Lecture Notes in Computer Science

Edited by G. Goos, J. Hartmanis and J. van Leeu

T0237789

Springer
Berlin
Heidelberg
New York
Barcelona
Hong Kong
London
Milan
Paris
Singapore
Tokyo

Ding-Zhu Du   Peter Eades   Vladimir Estivill-Castro
Xuemin Lin   Arun Sharma (Eds.)

Computing
and Combinatorics

6th Annual International Conference, COCOON 2000
Sydney, Australia, July 26-28, 2000
Proceedings

 Springer

Series Editors

Gerhard Goos, Karlsruhe University, Germany
Juris Hartmanis, Cornell University, NY, USA
Jan van Leeuwen, Utrecht University, The Netherlands

Volume Editors

Ding-Zhu Du
University of Minnesota, Department of Computer Science and Engineering
Minneapolis, MN 55455, USA
E-mail: dzd@cs.umn.edu

Peter Eades
Vladimir Estivill-Castro
University of Newcastle
Department of Computer Science and Software Engineering
Callaghan, NSW 2308, Australia
E-mail: {eades,vlad}@cs.newcastle.edu.au

Xuemin Lin
Arun Sharma
The University of New South Wales
School of Computer Science and Engineering
Sydney, NSW 2052, Australia
E-mail: {lxue,arun}@cse.unsw.edu.au

Cataloging-in-Publication Data applied for

Die Deutsche Bibliothek - CIP-Einheitsaufnahme

Computing and combinatorics : 6th annual international conference ;
proceedings / COCOON 2000, Sydney, Australia, July 26 - 28, 2000.
D.-Z. Du ... (ed.). - Berlin ; Heidelberg ; New York ; Barcelona ;
Hong Kong ; London ; Milan ; Paris ; Singapore ; Tokyo : Springer, 2000
  (Lecture notes in computer science ; Vol. 1858)
  ISBN 3-540-67787-9

CR Subject Classification (1998): F.2, G.2.1-2, I.3.5, C.2.3-4, E.1

ISSN 0302-9743
ISBN 3-540-67787-9 Springer-Verlag Berlin Heidelberg New York

Springer-Verlag Berlin Heidelberg New York
a member of BertelsmannSpringer Science+Business Media GmbH
© Springer-Verlag Berlin Heidelberg 2000
Printed in Germany

Typesetting: Camera-ready by author, data conversion by Boller Mediendesign
Printed on acid-free paper   SPIN: 10722191   06/3142   5 4 3 2 1 0

Preface

The papers in this volume were selected for presentation at the 6th Annual International Computing and Combinatorics Conference (COCOON 2000), in Sydney, Australia from July 26 - 28, 2000. The topics cover many areas in theoretical computer science and combinatorial optimization.

There were 81 high quality papers submitted to COCOON 2000. Each paper was reviewed by at least three program committee members, and the 44 papers were selected. It is expected that most of them will appear in a more complete form in scientific journals. In addition to the selected papers, the volume also contains the papers from two invited keynote speeches by Christos Papadimitriou and Richard Brent.

This year the *Hao Wang Award* was given to honor the paper judged by the program committee to have the greatest merit. The recipient is "Approximating Uniform Triangular Meshes in Polygons" by Franz Aurenhammer, Naoki Katoh, Hiromichi Kojima, Makoto Ohsaki, and Yinfeng Xu. The first *Best Young Researcher* paper award was given to William Duckworth for his paper "Maximum Induced Matchings of Random Cubic Graphs".

We wish to thank all who have made this meeting possible: the authors for submitting papers, the program committee members and the external referees, sponsors, the local organizers, ACM SIGACT for handling electronic submissions, Springer-Verlag for their support, and Debbie Hatherell for her assistance.

July 2000

Ding-Zhu Du
Peter Eades
Xuemin Lin

Sponsoring Institutions

University of New South Wales
University of Newcastle
University of Sydney
Computer Science Association of Australasia

Conference Organization

Program Committee Co-chairs

Ding-Zhu Du (University of Minnesota, USA)
Peter Eades (University of Sydney, Australia)
Xuemin Lin (University of New South Wales, Australia)

Program Committee

David Avis (McGill, Canada)
Jianer Chen (Texas A&M, USA)
Francis Chin (Hong Kong U., Hong Kong)
Vladimir Estivill-Castro (Newcastle, Australia),
George Havas (UQ, Australia)
Hiroshi Imai (Tokyo, Japan)
Tao Jiang (UC Riverside, USA)
Michael Juenger (Cologne, Germany)
Richard Karp (UC Berkeley, USA)
D. T. Lee (Academia Sinica, Taiwan)
Bernard Mans (Macquarie U., Australia)
Brendan McKay (ANU, Australia)
Maurice Nivat (Université de Paris VII, France)
Takeshi Tokuyama (Tohoku, Japan)
Roberto Tamassia (Brown, USA)
Jie Wang (UNC Greensboro, USA)
Shmuel Zaks (Technion, Israel)
Louxin Zhang (NUS, Singapore)
Shuzhong Zhang (CUHK, Hong Kong)
Binhai Zhu (City U. Hong Kong, Hong Kong)

Organizing Committee

Vladimir Estivill-Castro (Conference Co-chair, University of Newcastle)
Arun Sharma (Conference Co-chair, University of New South Wales)
Hai-Xin Lu (University of New South Wales)

Referees

Kazuyuki Amano
Joffroy Beauquier
Francois Bertault
Christoph Buchheim
S.W. Cheng
L. Devroye
Matthias Elf
Stanley Fung
Yehuda Hassin
Seokhee Hong
Tsan-Sheng Hsu
Wen-Lian Hsu
Ming-Tat Ko
Joachim Kupke
Thierry Lecroq
Weifa Liang
Frauke Liers
Chi-Jen Lu
Hsueh-I Lu

Marcus Oswald
Andrzej Pelc
David Peleg
Stephane Perennes
Z.P. Qin
Bruce Reed
Maurice Rojas
James L. Schwing
Rinovia M.G. Simanjuntak
Slamin
Rinovia Simanjuntak
Peter Tischer
Da-Wei Wang
John Watorus
Martin Wolff
Yinfeng Xu
Ding Feng Ye
Hong Zhu

Table of Contents

Graph Theory and Algorithms 1

Complexity, Discrete Mathematics, and Number Theory

Graph Theory and Algorithms 2

Online Algorithms

Parallel and Distributed Computing

Computational Geometry 2

Combinatorial Optimization

Data Structures and Computational Biology

Learning and Cryptography

Automata and Quantum Computing

Author Index

Theoretical Problems Related to the Internet

(Extended Abstract)

Christos H. Papadimitriou*

University of California, Berkeley, USA,
christos@cs.berkeley.edu
http://www.cs.berkeley.edu/~christos/

The Internet has arguably superseded the computer as the most complex cohesive artifact (if you can call it that), and is, quite predictably, the source of a new generation of foundational problems for Theoretical Computer Science. These new theoretical challenges emanate from several novel aspects of the Internet: (a) Its unprecedented size, diversity, and availability as an information repository; (b) its novel nature as a computer system that intertwines a multitude of economic interests in varying degrees of competition; and (c) its history as a shared resource architecture that emerged in a remarkably *ad hoc* yet gloriously successful manner. In this talk I will survey some recent research (done in collaboration with Joan Feigenbaum, Dick Karp, Elias Koutsoupias, and Scott Shenker, see [1,2]) on problems of the two latter kinds. See [3,5] for examples of theoretical work along the first axis.

To a large extent, the Internet owes its remarkable ascent to the surprising empirical success of TCP's congestion control protocol [4]. A flow between two points strives to determine the correct transmission rate by trying higher and higher rates, in a linearly increasing function. As links shared by several flows reach their capacity, one or more packets may be lost, and, in response, the corresponding flow *halves* its transmission rate. The fantastic empirical success of this crude mechanism raises important, and at times theoretically interesting, questions related to novel search algorithms (the process clearly contains a novel variant of binary search), on-line algorithms (when the true capacity that a flow can use changes in an unknown and, in a worst case, adversarial fashion), and algorithmic game theory (why is it that flows seem to be unwilling, or unmotivated, to move away from TCP?).

In a situation in which the same piece of information is multicast to millions of individuals, it is of interest to determine the price that each recipient must pay in return. The information in question is of different value to each individual, and this value is known only to this individual. The participants must engage in a *mechanism*, that is to say, a distributed protocol whereby they determine the individuals who will actually receive the information, and the price each will pay. Economists over the past fourty years have studied such situations arising in auctions, welfare economics, and pricing of public goods, and have come up with clever mechanisms —solutions that satisfy various sets of desiderata

---

* Research partially supported by the National Science Foundation.

D.-Z. Du et al. (Eds.): COCOON 2000, LNCS 1858, pp. 1–2, 2000.

("axioms")— or ingenious proofs that satisfying other sets of desiderata is impossible. In the context of the Internet, however, an important computational desideratum emerges: *The protocol should be efficient,* with respect to the total number of messages sent. It turns out that this novel condition changes the situation dramatically, and carves out a new set of pricing mechanisms that are appropriate and feasible.

References

1. Joan Feigenbaum, Christos H. Papadimitriou, Scott Shenker, "Sharing the Cost of Multicast Transmissions," *Proc. 2000 STOC.*
2. Richard M. Karp, Elias Koutsoupias, Christos H. Papadimitriou, Scott Shenker, "Optimization Problems in Congestion Control," manuscript, April 2000.
3. Jon Kleinberg, "Authoritative sources in a hyperlinked environment," *Proc. ACM-SIAM Symposium on Discrete Algorithms*, 1998.
4. Van Jacobson "Congestion Avoidance and Control," in *ACM SigComm Proceedings*, pp 314-329, 1988.
5. Christos H. Papadimitriou, Prabhakar Raghavan, Hisao Tamaki, Santosh Vempala, "Latent Semantic Indexing: A Probabilistic Analysis," Proc. 1998 *PODS*, to appear in the special issue of *JCSS*.

Recent Progress and Prospects for Integer Factorisation Algorithms

Richard P. Brent

Oxford University Computing Laboratory,
Wolfson Building, Parks Road,
Oxford OX1 3QD, UK
rpb@comlab.ox.ac.uk
http://www.comlab.ox.ac.uk/

Abstract. The integer factorisation and discrete logarithm problems
are of practical importance because of the widespread use of public key
cryptosystems whose security depends on the presumed difficulty of solving these problems. This paper considers primarily the integer factorisation problem. In recent years the limits of the best integer factorisation
algorithms have been extended greatly, due in part to Moore's law and in
part to algorithmic improvements. It is now routine to factor 100-decimal
digit numbers, and feasible to factor numbers of 155 decimal digits (512
bits). We outline several integer factorisation algorithms, consider their
suitability for implementation on parallel machines, and give examples
of their current capabilities. In particular, we consider the problem of
parallel solution of the large, sparse linear systems which arise with the
MPQS and NFS methods.

1 Introduction

There is no known deterministic or randomised polynomial-time[1] algorithm for
finding a factor of a given composite integer N. This empirical fact is of great
interest because the most popular algorithm for public-key cryptography, the
RSA algorithm [54], would be insecure if a fast integer factorisation algorithm
could be implemented.

In this paper we survey some of the most successful integer factorisation
algorithms. Since there are already several excellent surveys emphasising the
number-theoretic basis of the algorithms, we concentrate on the computational
aspects.

This paper can be regarded as an update of [8], which was written just
before the factorisation of the 512-bit number RSA155. Thus, to avoid duplica-
tion, we refer to [8] for primality testing, multiple-precision arithmetic, the use
of factorisation and discrete logarithms in public-key cryptography, and some
factorisation algorithms of historical interest.

---

[1] For a polynomial-time algorithm the expected running time should be a polynomial
in the length of the input, i.e. $O((\log N)^c)$ for some constant c.

D.-Z. Du et al. (Eds.): COCOON 2000, LNCS 1858, pp. 3–22, 2000.

1.1   Randomised Algorithms

All but the most trivial algorithms discussed below are *randomised algorithms*, i.e. at some point they involve pseudo-random choices. Thus, the running times are *expected* rather than *worst case*. Also, in most cases the expected running times are conjectured, not rigorously proved.

1.2   Parallel Algorithms

When designing parallel algorithms we hope that an algorithm which requires time T_1 on a computer with one processor can be implemented to run in time $T_P \sim T_1/P$ on a computer with P independent processors. This is not always the case, since it may be impossible to use all P processors effectively. However, it is true for many integer factorisation algorithms, provided P is not too large.

The *speedup* of a parallel algorithm is $S = T_1/T_P$. We aim for a linear speedup, i.e. $S = \Theta(P)$.

1.3   Quantum Algorithms

In 1994 Shor [57,58] showed that it is possible to factor in polynomial expected time on a quantum computer [20,21]. However, despite the best efforts of several research groups, such a computer has not yet been built, and it remains unclear whether it will ever be feasible to build one. Thus, in this paper we restrict our attention to algorithms which run on classical (serial or parallel) computers. The reader interested in quantum computers could start by reading [50,60].

1.4   Moore's Law

Moore's "law" [44,56] predicts that circuit densities will double every 18 months or so. Of course, Moore's law is not a theorem, and must eventually fail, but it has been surprisingly accurate for many years. As long as Moore's law continues to apply and results in correspondingly more powerful parallel computers, we expect to get a steady improvement in the capabilities of factorisation algorithms, without any algorithmic improvements. Historically, improvements over the past thirty years have been due both to Moore's law and to the development of new algorithms (e.g. ECM, MPQS, NFS).

2   Integer Factorisation Algorithms

There are many algorithms for finding a nontrivial factor f of a composite integer N. The most useful algorithms fall into one of two classes –

A. The run time depends mainly on the size of N, and is not strongly dependent on the size of f. Examples are –

- Lehman's algorithm [28], which has worst-case run time $O(N^{1/3})$.
- The Continued Fraction algorithm [39] and the Multiple Polynomial Quadratic Sieve (MPQS) algorithm [46,59], which under plausible assumptions have expected run time $O(\exp(\sqrt{c \ln N \ln \ln N}))$, where c is a constant (depending on details of the algorithm). For MPQS, $c \approx 1$.
- The Number Field Sieve (NFS) algorithm [29,30], which under plausible assumptions has expected run time $O(\exp(c(\ln N)^{1/3}(\ln \ln N)^{2/3}))$, where c is a constant (depending on details of the algorithm and on the form of N).

B. The run time depends mainly on the size of f, the factor found. (We can assume that $f \leq N^{1/2}$.) Examples are –

- The trial division algorithm, which has run time $O(f \cdot (\log N)^2)$.
- Pollard's "rho" algorithm [45], which under plausible assumptions has expected run time $O(f^{1/2} \cdot (\log N)^2)$.
- Lenstra's Elliptic Curve (ECM) algorithm [34], which under plausible assumptions has expected run time $O(\exp(\sqrt{c \ln f \ln \ln f}) \cdot (\log N)^2)$, where $c \approx 2$ is a constant.

In these examples, the time bounds are for a sequential machine, and the term $(\log N)^2$ is a generous allowance for the cost of performing arithmetic operations on numbers which are $O(N^2)$. If N is very large, then fast integer multiplication algorithms [19,24] can be used to reduce the $(\log N)^2$ term.

Our survey of integer factorisation algorithms is necessarily cursory. For more information the reader is referred to [8,35,48,53].

3   Lenstra's Elliptic Curve Algorithm

Lenstra's *elliptic curve* method/algorithm (abbreviated ECM) was discovered by H. W. Lenstra, Jr. about 1985 (see [34]). It is the best known algorithm in class B. To save space, we refer to [7,8,34] and the references there for a description of ECM, and merely give some examples of its successes here.

3.1   ECM Examples

1. In 1995 we completed the factorisation of the 309-decimal digit (1025-bit) Fermat number $F_{10} = 2^{2^{10}} + 1$. In fact

$$F_{10} = 45592577 \cdot 6487031809 \cdot$$
$$4659775785220018543264560743076778192897 \cdot p_{252}$$

where $46597 \cdots 92897$ is a 40-digit prime and $p_{252} = 13043 \cdots 24577$ is a 252-digit prime. The computation, which is described in detail in [7], took about 240 Mips-years. (A Mips-year is the computation performed by a hypothetical machine performing 10^6 instructions per second for one year, i.e. about 3.15×10^{13} instructions. Is is a convenient measure but should not be taken too seriously.)

2. The largest factor known to have been found by ECM is the 54-digit factor

$$484061254276878368125726870789180231995964870094916937$$

of $(6^{43} - 1)^{42} + 1$, found by Nik Lygeros and Michel Mizony with Paul Zimmermann's GMP-ECM program [63] in December 1999 (for more details see [9]).

3.2   Parallel/Distributed Implementation of ECM

ECM consists of a number of independent pseudo-random trials, each of which can be performed on a separate processor. So long as the expected number of trials is much larger than the number P of processors available, linear speedup is possible by performing P trials in parallel. In fact, if T_1 is the expected run time on one processor, then the expected run time on a MIMD parallel machine with P processors is

$$T_P = T_1/P + O(T_1^{1/2+\epsilon}) \tag{1}$$

3.3   ECM Factoring Records

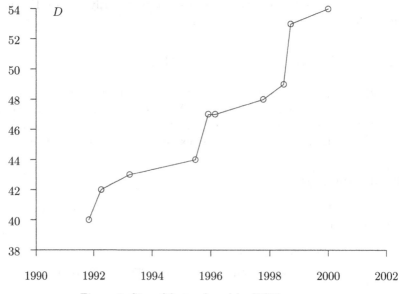

Figure 1: Size of factor found by ECM versus year

Figure 1 shows the size D (in decimal digits) of the largest factor found by ECM against the year it was done, from 1991 (40D) to 1999 (54D) (historical data from [9]).

3.4   Extrapolation of ECM Records

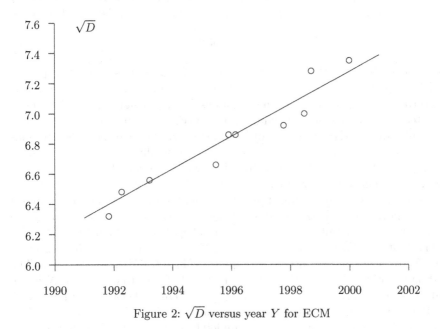

Figure 2: $\sqrt{D}$ versus year Y for ECM

Let D be the number of decimal digits in the largest factor found by ECM up to a given date. From the theoretical time bound for ECM, assuming Moore's law, we expect $\sqrt{D}$ to be roughly a linear function of calendar year (in fact $\sqrt{D \ln D}$ should be linear, but given the other uncertainties we have assumed for simplicity that $\sqrt{\ln D}$ is roughly a constant). Figure 2 shows $\sqrt{D}$ versus year Y.

The straight line shown in the Figure 2 is

$$\sqrt{D} = \frac{Y - 1932.3}{9.3} \quad \text{or equivalently} \quad Y = 9.3\sqrt{D} + 1932.3 \,,$$

and extrapolation gives $D = 60$ in the year $Y = 2004$ and $D = 70$ in the year $Y = 2010$.

4   Quadratic Sieve Algorithms

Quadratic sieve algorithms belong to a wide class of algorithms which try to find two integers x and y such that $x \neq \pm y \pmod{N}$ but

$$x^2 = y^2 \pmod{N} . \tag{2}$$

Once such x and y are found, then GCD $(x - y, N)$ is a nontrivial factor of N.

One way to find x and y satisfying (2) is to find a set of *relations* of the form

$$u_i^2 = v_i^2 w_i \pmod{N}, \tag{3}$$

where the w_i have all their prime factors in a moderately small set of primes (called the *factor base*). Each relation (3) gives a colunm in a matrix R whose rows correspond to the primes in the factor base. Once enough columns have been generated, we can use Gaussian elimination in $\mathbf{GF}(2)$ to find a linear dependency (mod 2) between a set of columns of R. Multiplying the corresponding relations now gives an expression of the form (2). With probability at least $1/2$, we have $x \neq \pm y \bmod N$ so a nontrivial factor of N will be found. If not, we need to obtain a different linear dependency and try again.

In quadratic sieve algorithms the numbers w_i are the values of one (or more) quadratic polynomials with integer coefficients. This makes it easy to factor the w_i by *sieving*. For details of the process, we refer to [11,32,35,46,49,52,59].

The best quadratic sieve algorithm (MPQS) can, under plausible assumptions, factor a number N in time $\Theta(\exp(c(\ln N \ln \ln N)^{1/2}))$, where $c \sim 1$. The constants involved are such that MPQS is usually faster than ECM if N is the product of two primes which both exceed $N^{1/3}$. This is because the inner loop of MPQS involves only single-precision operations.

Use of "partial relations", i.e. incompletely factored w_i, in MPQS gives a significant performance improvement [2]. In the "one large prime" (P-MPQS) variation w_i is allowed to have one prime factor exceeding m (but not too much larger than m). In the "two large prime" (PP-MPQS) variation w_i can have two prime factors exceeding m – this gives a further performance improvement at the expense of higher storage requirements [33].

4.1   Parallel/Distributed Implementation of MPQS

The sieving stage of MPQS is ideally suited to parallel implementation. Different processors may use different polynomials, or sieve over different intervals with the same polynomial. Thus, there is a linear speedup so long as the number of processors is not much larger than the size of the factor base. The computation requires very little communication between processors. Each processor can generate relations and forward them to some central collection point. This was demonstrated by A. K. Lenstra and M. S. Manasse [32], who distributed their program and collected relations via electronic mail. The processors were scattered around the world – anyone with access to electronic mail and a C compiler could volunteer to contribute[2]. The final stage of MPQS – Gaussian elimination to combine the relations – is not so easily distributed. We discuss this in §7 below.

4.2 MPQS Examples

MPQS has been used to obtain many impressive factorisations [10,32,52,59]. At the time of writing (April 2000), the largest number factored by MPQS is the 129-digit "RSA Challenge" [54] number RSA129. It was factored in 1994 by Atkins *et al* [1]. The relations formed a sparse matrix with 569466 columns, which was reduced to a dense matrix with 188614 columns; a dependency was then found on a MasPar MP-1. It is certainly feasible to factor larger numbers by MPQS, but for numbers of more than about 110 decimal digits GNFS is faster [22]. For example, it is estimated in [16] that to factor RSA129 by MPQS required 5000 Mips-years, but to factor the slightly larger number RSA130 by GNFS required only 1000 Mips-years [18].

5 The Special Number Field Sieve (SNFS)

The *number field sieve* (NFS) algorithm was developed from the *special number field sieve* (SNFS), which we describe in this section. The *general number field sieve* (GNFS or simply NFS) is described in §6.

Most of our numerical examples have involved numbers of the form

$$a^e \pm b, \tag{4}$$

for small a and b, although the ECM and MPQS factorisation algorithms do not take advantage of this special form.

The *special number field sieve* (SNFS) is a relatively new algorithm which does take advantage of the special form (4). In concept it is similar to the quadratic sieve algorithm, but it works over an algebraic number field defined by a, e and b. We refer the interested reader to Lenstra *et al* [29,30] for details, and merely give an example to show the power of the algorithm.

5.1 SNFS Examples

Consider the 155-decimal digit number

$$F_9 = N = 2^{2^9} + 1$$

as a candidate for factoring by SNFS. Note that $8N = m^5 + 8$, where $m = 2^{103}$. We may work in the number field $Q(\alpha)$, where α satisfies

$$\alpha^5 + 8 = 0,$$

---

[2] This idea of using machines on the Internet as a "free" supercomputer has been adopted by several other computation-intensive projects

and in the ring of integers of $Q(\alpha)$. Because

$$m^5 + 8 \equiv 0 \pmod{N},$$

the mapping $\phi : \alpha \mapsto m \bmod N$ is a ring homomorphism from $Z[\alpha]$ to Z/NZ.

The idea is to search for pairs of small coprime integers u and v such that both the algebraic integer $u + \alpha v$ and the (rational) integer $u + mv$ can be factored. (The factor base now includes prime ideals and units as well as rational primes.) Because

$$\phi(u + \alpha v) \equiv (u + mv) \pmod{N},$$

each such pair gives a relation analogous to (3).

The prime ideal factorisation of $u + \alpha v$ can be obtained from the factorisation of the *norm* $u^5 - 8v^5$ of $u + \alpha v$. Thus, we have to factor simultaneously two integers $u + mv$ and $|u^5 - 8v^5|$. Note that, for moderate u and v, both these integers are much smaller than N, in fact they are $O(N^{1/d})$, where $d = 5$ is the degree of the algebraic number field. (The optimal choice of d is discussed in §6.)

Using these and related ideas, Lenstra *et al* [31] factored F_9 in June 1990, obtaining

$$F_9 = 2424833 \cdot 7455602825647884208337395736200454918783366342657 \cdot p_{99},$$

where p_{99} is an 99-digit prime, and the 7-digit factor was already known (although SNFS was unable to take advantage of this). The collection of relations took less than two months on a network of several hundred workstations. A sparse system of about 200,000 relations was reduced to a dense matrix with about 72,000 rows. Using Gaussian elimination, dependencies (mod 2) between the columns were found in three hours on a Connection Machine. These dependencies implied equations of the form $x^2 = y^2 \bmod F_9$. The second such equation was nontrivial and gave the desired factorisation of F_9.

More recently, considerably larger numbers have been factored by SNFS, for example, the 211-digit number $10^{211} - 1$ was factored early in 1999 by a collaboration called "The Cabal" [13].

6   The General Number Field Sieve (GNFS)

The *general number field sieve* (GNFS or just NFS) is a logical extension of the special number field sieve (SNFS). When using SNFS to factor an integer N, we require two polynomials $f(x)$ and $g(x)$ with a common root $m \bmod N$ but no common root over the field of complex numbers. If N has the special form (4) then it is usually easy to write down suitable polynomials with small coefficients, as illustrated by the two examples given in §5. If N has no special form, but is just some given composite number, we can also find $f(x)$ and $g(x)$, but they no longer have small coefficients.

Suppose that $g(x)$ has degree $d > 1$ and $f(x)$ is linear[3]. d is chosen empirically, but it is known from theoretical considerations that the optimum value

---

[3] This is not necessary. For example, Montgomery found a clever way of choosing two quadratic polynomials.

is

$$d \sim \left(\frac{3 \ln N}{\ln \ln N} \right)^{1/3} .$$

We choose $m = \lfloor N^{1/(d+1)} \rfloor$ and write

$$N = \sum_{j=0}^{d} a_j m^j$$

where the a_j are "base m digits". Then, defining

$$f(x) = x - m, \quad g(x) = \sum_{j=0}^{d} a_j x^j ,$$

it is clear that $f(x)$ and $g(x)$ have a common root $m \bmod N$. This method of polynomial selection is called the "base m" method.

In principle, we can proceed as in SNFS, but many difficulties arise because of the large coefficients of $g(x)$. For details, we refer the reader to [36,37,41,47,48,62]. Suffice it to say that the difficulties can be overcome and the method works! Due to the constant factors involved it is slower than MPQS for numbers of less than about 110 decimal digits, but faster than MPQS for sufficiently large numbers, as anticipated from the theoretical run times given in §2.

Some of the difficulties which had to be overcome to turn GNFS into a practical algorithm are:

1. Polynomial selection. The "base m" method is not very good. Peter Montgomery and Brian Murphy [40,41,42] have shown how a very considerable improvement (by a factor of more than ten for number of 140–155 digits) can be obtained.

2. Linear algebra. After sieving a very large, sparse linear system over $\mathbf{GF}(2)$ is obtained, and we want to find dependencies amongst the columns. It is not practical to do this by structured Gaussian elimination [25, §5] because the "fill in" is too large. Odlyzko [43,17] and Montgomery [37] showed that the Lanczos method [26] could be adapted for this purpose. (This is non-trivial because a nonzero vector x over $\mathbf{GF}(2)$ can be orthogonal to itself, i.e. $x^T x = 0$.) To take advantage of bit-parallel operations, Montgomery's program works with blocks of size dependent on the wordlength (e.g. 64).

3. Square roots. The final stage of GNFS involves finding the square root of a (very large) product of algebraic numbers[4]. Once again, Montgomery [36] found a way to do this.

---

[4] An idea of Adleman, using quadratic characters, is essential to ensure that the desired square root exists with high probability.

6.1 RSA155

At the time of writing, the largest number factored by GNFS is the 155-digit RSA Challenge number RSA155. It was split into the product of two 78-digit primes on 22 August, 1999, by a team coordinated from CWI, Amsterdam. For details see [51]. To summarise: the amount of computer time required to find the factors was about 8000 Mips-years. The two polynomials used were

$$f(x) = x - 39123079721168000771313449081$$

and

$$g(x) = +119377138320x^5$$
$$-80168937284997582x^4$$
$$-662698522234118574445x^3$$
$$+1181684843007952188035 6852x^2$$
$$+7459661580071786443919743056x$$
$$-40679843542362159361913708405064 \, .$$

The polynomial $g(x)$ was chosen to have a good combination of two properties: being unusually small over the sieving region and having unusually many roots modulo small primes (and prime powers). The effect of the second property alone makes $g(x)$ as effective at generating relations as a polynomial chosen at random for an integer of 137 decimal digits (so in effect we have removed at least 18 digits from RSA155 by judicious polynomial selection). The polynomial selection took approximately 100 Mips-years or about 1.25% of the total factorisation time. For details of the polynomial selection, see Brian Murphy's thesis [41].

The total amount of CPU time spent on sieving was 8000 Mips-years on assorted machines (calendar time 3.7 months). The resulting matrix had about 6.7×10^6 rows and weight (number of nonzeros) about 4.2×10^8 (about 62 nonzeros per row). Using Montgomery's block Lanczos program, it took almost 224 CPU-hours and 2 GB of memory on a Cray C916 to find 64 dependencies. Calendar time for this was 9.5 days.

Table 1. RSA130, RSA140 and RSA155 factorisations

	RSA130	RSA140	RSA155
Total CPU time in Mips-years	1000	2000	8000
Matrix rows	3.5×10^6	4.7×10^6	6.7×10^6
Total nonzeros	1.4×10^8	1.5×10^8	4.2×10^8
Nonzeros per row	39	32	62
Matrix solution time (on Cray C90/C916)	68 hours	100 hours	224 hours

Table 1 gives some statistics on the RSA130, RSA140 and RSA155 factorisations. The Mips-year times given are only rough estimates. Extrapolation of these figures is risky, since various algorithmic improvements were incorporated in the later factorisations and the scope for further improvements is unclear.

7 Parallel Linear Algebra

At present, the main obstacle to a fully parallel and scalable implementation of GNFS (and, to a lesser extent, MPQS) is the linear algebra. Montgomery's block Lanczos program runs on a single processor and requires enough memory to store the sparse matrix (about five bytes per nonzero element). It is possible to distribute the block Lanczos solution over several processors of a parallel machine, but the communication/computation ratio is high [38].

Similar remarks apply to some of the best algorithms for the integer discrete logarithm problem. The main difference is that the linear algebra has to be performed over a (possibly large) finite field rather than over $\mathbf{GF}(2)$.

In this Section we present some preliminary ideas on how to implement the linear algebra phase of GNFS (and MPQS) in parallel on a network of relatively small machines. This work is still in progress and it is too early to give definitive results.

7.1 Assumptions

Suppose that a collection of P machines is available. It will be convenient to assume that $P = q^2$ is a perfect square. For example, the machines might be 400 Mhz PCs with 256 MByte of memory each, and $P = 16$ or $P = 64$.

We assume that the machines are connected on a reasonably high-speed network. For example, this might be multiple 100Mb/sec Ethernets or some higher-speed proprietary network. Although not essential, it is desirable for the physical network topology to be a $q \times q$ grid or torus. Thus, we assume that a processor $P_{i,j}$ in row i can communicate with another processor $P_{i,j'}$ in the same row or broadcast to all processors $P_{i,*}$ in the same row without interference from communication being performed in other rows. (Similarly for columns.) The reason why this topology is desirable is that it matches the communication patterns necessary in the linear algebra: see for example [4,5,6].

To simplify the description, we assume that the standard Lanczos algorithm is used. In practice, a block version of Lanczos [37] would be used, both to take advantage of word-length Boolean operations and to overcome the technical problem that $u^T u$ can vanish for $u \neq 0$ when we are working over a finite field. The overall communication costs are nearly the same for blocked and unblocked Lanczos, although blocking reduces startup costs.

If R is the matrix of relations, with one relation per column, then R is a large, sparse matrix over $\mathbf{GF}(2)$ with slightly more columns than rows. (Warning: in the literature, the transposed matrix is often considered, i.e. rows and columns are often interchanged.) We aim to find a nontrivial *dependency*, i.e. a nonzero vector x such that $Rx = 0$. If we choose $b = Ry$ for some random y, and solve $Rx = b$, then $x - y$ is a dependency (possibly trivial, if $x = y$). Thus, in the description below we consider solving a linear system $Rx = b$.

The Lanczos algorithm over the real field works for a positive-definite symmetric matrix. In our case, R is not symmetric (or even square). Hence we apply Lanczos to the *normal equations*

$$R^T R x = R^T b$$

rather than directly to $Rx = b$. We do not need to compute the matrix $A = R^T R$, because we can compute Ax as $R^T(Rx)$. To compute $R^T z$, it is best to think of computing $(z^T R)^T$ because the elements of R will be scattered across the processors and we do not want to double the storage requirements by storing R^T as well as R.

The details of the Lanczos iteration are not important for the description below. It is sufficient to know that the Lanczos recurrence has the form

$$w_{i+1} = Aw_i - c_{i,i} w_i - c_{i,i-1} w_{i-1}$$

where the multipliers $c_{i,i}$ and $c_{i,i-1}$ can be computed using inner products of known vectors. Thus, the main work involved in one iteration is the computation of $Aw_i = R^T(Rw_i)$. In general, if R has n columns, then w_i is a *dense* n-vector. The Lanczos process terminates with a solution after $n + O(1)$ iterations [37].

7.2   Sparse Matrix-Vector Multiplication

The matrix of relations R is sparse and almost square. Suppose that R has n columns, $m \approx n$ rows, and ρn nonzeros, i.e. ρ nonzeros per column. Consider the computation of Rw, where w is a dense n-vector. The number of multiplications involved in the computation of Rw is ρn. (Thus, the number of operations involved in the complete Lanczos process is of order $\rho n^2 \ll n^3$.)

For simplicity assume that $q = \sqrt{P}$ divides n, and the matrix R is distributed across the $q \times q$ array of processors using a "scattered" data distribution to balance the load and storage requirements [5]. Within each processor $P_{i,j}$ a sparse matrix representation will be used to store the relevant part $R_{i,j}$ of R. Thus the local storage of requirement is $O(\rho n/q^2 + n/q)$.

After each processor computes its local matrix-vector product with about $\rho n/q^2$ multiplications, it is necessary to sum the partial results (n/q-vectors) in each row to obtain n/q components of the product Rw. This can be done in several ways, e.g. by a systolic computation, by using a binary tree, or by each processor broadcasting its n/q-vector to other processors in its row and summing the vectors received from them to obtain n/q components of the product Rw. If broadcasting is used then the results are in place to start the computation of $R^T(Rw)$. If an inner product is required then communication along columns of the processor array is necessary to transpose the dense n-vector (each diagonal processor $P_{i,i}$ broadcasts its n/q elements along its column).

We see that, overall, each Lanczos iteration requires computation $O(\rho n/q^2 + n)$ on each processor, and communication of $O(n)$ bits along each row/column of the processor array. Depending on details of the communication hardware, the time for communication might be $O((n/q)\log q)$ (if the binary tree method is used) or even $O(n/q+q)$ (if a systolic method is used with pipelining). In the following discussion we assume that the communication time is proportional to the number of bits communicated along each row or column, i.e. $O(n)$ per iteration (not the total number of bits communicated per iteration, which is $O(nq)$).

The overall time for n iterations is

$$T_P \approx \alpha \rho n^2/P + \beta n^2 \,, \tag{5}$$

where the constants α and β depend on the computation and communication speeds respectively. (We have omitted the communication "startup" cost since this is typically much less than βn^2.)

Since the time on a single processor is $T_1 \approx \alpha \rho n^2$, the *efficiency E_P* is

$$E_P = \frac{T_1}{PT_P} \approx \frac{\alpha \rho}{\alpha \rho + \beta P}$$

and the condition for $E_P \geq 1/2$ is

$$\alpha \rho \geq \beta P \,.$$

Typically β/α is large (50 to several hundred) because communication between processors is much slower than computation on a processor. The number of nonzeros per column, ρ, is typically in the range 30 to 100. Thus, we can not expect high efficiency.

Fortunately, high efficiency is not crucial. The essential requirements are that the (distributed) matrix R fits in the local memories of the processors, and that the time for the linear algebra does not dominate the overall factorisation time. Because sieving typically takes *much* longer than linear algebra on a single processor, we have considerable leeway.

For example, consider the factorisation of RSA155 (see Table 1 above). We have $n \approx 6.7 \times 10^6$, $\rho \approx 62$. Thus $n^2 \approx 4.5 \times 10^{13}$. If the communications speed along each row/column is a few hundred Mb/sec then we expect β to be a small multiple of 10^{-8} seconds. Thus βn^2 might be (very roughly) 10^6 seconds or say 12 days. The sieving takes 8000 Mips-years or 20 processor-years if each processor runs at 400 Mips. With P processors this is $20/P$ years, to be compared with $12/365$ years for communication. Thus,

$$\frac{\text{sieving time}}{\text{communication time}} \approx \frac{600}{P}$$

and the sieving time dominates unless $P > 600$. For larger problems we expect the sieving time to grow at least as fast as n^2.

For more typical factorisations performed by MPQS rather than GNFS, the situation is better. For example, using Arjen Lenstra's implementation of PP-MPQS in the Magma package, we factored a 102-decimal digit composite c_{102} (a factor of $14^{213} - 1$) on a 4×250 Mhz Sun multiprocessor. The sieving time was 11.5 days or about 32 Mips-years. After combining partial relations the matrix R had about $n = 73000$ columns. Thus $n^2 \approx 5.3 \times 10^9$. At 100 Mb/sec it takes less than one minute to communicate n^2 bits. Thus, although the current Magma implementation performs the block Lanczos method on only one processor, there is no reason why it could not be performed on a network of small machines (e.g. a "Beowulf" cluster) and the communication time on such a network would only be a few minutes.

7.3   Reducing Communication Time

If our assumption of a grid/torus topology is not satisfied, then the communication times estimated above may have to be multiplied by a factor of order $q = \sqrt{P}$. In order to reduce the communication time (without buying better hardware) it may be worthwhile to perform some steps of Gaussian elimination before switching to Lanczos.

If there are k relations whose largest prime is p, we can perform Gaussian elimination to eliminate this large prime and one of the k relations. (The work can be done on a single processor if the relations are hashed and stored on processors according to the largest prime occurring in them.) Before the elimination we expect each of the k relations to contain about ρ nonzeros; after the elimination each of the remaining $k - 1$ relations will have about $\rho - 1$ additional nonzeros (assuming no "cancellation").

From (5), the time for Lanczos is about

$$T = n(\alpha z/P + \beta n) ,$$

where $z = \rho n$ is the weight (number of nonzeros) of the matrix R. The effect of eliminating one relation is $n \to n' = n - 1$, $z \to z' = z + (\rho - 1)(k - 1) - \rho$, $T \to T' = n'(\alpha z'/P + \beta n')$. After some algebra we see that $T' < T$ if

$$k < \frac{2\beta P}{\alpha \rho} + 3 .$$

For typical values of $\beta/\alpha \approx 100$ and $\rho \approx 50$ this gives the condition $k < 4P + 3$.

The disadvantage of this strategy is that the total weight z of the matrix (and hence storage requirements) increases. In fact, if we go too far, the matrix becomes full (see the RSA129 example in §4.2).

As an example, consider the c_{102} factorisation mentioned at the end of §7.2. The initial matrix R had $n \approx 73000$; we were able to reduce this to $n \approx 32000$, and thus reduce the communication requirements by 81%, at the expense of increasing the weight by 76%. Our (limited) experience indicates that this behaviour is typical – we can usually reduce the communication requirements by a factor of five at the expense of at most doubling the weight. However, if we try to go much further, there is an "explosion", the matrix becomes dense, and the advantage of the Lanczos method over Gaussian elimination is lost. (It would be nice to have a theoretical explanation of this behaviour.)

8 Historical Factoring Records

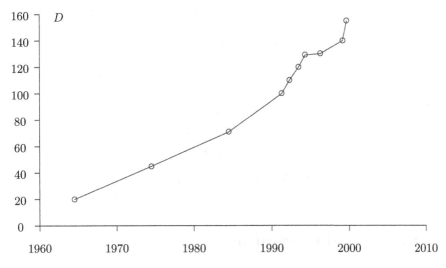

Figure 3: Size of "general" number factored versus year

Figure 3 shows the size D (in decimal digits) of the largest "general" number factored against the year it was done, from 1964 (20D) to April 2000 (155D) (historical data from [41,44,55]).

8.1 Curve Fitting and Extrapolation

Let D be the number of decimal digits in the largest "general" number factored by a given date. From the theoretical time bound for GNFS, assuming Moore's law, we expect $D^{1/3}$ to be roughly a linear function of calendar year (in fact $D^{1/3}(\ln D)^{2/3}$ should be linear, but given the other uncertainties we have assumed for simplicity that $(\ln D)^{2/3}$ is roughly a constant). Figure 4 shows $D^{1/3}$ versus year Y.

The straight line shown in the Figure 4 is

$$D^{1/3} = \frac{Y - 1928.6}{13.24} \quad \text{or equivalently} \quad Y = 13.24D^{1/3} + 1928.6 \,,$$

and extrapolation, for what it is worth, gives $D = 309$ (i.e. 1024 bits) in the year $Y = 2018$.

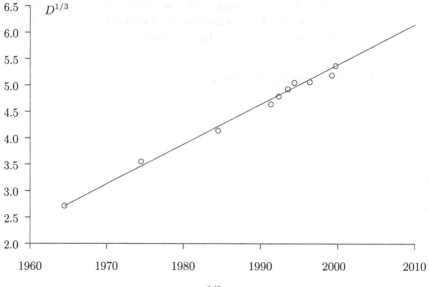

Figure 4: $D^{1/3}$ versus year Y

9   Summary and Conclusions

We have sketched some algorithms for integer factorisation. The most important are ECM, MPQS and NFS. As well as their inherent interest and applicability to other areas of mathematics, advances in public key cryptography have lent them practical importance.

Despite much progress in the development of efficient algorithms, our knowledge of the complexity of factorisation is inadequate. We would like to find a polynomial time factorisation algorithm or else prove that one does not exist. Until a polynomial time algorithm is found or a quantum computer capable of running Shor's algorithm [57,58] is built, large factorisations will remain an interesting challenge.

A survey similar to this one was written in 1990 (see [3]). Comparing the examples there we see that significant progress has been made. This is partly due to Moore's law, partly due to the use of many machines on the Internet, and partly due to improvements in algorithms (especially GNFS). The largest number factored by MPQS at the time of writing [3] had 111 decimal digits. According to [16], RSA110 was factored in 1992 (by MPQS). In 1996 GNFS was used to factor RSA130; in August 1999, RSA155 also fell to GNFS. Progress seems to be accelerating. This is due in large part to algorithmic improvements which seem unlikely to be repeated. On the other hand, it is very hard to anticipate algorithmic improvements!

From the predicted run time for GNFS, we would expect RSA155 to take 6.5 times as long as RSA140[5]. On the other hand, Moore's law predicts that circuit densities will double every 18 months or so. Thus, as long as Moore's law continues to apply and results in correspondingly more powerful parallel computers, we expect to get three to four decimal digits per year improvement in the capabilities of GNFS, without any algorithmic improvements. The extrapolation from historical figures is more optimistic: it predicts 6–7 decimal digits per year in the near future.

Similar arguments apply to ECM, for which we expect slightly more than one decimal digit per year in the size of factor found [9].

Although we did not discuss them here, the best algorithms for the integer discrete logarithm problem are analogous to NFS, requiring the solution of large, sparse linear systems over finite fields, and similar remarks apply to such algorithms.

Regarding cryptographic consequences, we can say that 512-bit RSA keys are already insecure. 1024-bit RSA keys should remain secure for at least fifteen years, barring the unexpected (but unpredictable) discovery of a completely new algorithm which is better than GNFS, or the development of a practical quantum computer. In the long term, public key algorithms based on the discrete logarithm problem for elliptic curves appear to be a promising alternative to RSA or algorithms based on the integer discrete logarithm problem.

Acknowledgements

Thanks are due to Peter Montgomery, Brian Murphy, Andrew Odlyzko, John Pollard, Herman te Riele, Sam Wagstaff, Jr. and Paul Zimmermann for their assistance.

References

1. D. Atkins, M. Graff, A. K. Lenstra and P. C. Leyland, The magic words are squeamish ossifrage, *Advances in Cryptology: Proc. Asiacrypt'94, LNCS* **917**, Springer-Verlag, Berlin, 1995, 263–277.
2. H. Boender and H. J. J. te Riele, *Factoring integers with large prime variations of the quadratic sieve,* Experimental Mathematics, **5** (1996), 257–273.
3. R. P. Brent, Vector and parallel algorithms for integer factorisation, *Proceedings Third Australian Supercomputer Conference* University of Melbourne, December 1990, 12 *pp.* http://www.comlab.ox.ac.uk/oucl/work/richard.brent/pub/pub122.html .
4. R. P. Brent, The LINPACK benchmark on the AP 1000, *Proceedings of Frontiers '92* (McLean, Virginia, October 1992), IEEE Press, 1992, 128-135. http://www.comlab.ox.ac.uk/oucl/work/richard.brent/pub/pub130.html

---

[5] It actually took only about 4.0 times as long, thanks to algorithmic improvements such as better polynomial selection.

5. R. P. Brent, Parallel algorithms in linear algebra, *Algorithms and Architectures: Proc. Second NEC Research Symposium* held at Tsukuba, Japan, August 1991 (edited by T. Ishiguro), SIAM, Philadelphia, 1993, 54-72. http://www.comlab.ox.ac.uk/oucl/work/richard.brent/pub/pub128.html

6. R. P. Brent and P. E. Strazdins, Implementation of the BLAS level 3 and Linpack benchmark on the AP 1000, *Fujitsu Scientific and Technical Journal* 29, 1 (March 1993), 61-70. http://www.comlab.ox.ac.uk/oucl/work/richard.brent/pub/pub136.html

7. R. P. Brent, Factorization of the tenth Fermat number, *Math. Comp.* **68** (1999), 429-451. Preliminary version available as *Factorization of the tenth and eleventh Fermat numbers*, Technical Report TR-CS-96-02, CSL, ANU, Feb. 1996, 25*pp.* http://www.comlab.ox.ac.uk/oucl/work/richard.brent/pub/pub161.html .

8. R. P. Brent, Some parallel algorithms for integer factorisation *Proc. Europar'99*, Toulouse, Sept. 1999. *LNCS* **1685**, Springer-Verlag, Berlin, 1–22.

9. R. P. Brent, *Large factors found by ECM*, Oxford University Computing Laboratory, March 2000. ftp://ftp.comlab.ox.ac.uk/pub/Documents/techpapers/Richard.Brent/champs.txt .

10. J. Brillhart, D. H. Lehmer, J. L. Selfridge, B. Tuckerman and S. S. Wagstaff, Jr., *Factorisations of $b^n \pm 1, b = 2, 3, 5, 6, 7, 10, 11, 12$ up to high powers*, American Mathematical Society, Providence, Rhode Island, second edition, 1988. Updates available from http://www/cs/purdue.edu/homes/ssw/cun/index.html .

11. T. R. Caron and R. D. Silverman, Parallel implementation of the quadratic sieve, *J. Supercomputing* **1** (1988), 273–290.

12. S. Cavallar, B. Dodson, A. K. Lenstra, P. Leyland, W. Lioen, P. L. Montgomery, B. Murphy, H. te Riele and P. Zimmermann, *Factorization of RSA-140 using the number field sieve*, announced 4 February 1999. Available from ftp://ftp.cwi.nl/pub/herman/NFSrecords/RSA-140 .

13. S. Cavallar, B. Dodson, A. K. Lenstra, P. Leyland, W. Lioen, P. L. Montgomery, H. te Riele and P. Zimmermann, *211-digit SNFS factorization*, announced 25 April 1999. Available from ftp://ftp.cwi.nl/pub/herman/NFSrecords/SNFS-211 .

14. D. V. and G. V. Chudnovsky, Sequences of numbers generated by addition in formal groups and new primality and factorization tests, *Adv. in Appl. Math.* **7** (1986), 385–434.

15. H. Cohen, *A Course in Computational Algebraic Number Theory*, Springer-Verlag, Berlin, 1993.

16. S. Contini, The factorization of RSA-140, RSA Laboratories Bulletin **10**, 8 (March 1999). Available from http://www.rsa.com/rsalabs/html/bulletins.html .

17. D. Coppersmith, A. Odlyzko and R. Schroeppel, Discrete logarithms in $\mathbf{GF}(p)$, *Algorithmica* **1** (1986), 1–15.

18. J. Cowie, B. Dodson, R. M. Elkenbracht-Huizing, A. K. Lenstra, P. L. Montgomery and J. Zayer, A world wide number field sieve factoring record: on to 512 bits, *Advances in Cryptology: Proc. Asiacrypt'96*, *LNCS* **1163**, Springer-Verlag, Berlin, 1996, 382–394.

19. R. Crandall and B. Fagin, Discrete weighted transforms and large-integer arithmetic, *Math. Comp.* **62** (1994), 305–324.

20. D. Deutsch, Quantum theory, the Church-Turing principle and the universal quantum computer, *Proc. Roy. Soc. London, Ser. A* **400** (1985), 97–117.

21. D. Deutsch, Quantum computational networks, *Proc. Roy. Soc. London, Ser. A* **425** (1989), 73–90.

22. M. Elkenbracht-Huizing, A multiple polynomial general number field sieve *Algorithmic Number Theory – ANTS III, LNCS* **1443**, Springer-Verlag, Berlin, 1998, 99–114.

23. K. F. Ireland and M. Rosen, *A Classical Introduction to Modern Number Theory*, Springer-Verlag, Berlin, 1982.

24. D. E. Knuth, *The Art of Computer Programming*, Vol. 2, Addison Wesley, third edition, 1997.

25. B. A. LaMacchia and A. M. Odlyzko, Solving large sparse systems over finite fields, *Advances in Cryptology, CRYPTO '90* (A. J. Menezes and S. A. Vanstone, eds.), *LNCS* **537**, Springer-Verlag, Berlin, 109–133.

26. C. Lanczos, Solution of systems of linear equations by minimized iterations, *J. Res. Nat. Bureau of Standards* **49** (1952), 33–53.

27. S. Lang, *Elliptic Curves – Diophantine Analysis*, Springer-Verlag, Berlin, 1978.

28. R. S. Lehman, Factoring large integers, *Math. Comp.* **28** (1974), 637–646.

29. A. K. Lenstra and H. W. Lenstra, Jr. (editors), The development of the number field sieve, *Lecture Notes in Mathematics* **1554**, Springer-Verlag, Berlin, 1993.

30. A. K. Lenstra, H. W. Lenstra, Jr., M. S. Manasse and J. M. Pollard, *The number field sieve*, Proc. 22nd Annual ACM Conference on Theory of Computing, Baltimore, Maryland, May 1990, 564–572.

31. A. K. Lenstra, H. W. Lenstra, Jr., M. S. Manasse, and J. M. Pollard, The factorization of the ninth Fermat number, *Math. Comp.* **61** (1993), 319–349.

32. A. K. Lenstra and M. S. Manasse, Factoring by electronic mail, *Proc. Eurocrypt '89, LNCS* **434**, Springer-Verlag, Berlin, 1990, 355–371.

33. A. K. Lenstra and M. S. Manasse, Factoring with two large primes, *Math. Comp.* **63** (1994), 785–798.

34. H. W. Lenstra, Jr., Factoring integers with elliptic curves, *Annals of Mathematics* (2) **126** (1987), 649–673.

35. P. L. Montgomery, A survey of modern integer factorization algorithms, *CWI Quarterly* **7** (1994), 337–366. `ftp://ftp.cwi.nl/pub/pmontgom/cwisurvey.psl.Z` .

36. P. L. Montgomery, Square roots of products of algebraic numbers, *Mathematics of Computation 1943 – 1993, Proc. Symp. Appl. Math.* **48** (1994), 567–571.

37. P. L. Montgomery, A block Lanczos algorithm for finding dependencies over $\mathbf{GF}(2)$, *Advances in Cryptology: Proc. Eurocrypt'95, LNCS* **921**, Springer-Verlag, Berlin, 1995, 106–120. `ftp://ftp.cwi.nl/pub/pmontgom/BlockLanczos.psa4.gz` .

38. P. L. Montgomery, *Parallel block Lanczos*, Microsoft Research, Redmond, USA, 17 January 2000 (transparencies of a talk presented at RSA 2000).

39. M. A. Morrison and J. Brillhart, A method of factorisation and the factorisation of F_7, *Math. Comp.* **29** (1975), 183–205.

40. B. A. Murphy, Modelling the yield of number field sieve polynomials, *Algorithmic Number Theory – ANTS III, LNCS* **1443**, Springer-Verlag, Berlin, 1998, 137–150.

41. B. A. Murphy, *Polynomial selection for the number field sieve integer factorisation algorithm*, Ph. D. thesis, Australian National University, July 1999.

42. B. A. Murphy and R. P. Brent, On quadratic polynomials for the number field sieve, *Australian Computer Science Communications* **20** (1998), 199–213. `http://www.comlab.ox.ac.uk/oucl/work/richard.brent/pub/pub178.html` .

43. A. M. Odlyzko, Discrete logarithms in finite fields and their cryptographic significance, *Advances in Cryptology: Proc. Eurocrypt '84, LNCS* **209**, Springer-Verlag, Berlin, 1985, 224–314.

44. A. M. Odlyzko, The future of integer factorization, *CryptoBytes* **1**, 2 (1995), 5–12. Available from `http://www.rsa.com/rsalabs/pubs/cryptobytes` .

45. J. M. Pollard, A Monte Carlo method for factorisation, *BIT* **15** (1975), 331–334.

46. C. Pomerance, The quadratic sieve factoring algorithm, *Advances in Cryptology, Proc. Eurocrypt '84, LNCS* **209**, Springer-Verlag, Berlin, 1985, 169–182.

47. C. Pomerance, The number field sieve, *Proceedings of Symposia in Applied Mathematics* **48**, Amer. Math. Soc., Providence, Rhode Island, 1994, 465–480.

48. C. Pomerance, A tale of two sieves, *Notices Amer. Math. Soc.* **43** (1996), 1473–1485.

49. C. Pomerance, J. W. Smith and R. Tuler, A pipeline architecture for factoring large integers with the quadratic sieve algorithm, *SIAM J. on Computing* **17** (1988), 387–403.

50. J. Preskill, *Lecture Notes for Physics 229: Quantum Information and Computation*, California Institute of Technology, Los Angeles, Sept. 1998. `http://www.theory.caltech.edu/people/preskill/ph229/` .

51. H. te Riele *et al*, *Factorization of a 512-bits RSA key using the number field sieve*, announcement of 26 August 1999, `http://www.loria.fr/~zimmerma/records/RSA155` .

52. H. J. J. te Riele, W. Lioen and D. Winter, Factoring with the quadratic sieve on large vector computers, *Belgian J. Comp. Appl. Math.* **27** (1989), 267–278.

53. H. Riesel, *Prime numbers and computer methods for factorization,* 2nd edition, Birkhäuser, Boston, 1994.

54. R. L. Rivest, A. Shamir and L. Adleman, A method for obtaining digital signatures and public-key cryptosystems, *Comm. ACM* **21** (1978), 120–126.

55. RSA Laboratories, Information on the RSA challenge, `http://www.rsa.com/rsalabs/html/challenges.html` .

56. R. S. Schaller, Moore's law: past, present and future, *IEEE Spectrum* **34**, 6 (June 1997), 52–59.

57. P. W. Shor, Algorithms for quantum computation: discrete logarithms and factoring, *Proc. 35th Annual Symposium on Foundations of Computer Science,* IEEE Computer Society Press, Los Alamitos, California, 1994, 124–134. CMP 98:06

58. P. W. Shor, Polynomial time algorithms for prime factorization and discrete logarithms on a quantum computer, *SIAM J. Computing* **26** (1997), 1484–1509.

59. R. D. Silverman, The multiple polynomial quadratic sieve, *Math. Comp.* **48** (1987), 329–339.

60. U. Vazirani, Introduction to special section on quantum computation, *SIAM J. Computing* **26** (1997), 1409–1410.

61. D. H. Wiedemann, Solving sparse linear equations over finite fields, *IEEE Trans. Inform. Theory* **32** (1986), 54–62.

62. J. Zayer, *Faktorisieren mit dem Number Field Sieve*, Ph. D. thesis, Universität des Saarlandes, 1995.

63. P. Zimmermann, *The ECMNET Project,* `http://www.loria.fr/~zimmerma/records/ecmnet.html`.

Approximating Uniform Triangular Meshes in Polygons

Franz Aurenhammer[1], Naoki Katoh[2] and Hiromichi Kojima[2], Makoto Ohsaki[2],
and Yinfeng Xu[3]

[1] Institute for Theoretical Computer Science, Graz University of Technology
Klosterwiesgasse 32/2, A-8010 Graz, Austria,
auren@igi.tu-graz.ac.at
[2] Department of Architecture and Architectural Systems, Kyoto University
Yoshida-Honmachi, Sakyo-ku, Kyoto, 606-8501 Japan
{naoki,kojima,ohsaki}@is-mj.archi.kyoto-u.ac.jp
[3] School of Management, Xi'an Jiaotong University
Xi'an, 710049 P.R.China,
yfxu@xjtu.edu.cn

1 Introduction

Given a convex polygon P in the plane and a positive integer n, we consider the problem of generating a length-uniform triangular mesh for the interior of P using n Steiner points. More specifically, we want to find both a set S_n of n points inside P, and a triangulation of P using S_n, with respect to the following minimization criteria: (1) ratio of the maximum edge length to the minimum one, (2) maximum edge length, and (3) maximum triangle perimeter.

These problems can be formalized as follows: Let V be the set of vertices of P. For an n-point set S_n interior to P let $\mathcal{T}(S_n)$ denote the set of all possible triangulations of $S_n \cup V$. Further, let $l(e)$ denote the (Euclidean) length of edge e, and let $peri(\Delta)$ be the perimeter of triangle Δ.

Problem 1 $\min_{S_n \subset P} \min_{T \in \mathcal{T}(S_n)} \max_{e,f \in T} \frac{l(e)}{l(f)}$

Problem 2 $\min_{S_n \subset P} \min_{T \in \mathcal{T}(S_n)} \max_{e \in T} l(e)$

Problem 3 $\min_{S_n \subset P} \min_{T \in \mathcal{T}(S_n)} \max_{\Delta \in T} peri(\Delta)$

Finding an optimal solution for any of the three problems seems to be difficult, in view of the NP-completeness of packing problems in the plane, see e.g. Johnson [10], or in view of the intrinsic complexity of Heilbronn's triangle problem, see [14]. For the case of a fixed point set, minimizing the maximum edge length is known to be solvable in $O(n^2)$ time; see Edelsbrunner and Tan [7]. Nooshin et al. [12] developed a potential-based heuristic method for Problem 2, but did not give a theoretical guarantee for the obtained solution.

In this paper, we offer an $O(n^2 \log n)$ heuristic capable of producing constant approximations for any of the three problems stated above. Respective approximation factors of 6, $4\sqrt{3}$, and $6\sqrt{3}$ are proven, provided n is reasonably

D.-Z. Du et al. (Eds.): COCOON 2000, LNCS 1858, pp. 23–33, 2000.

large. Our experiments reveal a much better behaviour, concerning the quality as well as the runtime. With minor modifications, our method works for *arbitrary* polygons (with possible holes), and yields the same approximation result for Problem 1. Concerning Problems 2 and 3, the approximation factors above can be guaranteed for a restricted class of non-convex polygons.

We first develop a heuristic we called *canonical Voronoi insertion* which approximately solves a certain extreme packing problem for point sets within P. The method is similar to the one used in Gonzalez [9] and Feder and Greene [8] developed for clustering problems. We then show how to modify the heuristic, in order to produce a set of n points whose Delaunay triangulation within P constitutes a constant approximation for the problems stated above. Note that the solution we construct is neccessarily a triangulation of constant vertex degree.

Generating triangular meshes is one of the fundamental problems in computational geometry, and has been extensively studied; see e.g. the survey article by Bern and Eppstein [3]. Main fields of applications are finite element methods and computer aided design. In finite element methods, for example, it is desirable to generate triangulations that do not have too large or too small angles. Along this direction, various algorithms have been reported [4,11,6,2,5,15]. Restricting angles means bounding the edge length ratio for the individual triangles, but not necessarily for a triangulation in global, which might be desirable in some applications. That is, the produced triangulation need not be uniform concerning, e.g., the edge length ratio of its triangles. Chew [6] and Melisseratos and Souvaine [11] construct uniform triangular meshes in the weaker sense that only upper bounds on the triangle size are required. To the knowledge of the authors, the problems dealt with in this paper have not been studied in the field of computational geometry. The mesh refinement algorithms in Chew [6] and in Ruppert [15] are similar in spirit to our Voronoi insertion method, but do not proceed in a canonical way and aim at different optimality criteria.

A particular application of length-uniform triangulation arises in designing structures such as plane trusses with triangular units, where it is required to determine the shape from aesthetic points of view under the constraints concerning stress and nodal displacement. The plane truss can be viewed as a triangulation of points in the plane by regarding truss members and nodes as edges and points, respectively. When focusing on the shape, edge lengths should be as equal as possible from the viewpoint of design, mechanics and manufacturing; see [12,13]. In such applications, the locations of the points are usually not fixed, but can be viewed as decision variables. In view of this application field, it is quite natural to consider Problems 1, 2, and 3.

The following notation will be used throughout. For two points x and y in the plane, let $l(x, y)$ denote their Euclidean distance. The minimum (non-zero) distance between two point sets X and Y is defined as $l(X, Y) = \min\{l(x, y) \mid x \in X, y \in Y, x \neq y\}$. When X is a singleton set $\{x\}$ we simply write $l(X, Y)$ as $l(x, Y)$. Note that $l(X, X)$ defines the minimum interpoint distance among the point set X.

2 Canonical Voronoi Insertion and Extreme Packing

In this section, we consider the following *extreme packing problem*. Let P be a (closed) convex polygon with vertex set V.

$$\text{Maximize } l(V \cup S_n, V \cup S_n)$$
$$\text{subject to a set } S_n \text{ of } n \text{ points within } P.$$

We shall give a 2-approximation algorithm for this problem using *canonical Voronoi insertion*. In Section 3 we then show that the point set S_n produced by this algorithm, as well as the Delaunay triangulation induced by S_n within P, can be modified to give an approximate solution for the three problems addressed in Section 1.

The algorithm determines the location of the point set S_n in a greedy manner. Namely, starting with an empty set S, it repeatedly places a new point inside P at the position which is farthest from the set $V \cup S$. The idea of the algorithm originates with Gonzalez [9] and Feder and Greene [8], and was developed for approximating minimax k-clusterings. Comparable insertion strategies are also used for mesh generation in Chew [6] and in Ruppert [15], there called *Delaunay refinement*. Their strategies aim at different quality measures, however, and insertion does not take place in a canonical manner.

The algorithm uses the Voronoi diagram of the current point set to select the next point to be inserted. We assume familiarity with the basic properties of a Voronoi diagram and its dual, the Delaunay triangulation, and refer to the survey paper [1].

Algorithm INSERT

Step 1: Initialize $S := \emptyset$.

Step 2: Compute the Voronoi diagram $\text{Vor}(V \cup S)$ of $V \cup S$.

Step 3: Find the set B of intersection points between edges of $\text{Vor}(V \cup S)$ and the boundary of P. Among the points in B and the vertices of $\text{Vor}(V \cup S)$ inside P, choose the point u which maximizes $l(u, V \cup S)$.

Step 4: Put $S := S \cup \{u\}$ and return to Step 2 if $|S| < n$.

Let p_j and S_j, respectively, denote the point chosen in Step 3 and the set obtained in Step 4 at the j-th iteration of the algorithm. For an arbitrary point $x \in P$ define the *weight* of x with respect to S_j as $w_j(x) = l(x, S_j \cup V)$. That is, $w_j(x)$ is the radius of the largest circle centered at x which does not enclose any point from $S_j \cup V$. By definition of a Voronoi diagram, the point p_j maximizes $w_{j-1}(x)$ over all $x \in P$. Let

$$d_n = l(S_n \cup V, S_n \cup V) \tag{1}$$

be the minimum interpoint distance realized by $S_n \cup V$. Furthermore, denote by S_n^* the optimal solution for the extreme packing problem for P and let d_n^* denote the corresponding objective value. The following approximation result might be of interest in its own right. Its proof is an adaptation of techniques in [9,8] and contains observations that will be used in our further analysis.

Theorem 1. *The solution S_n obtained by Algorithm INSERT is a 2-approximation of the extreme packing problem for $\boldsymbol{P}$. That is, $d_n \geq d_n^*/2$.*

Proof. We claim that p_n realizes the minimum (non-zero) distance from S_n to $S_n \cup V$. Equivalently, the claim is

$$w_{n-1}(p_n) = l(S_n, S_n \cup V). \tag{2}$$

To see this, assume that the minimum distance is realized by points p_k and p_j different from p_n. Let p_k be inserted after p_j by the algorithm. Then we get $w_{k-1}(p_k) \leq l(p_k, p_j) < l(p_n, S_{n-1} \cup V) = w_{n-1}(p_n)$. On the other hand, the sequence of weights chosen by the algorithm must be non-increasing. More exactly, $w_{k-1}(p_k) \geq w_{k-1}(p_n) \geq w_{n-1}(p_n)$. This is a contradiction.

The observations $d_n = \min\{l(S_n, S_n \cup V), l(V, V)\}$ and $l(V, V) \geq d_n^* \geq d_n$ now imply $d_n = \min\{w_{n-1}(p_n), d_n^*\}$ by (2). But p_n maximizes $w_{n-1}(x)$ for all $x \in \boldsymbol{P}$. So the lemma below (whose proof is omitted) completes the argument.

Lemma 1. *For any set $S \subset \boldsymbol{P}$ of $n-1$ points there exists a point $x \in \boldsymbol{P}$ with $l(x, S \cup V) \geq d_n^*/2$.*

3   Delaunay Triangulation of Bounded Edge Ratio

Our aim is to show that Algorithm INSERT is capable of producing a point set appropriate for Problems 1, 2, and 3. To this end, we first investigate the Delaunay triangulation $\mathrm{DT}(S_n \cup V)$ of $S_n \cup V$. This triangulation is implicitly constructed by the algorithm, as being the dual structure of $\mathrm{Vor}(S_n \cup V)$. However, $\mathrm{DT}(S_n \cup V)$ need not exhibit good edge length properties. We therefore prescribe the placement of the first k inserted points, and show that Algorithm INSERT completes them to a set of n points whose Delaunay triangulation has its edge lengths controlled by the minimum interpoint distance d_n for $S_n \cup V$.

3.1   Triangle types. For $1 \leq j \leq n$, consider the triangulation $\mathrm{DT}(S_j \cup V)$. Let us classify a triangles Δ of $\mathrm{DT}(S_j \cup V)$ as either *critical* or *non-critical*, depending on whether the Voronoi vertex dual to Δ (i.e., the circumcenter of Δ) lies outside of the polygon $\boldsymbol{P}$ or not. Whereas edges of critical triangles can be arbitrarily long, edge lengths are bounded in non-critical triangles.

Lemma 2. *No edge e of a non-critical triangle Δ of $DT(S_j \cup V)$ is longer than $2 \cdot w_{j-1}(p_j)$.*

Proof. Let $e = (p, q)$ and denote with x the Voronoi vertex dual to Δ. As x lies inside of $\boldsymbol{P}$, we get $l(x, p) = l(x, q) = w_{j-1}(x) \leq w_{j-1}(p_j)$, by the choice of point p_j in Step 3 of Algorithm INSERT. The triangle inequality now implies $l(p, q) \leq 2 \cdot w_{j-1}(p_j)$.

We make an observation on critical triangles. Consider some edge e of $DT(S_j \cup V)$ on the boundary of $\boldsymbol{P}$. Edge e cuts off some part of the diagram $Vor(S_j \cup V)$ that is outside of $\boldsymbol{P}$. If that part contains Voronoi vertices then we define the *critical region*, $R(e)$, for e as the union of all the (critical) triangles that are dual to these vertices. Notice that each critical triangle of $DT(S_j \cup V)$ belongs to a unique critical region.

Lemma 3. *No edge f of a critical triangle in $R(e)$ is longer than $l(e)$.*

Proof. Let p be an endpoint of f. Then the region of p in $Vor(S_j \cup V)$ intersects e. Let x be a point in this region but outside of $\boldsymbol{P}$. There is a circle around x that encloses p but does not enclose any endpoint of e. Within $\boldsymbol{P}$, this circle is completely covered by the circle C with diameter e. This implies that p lies in C. As the distance between any two points in C is at most $l(e)$, we get $l(f) \leq l(e)$.

Let us further distinguish between *interior* triangles and *non-interior* ones, the former type having no two endpoints on the boundary of $\boldsymbol{P}$. The shortest edge of an interior triangle can be bounded as follows.

Lemma 4. *Each edge e of an interior triangle Δ of $DT(S_j \cup V)$ has a length of at least $w_{j-1}(p_j)$.*

Proof. We have $l(e) \geq l(S_j, S_j \cup V)$, because Δ has no two endpoints on $\boldsymbol{P}$'s boundary. But from (2) we know $l(S_j, S_j \cup V) = w_{j-1}(p_j)$.

3.2 Edge length bounds.

We are now ready to show how a triangulation with edge lengths related to d_n can be computed. First, Algorithm INSERT is run on $\boldsymbol{P}$, in order to compute the value d_n. We assume than n is chosen sufficiently large to assure $d_n \leq l(V, V)/2$. This assumption is not unnatural as the shortest edge of the desired triangulation cannot be longer than the shortest edge of $\boldsymbol{P}$. After having d_n available, k points $p'_1, \ldots, p'_k$ are placed on the boundary of $\boldsymbol{P}$, with consecutive distances between $2 \cdot d_n$ and $3 \cdot d_n$, and such that $l(V', V') \geq d_n$ holds, for $V' = V \cup \{p'_1, \ldots, p'_k\}$. Notice that such a placement is always possible. Finally, $n - k$ additional points $p'_{k+1}, \ldots, p'_n$ are produced by re-running Algorithm INSERT after this placement.

For $1 \leq j \leq n$, let $S'_j = \{p'_1, \ldots, p'_j\}$. Define $w(x) = l(x, S'_n \cup V)$ for a point $x \in \boldsymbol{P}$. The value of $w(p'_n)$ will turn out to be crucial for analyzing the edge length behavior of the triangulation $DT(S'_n \cup V)$. The lemma below asserts that $w(p'_n)$ is small if n exceeds twice the number k of prescribed points.

Lemma 5. *Suppose $n \geq 2k$. Then $w(p'_n) \leq 3 \cdot d_n$.*

Proof. The point set S_n produced by Algorithm INSERT in the first run is large enough to ensure $d_n < l(V, V)$. So we get $d_n = w_{n-1}(p_n)$ from (2). As point p_n maximizes $w_{n-1}(x)$ for all $x \in \boldsymbol{P}$, the $n + |V|$ circles centered at the points in $S_n \cup V$ and with radii d_n completely cover the polygon $\boldsymbol{P}$. Let $d_n = 1$ for the moment. Then

$$A(\boldsymbol{P}) \leq \pi(n + |V|) - A' \tag{3}$$

where $A(P)$ is the area of P, and A' denotes the area outside of P which is covered by the circles centered at V.

Assume now $w(p'_n) > 3 \cdot d_n$. Draw a circle with radius $\frac{3}{2} d_n$ around each point in $S'_n \setminus S'_k$. Since $w(p'_n) = l(S'_n \setminus S'_k, S'_n \cup V)$ by (2), these circles are pairwise disjoint. By the same reason, and because boundary distances defined by $V' = V \cup S'_k$ are at most $3 \cdot d_n$, these circles all lie completely inside P. Obviously, these circles are also disjoint from the $|V|$ circles of radius d_n centered at V. Finally, the latter circles are pairwise disjoint, since $d_n \leq l(V, V)/2$. Consequently,

$$A(P) \geq \frac{9}{4}\pi(n - k) + A''$$

(4)

where A'' denotes the area inside of P which is covered by the circles centered at V. Combining (3) and (4), and observing $A' + A'' = \pi \cdot |V|$ now implies $n < 2k$, a contradiction.

It has to be observed that the number k depends on n. The following fact guarantees the assumption in Lemma 5, provided n is sufficiently large. Let $B(P)$ denote the perimeter of P.

Lemma 6. *The condition $d_n \leq A(P)/(\pi \cdot B(P))$ implies $n \geq 2k$.*

Proof. By (3) we have

$$n \geq \frac{A(P)}{\pi \cdot (d_n)^2} - |V|.$$

To get a bound on k, observe that at most $l(e)/2d_n - 1$ points are placed on each edge e of P. This sums up to

$$k \leq \frac{B(P)}{2d_n} - |V|.$$

Simple calculations now show that the condition on d_n stated in the lemma implies $n \geq 2k$.

The following is a main theorem of this paper.

Theorem 2. *Suppose n is large enough to assure the conditions $d_n \leq l(V, V)/2$ and $d_n \leq A(P)/(\pi \cdot B(P))$. Then no edge in the triangulation $T^+ = DT(S'_n \cup V)$ is longer than $6 \cdot d_n$. Moreover, T^+ exhibits an edge length ratio of 6.*

Proof. Two cases are distinguished, according to the value of $w(p'_n)$.

Case 1: $w(p'_n) < d_n$. Concerning upper bounds, Lemma 2 implies $l(e) \leq 2 \cdot w(p'_n) < 2 \cdot d_n$ for all edges e belonging to non-critical triangles of T^+. If e belongs to some critical triangle, Lemma 3 shows that $l(e)$ cannot be larger than the maximum edge length on the boundary of P, which is at most $3 \cdot d_n$ by construction. Concerning lower bounds, Lemma 4 gives $l(e) \geq w(p'_n)$ for edges of interior triangles. We know $w(p'_n) \geq d^*_n/2$ from Lemma 1, which implies $l(e) \geq d_n/2$ because $d^*_n \geq d_n$. For edges spanned by V', we trivially obtain $l(e) \geq d_n$ as $l(V', V') \geq d_n$ by construction.

Case 2: $w(p'_n) \geq d_n$. The upper bound $2 \cdot w(p'_n)$ for non-critical triangles now gives $l(e) \leq 6 \cdot d_n$, due to Lemmas 5 and 6. The lower bound for interior triangles becomes $l(e) \geq w(p'_n) \geq d_n$. The remaining two bounds are the same as in the former case.

3.3 Computational issues.

The time complexity of computing the triangulation T^+ is dominated by Steps 2 and 3 of Algorithm INSERT. In the very first iteration of the algorithm, both steps can be accomplished in $O(|V| \log |V|)$ time. In each further iteration j we update the current Voronoi diagram under the insertion of a new point p_j in Step 2, as well as a set of weights for the Voronoi vertices and relevant polygon boundary points in Step 3.

Since we already know the location of the new point p_j in the current Voronoi diagram, the region of p_j can be integrated in time proportional to the degree of p_j in the corresponding Delaunay triangulation, $\deg(p_j)$. We then need to calculate, insert, and delete $O(\deg(p_j))$ weights, and then select the largest one in the next iteration. This gives a runtime of $O(\deg(p_j) \cdot \log n)$ per iteration.

The following lemma bounds the number of constructed triangles, of a certain type. Let us call a triangle *good* if it is both interior and non-critical.

Lemma 7. *The insertion of each point p_j creates only a constant number of good triangles.*

Proof. Consider the endpoints of all good triangles incident to p_j in DT$(S_j \cup V)$, and let X be the set of all such endpoints interior to $\mathbf{P}$. Then $l(X, X) \geq l(S_j, S_j) \geq w_{j-1}(p_j)$, due to (2). On the other hand, by Lemma 2, X lies in the circle of radius $2 \cdot w_{j-1}(p_j)$ around p_j. As a consequence, $|X|$ is constant. The number of good triangles incident to p_j is at most $2 \cdot |X|$, as one such triangle would have two endpoints on $\mathbf{P}$'s boundary, otherwise.

For most choices of $\mathbf{P}$ and n, the good triangle type will be most frequent. Note also that the degree of *all* points in the final triangulation T^+ has to be constant.

In conclusion, we obtain a runtime bound of $O(n^2 \log n)$ and a space complexity of $O(n)$. However, Lemma 7 suggests a runtime of $O(\log n)$ in most iterations.

4   Approximation Results

Let us now return to the three optimization problems for the polygon $\mathbf{P}$ posed in the introduction. We will rely on Theorem 2 in the following. Recall that, in order to make the theorem hold, we have to choose n sufficiently large.

Theorem 3. *The triangulation T^+ approximates the optimal solution for Problem 1 by a factor of 6.*

Proof. Theorem 2 guarantees for T^+ an edge length ratio of 6, and for no triangulation this ratio can be smaller than 1.

We now turn our attention to Problem 2. Let the point set $\tilde{S}$ in conjunction with the triangulation $\tilde{T}$ of $\tilde{S} \cup V$ be the corresponding optimum solution. Let d_{long} denote the optimum objective value, that is, d_{long} measures the longest edge in $\tilde{T}$. The lemma below relates d_{long} to the optimum value d_n^* for the extreme packing problem for $\boldsymbol{P}$. The proof is omitted in this extended abstract.

Lemma 8. $$d_{long} \geq \tfrac{\sqrt{3}}{2} d_n^*.$$

We strongly conjecture that the statement of Lemma 8 can be strengthened to $d_{long} \geq d_n^*$, which would improve the bounds in Theorems 4 and 5 below.

Theorem 4. *The triangulation T^+ constitutes a $4\sqrt{3}$–approximation for Problem 2.*

Proof. Let e_{max} denote the longest edge in T^+. By Theorem 2 we have $l(e_{max}) \leq 6 \cdot d_n$. Trivially $d_n \leq d_n^*$ holds, and Lemma 8 implies the theorem, $l(e_{\max})/d_{long} \leq 4\sqrt{3}$.

Finally let us consider Problem 3. Let d_{peri} denote the optimum objective value for this problem. We show the following:

Theorem 5. *The triangulation T^+ gives a $6\sqrt{3}$–approximation for Problem 3.*

Proof. For any triangulation of $\boldsymbol{P}$ with n Steiner points, its longest edge cannot be shorter than $\tfrac{\sqrt{3}}{2} \cdot d_n^*$ by Lemma 8. This implies $d_{peri} \geq \sqrt{3} \cdot d_n^*$ by the triangle inequality. On the other hand, for the longest edge e_{max} of T^+ we have $l(e_{max}) \leq 6 \cdot d_n^*$ due to Theorem 2. The longest triangle perimeter δ_{max} that occurs in T^+ is at most $3 \cdot l(e_{max})$. In summary, $\delta_{max}/d_{peri} \leq 6 \cdot \sqrt{3}$.

We conclude this section by mentioning an approximation result concerning minimum-weight triangulations. The easy proof is omitted.

Theorem 6. *Let S^+ be the vertex set of T^+ and let $MWT(S^+)$ denote the minimum-weight triangulation of S^+. Then T^+ is a 6–length approximation for $MWT(S^+)$.*

5   Experimental Results

We have performed computational experiments in order to see the effectiveness of the proposed algorithm. For space limitations, we focus on Problem 1 and we only give detailed results for two typical convex polygons. The first polygon is rather fat while the second one is skinny and has a very long edge. The length of the shortest edge in both polygons is roughly equal to d_n for $n = 50$. We have tested the four cases of $n = 50, 100, 200, 300$.

As we described in Section 3, Algorithm INSERT places points on the boundary in the first run, and then is restarted to produce the triangulation. Thereby, the consecutive distance between placed points need to be set to $2d_n \sim 3d_n$ in

points	consec_dist	edge ratio	flips run 1	flips run 2	edge ratio	flips run 1	flips run 2
50	$1 \sim 2$	2.67	81	97	5.08	65	65
	$2 \sim 3$	3.96	81	86	5.43	65	64
100	$1 \sim 2$	2.00	180	214	3.18	147	191
	$2 \sim 3$	2.61	180	174	3.62	147	158
200	$1 \sim 2$	1.96	371	475	2.08	335	427
	$2 \sim 3$	3.23	371	392	2.71	335	343
300	$1 \sim 2$	2.04	567	700	2.1	531	657
	$2 \sim 3$	3.85	567	580	2.5	531	563

order to ensure the desired results theoretically. However, the case of $d_n \sim 2d_n$ is also tested in our experiments. The computational results are summarized below (fat polygon left, skinny polygon right).

We observe that (1) the actual edge length ratio is much better than the worst-case ratio of 6 given by Theorem 3, and (2) the number of flips per insertion of a point is very small (roughly two). Although the example polygons do not satisfy $d_n < l(V, V)/2$ (as required by Theorem 3) we obtained much better ratios. This exhibits the practical effectiveness of the proposed algorithm.

Note also that for a fat polygon (left) the choice of consecutive distance of $d_n \sim 2d_n$ produces a better ratio, while for a skinny polygon (right) the contrary can be observed.

6   Discussion and Extensions

We have considered the problem of generating length-uniform triangular meshes for the interior of convex polygons. A unifying algorithm capable of computing constant approximations for several criteria has been developed. The basic idea has been to relate the length of triangulation edges to the optimum extreme packing distance. The method is easy to implement and seems to produce acceptably good triangular meshes as far as computational experiments are concerned.

In practical applications, more general input polygons need to be triangulated. We stress that our algorithm works with minor modification for arbitrary polygons with possible holes. Convexity is used solely in the proof of Lemma 8. As a consequence, Theorems 1 and 2, the approximation result for Problem 1, and Theorem 6 still hold. The modification needed is that *visible distances* in a non-convex polygon P should be considered only, in the proofs as well as concerning the algorithm. That is, for the point sets $S \subset P$ in question, the Delaunay triangulation of $S \cup V$ *constrained by* P has to be utilized rather than $DT(S \cup V)$.

The proof of Lemma 8 (and with it the approximation results for Problems 2 and 3) still go through for non-convex polygons P with interior angles of at most $\frac{3\pi}{2}$, provided n is large enough to make the value $\frac{2}{\sqrt{3}}d_{long}$ fall short of the minimum distance between non-adjacent edges of P. We pose the question of establishing a version of Lemma 8 for general non-convex polygons, and of improving the respective bound $\frac{\sqrt{3}}{2}$ for the convex case.

Viewed from the point of applications to the design of structures, it is also important to generate a triangular mesh for approximating surfaces such as large-span structures. For this direction, our result concerning Problem 1 can be extended to spherical polygons. More precisely, given a convex polygon whose vertices lie on a hemisphere (or a smaller region of a sphere cut by a plane), the problem is to find a triangular mesh whose points are on the hemisphere that minimizes the objective function of Problem 1. It can be shown that the algorithm obtained by appropriately modifying Algorithm INSERT attains a $3\sqrt{5}$ approximation ratio.

Acknowledgements. This research was partially supported by the Austrian Spezial-forschungsbereich F003, Optimierung und Kontrolle, and the Grant-in-Aid for Scientific Research on Priority Areas (B) (No. 10205214) by the Ministry of Education, Science, Sports and Culture of Japan. In addition, the fifth author was also supported by NSF of China (No. 19731001) and the Japan Society for the Promotion of Science of Japan. We gratefully acknowledge all these supports.

References

1. F. Aurenhammer, "Voronoi diagrams – a survey of a fundamental geometric data structure", *ACM Computing Surveys* 23 (1991), 345-405.
2. M. Bern, D. Dobkin and D. Eppstein, "Triangulating polygons without large angles", *Intl. J. Comput. Geom. and Appl.* 5 (1995), 171-192.
3. M. Bern and D. Eppstein, "Mesh generation and optimal triangulation", in D.-Z. Du (ed.), *Computing in Euclidean Geometry*, World Scientific Publishing, 1992, 47-123.
4. M. Bern, D. Eppstein and J.R. Gilbert, "Provably good mesh generation", *Journal of Computer and System Sciences* 48 (1994), 384-409.
5. M. Bern, S. Mitchell and J. Ruppert, "Linear-size nonobtuse triangulation of polygons", Proceedings of the 10th Ann. ACM Symposium on Computational Geometry (1994), 221-230.
6. P. Chew, "Guaranteed-Quality Mesh Generation for Curved Surfaces", Proceedings of the 9th Ann. ACM Symposium on Computational Geometry (1993), 274-280.
7. H. Edelsbrunner and T.S. Tan, "A quadratic time algorithm for the minmax length triangulation", *SIAM Journal on Computing* 22 (1993), 527-551.
8. T. Feder and D.H. Greene, "Optimal Algorithms for Approximate Clustering", Proceedings of the 20th Ann. ACM Symposium STOC (1988), 434-444.
9. T. Gonzalez, "Clustering to minimize the maximum intercluster distance", *Theoretical Computer Science* 38 (1985), 293-306.
10. D.S. Johnson, "The NP-completeness column: An ongoing guide", *Journal of Algorithms* 3 (1982), 182-195.
11. E. Melisseratos and D. Souvaine, "Coping with inconsistencies: A new approach to produce quality triangulations of polygonal domains with holes", Proceedings of the 8th Ann. ACM Symposium on Computational Geometry (1992),202-211.
12. H. Nooshin, K. Ishikawa, P.L. Disney and J.W. Butterworth, "The traviation process", *Journal of the International Association for Shell and Spatial Structures* 38 (1997), 165-175.

13. M. Ohsaki, T. Nakamura and M. Kohiyama, "Shape optimization of a double-layer space truss described by a parametric surface", *International Journal of Space Structures* 12 (1997), 109-119.
14. K.F. Roth, "On a problem of Heilbronn", *Proc. London Mathematical Society* 26 (1951), 198-204.
15. J. Ruppert, "A Delaunay Refinement Algorithm for Quality 2-Dimensional Mesh Generation", *Journal of Algorithms* 18 (1995), 548-585.

Maximum Induced Matchings
of Random Cubic Graphs

William Duckworth[1], Nicholas C. Wormald[1]*, and Michele Zito[2]**

[1] Department of Mathematics and Statistics, University of Melbourne, Australia
{billy,nick}@ms.unimelb.edu.au
[2] Department of Computer Science, University of Liverpool, UK
michele@csc.liv.ac.uk

Abstract. In this paper we present a heuristic for finding a large induced matching $\mathcal{M}$ of cubic graphs. We analyse the performance of this heuristic, which is a random greedy algorithm, on random cubic graphs using differential equations and obtain a lower bound on the expected size of the induced matching returned by the algorithm. The corresponding upper bound is derived by means of a direct expectation argument. We prove that $\mathcal{M}$ asymptotically almost surely satisfies $0.2704n \leq |\mathcal{M}| \leq 0.2821n$.

1 Introduction

An *induced matching* of a graph $G = (V, E)$ is a set of vertex disjoint edges $\mathcal{M} \subseteq E$ with the additional constraint that no two edges of $\mathcal{M}$ are connected by an edge of $E \setminus \mathcal{M}$. We are interested in finding induced matchings of large cardinality.

Stockmeyer and Vazirani [11] introduced the problem of finding a maximum induced matching of a graph, motivating it as the "risk-free marriage problem" (find the maximum number of married couples such that each person is compatible only with the person (s)he is married to). This in turn stimulated much interest in other areas of theoretical computer science and discrete mathematics as finding a maximum induced matching of a graph is a sub-task of finding a strong edge-colouring of a graph (a proper colouring of the edges such that no edge is incident with more than one edge of the same colour as each other, see (for example) [5,6,9,10]).

The problem of deciding whether for a given integer k a given graph G has an induced matching of size at least k is NP-Complete [11], even for bipartite graphs of maximum degree 4. It has been shown [3,13] that the problem is APX-complete even when restricted to ks-regular graphs for $k \in \{3, 4\}$ and any integer $s \geq 1$. The problem of finding a maximum induced matching is polynomial-time solvable for chordal graphs [2] and circular arc graphs [7].

---

* Supported by the Australian Research Council
** Supported by EPSRC grant GR/L/77089

D.-Z. Du et al. (Eds.): COCOON 2000, LNCS 1858, pp. 34–43, 2000.

Recently Golumbic and Lewenstein [8] have constructed polynomial-time algorithms for maximum induced matching in trapezoid graphs, interval-dimension graphs and co-comparability graphs, and have given a linear-time algorithm for maximum induced matching in interval graphs.

In this paper we present a heuristic for finding a large induced matching $\mathcal{M}$ of cubic graphs. We analyse the performance of this heuristic, which is a random greedy algorithm, on random cubic graphs using differential equations and obtain a lower bound on the expected size of the induced matching returned by the algorithm. The corresponding upper bound is derived by means of a direct expectation argument. We prove that $\mathcal{M}$ asymptotically almost surely satisfies $0.2704n \leq |\mathcal{M}| \leq 0.2821n$. Little is known on the complexity of this problem under the additional assumption that the input graphs occur with a given probability distribution. Zito [14] presented some simple results on dense random graphs. The algorithm we present was analysed deterministically in [3] where it was shown that, given an n-vertex cubic graph, the algorithm returns an induced matching of size at least $3n/20 + o(1)$, and there exist infinitely many cubic graphs realising this bound.

Throughout this paper we use the notation $\mathbf{P}$ (probability), $\mathbf{E}$ (expectation), u.a.r. (uniformly at random) and a.a.s. (asymptotically almost surely). When discussing any cubic graph on n vertices, we assume n to be even to avoid parity problems.

In the following section we introduce the model used for generating cubic graphs u.a.r. and in Section 3 we describe the notion of analysing the performance of algorithms on random graphs using a system of differential equations. Section 4 gives the randomised algorithm and Section 5 gives its analysis showing the a.a. sure lower bound. In Section 6 we give a direct expectation argument showing the a.a. sure upper bound.

2   Uniform Generation of Random Cubic Graphs

The model used to generate a cubic graph u.a.r. (see for example Bollobás [1]) can be summarised as follows. For an n vertex graph

- take $3n$ points in n buckets labelled $1 \ldots n$ with three points in each bucket and
- choose u.a.r. a perfect matching of the $3n$ points.

If no edge of the matching contains two points from the same bucket and no two edges contain four points from just two buckets then this represents a cubic graph on n vertices with no loops and no multiple edges. With probability bounded below by a positive constant, loops and parallel edges do not occur [1]. The edges of the random cubic graph generated are represented by the edges of the matching and its vertices are represented by the buckets.

We may consider the generation process as follows. Initially, all vertices have degree 0. Throughout the execution of the generation process, vertices will increase in degree until the generation is complete and all vertices have degree 3. We refer to the graph being generated by this process as the *evolving graph*.

3   Analysing Algorithms Using Differential Equations

We incorporate the algorithm as part of a matching process generating a random cubic graph. During the generation of a random cubic graph, we choose the matching edges sequentially. The first end-point of a matching edge may be chosen by any rule, but in order to ensure that the cubic graph is generated u.a.r., the second end-point of that edge must be selected u.a.r. from all the remaining free points. The freedom of choice of the first end-point of a matching edge enables us to select it u.a.r. from the vertices of given degree in the evolving graph.

The algorithm we use to generate an induced matching of cubic graphs is a greedy algorithm based on choosing a vertex of a given degree and performing some edge and vertex deletions. In order to analyse our algorithm using a system of differential equations, we generate the random graph in the order that the edges are examined by the algorithm.

At some stage of the algorithm, a vertex is chosen of given degree based upon the number of free points remaining in each bucket. The remaining edges incident with this vertex are then *exposed* by randomly selecting a mate for each spare point in the bucket. This allows us to determine the degrees of the neighbours of the chosen vertex and we refer to this as *probing* those vertices. This is done without exposing the other edges incident with these probed vertices. Once the degrees of the neighbours of the chosen vertex are known, an edge is selected to be added to the induced matching. Further edges are then exposed in order to ensure the matching is induced. More detail is given in the following section.

In what follows, we denote the set of vertices of degree i of the evolving graph by V_i and let $Y_i(= Y_i(t))$ denote $|V_i|$ (at time t). The number of edges in the induced matching at any stage of the algorithm (time t) is denoted by $M(= M(t))$ and we let $s(= s(t))$ denote the number of free points available in buckets at any stage of the algorithm (time t). Note that $s = \sum_{i=0}^{2}(3-i)Y_i$. Let $\mathbf{E}(\Delta X)$ denote the expected change in a random variable X conditional upon the history of the process.

One method of analysing the performance of a randomised algorithm is to use a system of differential equations to express the expected changes in variables describing the state of an algorithm during its execution (see [12] for an exposition of this method).

We can express the state of the evolving graph at any point during the execution of the algorithm by considering Y_0, Y_1 and Y_2. In order to analyse our randomised algorithm for finding an induced matching of cubic graphs, we calculate the expected change in this state over one time step in relation to the expected change in the size M of the induced matching. We then regard $\mathbf{E}(\Delta Y_i)/\mathbf{E}(\Delta M)$ as the derivative $\mathrm{d}Y_i/\mathrm{d}M$, which gives a system of differential equations. The solution to these equations describes functions which represent the behaviour of the variables Y_i. There is a general result which guarantees that the solution of the differential equations almost surely approximates the variables Y_i. The expected size of the induced matching may be deduced from these results.

4   The Algorithm

In order to find an induced matching of a cubic graph, we use the following algorithm. We assume the generated graph G to be connected (since random cubic graphs are connected a.a.s.). Then, at any stage of the algorithm after the first step and before its completion, $Y_1 + Y_2 > 0$. The degree of a vertex v in the evolving graph is denoted by $deg(v)$. We denote the initial set of $3n$ points by P. $B(p)$ denotes the bucket that a given point p belongs to and we use $q(b)$ to denote the set of free points in bucket b. The set of induced matching edges returned is denoted by $\mathcal{M}$. Here is the algorithm; a description is given below.

select(p_1, P);
expose(p_1, p_2);
isolate$(B(p_1), B(p_2))$;
$\mathcal{M} \leftarrow (B(p_1), B(p_2))$;
while $(Y_1 + Y_2 > 0)$
  { **if** $(Y_2 > 0)$
    { **select**(u, V_2); $\{p_1\} \leftarrow q(u)$;
      expose(p_1, p_2); $v \leftarrow B(p_2)$;}
    else
    { **select**(u, V_1); $\{p_1, p_2\} \leftarrow q(u)$;
      expose(p_1, p_3); $a \leftarrow B(p_3)$;
      expose(p_2, p_4); $b \leftarrow B(p_4)$;
      if $(deg(a) > deg(b))$ $v \leftarrow a$;
      else if $(deg(b) > deg(a))$ $v \leftarrow b$;
      else   **select**$(v, \{a, b\})$;}
    isolate(u, v);
    $\mathcal{M} \leftarrow \mathcal{M} \cup (u, v)$;
  }

The function **select**(s, S) involves the process of selecting the element s u.a.r. from the set S. The function **expose**(p_i, p_j) involves the process of deleting the selected point p_i from P, exposing the edge $(B(p_i), B(p_j))$ by randomly selecting the point p_j from P and deleting the point p_j from P. The function **isolate**(B_1, B_2) involves the process of randomly selecting a mate for each free point in the buckets B_1 and B_2 and then exposing all edges incident with these selected mates. This ensures that the matching is induced.

The first step of the algorithm involves randomly selecting the first edge of the induced matching and exposing the appropriate edges. We split the remainder of the algorithm into two distinct phases. We informally define Phase 1 as the period of time where any vertices in V_2 that are created are used up almost immediately and Y_2 remains small. Once the rate of generating vertices in V_2 becomes larger than the rate that they are used up, the algorithm moves into Phase 2 and the majority of operations involve selecting a vertex from V_2 and including its incident edge in the induced matching.

There are two basic operations performed by the algorithm. Type 1 refers to a step when $Y_2 > 0$ and a vertex is chosen from V_2. Similarly, Type 2 refers to a step where $Y_2 = 0$ and a vertex is chosen from V_1.

Figs. 1 and 2 show the configurations that may be encountered performing operations of Type 1 and Type 2 respectively (a.a.s.). For Type 1, we add the edge incident with the chosen vertex. For Type 2, if (after exposing the edges incident with the chosen vertex u) exactly one of the neighbours of u has exactly two free points, we add the edge incident with this vertex and the chosen vertex. Otherwise we randomly choose an edge incident with the chosen vertex.

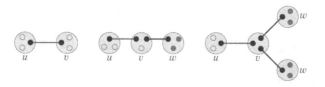

Fig. 1. Selecting a vertex from V_2 and adding its incident edge to $\mathcal{M}$.

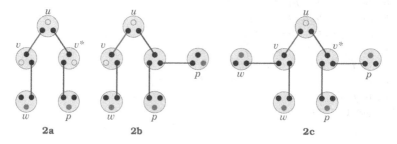

Fig. 2. Selecting a vertex from V_1 and choosing an edge to add to $\mathcal{M}$.

The larger circles represent buckets each containing 3 smaller circles representing the points of that bucket. Smaller circles coloured black represent points that are free. Used points are represented by white circles. Points which are not known to be free at this stage of the algorithm are shaded.

In all cases, the selected vertex is labelled u and the other end-point of the induced matching edge chosen is labelled v. A vertex labelled v^* denotes that a random choice has been made between 2 vertices and this one was not selected. After selecting a vertex u of given degree, the edges incident with this vertex are exposed. Once we probe the degrees of the neighbours of this vertex, we then make the choice as to which edge to add to the induced matching. Only then are other edges exposed. Therefore, at this stage, we do not know the degrees of all the vertices at distance at most two from the end-points of the selected induced matching edge. A vertex whose degree is unknown is labelled either w

or p. A vertex labelled p will have one of its incident edges exposed and will subsequently have its degree increased by one. We refer to these vertices as *incs*. A vertex labelled w will have all of its incident edges exposed and we refer to these vertices as *rems*. Should any *rem* be incident with other vertices of unknown degree then these vertices will be *incs*.

Once the choice of induced matching edge has been made, all edges incident with the endpoints of the induced matching edge are exposed and subsequently the edges of the neighbours of the end-points of the chosen edge are exposed. This ensures the matching is induced.

5   The Lower Bound

Theorem 1. *For a cubic graph on n vertices the size of a maximum induced matching is asymptotically almost surely greater than $0.2704n$*

Proof. We define a *clutch* to be a series of operations in Phase i involving the selection a vertex from V_i and all subsequent operations up to but not including the next selection of a vertex from V_i. Increment time by 1 step for each clutch of vertices processed. We calculate $\mathbf{E}(\Delta Y_i)$ and $\mathbf{E}(\Delta M)$ for a clutch in each Phase.

5.1   Preliminary Equations for Phase 1

The initial step of Phase 1 is of Type 2 (at least a.a.s.). We consider operations of Type 1 first and then combine the equations given by these operations with those given by the operations of Type 2.

Operations of Type 1 involve the selection of a vertex u from V_2 (which has been created from processing a vertex from V_1). The expected change in Y_2 is negligible and can assumed to be 0 since this is phase 1 and Y_2 remains small as vertices of V_2 are used up almost as fast as they are created. The desired equation may be formulated by considering the contribution to $\mathbf{E}(\Delta Y_i)$ in three parts; deleting v (the neighbour of u), deleting the neighbour(s) of v and increasing the degree of the other neighbours of these neighbours by one.

Let ρ_i denote the contribution to $\mathbf{E}(\Delta Y_i)$ due to changing the degree of an *inc* from i to $i+1$ and we have

$$\rho_i = \frac{(i-3)Y_i + (4-i)Y_{i-1}}{s}.$$

This equation is valid under the assumption that $Y_{-1} = 0$.

We let $\mu_i = \mu_i(t)$ denote the contribution to $\mathbf{E}(\Delta Y_i)$ due to a *rem* and all its edges incident with *incs* and we have

$$\mu_i = \frac{(i-3)}{s} Y_i + \frac{(6Y_0 + 2Y_1)}{s} \rho_i.$$

Let $\alpha_i = \alpha_i(t)$ denote the contribution to $\mathbf{E}(\Delta Y_i)$ for an operation of Type 1 in Phase 1. We have

$$\alpha_i = \frac{(i-3)}{s} Y_i + \left(\frac{6Y_0 + 2Y_1}{s} \right) \mu_i.$$

We now consider operations of Type 2. Let $\beta_{h,i}(=\beta_{h,i}(t))$ denote the contribution to $\mathbf{E}(\Delta Y_i)$ for operation $2h$ given in Fig. 2 (at time t). We will also use $\delta_{ij} = 1$ if $i = j$ and $\delta_{ij} = 0$ otherwise.

We have

$$\beta_{a,i} = -3\delta_{i1} + \mu_i + \rho_i,$$

$$\beta_{b,i} = -\delta_{i0} - 2\delta_{i1} + \mu_i + 2\rho_i,$$

$$\beta_{c,i} = -2\delta_{i0} - \delta_{i1} + 2\mu_i + 2\rho_i.$$

These equations are formulated by considering the contribution to $\mathbf{E}(\Delta Y_i)$ by deletion of vertices of known degree and the contribution given by the expected degree of vertices of unknown degree.

For an operation of Type 2 in Phase 1, neighbours of u (the vertex selected at random from V_1) are in $\{V_0 \cup V_1\}$, since $Y_2 = 0$ when the algorithm performs this type of operation. The probability that these neighbours are in V_0 or V_1 are $3Y_0/s$ and $2Y_1/s$ respectively. Therefore the probabilities that, given we are performing an operation of Type 2 in Phase 1, the operation is of type 2a, 2b or 2c are given by $\mathbf{P}(2a) = \frac{4Y_1^2}{s^2}$, $\mathbf{P}(2b) = \frac{12Y_0Y_1}{s^2}$ and $\mathbf{P}(2c) = \frac{9Y_0^2}{s^2}$ respectively.

We define a birth to be the generation of a vertex in V_2 by processing a vertex of V_1. Let $\nu_1(=\nu_1(t))$ be the expected number of births from processing a vertex from V_1 (at time t). Then we have

$$\nu_1 = \mathbf{P}(2a)\left(\mu_2 + \frac{2Y_1}{s}\right) + \mathbf{P}(2b)\left(\mu_2 + \frac{4Y_1}{s}\right) + \mathbf{P}(2c)\left(2\mu_2 + \frac{4Y_1}{s}\right).$$

Here, for each case, we consider the probability that vertices of degree one (in the evolving graph) become vertices of degree two by exposing an edge incident with the vertex.

Similarly, we let $\nu_2 = \nu_2(t)$ be the expected number of births from processing a vertex from V_2. Then we have

$$\nu_2 = \left(\frac{6Y_0 + 2Y_1}{s}\right)\mu_2,$$

giving the expected number of births in a clutch to be $\nu_1/(1 - \nu_2)$.

For Phase 1, the equation giving the expected change in Y_i for a clutch is therefore given by

$$\mathbf{E}(\Delta Y_i) = \mathbf{P}(2a)\beta_{a,i} + \mathbf{P}(2b)\beta_{b,i} + \mathbf{P}(2c)\beta_{c,i} + \frac{\nu_1}{1 - \nu_2}\alpha_i.$$

The equation giving the expected increase in M for a clutch is given by

$$\mathbf{E}(\Delta M) = 1 + \frac{\nu_1}{1 - \nu_2}$$

since the contribution to the increase in the size of the induced matching by the Type 2 operation in a clutch is 1.

5.2 Preliminary Equations for Phase 2

In Phase 2, all operations are considered to be of Type 1 and therefore a clutch consists of one step, but we must also consider the expected change in Y_2 since this is no longer a negligible amount. The expected change in Y_i is given by

$$\mathbf{E}(\Delta Y_i) = \alpha_i - \delta_{i2}$$

where α_i remains the same as that given for Phase 1 and the expected increase in M is 1 per step.

5.3 The Differential Equations

The equation representing $\mathbf{E}(\Delta Y_i)$ for processing a clutch of vertices in Phase 1 forms the basis of a differential equation. Write $Y_i(t) = nz_i(t/n)$, $\mu_i(t) = n\tau_i(t/n)$, $\beta_{j,i}(t) = n\psi_{j,i}(t/n)$, $s(t) = n\xi(t/n)$, $\alpha_i(t) = n\chi_i(t/n)$ and $\nu_j(t) = n\omega_j(t/n)$. The differential equation suggested is

$$z_i' = \frac{4z_1^2}{\xi^2}\psi_{a,i} + \frac{12z_0z_1}{\xi^2}\psi_{b,i} + \frac{9z_0^2}{\xi^2}\psi_{c,i} + \frac{\omega_1}{1-\omega_2}\chi_i \qquad (i \in \{0,1\})$$

where differentiation is with respect to x and xn represents the number of clutches. From the definitions of μ, β, s, α and ν we have

$$\tau_i = \frac{(i-3)}{\xi}z_i + \frac{(6z_0+2z_1)((i-3)z_i+(4-i)z_{i-1})}{\xi^2},$$

$$\psi_{a,i} = -3\delta_{i1} + \tau_i + \frac{(i-3)z_i+(4-i)z_{i-1}}{\xi},$$

$$\psi_{b,i} = -\delta_{i0} - 2\delta_{i1} + \tau_i + 2\frac{(i-3)z_i+(4-i)z_{i-1}}{\xi},$$

$$\psi_{c,i} = -2\delta_{i0} - \delta_{i1} + +2\tau_i + 2\frac{(i-3)z_i+(4-i)z_{i-1}}{\xi},$$

$$\xi = \sum_{i=0}^{2}(3-i)z_i,$$

$$\chi_i = \frac{(i-3)}{\xi}z_i + \frac{6z_0+2z_1}{\xi}\tau_i,$$

$$\omega_1 = \frac{4z_1^2}{\xi^2}\left(\tau_2 + \frac{2z_1}{\xi}\right) + \frac{12z_0z_1}{\xi^2}\left(\tau_2 + \frac{4z_1}{\xi}\right) + \frac{4z_1^2}{\xi^2}\left(2\tau_2 + \frac{4z_1}{\xi}\right), \quad \text{and}$$

$$\omega_2 = \frac{6z_0+2z_1}{\xi}\tau_2.$$

Using the equation representing the expected increase in the size of M after processing a clutch of vertices in Phase 1 and writing $M(t) = nz(t/n)$ suggests the differential equation for z as

$$z' = 1 + \frac{\omega_1}{1-\omega_2}.$$

We compute the ratio $\frac{dz_i}{dz} = \frac{z_i'(x)}{z'(x)}$ and we have

$$z_i' = \frac{\frac{4z_1^2}{\xi^2}\psi_{a,i} + \frac{12z_0z_1}{\xi^2}\psi_{b,i} + \frac{9z_0^2}{\xi^2}\psi_{c,i} + \frac{\omega_1}{1-\omega_2}\chi_i}{1 + \frac{\omega_1}{1-\omega_2}} \qquad (i \in \{0,1\})$$

where differentiation is with respect to z and all functions can be taken as functions of z.

For Phase 2 the equation representing $\mathbf{E}(\Delta Y_i)$ for processing a clutch of vertices suggests the differential equation

$$z_i' = \chi_i - \delta_{i2} \qquad (0 \le i \le 2). \tag{1}$$

The increase in the size of the induced matching per clutch of vertices processed in this Phase is 1, so computing the ratio $\frac{dz_i}{dz} = \frac{z_i'(x)}{z'(x)}$ gives the same equation as that given in (1). Again, differentiation is with respect to z and all functions can be taken as functions of z.

The solution to this system of differential equations represents the cardinalities of the sets V_i for given M (scaled by $\frac{1}{n}$). For Phase 1, the initial conditions are

$$z_0(0) = 1, \quad z_i(0) = 0 \ (i > 0).$$

The initial conditions for Phase 2 are given by the final conditions for Phase 1.

Wormald [12] describes a general result which ensures that the solutions to the differential equations almost surely approximate the variables Y_i. It is simple to define a domain for the variables z_i so that Theorem 5.1 from [12] may be applied to the process within each phase. An argument similar to that given for independent sets in [12] or that given for independent dominating sets in [4] ensures that a.a.s. the process passes through phases as defined informally, and that Phase 2 follows Phase 1. Formally, Phase 1 ends at the time corresponding to $\omega_2 = 1$ as defined by the equations for Phase 1. Once in Phase 2, vertices in V_2 are replenished with high probability which keeps the process in Phase 2.

The differential equations were solved using a Runge-Kutta method, giving $\omega_2 = 1$ at $z = 0.1349$ and in Phase 2 $z_2 = 0$ at $z > 0.2704$. This corresponds to the size of the induced matching (scaled by $\frac{1}{n}$) when all vertices are used up, thus proving the theorem. □

6   The Upper Bound

Theorem 2. *For a cubic graph on n vertices the size of a maximum induced matching is asymptotically almost surely less than $0.2821n$*

Proof. Consider a random cubic graph G on n vertices. Let $M(G, k)$ denote the number of induced matchings of G of size k. We calculate $\mathbf{E}(M(G, k))$ and show that when $k > 0.2821n$, $\mathbf{E}(M(G, k)) = o(1)$ thus proving the theorem. Let $N(x) = \frac{(2x)!}{x!2^x}$.

$$\mathbf{E}(M(G,k)) \leq \binom{n}{2k} N(k)\, 3^{2k}\, \frac{(3(n-2k))!}{(3n-10k)!}\, \frac{N(\frac{3n}{2}-5k)}{N(\frac{3n}{2})}$$

$$= \frac{n!\, 3^{2k}\, (3(n-2k))!\, (\frac{3n}{2})!\, 2^{4k}}{k!\, (n-2k)!\, (\frac{3n}{2}-5k)!\, (3n)!}.$$

Approximate using Stirling's formula and re-write using $f(x) = x^x$, $\kappa = k/n$ and we have

$$\mathbf{E}(M(G,k))^{\frac{1}{n}} \sim \frac{3^{2\kappa}\, f(3(1-2\kappa))\, f(\frac{3}{2})\, 2^{4\kappa}}{f(\kappa)\, f(1-2\kappa)\, f(\frac{3}{2}-5\kappa)\, f(3)}.$$

Solving this we find that for $\kappa \geq 0.2821$ the expression on the right tends to 0.

$\square$

Note that this bound may be improved by counting only maximal matchings. However the improvement is slight and we do not include the details here for reasons of brevity.

References

1. Bollobás, B.: Random Graphs. Academic Press, London, 1985.
2. Cameron, K.: Induced Matchings. Discrete Applied Math., **24**:97–102, 1989.
3. Duckworth, W., Manlove, D. and Zito, M.: On the Approximability of the Maximum Induced Matching Problem. J. of Combinatorial Optimisation (Submitted).
4. Duckworth, W. and Wormald, N.C.: Minimum Independent Dominating Sets of Random Cubic Graphs. Random Structures and Algorithms (Submitted).
5. Erdős, E.: Problems and Results in Combinatorial Analysis and Graph Theory. Discrete Math., **72**:81–92, 1988.
6. Faudree, R.J., Gyárfás, A., Schelp, R.H. and Tuza, Z.: Induced Matchings in Bipartite Graphs. Discrete Math., **78**:83–87, 1989.
7. Golumbic, M.C. and Laskar, R.C.: Irredundancy in Circular Arc Graphs. Discrete Applied Math., **44**:79–89, 1993.
8. Golumbic, M.C. and Lewenstein, M.: New Results on Induced Matchings. Discrete Applied Math., **101**:157–165, 2000.
9. Liu, J. and Zhou, H.: Maximum Induced Matchings in Graphs. Discrete Math., **170**:271–281, 1997.
10. Steger, A. and Yu, M.: On Induced Matchings. Discrete Math., **120**:291–295, 1993.
11. Stockmeyer, L.J. and Vazirani, V.V.: NP-Completeness of Some Generalizations of the Maximum Matching Problem. Inf. Proc. Lett., **15**(1):14–19, 1982.
12. Wormald, N.C.: Differential Equations for Random Processes and Random Graphs. In Lectures on Approximation and Randomized Algorithms, 73–155, PWN, Warsaw, 1999. Michał Karoński and Hans-Jürgen Prömel (editors).
13. Zito, M.: Induced Matchings in Regular Graphs and Trees. In Proceedings of the 25th International Workshop on Graph Theoretic Concepts in Computer Science, volume **1665** of Lecture Notes in Computer Science , 89–100. Springer-Verlag, 1999.
14. Zito, M.: Randomised Techniques in Combinatorial Algorithmics. PhD thesis, Department of Computer Science, University of Warwick, UK, November 1999.

A Duality between Small-Face Problems in Arrangements of Lines and Heilbronn-Type Problems

Gill Barequet

The Technion—Israel Institute of Technology, Haifa 32000, Israel,
barequet@cs.technion.ac.il,
http://www.cs.technion.ac.il/~barequet

Abstract. Arrangements of lines in the plane and algorithms for computing extreme features of arrangements are a major topic in computational geometry. Theoretical bounds on the size of these features are also of great interest. Heilbronn's triangle problem is one of the famous problems in discrete geometry. In this paper we show a duality between extreme (small) face problems in line arrangements (bounded in the unit square) and Heilbronn-type problems. We obtain lower and upper combinatorial bounds (some are tight) for some of these problems.

1 Introduction

The investigation of arrangements of lines in the plane has attracted much attention in the literature. In particular, certain extremal features of such arrangements are of great interest. Using standard duality between lines and points, such arrangements can be mapped into sets of points in the plane, which have also been studied intensively. In this dual setting, distributions in which certain features (defined by triples of points) assume their maxima are often sought. In this paper we show a connection between these two classes of problems and summarize the known bounds for some extremal-feature problems.

Let $\mathcal{A}(\mathcal{L})$ be an arrangement of a set $\mathcal{L}$ of n lines. We assume the lines of $\mathcal{A}$ to be in general position, in the sense that no two lines have the same slope. Thus every triple of lines define a triangle. Let $U = [0,1]^2$ be the unit square. An arrangement $\mathcal{A}$ is called *narrow* if all its lines intersect the two vertical sides of U. A narrow arrangement $\mathcal{A}$ is called *transposed* if the lines of $\mathcal{A}$ intersect the vertical sides of U in two sequences that, when sorted in increasing y order, are the reverse of each other. Clearly, every transposed arrangement is also narrow. Note that *all* the vertices of a transposed arrangement lie in U. Later in the paper we will also define the set of *convex* arrangements which is a proper subset of the set of transposed arrangements.

In this paper we investigate the "size" of triangles defined by the lines of narrow arrangements, according to several measures of the size of a triangle. We consider the arrangements in which the minimum size of a triangle assumes its maximum, and attempt to bound this value.

D.-Z. Du et al. (Eds.): COCOON 2000, LNCS 1858, pp. 44–53, 2000.

Heilbronn's triangle problem is the following:

Let $\{P_1, P_2, \ldots, P_n\}$ be a set of n points in U, such that the minimum of the areas of the triangles $P_i P_j P_k$ (for $1 \leq i < j < k \leq n$) assumes its maximum possible value $\mathcal{G}_0(n)$. Estimate $\mathcal{G}_0(n)$.

Heilbronn conjectured that $\mathcal{G}_0(n) = O(\frac{1}{n^2})$. The first nontrivial upper bound (better than $O(\frac{1}{n})$), namely, $O(\frac{1}{n\sqrt{\log \log n}})$, was given by Roth [6]. Schmidt [9] improved this result twenty years later and obtained $\mathcal{G}_0(n) = O(\frac{1}{n\sqrt{\log n}})$. Then Roth improved the upper bound twice to $O(\frac{1}{n^{1.105\ldots}})$ [7] and $O(\frac{1}{n^{1.117\ldots}})$ [8]. The best currently-known upper bound, $\mathcal{G}_0(n) = O(\frac{1}{n^{8/7-\varepsilon}}) = O(\frac{1}{n^{1.142\ldots}})$ (for any $\varepsilon > 0$), is due to Komlós, Pintz, and Szemerédi [3] by a further refinement of the method of [7,8]. A simple probabilistic argument by Alon et al. [1, p. 30] proves a lower bound of $\Omega(\frac{1}{n^2})$. Erdős [6, appendix] showed the same lower bound by an example. However, Komlós, Pintz, and Szemerédi [4] show by a rather involved probabilistic construction that $\mathcal{G}_0(n) = \Omega(\frac{\log n}{n^2})$. In a companion paper [2] we show a lower bound for the generalization of Heilbronn's triangle problem to higher dimensions.

In this paper we consider some variants of Heilbronn's problem, in which other measures of triangles defined by triples of points are considered, and/or some restrictions are imposed on the locations of the points in U. Specifically, in a *monotone decreasing* distribution of points, the x-coordinates and the y-coordinates of the points appear in opposite permutations, and a *convex monotone decreasing* distribution is a monotone decreasing distribution in which the points form a convex (or concave) chain according to the respective permutation.

2   The Duality and the Problems

Let $\mathcal{A}$ be a narrow arrangement of n lines. We define two measures of triangles. Let $F_1(\tau)$ be the vertical height of a triangle τ, that is, the maximum length of a vertical segment contained in τ. Let $F_2(\tau)$ be the area of τ. Our goal is to bound $\mathcal{F}_1(n) = \max_{\mathcal{A}} \min_{\tau \in \mathcal{A}} F_1(\tau)$ and $\mathcal{F}_2(n) = \max_{\mathcal{A}} \min_{\tau \in \mathcal{A}} F_2(\tau)$, where the maximum is taken over all narrow arrangements $\mathcal{A}$ of n lines, and where the minimum is taken over all triangles τ formed by triples of lines in $\mathcal{A}$. We use the superscripts "(mon)" and "(conv)" for denoting the monotone decreasing and the convex monotone decreasing variants, respectively.

Denote by l_i (resp., r_i) the y-coordinate of the intersection point of the line $\ell_i \in \mathcal{L}$ with the left (resp., right) vertical side of U. We dualize the line ℓ_i to the point $P_i = (l_i, r_i)$. This dualization maps the measures F_1 and F_2 of lines into measures of triples of points in U. The two optimization problems are mapped into generalizations of Heilbronn's problem, where the difference is in the definition of the measure of a triple of points. Note that the dual of a transposed arrangement of lines is a monotone *decreasing* set of points. The measures G_1 and G_2 (defined below) crucially rely on the *decreasing* monotonicity of the points. A "convex" arrangement is an arrangement whose dual set of points lies in convex position.

It is easy to see that the vertical height of a triangle is the minimum length of a vertical segment that connects a vertex of the triangle to the line supporting the opposite edge. This follows from the convexity of a triangle, and indeed that segment always lies inside the triangle. We now specify (in terms of the dual representation) the vertical distance between the intersection of two lines to a third line of $\mathcal{A}$ (in the primal representation). Refer to Figure 1. The equation of ℓ_i is

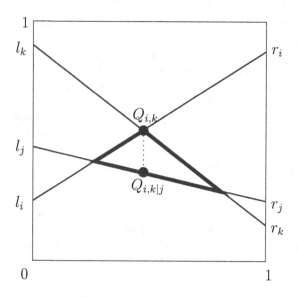

Fig. 1. Vertical distance

$y = (r_i - l_i)x + l_i$. We compute the distance between $Q_{i,k} = (x_{i,k}, y_{i,k})$, the intersection point of ℓ_i and ℓ_k, and $Q_{i,k|j} = (x_{i,k}, y_{i,k|j})$, the vertical projection of $Q_{i,k}$ on ℓ_j. A simple calculation shows that $Q_{i,k} = (\frac{l_k - l_i}{(l_k - l_i) - (r_k - r_i)}, \frac{l_k r_i - l_i r_k}{(l_k - l_i) - (r_k - r_i)})$. By substituting $x_{i,k}$ in the equation of ℓ_j we find that $y_{i,k|j} = \frac{r_j(l_k - l_i) - l_j(r_k - r_i)}{(l_k - l_i) - (r_k - r_i)}$. Finally,

$$\mathrm{Dist}(Q_{i,k}, Q_{i,k|j}) = |y_{i,k} - y_{i,k|j}| = \left| \frac{r_i(l_k - l_j) - r_j(l_k - l_i) + r_k(l_j - l_i)}{(l_k - l_i) + (r_i - r_k)} \right|$$

$$= 4 \, \mathrm{abs} \left(\frac{\frac{1}{2} \begin{vmatrix} l_i & r_i & 1 \\ l_j & r_j & 1 \\ l_k & r_k & 1 \end{vmatrix}}{2((l_k - l_i) + (r_i - r_k))} \right).$$

The numerator of the last term is the area of the triangle defined by the points P_i, P_j, and P_k. In case P_i and P_k are in monotone decreasing position, the denominator is the perimeter of the axis-aligned box defined by P_i and P_k. When setting $G_1(P_i, P_j, P_k) = \text{Dist}(Q_{i,k}, Q_{i,k|j})$, maximizing the smallest value of F_1 over the triangles defined by triples of lines of $\mathcal{A}$ dualizes to maximizing the smallest value of G_1 over triples of points in U. We denote the asymptotic value by $\mathcal{G}_1(n) = \mathcal{F}_1(n)$.

We now compute (in the dual representation) the area of the triangle defined by three lines of $\mathcal{A}$ (in the primal representation). Recall that

$$Q_{i,j} = (x_{i,j}, y_{i,j}) = \left(\frac{l_j - l_i}{(l_j - l_i) - (r_j - r_i)}, \frac{l_j r_i - l_i r_j}{(l_j - l_i) - (r_j - r_i)} \right).$$

Tedious computation shows that

$$\text{Area}(Q_{i,j}, Q_{j,k}, Q_{k,i}) = \frac{1}{2} \text{ abs} \begin{vmatrix} x_{i,j} & y_{i,j} & 1 \\ x_{j,k} & y_{j,k} & 1 \\ x_{k,i} & y_{k,i} & 1 \end{vmatrix}$$

$$= \frac{1}{2} \text{ abs}(x_{j,k} y_{k,i} - y_{j,k} x_{k,i} - x_{i,j} y_{k,i} + y_{i,j} x_{k,i} + x_{i,j} y_{j,k} - y_{i,j} x_{j,k})$$

$$= \frac{1}{2} \text{ abs}\left(\frac{(l_i r_j - l_j r_i + l_k r_i - l_i r_k + l_j r_k - l_k r_j)^2}{((l_j - l_i) + (r_i - r_j))((l_k - l_i) + (r_i - r_k))((l_k - l_j) + (r_j - r_k))} \right)$$

$$= 16 \text{ abs}\left(\frac{\left(\frac{1}{2} \begin{vmatrix} l_i & r_i & 1 \\ l_j & r_j & 1 \\ l_k & r_k & 1 \end{vmatrix} \right)^2}{2((l_j - l_i) + (r_i - r_j)) \cdot 2((l_k - l_i) + (r_i - r_k)) \cdot 2((l_k - l_j) + (r_j - r_k))} \right).$$

The numerator of the last term is the square of the area of the triangle defined by the points P_i, P_j, and P_k. In case P_i, P_j, and P_k are in monotone decreasing position, the denominator is the product of the perimeters of the axis-aligned boxes defined by P_i, P_j, and P_k. When setting $G_2(P_i, P_j, P_k) = \text{Area}(Q_{i,j}, Q_{j,k}, Q_{k,i})$, maximizing the smallest value of F_2 over the triangles defined by triples of lines of $\mathcal{A}$ dualizes to maximizing the smallest value of G_2 over triples of points in U. We denote the asymptotic value by $\mathcal{G}_2(n) = \mathcal{F}_2(n)$.

In summary, we define a duality between narrow arrangements of lines and distributions of points in the unit square. We define measures of triangles defined by triples of lines of the arrangements, and find their dual measures of triples of points. We consider some special structures of the arrangements and their dual distributions of points. In all cases we look for the arrangement of lines (or distribution of points) that maximizes the minimum value of the measure over all triples of lines (or points), and attempt to asymptotically bound this value.

3   Heilbronn's Triangle Problem

As noted in the introduction, the best known lower bound for the original Heilbronn's triangle problem, $\mathcal{G}_0(n) = \Omega(\frac{\log n}{n^2})$, is obtained by Komlós, Pintz, and Szemerédi [4]. The same authors obtain in [3] the best known upper bound, $O(\frac{1}{n^{1.142\ldots}})$.

Unlike the measures G_1 and G_2, it is not essential to require the set to be *decreasing* in the monotone case. Here we have only an upper bound:

Theorem 1. $\mathcal{G}_0^{(mon)}(n) = O(\frac{1}{n^2})$.

Proof. Refer to the sequence of segments that connect pairs of consecutive points. The total length of the sequence is at most 2. Cover the whole sequence of segments by at most $\frac{n}{3}$ squares with side $\frac{6}{n}$. Thus at least one square contains three points. The area of the triangle that they define is at most $\frac{18}{n^2}$.

For the convex case the bound is tight:

Theorem 2. $\mathcal{G}_0^{(conv)}(n) = \Theta(\frac{1}{n^3})$.

Proof. The lower bound is obtained by a simple example: Put n points equally spaced on an arc of radius 1 and of length $\frac{\pi}{2}$ (see Figure 2). The points are

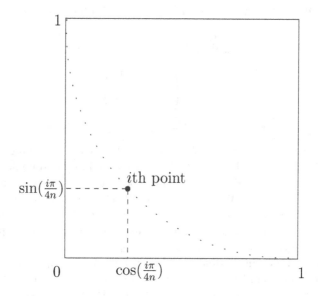

Fig. 2. Uniform distribution of points on an arc

ordered from bottom-right to top-left. The coordinates of the ith point are $(\cos(2i\delta), \sin(2i\delta))$, for $1 \leq i \leq n$, where $\delta = \pi/(4n)$. The area of the triangle defined by every three consecutive points is $4\cos(\delta)\sin^3(\delta) = \Theta(\frac{1}{n^3})$.

It is also easy to prove the upper bound. Refer to the sequence of segments that connect pairs of consecutive points. Drop all the segments of length greater than $8/n$. Since the maximum length of such a convex and monotone chain is 2, less than $n/4$ segments are dropped. Now drop all pairs of consecutive segments whose external angle is greater than $4\pi/n$. Since the maximum turning angle of such a convex chain is $\pi/2$, less than $n/8$ pairs are dropped. That is, less than $n/4$

segments are now dropped. In total we have dropped less than $n/2$ segments; therefore two consecutive segments have not been dropped. The area of the triangle which these two segments define is upper bounded by $\frac{1}{2}(\frac{8}{n})^2 \sin(\frac{4\pi}{n}) = \Theta(\frac{1}{n^3})$.

4   Vertical Height

Theorem 3. $\mathcal{G}_1(n) = \Theta(\frac{1}{n})$.

Proof. The lower bound is shown by an example.[1] Set $l_i = \frac{i}{n} - \frac{1}{n^i}$ and $r_i = \frac{i}{n}$, for $1 \leq i \leq n$ (a set of almost-parallel lines). In this example

$$G_1(P_i, P_j, P_k) = \left| \frac{(j-i)n^{i+j} - (k-i)n^{i+k} + (k-j)n^{j+k}}{n^{j+1}(n^k - n^i)} \right|,$$

which is minimized when i, j, and k are consecutive integers. It is easy to verify that in this example, for every i, $G_1(P_i, P_{i+1}, P_{i+2}) = \frac{n-1}{n(n+1)}$. Hence $\mathcal{G}_1(n) = \Omega(\frac{1}{n})$.

For the upper bound we return to the primal representation of the problem by an arrangement $\mathcal{A}$ of n lines. Since $\mathcal{A}$ is narrow, there exists a vertical segment of length 1 which stabs all the n lines. Hence there exists a triple of lines which are stabbed by a vertical segment of length at most $\frac{2}{n-1}$. Such a triple cannot define a triangle whose vertical height exceeds $\frac{2}{n-1}$, therefore $\mathcal{F}_1(n) = O(\frac{1}{n})$.

We now refer to transposed arrangements of lines (the dual of monotone decreasing Heilbronn's sets of points). The best upper bound of which we are aware is Mitchell's [5]. Here is a simplified version of the proof of this bound. Let θ_i denote the slope of ℓ_i (for $1 \leq i \leq n$). Assume without loss of generality that at least $n/2$ lines of $\mathcal{L}$ have positive slopes. For the asymptotic analysis we may assume that all the lines of $\mathcal{L}$ are ascending. Denote by h the minimum vertical height of a triangle in $\mathcal{A}(\mathcal{L})$. Then each pair of lines $\ell_i, \ell_j \in \mathcal{L}$ induces a hexagon of area $\frac{3h^2 \cos(\theta_i)\cos(\theta_j)}{4\sin(\theta_j - \theta_i)}$ (see Figure 3) through which no other line of $\mathcal{L}$ can pass. This is since such a line would form with ℓ_i and ℓ_j a triangle whose vertical height is at most $h/2$. The intersection of every pair of such hexagons is empty, for otherwise there would be a triangle whose vertical height is less than h. Denote by S the total area of the $\binom{n}{2}$ hexagons. On one hand,

$$S = \sum_{1 \leq i < j \leq n} \frac{3h^2 \cos(\theta_i)\cos(\theta_j)}{4\sin(\theta_j - \theta_i)} \geq \frac{3h^2}{8} \sum_{1 \leq i < j \leq n} \frac{1}{\sin(\theta_j - \theta_i)} \geq \frac{3h^2}{8} \sum_{1 \leq i < j \leq n} \frac{1}{\theta_j - \theta_i},$$

where we use the facts that $0 < \theta_i \leq \pi/4$ for $1 \leq i \leq n$ and $\sin(\theta) < \theta$.

---

[1]   Note that this construction is monotone and even convex, and yet it beats the $O(\frac{1}{n^2})$ upper bound of $\mathcal{G}_1^{(\mathrm{conv})}(n)$. This is because it is an *increasing* construction.

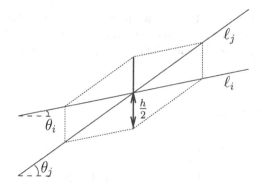

Fig. 3. Hexagonal forbidden zones induced by intersection of lines

Lemma 1. $\sum_{1 \leq i < j \leq n} \frac{1}{\theta_j - \theta_i} = \Omega(n^2 \log n).$

Proof. Omitted in this short version of the paper.

We thus have $S = \Omega(h^2 n^2 \log n)$. On the other hand, the total area of all the hexagons along one line of $\mathcal{L}$ cannot exceed h (the area of a strip of vertical height h clipped to the unit square). By summing up for all the lines of $\mathcal{L}$ we obtain $2S \leq hn$. The combination of the two inequalities implies that $h = O(\frac{1}{n \log n})$. (Note that the "dropped-segments" argument gives in this case a weaker upper bound of $O(\frac{1}{n})$.) Thus we have the following:

Theorem 4. $\mathcal{G}_1^{(mon)}(n) = O(\frac{1}{n \log n}).$

Finally we consider convex arrangements (the dual of convex monotone decreasing Heilbronn's sets). As with the convex case of the original Heilbronn's triangle problem, the bound in this case is tight:

Theorem 5. $\mathcal{G}_1^{(conv)}(n) = \Theta(\frac{1}{n^2}).$

Proof. For the lower bound we use the same example (points on an arc) as in the proof of Theorem 2. Clearly, the perimeter of the bounding box of every triangle defined by three consecutive points in this example is $\Theta(\frac{1}{n})$. Since the area of each such triangle is $\Theta(\frac{1}{n^3})$ we obtain the lower bound. For the upper bound we follow again the argument of dropping segments. The perimeter of the bounding box of the triangle defined by the two remaining consecutive segments is linear in the length s of the longer segment. The quotient of the area of the triangle and s is upper bounded by $\frac{1}{2}(\frac{8}{n}) \sin(\frac{2\pi}{n}) = \Theta(\frac{1}{n^2}).$

Table 1 shows a few examples which show the lower bound $\mathcal{G}_1^{(conv)}(n) = \Omega(\frac{1}{n^2}).$

l_i	r_i	$G_1(P_i, P_j, P_k)$
$\frac{i}{n}$	$1 - \left(\frac{i}{n}\right)^2$	$\left\lvert \frac{(i-j)(k-j)}{n(i+k+n)} \right\rvert$
$\frac{i}{n}$	$1 - \sqrt{\frac{i}{n}}$	$\left\lvert \frac{(\sqrt{j}-\sqrt{i})(\sqrt{k}-\sqrt{j})}{(\sqrt{i}+\sqrt{k}+\sqrt{n})\sqrt{n}} \right\rvert$
$\sin(\frac{i\pi}{n})$	$\cos(\frac{i\pi}{n})$	$\left\lvert \frac{\sin(\frac{(i-j)\pi}{2n})-\sin(\frac{(i-k)\pi}{2n})+\sin(\frac{(i-k)\pi}{2n})}{\cos(\frac{i\pi}{2n})-\cos(\frac{k\pi}{2n})-\sin(\frac{i\pi}{2n})+\sin(\frac{k\pi}{2n})} \right\rvert$

Table 1. Convex line arrangements which show that $\mathcal{G}_1^{(\mathrm{conv})}(n) = \Omega(\frac{1}{n^2})$

5   Area

The function $\mathcal{G}_2(n)$ is as high as we like:

Theorem 6. $\mathcal{G}_2(n) = \Omega(f(n))$ for any $f(n)$.

Proof. We use the construction $l_i = \frac{i}{n} - \frac{1}{n^{Mi}}$ and $r_i = \frac{i}{n}$ (for $1 \leq i \leq n$ and for an arbitrarily large M). In this example

$$G_2(P_i, P_j, P_k) = \left\lvert \frac{[(j-i)n^{M(i+j)} - (k-i)n^{M(i+k)} + (k-j)n^{M(j+k)}]^2}{2n^2(n^{Mj} - n^{Mi})(n^{Mk} - n^{Mi})(n^{Mk} - n^{Mj})} \right\rvert ,$$

which is minimized by $i = 1$, $j = 2$, and $k = 3$. In this example $G_2(P_1, P_2, P_3) = \frac{n^{2(M-1)}(n^M-1)}{2n^M+1}$. Set $M = \frac{\log f(n)}{2\log n} + 1$, and the claim follows.

We now refer to transposed arrangements of lines (the dual of monotone decreasing Heilbronn's sets of points).

Theorem 7. $\mathcal{G}_2^{(mon)}(n) = O(\frac{1}{n\sqrt{\log n}})$.

Proof. We follow an argument similar to that used in the proof of Theorem 4. Let again θ_i denote the slope of ℓ_i (for $1 \leq i \leq n$), and assume without loss of generality that all the lines of $\mathcal{L}$ are ascending. Denote by A the minimum area of a triangle in $\mathcal{A}(\mathcal{L})$. Here the "forbidden zone" induced by each pair of lines $\ell_i, \ell_j \in \mathcal{L}$ is a rectangle whose diagonal is of length $\frac{2A}{\sin(\theta_j - \theta_i)}$ (refer to Figure 4). Indeed, if any other line of $\mathcal{L}$ passed through this rectangle, then together with ℓ_i and ℓ_j it would form a triangle whose area is less then $A/2$. The intersection of every pair of such rectangles is empty, for otherwise there would be a triangle whose area is less than A. The area of the forbidden rectangle is $\frac{2A^2}{\sin(\theta_j - \theta_i)}$. Let again S denote the total area of the $\binom{n}{2}$ forbidden zones. Here we have:

$$S = \sum_{1 \leq i < j \leq n} \frac{2A^2}{\sin(\theta_j - \theta_i)} = \Omega(A^2 n^2 \log n),$$

by using again Lemma 1. Since $S \leq 1$, we obtain $A = O(\frac{1}{n\sqrt{\log n}})$.

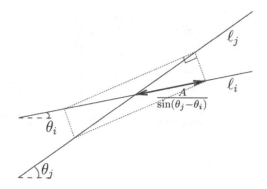

Fig. 4. Forbidden zones for $\mathcal{G}_2^{(\mathrm{mon})}$

Finally we consider convex arrangements (the dual of convex monotone decreasing Heilbronn's sets).

Theorem 8. $\mathcal{G}_2^{(conv)}(n) = \Theta(\frac{1}{n^3})$.

Proof. For the lower bound we use the construction $l_i = \frac{i}{n}$ and $r_i = 1 - (\frac{i}{n})^2$ (for $1 \leq i \leq n$). A simple calculation shows that in this example

$$G_2^{(conv)}(P_i, P_j, P_k) = \left| \frac{(j-i)(k-i)(k-j)}{2(i+j+n)(i+k+n)(j+k+n)} \right|,$$

whose minimum is easily verified to be $\Theta(\frac{1}{n^3})$.

For the upper bound we follow again the argument of dropping segments. Then $\mathcal{G}_2(n)$ is upper bounded by A^2/xy^2, where x and y are the lengths of the two remaining consecutive segments, and A is the area of the triangle spanned by these two segments. But $A^2/xy^2 \leq \frac{1}{4}(\frac{8}{n}) \sin^2(\frac{2\pi}{n}) = \Theta(\frac{1}{n^3})$.

6   Summary

In this paper we show a relation between small-face problems in arrangements of lines and Heilbronn-type problems in point sets. We use a duality between the two classes for obtaining bounds for these problems. We summarize in Table 2 the best bounds that we are aware of. (There is a slight abuse of notation in the synonyms of $\mathcal{G}_1$ and $\mathcal{G}_2$.) Note that every lower (resp., upper) bound trivially applies for columns on the left (resp., right) of it. We mention such trivial bounds in square brackets. The main open problem is to obtain tight bounds for some of the problems.

Acknowledgment

The author wishes to thank Noga Alon (who suggested the "dropped-segments" argument) and Micha Sharir for helpful discussions. The original proof that

Measure		Construction		
Arrangement of Lines	Point Set	Narrow (Arng.) General (Heil.)	Transp. (Arng.) Mon. Dec. (Heil.)	"Convex" (Arng.) Conv. Dec. (Heil.)
	$\mathcal{G}_0$: A	$\Omega(\frac{\log n}{n^2})$, $O(\frac{1}{n^{8/7-\varepsilon}})$	$[\Omega(\frac{1}{n^3})]$, $O(\frac{1}{n^2})$	$\Theta(\frac{1}{n^3})$
$\mathcal{F}_1$: V.H.	$\mathcal{G}_1$: A/P	$\Theta(\frac{1}{n})$	$[\Omega(\frac{1}{n^2})]$, $O(\frac{1}{n\log n})$	$\Theta(\frac{1}{n^2})$
$\mathcal{F}_2$: Area	$\mathcal{G}_2$: A^2/P^3	Unbounded	$[\Omega(\frac{1}{n^3})]$, $O(\frac{1}{n\sqrt{\log n}})$	$\Theta(\frac{1}{n^3})$

Table 2. Summary of known bounds

$\mathcal{G}_1^{(\mathrm{mon})}(n) = O(\frac{1}{n\log n})$ is given in the unpublished manuscript [5]. It inspired the simplified proof given in this paper with the kind permission of Joseph Mitchell.

References

1. Alon, N., Spencer, J.H.: The Probabilistic Method. John Wiley & Sons, 1992
2. Barequet, G.: A lower bound for Heilbronn's triangle problem in d dimensions. Proc. 10th Ann. ACM-SIAM Symp. on Discrete Algorithms. Baltimore, MD, 76–81, January 1999
3. Komlós, J., Pintz, J., Szemerédi, E.: On Heilbronn's triangle problem. J. London Mathematical Society (2) **24** (1981) 385–396
4. Komlós, J., Pintz, J., Szemerédi, E.: A lower bound for Heilbronn's problem. J. London Mathematical Society (2) **25** (1982) 13–24
5. Mitchell, J.S.B.: On the existence of small faces in arrangements of lines. Manuscript, Dept. of Applied Mathematics, SUNY Stony Brook, NY, 1995. (available at http://www.ams.sunysb.edu/~jsbm/publications.html)
6. Roth, K.F.: On a problem of Heilbronn. Proc. London Mathematical Society **26** (1951) 198–204
7. Roth, K.F.: On a problem of Heilbronn, II. Proc. London Mathematical Society (3) **25** (1972) 193–212
8. Roth, K.F.: On a problem of Heilbronn, III. Proc. London Mathematical Society (3) **25** (1972) 543–549
9. Schmidt, W.M.: On a problem of Heilbronn. J. London Mathematical Society (2) **4** (1971) 545–550

On Local Transformation of Polygons with Visibility Properties

Carmen Hernando[1]*, Michael E. Houle[2], and Ferran Hurtado[3]*

[1] Departament de Matemàtica Aplicada I, Universitat Politècnica de Catalunya
(UPC), Avda. Diagonal 647, 08028 Barcelona, SPAIN
(hernando@ma1.upc.es)
[2] Basser Dept. of Computer Science, The University of Sydney, Sydney, NSW 2006,
AUSTRALIA
(meh@cs.usyd.edu.au)
[3] Departament de Matemàtica Aplicada II, Universitat Politècnica de Catalunya
(UPC), Pau Gargallo 5, 08028 Barcelona, SPAIN
(hurtado@ma2.upc.es)

Abstract. One strategy for the enumeration of a class of objects is
local transformation, in which new objects of the class are produced by
means of a small modification of a previously-visited object in the same
class. When local transformation is possible, the operation can be used
to generate objects of the class via random walks, and as the basis for
such optimization heuristics as simulated annealing.
For general simple polygons on fixed point sets, it is still not known
whether the class of polygons on the set is connected via a constant-size
local transformation. In this paper, we exhibit a simple local transformation for which the classes of (weakly) edge-visible and (weakly) externally
visible polygons are connected. The latter class is particularly interesting
as it is the most general polygon class known to be connected under local
transformation.

1 Introduction

Even for small instances, strategies for combinatorial enumeration must be very
efficient in order to handle the huge numbers of objects which are often produced.
In particular, they must avoid both the excessive recomputation of individual
objects, and the excessive computation of objects which are outside the class
under consideration. One way of limiting the construction of invalid objects is
through the use of *local transformation*, in which new objects of the class are
generated only from previously-examined objects by means of small changes. The
class must be *connected* with respect to the local transformation in question: that
is, every valid object must be reachable from some initial valid object by means
of a finite sequence of transformations. Local search and optimization methods
such as reverse search enumeration [1], random walks and simulated annealing
all make use of local transformation to visit new objects.

---

* Partially supported by CUR Gen. Cat. 1999SGR00356 and MEC-DGES-SEUID
PB98-0933

D.-Z. Du et al. (Eds.): COCOON 2000, LNCS 1858, pp. 54–63, 2000.

Within the discipline of computational geometry, interest in local transformation can be traced back to the investigation of the connectivity of Delaunay triangulations under the diagonal flip operation [12,4]. Recent connectivity results have been established for triangulations, including triangulations of polygons [9,10], topological triangulations [3,14], and triangulations in higher dimensions [11]. Other classes of objects, such as non-crossing trees and Euclidean matchings have also been studied [5,6].

One class of great importance to computational geometry, but for which no satisfactory enumeration method is yet known, is that of simple polygons. Given a set of n points S, one can generate all simple polygons on S by considering each of the $\frac{(n-1)!}{2}$ possible Hamiltonian circuits of S, and then testing each for self-intersection. However, this approach is hardly practical, as the proportion of circuits which are simple diminishes (at least) exponentially with n. Zhu, Sundaram, Snoeyink, and Mitchell have provided a method for generating x-monotone polygons in $O(n^2)$ time which does not use local transformation; unfortunately, there is no obvious way of extending this to other classes of polygons.

As in the case of triangulations, if any simple polygon on S could be transformed into any other by means of a finite sequence of 'local' transformations, then methods such as reverse search enumeration could be applied. The time complexity of the method would depend heavily on the number of edges modified by the transformation.

1.1 Edge Flips

Let S be a set of n points not all collinear. If two distinct polygons P and P' on S differ in exactly k edges, then we shall say that P is (edge) k-flippable from P to P'. Alternatively, we say that P is transformable to P' by means of a k-flip. The edges of P not contained in P' are called the *leaving edges*, and the edges of P' not contained in P are called the *entering edges*.

Figure 1 gives a characterization of k-flips for $k = 2$ and $k = 3$. A few special cases are worth mentioning:

- If the vertices of the internal chain of a 2-flip are collinear (or if the chain consists of a single edge), the 2-flip is called a *line flip* (L-flip).
- If two leaving edges of a 3-flip share a vertex in common, the flip will be referred to as a *vertex-edge flip* (VE-flip).
- A flip is called *planar* if no leaving edge intersects any entering edge, except perhaps at their endpoints.

Planar flips are of computational interest, in that the entering edges are edges of the visibility graph of P. Visibility graphs of simple polygons can be computed in $\Theta(M)$ time, where M is the number of edges of the graph [7]. Using the visibility graph, all candidate L-flips and planar VE-flips of a given polygon can be generated in a straightforward manner, in $O(M)$ time.

In 1991, David Avis posed the following question: can any simple polygon on S be transformed to any other simple polygon on S by means of a finite

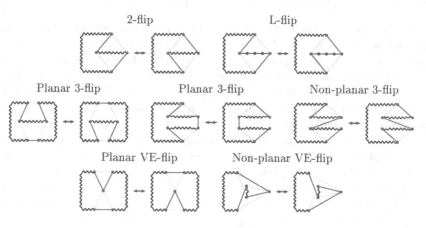

Fig. 1. 2-flips and 3-flips.

Fig. 2. A non-2-flippable, non-VE-flippable polygon.

sequence of 2-flips? Unfortunately, the answer to Avis' question turns out to be 'no' [8]. The 19-vertex polygon shown in Figure 2 cannot be 2-flipped or VE-flipped to any other polygon on S (although it is 3-flippable). Arbitrarily large non-2-flippable and non-VE-flippable polygons may be exhibited.

The natural question to ask at this point is, is there any *constant k* such that any polygon P on a given set of vertices S can be transformed into any other polygon P' via a finite sequence of k-flips? For general polygons, the question is still very much open. In this paper, we show that two of the simplest operations, namely L-flips and planar VE-flips, together are sufficient to connect polygon classes with certain visibility properties: (weakly) edge-visible and (weakly) externally visible. These polygon classes are formally defined in the next section, although more information concerning visibility within polygons can be found in [13]. In Section 3, we prove the connectivity of edge-visible polygons under the *VE-L-flip* operation, in which either an L-flip or a planar VE-flip can be performed. The edge-visibility is taken with respect to a fixed edge. In Section 4, we prove the connectivity of general edge-visible polygons, and of externally visible polygons.

2 Polygon Classes

Let $P = (p_0, p_1, p_2, \ldots, p_{n-1})$ be a simple polygon in the plane. With respect to P, points $s, t \in \mathbf{R}^2$ are said to be mutually (clearly) *visible* if the open line segment $\overline{st}$ joining s and t does not intersect P. The visibility is said to be *internal* if $\overline{st}$ lies in the interior of P, and *external* otherwise.

Let e be an edge of P. P is said to be (weakly) *edge-visible* from e if for every point s on P there exists a point t on e that is internally visible from s. One variant of edge visibility requires that t be in the relative interior of e – in such cases, it is not hard to see that e must be an edge of the convex hull of P. In this paper, we will assume that e coincides with an edge of the boundary of the convex hull, but will also consider some cases in which visibility to the endpoints of e is allowed.

Algorithms for edge-visible polygons tend to be simpler and more efficient than their counterparts for general simple polygons, due to their special visibility properties. For example, the most important property for our purposes also allows edge-visible polygons to be triangulated in linear time, simply by cutting off convex vertices one by one:

Property 1. [2] Every convex vertex of P other than the endpoints of e is an ear of P.

Closely related to edge-visible polygons are (weakly) *externally visible* polygons. A polygon P is (clearly) externally visible if for each point s on the boundary, there exists a ray ρ_s originating at s which avoids the interior of P, and which intersects P only at s. If s is on the boundary of the convex hull of P, this condition is trivially satisfied. Otherwise, s must be on some chain of P through the interior of the hull, with only its endpoints p_i and p_j on the hull boundary. Together with the *lid* segment (p_i, p_j), the chain forms what is referred to as a *pocket* polygon. As any external ray originating inside the pocket must intersect the lid in order to escape, every pocket of an externally visible polygon is edge-visible with respect to its lid (see Figure 3).

Two well-known classes of polygons, *monotone* and *shar-shaped*, are subclasses of the class of externally visible polygons. Although we have established connectivity for these classes as well, the details are omitted in this version of the paper, due to space limitations.

3 Edge-Visible Polygons

Instead of proving directly that a polygon of a given class can be transformed into another polygon from the same class, our strategy is to show that it can always be transformed into a unique 'canonical' polygon K. If this is so, polygon P could be transformed into another polygon P' by first transforming P into K, and then reversing the operations that transform P' into K. Before showing how this can be done for edge-visible polygons, we present a technical lemma concerning the visibility of subchains within a polygon.

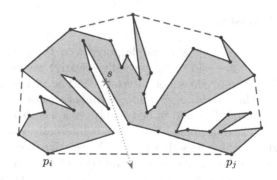

Fig. 3. An externally visible polygon.

3.1   Clearly Visible Chains

Let S be a set of n planar points, not all collinear. Let e be an edge of the convex hull boundary of S, and q be a point on e, such that the endpoints of e are the only pair of points of S forming a collinearity with q. Let $P = (p_0, p_1, p_2, \ldots, p_{n-1})$ be an edge-visible polygon on S from $e = (p_0, p_{n-1})$, with vertices listed in clockwise order about the interior.

With respect to q, for $0 < i < m - 1$, vertex p_i is a *left cusp* if both p_{i-i} and p_{i+1} appear after p_i in clockwise angular order about q. If instead both p_{i-1} and p_{i+1} appear before p_i, then p_i is a *right cusp*. The notion of a cusp extends to polygons which are not edge-visible, and even to polygonal chains, provided that adjacent vertices are not collinear with q.

The following lemma shows the existence of a chain of P that is entirely visible from q, and whose endpoints are either endpoints of e or cusps.

Lemma 1. *Let P, e and q be as defined above. Then there exist indices $0 \le a < b < n$ such that:*

1. *every point of subchain $(p_a, p_{a+1}, \ldots, p_b)$ is visible from q, with the possible exceptions of p_a if $a = 0$, and p_b if $b = n - 1$;*
2. *if $a > 0$, then p_a is a left cusp;*
3. *if $b < n - 1$, then p_b is a right cusp.*

Proof. Omitted.

3.2   Transformation to a Canonical Polygon

Let us now consider the polygon K^q spanning the points of S in such a way that its vertices appear consecutively in (clockwise) angular order about q. Polygon K^q shall be called the *canonical edge-visible polygon* on S with respect to q. Note that K^q is uniquely defined, and is simple.

Any edge-visible polygon P on S with visibility to e can be given a *score* depending on the visibility of its vertices from q. Every vertex s_i of P contributes

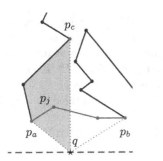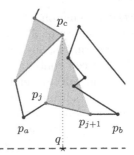

Fig. 4. Searching for a cusp behind chain P^*, and attempting VE-flip with vertex P_c.

either 0 or 1 to the score: s_i contributes 1 if s_i is an endpoint of e or if s_i is visible from q; otherwise, the contribution of s_i is 0. Every polygon on S with edge-visibility to e can be scored in this way. It is not hard to see that the score of the canonical polygon K^q is n, and that K^q is the only polygon on S which achieves this maximum score. Also, the minimum possible score is 3: a contribution of 1 from each endpoint of e, and 1 from that vertex of P closest to the line containing e (this vertex must be visible from q).

Lemma 2. *Let S be a set of n planar points, not all collinear. Let e be an edge of the convex hull boundary of S, and q be a point on e, such that the endpoints of e are the only pair of points of S forming a collinearity with q. If P is an edge-visible polygon spanning S with visibility to e, then P can be transformed into canonical edge-visible polygon K^q via a sequence of at most $n-3$ VE-L-flips.*

Proof. Let P be the polygon $(p_0, p_1, p_2, \ldots, p_{n-1})$, with $e = (p_0, p_{n-1})$, and vertices listed in clockwise order about the interior.

It suffices to show that if $P \neq K^q$, there exists an *improving* VE-flip; that is, one which transforms P into another edge-visible polygon P' such that the score of P' is greater than that of P.

Lemma 1 implies the existence of indices $0 \leq a < b \leq n - 1$ such that

- $a = 0$ or p_a is a left cusp,
- $b = n - 1$ or p_b is a right cusp, and
- $P^* = (p_a, p_{a+1}, \ldots, p_b)$ is entirely visible from q — the only possible exceptions being at p_a if $a = 0$, or p_b if $b = n - 1$.

The only situation in which a could equal 0 and b could equal $n - 1$ is if P were the canonical polygon K^q. Since by assumption $P \neq K^q$, either $a > 0$ or $b < n - 1$. We will consider only the case where $a > 0$, as the arguments for the case where $b < n - 1$ are virtually identical.

Starting from p_a, we search for a vertex and edge to participate in a VE-flip, by performing a radial sweep clockwise about q. As the sweep ray progresses, the closest intersection points (to q) with both P^* and $P \setminus P^*$ are maintained, the first intersections being on edge (p_a, p_{a+1}) of P^* and edge (p_a, p_{a-1}) of $P \setminus P^*$. Note that the visibility of P^* to q implies that the chain is radially monotone with respect to q; that is, any ray from q intersects P^* in at most one point (see Figure 4). The sweep continues until a cusp p_c of $P \setminus P^*$ is encountered. We claim that this must occur before the sweep ray reaches p_b.

Since p_c is a cusp, it must be a convex vertex. From the edge-visible property of P, it must therefore be an ear. Let (p_j, p_{j+1}) be the edge of P^* intersected by the ray through p_c, where $a \leq j < b$. If p_c is visible from both p_j and p_{j+1}, a VE-flip is possible: edges (p_{c-1}, p_c), (p_c, p_{c+1}) and (p_j, p_{j+1}) could be replaced by edges (p_{c-1}, p_{c+1}), (p_c, p_j) and (p_c, p_{j+1}). Due to the radial monotonicity of P^* with respect to q, the removal of edge (p_j, p_{j+1}) would result in p_c becoming visible from q. The internal visibility to e and to q of all other vertices would remain unchanged; the VE-flip would thus yield an edge-visible polygon of higher score than P. However, it is possible that some part of the polygon interferes with the visibility from p_c to p_{j+1} (see Figure 4).

If p_c is not visible from both p_j and p_{j+1}, we search for an alternative to p_c for the flip with (p_j, p_{j+1}). Let l be the line segment joining q and p_c, and let x be a point of l between (p_j, p_{j+1}) and p_c. Imagine a second sweep in which the point x begins on (p_j, p_{j+1}) and slowly moves along l towards p_c. As x moves, the empty triangular area Δ^x bounded by x, p_j and p_{j+1} would grow, until one or both of its sides comes into contact with at least one vertex of $P \setminus P^*$. Let us assume that contact occurs along the side (x, p_{j+1}) of the triangle; the case in which the only contact is along (x, p_j) is handled symmetrically.

For contact to occur along (x, p_{j+1}), there must exist $b < d \leq d' < n - 1$ such that

- p_i is collinear with (x, p_{j+1}) for all $d \leq i \leq d'$,
- $p_{d'+1}$ is not collinear with (x, p_{j+1}), and
- p_i is not collinear with (x, p_{j+1}) for all $j + 1 < i < d$.

There are two cases remaining to be considered.

1. $j + 2 < d$.
   The segment (p_j, p_d) passes through the interior of Δ^x, and thus p_j and p_d are mutually visible. The construction guarantees that p_d and p_{j+1} are also visible, and that p_d is a convex vertex. Here, a VE-flip is possible: edges (p_{d-1}, p_d), (p_d, p_{d+1}) and (p_j, p_{j+1}) could be replaced by edges (p_{d-1}, p_{d+1}), (p_d, p_j) and (p_d, p_{j+1}). By the same arguments as when p_c was visible from p_j and p_{j+1}, the VE-flip would yield an edge-visible polygon of higher score than P (see Figure 5).
2. $j + 2 = d$.
   The segment $(p_{d'}, p_j)$ passes through the interior of Δ^x, and thus $p_{d'}$ and p_j are mutually visible. The collinearity of $(p_{j+1}, p_{j+2}, \ldots, p_{d'})$ and the fact that $p_{d'}$ is a convex vertex together imply that $p_{d'+1}$ and p_{j+1} are mutually visible. To see this, consider the polygon that would result if the collinear chain $(p_{j+1}, p_{j+2}, \ldots, p_{d'})$ were replaced by the new edge $(p_{j+1}, p_{d'})$. This resulting polygon would also be edge-visible with respect to e, and $p_{d'}$ would

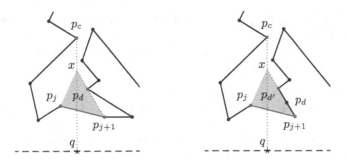

Fig. 5. Two cases: no contact with p_{j+2}; contact with p_{j+2}.

be an ear. Here, an L-flip is possible: edges $(p_{d'}, p_{d'+1})$ and (p_j, p_{j+1}) could be replaced by edges $(p_j, p_{d'})$ and $(p_{j+1}, p_{d'+1})$. With the deletion of (p_j, p_{j+1}), each of the vertices $(p_{j+2}, p_{j+3}, \ldots, p_{d'})$ would become visible to q without disturbing the visibility to q of any other vertices. The L-flip would thus yield an edge-visible polygon of higher score than P (see Figure 5).

It should be observed that as long as no other vertex of P is collinear with the endpoints of e, it is always possible to choose some point q on e satisfying the non-collinearity condition of the lemma. The main theorem of this section follows as an immediate corollary.

Theorem 1. *Let S be a set of n planar points, not all collinear. Let P and P' be edge-visible polygons spanning S, with visibility to a common convex hull edge e. Then P can be transformed into P' via a sequence of no more than $2n - 6$ VE-L-flips.*

4   Other Polygon Classes

The methods of the previous section for proving connectivity under the VE-L-flip operation of edge-visible polygons can be extended to other classes as well. The general strategy is the same: define a score function and a unique canonical polygon which maximizes it, and then show that there always exists an improving VE-L-flip for every polygon other than the canonical.

4.1   Externally Visible Polygons

Let point set S, convex hull edge e, and point q on e be defined as in the previous section. Let h be the number of points of S on the boundary of its convex hull. The *canonical externally visible polygon* $K^q = (k_0, k_1, \ldots, k_{h-1}, k_h, \ldots, k_{n-1})$ is defined as follows:

- the endpoints of e are k_0 and k_{h-1};
- the subchain $(k_0, k_1, \ldots, k_{h-1})$ of K^q spans the points of S on its convex hull boundary, in counterclockwise order about the interior;

- the subchain $(k_h, k_{h+1}, \ldots, k_{n-1})$ of K^q spans the points of S interior to the convex hull;
- the vertices of subchain $(k_{h-1}, \ldots, k_{n-1}, k_0)$ appear consecutively in clockwise angular order about q.

Again, K^q is uniquely defined, and is simple

An externally visible polygon P on S will be scored as follows: vertex s_i in the interior of the convex hull of S contributes 1 if s_i has external visibility to q; otherwise, if s_i is not externally visible to q or if s_i is on the convex hull boundary, the contribution of s_1 is 0. The score of the canonical polygon K^q is $n - h$, and no other externally visible polygon on S achieves this maximum score. The minimum possible score is 0, occurring whenever e is an edge of P.

The following lemma is a corollary of Lemma 2.

Lemma 3. *Let S be a set of n planar points, not all collinear. Let e be an edge of the convex hull boundary of S, and q be a point on e, such that no pair of points form a collinearity with q unless they are both collinear with e. If P is an externally visible polygon spanning S, then P can be transformed into canonical externally visible polygon K^q via a sequence of at most $n - h$ VE-L-flips.*

Proof. The details have been omitted, as the proof is virtually identical to that of Lemma 2.

Lemma 3 immediately implies the connectivity result for externally visible polygons.

Theorem 2. *Let S be a set of n planar points, not all collinear. Let P and P' be externally visible polygons spanning S. Then P can be transformed into P' via a sequence of no more than $2n - 2h$ VE-L-flips.*

4.2   Edge-Visible Polygons with Differing Visibility Edges

Theorem 3. *Let S be a set of n planar points, and let S^* be the subset of S lying on the boundary of the convex hull of S. Assume that no point s of S^* is collinear with any pair of points of $S \setminus \{s\}$. Let P and P' be edge-visible polygons spanning S, with visibility to convex hull edges e and e', respectively. Then P can be transformed into P' via a sequence of no more than $\lceil \frac{h+1}{2} \rceil (n - 3)$ VE-L-flips, where h is the cardinality of S^*.*

Proof. Consider the shorter of the two sequences of edges between e and e' along the boundary of the convex hull of S. If both sequences have the same number of edges, then choose one arbitrarily. Let the chosen sequence be $E = (e_1, e_2, \ldots, e_m)$, where $e_1 = e$ and $e_m = e'$. Let the endpoints of e_i be (s_{i-1}, s_i) for all $1 \le i \le m$.

To transform P into P', we first transform P to the canonical edge-visible polygon K^{q_1} with respect to a point q_1 on e_1 very close to s_1: q_1 is chosen so that no line passing through a pair of points of $S \setminus \{s_0, s_1\}$ separates it from s_1. This can be done via a sequence of at most $n - 3$ VE-L-flips, by Lemma 2.

Since s_1 is assumed not to be involved in any collinearity among the points of S, there exists a point q on e_2 from which no line of collinearity separates q_1. This implies that q and q_1 are both visible from every point on the boundary of K^{q_1}, which in turn implies that $K^q = K^{q_1}$ is edge-visible with respect to e_2. K^q can then be transformed into a new canonical edge-visible polygon K^{q_2} for some point arbitrarily close to s_2, again by Lemma 2.

This process may be continued, choosing points q_i on e_i sufficiently close to s_i for all $1 \le i < m$. The result, after $m - 1$ applications of Lemma 2, is a polygon edge-visible with respect to $e' = e_m$. A final application of Lemma 2 shows that this polygon can then be transformed to P'.

The total number of flips required is bounded by $m(n-3)$. Since E is chosen to be the smaller of the two sequences available, it is easy to show that $m \le \lceil \frac{h+1}{2} \rceil$. From this, the result follows.

References

1. D. Avis and K. Fukuda, 'Reverse search for enumeration,' *Discr. Appl. Math.* 65:21–46, 1996.
2. D. Avis and G. T. Toussaint, 'An optimal algorithm for determining the visibility of a polygon from an edge,' *IEEE Trans. Comput.* C-30:910–914, 1981.
3. H. Edelsbrunner and N. R. Shah, 'Incremental topological flipping works for regular triangulations,' *Algorithmica* 15:223–241, 1996.
4. S. Fortune, 'Voronoi diagrams and Delaunay triangulations,' in *Computing in Euclidean Geometry*, D. Z. Du and F. K. Hwang, eds., World Scientific, pp. 193–234, 1992.
5. C. Hernando, F. Hurtado, A. Márquez, M. Mora and M. Noy, 'Geometric tree graphs of points in convex position,' *Discr. Appl. Math.* 93:51–66, 1999.
6. C. Hernando, F. Hurtado and M. Noy, 'Graphs of non-crossing matchings,' *Proc. 15th European Conference on Computational Geometry*, pp. 97–100, 1999.
7. J. Hershberger, 'Finding the visibility graph of a simple polygon in time proportional to its size,' *Algorithmica* 4:141–155, 1989.
8. M. E. Houle, 'On local transformations of simple polygons,' *Australian Comp. Sci. Commun.* 18(3):64–71 (Proc. CATS'96, Melbourne), 1996.
9. F. Hurtado and M. Noy, 'Graphs of triangulations of a convex polygon and tree of triangulations,' *Comput. Geom. Theory Appl.* 13:179–188, 1999.
10. F. Hurtado, M. Noy and J. Urrutia, 'Flipping edges in triangulations,' *Discr. Comp. Geom.* 22:333–346, 1999.
11. B. Joe, 'Construction of three-dimensional Delaunay triangulations using local transformations', *Computer Aided Geom. Design* 8:123–142, 1991.
12. C. L. Lawson, 'Transforming triangulations,' *Discr. Math.* 3:365–372, 1972.
13. J. O'Rourke, *Art Gallery Theorems and Algorithms*, Oxford University Press, New York, 1987.
14. M. Pocchiola and G. Vegter, 'Computing the visibility graph via pseudo-triangulations' *Proc. 11th ACM Symp. Comp. Geom.*, 1995, pp. 248–257.
15. C. Zhu, G. Sundaram, J. Snoeyink, and J. S. B. Mitchell, 'Generating random polygons with given vertices,' *Comput. Geom. Theory Appl.* 6:277–290, 1996.

Embedding Problems for Paths with Direction Constrained Edges[*]

Giuseppe Di Battista, Giuseppe Liotta[2], Anna Lubiw[3], and Sue Whitesides[4]

[1] Dipartimento di Informatica ed Automazione, Università di Roma Tre, Roma, Italy.
gdb@dia.uniroma3.it
[2] Dipartimento di Ingegneria Elettronica e dell'Informazione,
Università di Perugia, Perugia, Italy.
liotta@diei.unipg.it
[3] Department of Computer Science,
University of Waterloo,Waterloo, Canada.
alubiw@daisy.uwaterloo.ca
[4] School of Computer Science,
McGill University,Montreal, Canada.
sue@cs.mcgill.ca

Abstract. We determine the reachability properties of the embeddings in R^3 of a directed path, in the graph theoretic sense, whose edges have each been assigned a desired direction (East, West, North, South, Up, or Down) but no length. We ask which points of R^3 can be reached by the terminus of an embedding of such a path, by choosing appropriate positive lengths for the edges, if the embedded path starts at the origin, does not intersect itself, and respects the directions assigned to its edges. This problem arises in the context of extending planar graph embedding techniques and VLSI rectilinear layout techniques from $2D$ to $3D$. We give combinatorial characterizations of reachability that yield linear time recognition and layout algorithms.

1 Introduction

The *shape* of a directed, polygonal curve in $3D$ that consists of n axis-parallel straight line segments is the sequence σ of n direction labels *East*, *West*, *North*, *South*, *Up*, or *Down* determined by the directions of the n segments of the curve.

A *shape* for a directed, graph theoretic path P is an ordered sequence σ of n labels *East*, *West*, *North*, *South*, *Up*, or *Down*, one for each of the n edges of P. Each label in the sequence specifies a direction for the corresponding directed edge of P when that edge is realized as a straight line segment in $3D$.

[*] Research partially supported by operating grants from the Natural Sciences and Engineering Research Council (NSERC) of Canada, by the project "Algorithms for Large Data Sets: Science and Engineering" of the Italian Ministry of University and Scientific and Technological Research (MURST 40%), and by the project "Geometria Computazionale Robusta con Applicazioni alla Grafica ed a CAD" of the Italian National Research Council (CNR).

D.-Z. Du et al. (Eds.): COCOON 2000, LNCS 1858, pp. 64–73, 2000.
© Springer-Verlag Berlin Heidelberg 2000

A realization of a graph theoretic directed path P as an orthogonal polygonal curve in $3D$ is specified by giving a start point, a shape σ, and an assignment of positive lengths to the edges.

Given a point p in $3D$ and a shape σ for a directed, graph theoretic path P, the $3D$ *shape path reachability problem* is to find an assignment (if one exists) of positive integer lengths to the directed edges of P so that the resulting realization of P as an orthogonal, polygonal curve

- ends at p if it starts at the origin;
- is *simple*, that is, does not intersect itself; and
- satisfies the direction constraints on its segments as specified by σ.

The $3D$ shape path reachability problem has instances that do not admit solution. For example, consider a point p that lies East of the origin and the shape σ given by the sequence of labels $UWDESWN$, where U stands for Up, W stands for $West$, and so on. Shape σ cannot be the shape of any simple, orthogonal, polygonal realization of a path that starts at the origin and terminates at p, even though σ contains an E label (see Figure 1).

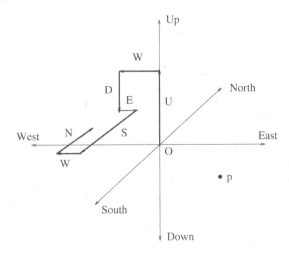

Fig. 1. No path of shape $\sigma = UWDESWN$ can start at the origin O and reach a point p East of O.

The main result of this paper is a combinatorial characterization of those $3D$ shape path reachability problem instances that admit solution, together with linear time recognition and embedding algorithms based on this characterization.

Various $2D$ versions of our $3D$ shape path problem have been studied in several papers, including [10, 9]. The problem has also been generalized to non-orthogonal polygons and graphs in [11, 4, 7].

A basic technique for $2D$ orthogonal layout in VLSI and in graph drawing is to give a "shape" for a graph, that is, an assignment of direction labels E,

W, N, and S to the edges, then to determine lengths for the edges so that the layout is non-crossing [Vijayan and Widgerson] ([10]). The graph must have maximum degree 4. In the VLSI context, each vertex represents either a corner of a bounding box containing components, or a connection pin on the side of such a box. The edges at a vertex connect it to its two neighbors (pins or box corners) on the bounding box and, in the case of a pin vertex, to a pin on another box.

The well-known *topology-shape-metrics approach* [3] for constructing a rectilinear embedding of a planar graph consists of three main steps, called planarization, orthogonalization, and compaction. The planarization step determines an embedding, i.e., the face cycles, for the graph in the plane. The orthogonalization step then specifies for each edge (u, v), an ordered list of the bends that the orthogonal polygonal line representing (u, v) is to have in the final embedding. For example, (u, v) could be labeled $NESNE$, which would say "starting from u first go North, then go East, etc." Finally, the compaction step computes the final embedding, giving coordinates to vertices and bends.

While the literature on $3D$ orthogonal embeddings of graphs is quite rich (see e.g. [1, 5, 6, 8, 12, 13]), the extension of the topology-shape-metrics approach to $3D$ remains, as far as we know, to be carried out. The $3D$ shape path reachability results we present here are an essential prerequisite for such a program.

2   Overview of the Main Results

In order to state our characterization result precisely, we introduce the concepts of a *flat* and of a *canonical sequence* in a shape.

Let σ be a shape. A *flat* of σ is a consecutive subsequence σ' of σ that is maximal with respect to the property that its labels come from the union of two oppositely directed pairs of directions, i.e., either from the set $\{N, S, E, W\}$, or from the set $\{N, S, U, D\}$, or from the set $\{U, D, E, W\}$. Thus any realization of the shape sequence σ' must consist of segments that lie on the same axis-aligned plane. For example, the shape $\sigma = UWDESWN$ depicted in Figure 1 contains two flats: $F_1 = UWDE$, and $F_2 = ESWN$. Observe that two consecutive flats of σ share a label and that they must be drawn on perpendicular planes. For example, the last label, E, of F_1 coincides with the first label, E, of F_2.

A not necessarily consecutive subsequence $\tau \subseteq \sigma$, where τ consists of k elements, is a *canonical sequence* provided that:

- $1 \leq k \leq 3$;
- any two elements of τ indicate mutually orthogonal directions; and
- the elements of τ that appear on any given flat of σ form a possibly empty consecutive subsequence of σ.

The *type* of a canonical sequence τ is given by the (unordered) set of labels it contains.

For example, the shape $\sigma = UWDESWN$ depicted in Figure 1 contains two canonical sequences of type $\{U, N, W\}$. The first two elements of σ are a U and a

W. These elements belong to the same flat, F_1, where they appear consecutively; they do not belong to F_2. The last element of σ is an N, which belongs to flat F_2 but not to F_1. Thus the conditions for a canonical sequence of length $k = 3$ and type $\{U, N, W\}$ are satisfied by the subsequence of σ consisting of its first, second, and last elements. The first, next-to-last, and last elements of σ form a second canonical sequence of type $\{U, N, W\}$. Note, however, that σ does not contain a canonical subsequence of type $\{U, E, N\}$: these labels occur in unique positions in σ, and in particular, the unique U and the unique E both belong to the same flat F_1, but they are not consecutive elements of σ.

The essence of our combinatorial characterization of solvable shape path reachability problems is given below. For concreteness, the characterization is given with respect to the UNE octant. As explained in the next section, the results for other octants can be obtained by a suitable permutation of the labels.

Theorem 1. *Let σ be a shape and let p be a point in the UNE octant. Then there exists a simple, orthogonal, polygonal curve of shape σ that starts at the origin and that terminates at p if and only if σ contains a canonical sequence of type $\{U, N, E\}$.*

In other words, p can be reached, starting from the origin, by a simple polygonal curve of shape σ if and only if it is possible to choose from σ a U element, an N element and an E element, not necessarily in that order, such that if any flat of σ contains two of these elements, then the two elements are consecutive in σ.

3   Preliminaries

We regard each coordinate axis in the standard $3D$ coordinate system as consisting of the origin plus two open semi-axes, directed away from the origin and labelled with a pair of opposite direction labels from the set $\{N, W, S, E, U, D\}$. A triple XYZ of distinct unordered labels no two of which are opposite defines the XYZ *octant*. Note that unless stated otherwise, we consider octants to be open sets. Similarly, a pair XY of distinct orthogonal labels defines the XY *quadrant* in $2D$ or $3D$. Finally, a label X defines the X *semi-axis*, which consists of those points that are positive multiples of the unit vector in the X direction. For short we call a semi-axis an axis. Thus, $3D$ space is partitioned into eight octants, twelve quadrants, six axes, and the origin.

Let σ be a shape consisting of n elements or labels[1]. An *embedding* or *drawing* of σ, denoted as $\Gamma(\sigma)$, is a *non-self-intersecting*, directed, orthogonal polygonal curve consisting of n segments such that the k-th segment ($k = 1, \ldots, n$) of $\Gamma(\sigma)$ has positive length and has the orientation specified by the k-th label of σ. Unless stated otherwise, we assume that the start of the directed curve $\Gamma(\sigma)$ lies at the origin of the reference system.

---

[1] Since the elements of a shape sequence have direction labels as values, we often refer to the elements as labels.

Assumptions. Since we are interested in shapes that admit embeddings, we assume from now on that shapes do not contain adjacent labels that are oppositely directed. Furthermore, we assume that shapes do not contain adjacent labels that are identical, since the reachability properties of shapes with adjacent, identical labels are not changed by coalescing such labels. Finally, we assume that the point p to be reached in an instance of the shape path reachability problem lies in an octant.

Remark. Let $\phi()$ be a permutation of the six direction labels that maps opposite pairs of labels to possibly different opposite pairs (for example, ϕ might map N, S, E, W, U, D to E, W, N, S, D, U, respectively). Note that $\phi()$ defines a linear transformation of $3D$ space that is not necessarily a rotation. Nevertheless, this transformation defines a bijection between embeddings of σ and embeddings of $\phi(\sigma)$. Here, $\phi(\sigma)$ denotes the sequence of labels obtained by applying ϕ to the labels of σ. For concreteness, we often state our results and proofs referring to some given octant, quadrant, or axis where points of embeddings of σ can lie. However, the results can also be stated with respect to any other octant, quadrant, or axis since they are preserved under the $\phi()$ transformation.

Definition. A drawing $\Gamma(\sigma)$ of a shape σ is an *expanding drawing* if each segment travels one unit farther in its direction than the extreme points, with respect to that direction, of the previous segments of $\Gamma(\sigma)$.

Expanding drawings are useful because a new segment of arbitrary length, perpendicular to the last segment of the drawing, may be appended to the last segment without causing any collisions. The definition above is a simple, technical way to achieve this property.

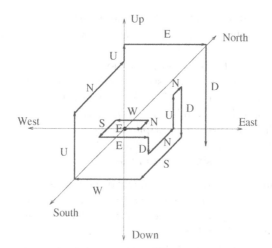

Fig. 2. The expanding drawing for $\sigma = ENWSEDNUNDSWUNUED$. The drawing starts at the origin, and each segment travels one unit farther in its direction than any preceding segment has gone.

Lemma 1. *Every shape admits an expanding drawing.*

In the next sections we shall characterize the reachabiity properties of shapes by means of canonical sequences.

4   The Shape Path Reachability Problem

We say that a drawing $\Gamma(\sigma)$ of a shape σ *reaches point* p if $\Gamma(\sigma)$ terminates at p when it starts at the origin. A shape σ *reaches a point* p if it admits a drawing $\Gamma(\sigma)$ that reaches p. A shape σ *reaches a set of points* if it reaches each point in the set. For example, shapes NEU, NUE, and $NWUE$ all reach the UNE octant; that is, for each point of the UNE octant, each of these shapes has a drawing that terminates at that point.

Lemma 2. *Let σ be a shape. If σ reaches a point p in an axis, quadrant or octant, then it reaches that entire axis, quadrant or octant, respectively.*

Observe that if a shape reaches the UNE octant, then it must contain a U label, an N label and an E label. Notice, however, that the converse is not always true. As previously observed, the shape $UWDESWN$ does not reach the UNE octant, even though it contains a U label, an E label, and an N label.

In the rest of this section, we investigate the problem of determining whether a shape can reach a given portion of $2D$ or $3D$ space. We distinguish between shapes that consists of only one flat and shapes that have more than one flat. A shape of the first type is a $2D$ *shape*, while a shape of the second type is a $3D$ *shape*.

4.1   Reachability in $2D$

In this section we answer the following question: Given a point p in a quadrant or on an axis and a $2D$ shape σ, can σ reach p? By Lemma 2, this question can be answered by characterizing when σ can reach the quadrant or the axis containing p.

We start by investigating a special type of reachability. Let $\Gamma(\sigma)$ be a drawing of σ that reaches the NE quadrant. We say that $\Gamma(\sigma)$ reaches the NE quadrant *for the first time with label* X if $X \in \{N, S, E, W\}$ is the direction associated with the first segment of $\Gamma(\sigma)$ entering the NE quadrant when walking along $\Gamma(\sigma)$ starting at the origin.

Lemma 3. *Let $\Gamma(\sigma)$ be a drawing of a $2D$ shape $\sigma = \sigma'E$ such that $\Gamma(\sigma)$ reaches the NE quadrant for the first time with its last label, E. Then σ contains a canonical sequence NE.*

Lemma 4. *Let $\sigma = \sigma'\sigma''$ be a $2D$ shape such that σ' reaches the NE quadrant. Then σ reaches the NE quadrant.*

We are now ready to consider the more general problem of quadrant reachability.

Theorem 2. *A 2D shape σ reaches the NE quadrant if and only if σ contains a canonical sequence of type $\{N, E\}$.*

Proof. We prove first that if σ contains a canonical sequence τ of type $\{N, E\}$, then it reaches the NE quadrant. Let $\sigma = \sigma_1 NE\sigma_2$, where σ_1 or σ_2 may be the empty string. By Lemma 1, σ_1 has an expanding drawing $\Gamma(\sigma_1)$. Extend $\Gamma(\sigma_1)$ to a drawing for σ as follows. Append to the terminal point of $\Gamma(\sigma_1)$ a segment oriented N that is so long that i) its terminal point projects to the N axis and ii) the resulting drawing is still an expanding drawing. Because the new drawing is expanding we can append to its terminal point a segment oriented E that is long enough to enter the NE quadrant. Thus $\sigma_1 NE \subseteq \sigma$ reaches the NE quadrant, and hence by Lemma 4, so does σ.

Suppose now that $\sigma = X_1, \ldots, X_n$ reaches the NE quadrant, and let $\Gamma(\sigma)$ be a drawing of σ that terminates at a point in the NE quadrant. Suppose the first segment of $\Gamma(\sigma)$ to intersect the NE quadrant is the ith segment, whose associated direction is X_i. Let σ' denote the initial sequence $X_1, \ldots X_i$ of σ. Then $\Gamma(\sigma)$ contains a drawing $\Gamma(\sigma')$ of σ'. If $X_i = E$, then by Lemma 3, σ' and hence σ contains the consecutive subsequence NE. Similarly, if $X_i = N$, then σ contains the consecutive subsequence EN.

Based on the results above and on Lemma 2 it is possible to design a linear time algorithm for deciding whether a $2D$ shape σ can reach a point of a quadrant.

Theorem 3. *Let σ be a 2D shape, and let p be a point of a quadrant. There exists a linear time and space algorithm that decides whether σ reaches p.*

4.2   Reachability in $3D$

In this section we answer the following question: Given a point p in an octant and a 3D shape σ, can σ reach p? Again, by Lemma 2, this question can answered by characterizing when σ can reach a given octant. To this aim, we introduce the key concept of a *doubly expanding drawing*. This concept will later be used for the study of shape path embeddability.

A *doubly expanding drawing* is one for which the first and last segments can be replaced by arbitrarily long segments without creating any intersections within the drawing. Figure 3 shows an example of a doubly expanding drawing.

Lemma 5. *Let σ be a shape with n labels such that it either consists of exactly two labels or it contains at least two flats. Then σ has a double expanding drawing that can be computed in $O(n)$ time.*

The next lemma gives a sufficient condition for octant reachability; the proof (omitted due to space constraints) is based on connecting together singly and doubly expanding drawings for subsequences of shapes and gives rise to a linear time embedding algorithm.

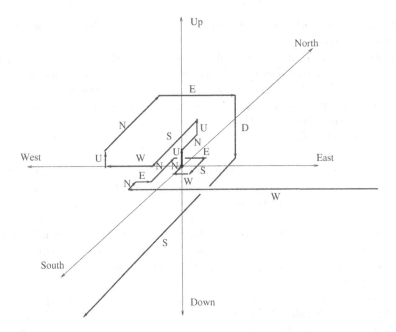

Fig. 3. A doubly expanding drawing for $\sigma = WNENESWNUNUSWUNEDS$.

Lemma 6. *Let σ be a 3D shape that contains a canonical sequence of type $\{U, N, E\}$. Then σ reaches the UNE octant.*

The next lemma proves that the condition of Lemma 6 is also necessary for octant reachability.

Lemma 7. *Let σ be a 3D shape that reaches the UNE octant. Then σ contains a canonical sequence of type $\{U, N, E\}$.*

Proof. First, we begin by proving a weaker form of this:
If σ is a 3D shape that can reach the NE quarter-space (i.e., the union of the UNE octant, the DNE octant, and the NE quadrant), then one can choose one N and one E in σ that either are consecutive or do not share a flat.

To prove this, consider a drawing $\Gamma(\sigma)$ that reaches the (open) NE quarter-space. Let (a, b) be the first edge of Γ that enters this quarter-space. Suppose without loss of generality that (a, b) is an E edge. Since this E edge lies in the N half-space, there must be a previous N edge, (c, d), that enters this half-space. If the chosen E and N edges do not share a flat, we are done. So assume that they do share a flat. Let Π be the $NSEW$ plane containing the flat. Note that b is in the (open) NE quadrant of Π. We claim that c is in the closure $\bar{S}\bar{W}$ of the W quadrant. This is because d, which is in the N half-plane but not in the NE quadrant, must lie in the $N\bar{W}$ quadrant. The portion of Γ from c to b thus travels to the NE quadrant. By the Lemma 3 one can find a consecutive pair

NE or EN in the corresponding portion of σ. This completes the proof of the preliminary result.

To obtain a proof of the original statement, consider a drawing $\Gamma(\sigma)$ that reaches the ENU octant. Let (p, q) be the first edge of Γ that enters this octant, and suppose without loss of generality that (p, q) is a U edge. Since this U edge lives in the NE quarter-space, there must be a previous edge, (r, s), that enters this quarter-space. Assume without loss of generality that (r, s) is an N edge. Note that q is in the ENU octant, s is in the $EN\bar{D}$ octant, and r is in the $E\bar{S}\bar{D}$ octant.

The path from the origin to s travels to the EN quarter-space, and thus by the preliminary result just proved, one can choose in the corresponding portion of σ an E and an N that either are consecutive or do not share a flat. The path from r to q travels to the NU quarter-space, and thus by the preliminary result above, one can choose in the corresponding portion of σ an N and a U that are either consecutive or do not share a flat. Note that the second pair of chosen labels follows the first pair in σ, except that the N edge (r, s) may be in both pairs. If it is, we have what we want.

Now consider the other possibilities. The chosen letters may appear in the order EN_1N_2U, with the N edge (r, s) in the closed interval from N_1 to N_2. We claim that either N_1 and U do not share a flat, or E and N_2 do not share a flat, for if both occurred then all four letters would share a flat. Thus at least one of N_1 or N_2 will suffice to establish the lemma.

As a next possibility, the chosen letters may appear in the order N_1EN_2U, with the N edge (r, s) in the closed interval from E to N_2. E and U cannot share a flat, because such a flat would also have to contain N_2, and thus N_1 suffices. The case EN_1UN_2 is similar—N_2 suffices.

The final possibility is that the chosen letters may appear in the order N_1EUN_2 with the N edge (r, s) in the interval from E to U. Because the N edge intervenes, E and U cannot share a flat, and thus either of N_1 or N_2 will suffice.

Lemmas 6 and 7 can be summarized as follows.

Theorem 4. *Let σ be a 3D shape. Shape σ reaches the UNE octant if and only if it contains a canonical sequence of type $\{U, N, E\}$.*

Theorem 4 and Lemma 2 yield a linear time algorithm for deciding whether a $3D$ shape σ can reach a point in an octant. Given a point p that can be reached, the linear time algorithm in the proof of Lemma 6 then gives an embedding that terminates at p.

Theorem 5. *Let σ be a 3D shape with n labels, and let p be a point of an octant. There exists an algorithm that decides whether σ reaches p, that runs in $O(n)$ time, and that computes an embedding for σ that reaches p when such an embedding exists.*

5 Open Problems

Several issues remain open. We mention three that in our opinion are interesting.

1. Study the reachability problem under the additional constraint that the lengths of some of the segments are given.
2. Study the reachability problem in the presence of obstacles.
3. Study the reachability problem for paths that must be embedded in a fixed grid and that are to reach specified points in this grid.

In the case of the last item, it is no longer true that a shape can reach a point if and only if it can reach all the other points in the same octant. One can no longer construct a drawing and then scale it so that its terminus is located at some specified point p, as scaling would move some edges off grid lines. For drawings in a fixed grid, volume becomes a meaningful issue to study.

References

[1] T. Biedl, T. Shermer, S. Wismath, and S. Whitesides. Orthogonal 3-D graph drawing. *J. Graph Algorithms and Applications*, 3(4):63–79, 1999.

[2] R. F. Cohen, P. Eades, T. Lin and F. Ruskey. Three-dimensional graph drawing. *Algorithmica* , 17(2):199–208, 1997.

[3] G. Di Battista, P. Eades, R. Tamassia, and I. Tollis. Graph Drawing. Prentice Hall, 1999.

[4] G. Di Battista and L. Vismara. Angles of planar triangular graphs. *SIAM J. Discrete Math.*, 9(3):349–359, 1996.

[5] P. Eades, C. Stirk, and S. Whitesides. The techniques of Kolmogorov and Bardzin for three dimensional orthogonal graph drawings. *Inform. Process. Lett.*, 60:97–103, 1996.

[6] P. Eades, A. Symvonis, and S. Whitesides. Two algorithms for three dimensional orthogonal graph drawing. In S. North, editor, *Graph Drawing (Proc. GD '96)*, volume 1190 of *Lecture Notes Comput. Sci.*, pages 139–154. Springer-Verlag, 1997.

[7] A. Garg. New results on drawing angle graphs. *Comput. Geom. Theory Appl.*, 9(1–2):43–82, 1998. Special Issue on Geometric Representations of Graphs, G. Di Battista and R. Tamassia, editors.

[8] A. Papakostas and I. Tollis. Incremental orthogonal graph drawing in three dimensions. In G. Di Battista, editor, *Graph Drawing (Proc. GD '97)*, volume 1353 of *Lecture Notes Comput. Sci.*, pages 139–154. Springer-Verlag, 1997.

[9] R. Tamassia. On embedding a graph in the grid with the minimum number of bends. *SIAM J. Comput.*, 16(3):421–444, 1987.

[10] G. Vijayan and A. Wigderson. Rectilinear graphs and their embeddings. *SIAM J. Comput.*, 14:355–372, 1985.

[11] V. Vijayan. Geometry of planar graphs with angles. In *Proc. 2nd Annu. ACM Sympos. Comput. Geom.*, pages 116–124, 1986.

[12] D. R. Wood. Two-bend three-dimensional orthogonal grid drawing of maximum degree five graphs. Technical Report 98/03, School of Computer Science and Software Engineering, Monash University, 1998.

[13] D. R. Wood. An algorithm for three-dimensional orthogonal graph drawing. In *Graph Drawing (Proc. GD '98)*, volume 1547 of *Lecture Notes Comput. Sci.*, pages 332-346. Springer-Verlag, 1998.

Characterization of Level Non-planar Graphs by Minimal Patterns

Patrick Healy[1], Ago Kuusik[1]*, and Sebastian Leipert[2]

[1] Dept. of Comp. Science and Info. Systems, University of Limerick, Ireland
[2] Institut für Informatik, Universität zu Köln, Germany

Abstract. In this paper we give a characterization of level planar graphs in terms of minimal forbidden subgraphs called minimal level non-planar subgraph patterns (MLNP). We show that a MLNP is completely characterized by either a tree, a level non-planar cycle or a level planar cycle with certain path augmentations. These characterizations are an important first step towards attacking the $\mathcal{NP}$-hard level planarization problem.

1 Introduction

Level graphs are an important class of graphs that generalize bipartite graphs. Such graphs are used to model hierarchical relationships or workflow diagrams. From the point of view of representing the relationships graphically, it is desirable that the graphs are drawn with as few edge crossings as possible. Jünger et al. [5] have presented a linear-time algorithm to determine if a level graph is planar; this algorithm can be modified to determine a planar embedding of the graph as well.

Clearly, level graphs that need to be visualized are not level planar in general. Mutzel [6] therefore studies the level planarization problem for the case $k = 2$ levels, removing a minimum number of edges such that the resulting subgraph is level planar. For the final diagram the removed edges are reinserted into a level planar drawing. However, the level planarization problem is $\mathcal{NP}$-hard [3]. Mutzel [6] gives an integer linear programming formulation for the 2-level planarization problem, studying the polytope associated with the set of all level planar subgraphs of a level graph with 2 levels.

In order to attack the level planarization problem for $k \geq 2$ levels, an integer linear programming formulation has to be found, and the polytope associated with the set of all level planar subgraphs of a level graph needs to be described. Besides, polynomial time separation algorithms need to be developed for practical application. One important step in the study of the polytope associated with the set of all level planar subgraphs is the characterization of level planar graphs in terms of minimal forbidden subgraphs called *minimal level non-planar subgraph patterns* (MLNP-patterns) These graphs are the analogue of K_5 and $K_{3,3}$ in the case of general graph planarity.

---

* Funded by Forbairt Basic Research Grant SC/97/611.

D.-Z. Du et al. (Eds.): COCOON 2000, LNCS 1858, pp. 74–84, 2000.
© Springer-Verlag Berlin Heidelberg 2000

This paper is organized as follows. After summarizing the necessary preliminaries in the next section, we introduce the minimal level non-planar subgraph patterns in Sect. 3. We prove that the patterns are minimal level non-planar and that the set of patterns is complete for the special case of hierarchies. This result is generalized to level graphs in Sect. 4.

2 Preliminaries

A *level graph* $G = (V, E, \phi)$ is a directed acyclic graph with a mapping $\phi : V \to \{1, 2, \ldots, k\}$, $k \geq 1$, that partitions the vertex set V as $V = V_1 \cup V_2 \cup \cdots \cup V_k$, $V_j = \phi^{-1}(j)$, $V_i \cap V_j = \emptyset$ for $i \neq j$, such that $\phi(v) = \phi(u) + 1$ for each edge $(u, v) \in E$.

A drawing of a level graph G in the plane is a *level drawing* if the vertices of every V_j, $1 \leq j \leq k$, are placed on a horizontal line $l_j = \{(x, k - j) \mid x \in \mathbb{R}\}$, and every edge $(u, v) \in E$, $u \in V_j$, $v \in V_{j+1}$, $1 \leq j < k$, is drawn as a straight line segment between the lines l_i and l_{i+1}. A level drawing of G is called *level planar* if no two edges cross except at common endpoints. A level graph is *level planar* if it has a level planar drawing.

A *hierarchy* is a level graph $G(V, E, \phi)$ where for every $v \in V_j$, $j > 1$, there exists at least one edge (w, v) such that $w \in V_{j-1}$. That is, all sources appear on the first level.

The characterization of level non-planarity by patterns of subgraphs has been suggested earlier by Di Battista and Nardelli who have identified three (not necessarily minimal) patterns of level non-planar subgraphs for hierarchies [1]. We call these patterns *Level Non-Planar* (LNP) patterns. To describe the LNP patterns, we give some terminology similar to theirs. A *path* is an ordered sequence of vertices $(v_1, v_2, \ldots, v_n)$, $n > 1$ such that for each pair (v_i, v_{i+1}), $i = 1, 2, \ldots, n - 1$ either (v_i, v_{i+1}) or (v_{i+1}, v_i) belongs to E. Let i and j, $i < j$ be two levels of a level graph $G = (V, E, \phi)$. $\mathrm{LACE}(i, j)$ denotes the set of paths C connecting any two vertices $x \in V_i$ and $y \in V_j$ such that $\{z \in C \mid z \in V_t, i \leq t \leq j\}$. If C_1 and C_2 are completely distinct paths belonging to $\mathrm{LACE}(i, j)$ then a *bridge* is a path connecting vertices $x \in C_1$ and $y \in C_2$. Vertices x and y are thus called the *endpoints* of a bridge. The next theorem gives a characterization of LNP patterns for hierarchies, as opposed to level graphs.

Theorem 1 (Di Battista and Nardelli). *Let $G = (V, E, \phi)$ be a hierarchy with $k > 1$ levels. G is level planar if and only if there is no triple $L_1, L_2, L_3 \in \mathrm{LACE}(i, j)$, $0 < i < j \leq k$, that satisfies one of the following conditions:*

(a) L_1, L_2 and L_3 are completely disjoint and pairwise connected by bridges. Bridges do not share a vertex with L_1, L_2 and L_1, except in their endpoints (see Fig. 1(a));

(b) L_1 and L_2 share an endpoint p and a path C (possibly empty) starting from p, $L_1 \cap L_3 = L_2 \cap L_3 = \emptyset$; there is a bridge b_1 between L_1 and L_3 and a bridge b_2 between L_2 and L_3, $b_1 \cap L_2 = b_2 \cap L_1 = \emptyset$ (see Fig. 1(b));

(c) L_1 and L_2 share an endpoint p and a path C_1 (possibly empty) starting from
   p; L_1 and L_3 share an endpoint q ($q \neq p$) and a path C_2 (possibly empty)
   starting from q, $C_2 \cap C_1 = \emptyset$; L_2 and L_3 are connected by a bridge b, $b \cap L_1 = \emptyset$
   (see Fig. 1(c)).

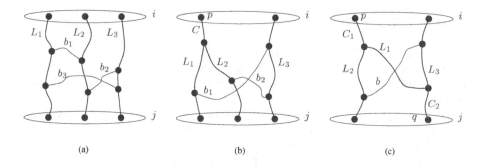

(a)   (b)   (c)

Fig. 1. Di Battista-Nardelli's Level Non-Planar Patterns

3   Minimal Level Non-planar Patterns in Hierarchies

Di Battista and Nardelli, as we have remarked, have identified three level non-
planar patterns for hierarchies. From the recognition of an LNP pattern in a
hierarchy, one can decide that the hierarchy is not level planar, but one cannot
guarantee that a removal of an edge from the subgraph matching an LNP pattern
leads to the level planarity of the graph.

   MLNP patterns are defined to have the following property: If a level graph
$G = (V, E, \phi)$ matches an MLNP pattern then any subgraph $G'(V, E', \phi)$ of G,
with $E' = E \setminus \{e\}, e \in E$, is embeddable without crossings on levels. The MLNP
patterns are divided into three categories: trees, level non-planar cycles, and level
planar cycles with incident paths. We give a comprehensive description of each
of these categories and show that the categories are complete for hierarchies.

   The terminology that is used to describe the MLNP patterns is compatible
with Harary [4], except that we denote by a *chain* a tree $T(V, E)$ where $E =
\{(v_1, v_2), (v_2, v_3), \dots, (v_{|V|-1}, v_{|V|})\}$. Furthermore, we define some terms that are
common to all of the patterns. The upper- and lower-most levels that contain
vertices of a pattern P are called *extreme levels* of P. The extreme levels of a
pattern are not necessarily the same as the extreme levels 1 and k of the input
graph G. If pattern P is instantiated in G as subgraph G_P and the uppermost
level of G_P in G is i and the lower most level is j ($i < j$) then the extreme levels
of P correspond to levels i and j in G. If vertex v lies on an extreme level then
we call this extreme level the *incident* extreme level and the other extreme level
the *opposite* extreme level of v.

3.1   Trees

Characterization We can characterize a MLNP tree pattern as follows. Let i and j be the extreme levels of a pattern and let x denote a root vertex with degree 3 that is located on one of the levels $i, \ldots, j$. From the root vertex emerge 3 subtrees that have the following common properties (see Fig. 2 for an illustration):

- each subtree has at least one vertex on both extreme levels;
- a subtree is either a chain or it has two branches which are chains;
- all the leaf vertices of the subtrees are located on the extreme levels, and if there is a leaf vertex v of a subtree S on an extreme level $l \in \{i, j\}$ then v is the only vertex of S on the extreme level l;
- those subtrees which are chains have one or more non-leaf vertices on the extreme level opposite to the level of their leaf vertices.

The location of the root vertex distinguishes the two characterizations.

T1 The root vertex x is on an extreme level $l \in \{i, j\}$ (see Fig. 2(a)):
  - at least one of the subtrees is a chain starting from x, going to the opposite extreme level of x and finishing on x's level;
T2 The root vertex x is on one of the intermediate levels l, $i < l < j$ (see Fig. 2(b)):
  - at least one of the subtrees is a chain that starts from the root vertex, goes to the extreme level i and finishes on the extreme level j;
  - at least one of the subtrees is a chain that starts from the root vertex, goes to the extreme level j and finishes on the extreme level i.

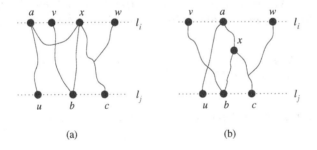

(a)   (b)

Fig. 2. Minimal Level Non-Planar Trees

Theorem 2. *A subgraph matching either of the two tree characterizations* T1 *or* T2 *is minimal level non-planar.*

Proof. The proof of level non-planarity of *T1* and *T2* is straightforward by matching *T1* and *T2* to the LNP pattern (a).

To prove minimality, we consider the two forms of the tree patterns separately. Consider *T1* where the root vertex x and vertices a, v and w of the chains are located on the same extreme level. Every subtree detached from the root-vertex x has a level planar layout. Thus if we remove one of the edges incident upon the leaf vertices on the extreme level of the root vertex (like the vertices v or w in Fig. 2(a)) then the corresponding subtree can be embedded under the root vertex x and between the other subtrees without any crossings. If we remove an edge incident upon the leaf vertices near the opposite level of the root vertex (for example, the path from vertex c to the branching point in Fig. 2(a)) then the modified subtree can be embedded on top of the chain-shaped subtree (according to the characterization there has to be one). Next, if we remove any other edge, we will have two disconnected subgraphs: one which contains the root vertex and the other which does not contain the root vertex. The former is a reduced case of the removal of an edge incident to a leaf vertex and the other component can be embedded.

In the case $T2$ when the root vertex is not on an extreme level, we consider two cases: the removal of an edge connecting the leaf vertex of a chain and the removal of an edge connecting a leaf vertex of a non-chain subtree. In the former case, the two chain subtrees can be embedded on top of each other. In the latter case, the path can be embedded under or on top of a chain by repositioning either vertices v or u as appropriate. If we remove any other edge then, again, we will have two disconnected subgraphs from which the subgraph containing the root vertex is a reduced case of the removal of an edge incident to a leaf vertex and the other subgraph can be embedded.

The following three lemmas (stated without proof) and theorem prove that the two tree patterns in our characterization are unique.

Lemma 1. *If LNP pattern (a) matches a tree then each one of the paths L_1, L_2, L_3 contains only one vertex where the bridges are connected.*

Lemma 2. *If LNP pattern (a) matches a tree then its bridges must form a subgraph homeomorphic to $K_{1,3}$.*

Lemma 3. *Only LNP pattern (a) can be matched to a tree.*

Theorem 3. *Let T be a tree. T is level non-planar if and only if it matches either of the two tree characterizations.*

Proof. From the previous lemmas it is possible to derive a level non-planar tree pattern (not necessarily minimal) from LNP pattern (a) only. Consider a tree matching pattern (a). If the pattern is bounded by levels i and j, but the vertices of bridges occur on levels $l_1, \ldots, l_2$, where $i < l_1$ and $l_2 < j$ then we can remove all the edges of the paths L_i which connect vertices on levels $i, \ldots, l_1$ and $l_2, \ldots, j$ without affecting level planarity. Moreover we can narrow the range of levels $l_1, \ldots, l_2$ even more, until both levels l_1 and l_2 contain at least one vertex v whose degree in the subgraph bounded by levels l_1 and l_2 is greater than 1.

Then, it can be shown that the tree between levels $l_1, \ldots, l_2$ is homeomorphic to one of the MLNP trees. From Lemmas 1 and 2 each of the paths L_i has exactly one vertex c_i to connect a bridge and the bridges form a subgraph homeomorphic to $K_{1,3}$. Consequently, after narrowing the levels to $l_1, \ldots, l_2$, each of the new extreme levels l_1, l_2 contains at least one of the following:

- a root vertex (x);
- a vertex of a path from x to c_i (c_i included).

In the latter case, if the vertex, say d, on level l_1 or l_2 is not identical to vertices x or c_i, we can remove the part of the upward path L_i from the extreme level of d to the vertex c_i. The tree maintains level non-planarity since the path L_i of LNP pattern (a) starts from the vertex d now. After performing this operation on each path L_i, we obtain a tree that matches either of our characterizations.

3.2   Cycles

We now study cycles that are bounded by the extreme levels of the pattern. A cycle must then contain at least two distinct paths between the extreme levels having vertices of the extreme levels only in their endpoints. These paths are called *pillars*.

Level Non-planar Cycles

Theorem 4. *If a cycle has more than two distinct paths connecting the vertices on the extreme levels of a pattern, it is minimal level non-planar.*

Proof. The number of such paths must be even. So, following our assumption of more than two paths, the number of paths must be at least 4 in a level non-planar cycle. Without loss of generality, consider the 4-path case first. Let the extreme levels be i and j. Let us denote a sequence of paths along the cycle $A = (v_a, \ldots, v_b)$, $B = (v_b, \ldots, v_c)$, $C = (v_c, \ldots, v_d)$, $D = (v_d, \ldots, v_a)$, and $v_a, v_c \in V_i$, $v_b, v_d \in V_j$. Consider LNP pattern (c). The paths A, B, C can be mapped always to the paths L_2, L_1, L_3 of the pattern, respectively. The remaining path D can be then mapped to the bridge in LNP pattern (c). If the number of paths is greater then 4, the first three paths can be mapped as in the case of 4 paths, and the remaining paths can be mapped to the bridge.

Such a cycle is minimal since any edge that is removed from a level non-planar cycle results in a chain that can be drawn as a level planar graph.

Level Planar Cycles
Level planar cycles can be augmented by a set of chains to obtain minimal level non-planarity. First, we give some terminology related to level planar cycles. A vertex that lies on a pillar is called an *outer* vertex; all the remaining vertices are *inner* vertices. The endpoints of pillars are *corner* vertices; if an extreme level i has only one vertex it is called a *single corner* vertex. A *bridge* in the context of a planar cycle is the shortest walk between

corner vertices on the same level; a bridge contains two corner vertices as its endpoints and the remainder are inner vertices. A pillar is *monotonic* if, in a walk of the cycle, the level numbers of subsequent vertices of the pillar are monotonically increasing or decreasing, depending on the direction of traversal. We call two paths or chains *parallel* if they start on the same pillar and end on the same extreme level. If a chain is connected to a cycle by one of its vertices having degree 1 (considering only edges of the chain) then this vertex is called the *starting vertex* of the chain and the level where this vertex lies, the *starting level*. The other vertex of degree 1 of the chain is then the *ending vertex* and corresponding level, the *ending level*.

Characterization Given a level planar cycle whose extreme levels are i and j, there are four cases to consider where augmentation of the level planar cycle by paths results in minimal level non-planarity. The pattern cannot contain one of the tree patterns given earlier. We enumerate these augmenting paths below. In all cases the paths start at a vertex on the cycle and end on an extreme level.

C1 A single path p_1 starting from an inner vertex and ending on the opposite extreme level of the inner vertex; p_1 and the cycle share only one vertex. The path will have at least one vertex on an extreme level, the end vertex, and at most two, the start and end vertices. An example of this is illustrated in Fig. 3 (a);

C2 Two paths p_1 and p_2, starting, respectively, from vertices v_{p_1} and v_{p_2}, $v_{p_1} \neq v_{p_2}$, of the same pillar $L = (v_i, \ldots, v_{p_1}, \ldots, v_{p_2}, \ldots v_j)$ terminating on extreme levels j and i, respectively. Vertices v_{p_1} or v_{p_2} may be identical to corner vertices of L ($v_{p_1} = v_i$ or $v_{p_2} = v_j$) only if the corner vertices are not single corner vertices on their extreme levels. Paths p_1 and p_2 do not have any vertices other than their start (if corner) and end vertices on the extreme levels. There are two subcases according to the levels of v_{p_1} and v_{p_2}:
  - $\phi(v_{p_1}) < \phi(v_{p_2})$;
  - $\phi(v_{p_1}) \geq \phi(v_{p_2})$, this means, L must be a non-monotonic pillar.
  Figures 3 (b) and (c) illustrate typical subgraphs matching the two subcases, respectively;

C3 Three paths, p_1, p_2 and p_3. Path p_1 starts from a single corner vertex and ends on the opposite extreme level; paths p_2 and p_3 start from opposite pillars and end on the extreme level where the single corner vertex is. Neither p_2 nor p_3 can start from a single corner vertex. Figure 3 (d) illustrates a level planar cycle augmented by three paths causing level non-planarity;

C4 Four paths, p_1, p_2, p_3 and p_4. The cycle comprises a single corner vertex on each of the extreme levels. Paths p_1 and p_2 start from different corner vertices and end on the opposite extreme level to their start with the paths embedded on either side of the cycle such that they do not intersect; paths p_3 and p_4 start from distinct non-corner vertices of the same pillar and finish on different extreme levels. The level numbers of starting vertices are such that they do not cause crossing of the last two paths. See Fig. 3 (e) for an illustration.

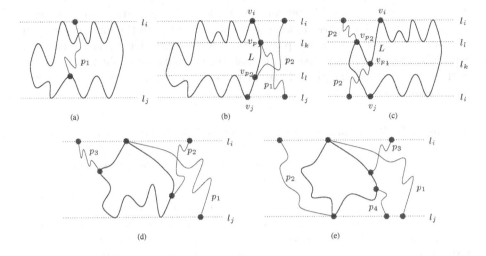

Fig. 3. Level Planar Cycles with Paths

We will now prove that each of the path-augmented cycles is minimal level non-planar and, in Theorems 6 to 10, prove that this set is complete for hierarchies.

Theorem 5. *Each of the four path-augmented cycles is minimal level non-planar.*

Proof. The augmented cycles are level non-planar because it can be shown that each can be mapped to one of Di Battista and Nardelli's LNP patterns. To see minimality we consider the three cases of the starting position of the path-augmentation on the cycle.

Suppose the start vertex is an inner vertex of a cycle. Since no subgraph matches an MLNP tree, breaking either an edge of the path or an edge of the cycle yields a level planar embedding.

In case of a path-augmented cycle of type $C2$, the removal of any edge of the cycle allows one of the augmenting paths to be embedded through the "gap" left by that edge. The removal of any edge of an augmenting path allows that path to be embedded on the internal face of the cycle. In both cases no crossings will remain.

For paths starting from corner vertices similar reasoning holds.

Theorem 6. *If a minimal level non-planar graph G comprises a level planar cycle and a single path p_1 connected to the cycle, then p_1 starts from an inner vertex of the cycle and ends on the opposite extreme level.*

Proof. For a path to cause level non-planarity, the path must start from an inner vertex of the cycle. Otherwise, the path can be embedded on the external face.

There are only two possibilities for causing level non-planarity: crossing with the incident bridge or, crossing with the opposite bridge. In the former, the path in combination with the lower part of the cycle forms a level non-planar tree. Since this level non-planar tree is minimal, the combination of the cycle and the path is not minimal. Therefore, the latter is the only remaining minimal level non-planar case.

Theorem 7. *If a minimal level non-planar graph G comprises a level planar cycle and two paths p_1 and p_2 connected to the cycle, then p_1 and p_2 start from the same pillar and end on an extreme level and either they cross or they start from a non-monotonic sub-chain of the pillar.*

Proof. Neither of the two paths may start from an inner vertex of the cycle because otherwise either they can be embedded on the internal face, or at least one of them matches type $C1$ above, or they form a level non-planar tree. Since the latter two cases are not minimal both paths must start from a pillar. Moreover, they must start from the same pillar, otherwise, the paths can be both embedded on the external face. The paths must finish on extreme levels, otherwise they can be embedded on the internal face. Moreover, the extreme levels must be different for if they are the same, the pattern – although it can be made level non-planar by introducing non-monotonic paths and a non-monotonic pillar – will not be minimal since it can be shown to match a minimal level non-planar tree pattern. If the extreme levels are different, then either the paths cross or there is a non-monotonic pillar that causes a crossing of the cycle and a path.

Theorem 8. *If a minimal level non-planar graph G comprises a level planar cycle and three paths p_1, p_2 and p_3 connected to the cycle, then G has a single corner vertex c_1 with p_1 starting at c_1 and extending to the opposite extreme level and p_2 and p_3 starting on opposite pillars and ending on the extreme level that contains c_1.*

Proof. As in the case of two paths, none of the paths may start from an inner vertex. Hence, all the paths should start from pillars. Additionally, all the paths must end on extreme levels, otherwise, they can be embedded on the internal face. No pair of paths can create minimal level non-planarity of the type $C2$ above. These conditions are met if one of the paths starts from a single corner vertex. If there were no other paths, the path starting from the single corner could be embedded on the external face on both sides of the cycle. However, if we have two paths starting from different pillars, not from a single corner vertex, and ending on the extreme level of the single corner vertex, a level planar embedding is not possible.

Theorem 9. *If a minimal level non-planar graph G comprises a level planar cycle and four paths $p_1, \ldots, p_4$ connected to the cycle, then G has two single corner vertices c_1 and c_2 with p_1 starting at c_1 and extending to the opposite* *extreme level of c_1, p_2 starting at c_2 and extending to the opposite extreme level of c_2, and p_3 and p_4 starting on the same pillar and diverging to end on opposite extreme levels such that they do not cross.*

Proof. This is proved analogously to the previous theorem, considering two corner vertices instead of one.

Theorem 10. *If a level non-planar graph G comprises a level planar cycle and five or more path augmentations that extend to extreme levels, then G cannot be minimal level non-planar.*

Proof. A level non-planar pattern with two parallel paths cannot be minimal (since any path that crosses one will cross both or one of the parallel paths is redundant) unless one of the parallel paths can be embedded on the other side of the cycle, in which case this path starts from a single corner vertex.

Suppose we have a minimal level non-planar graph comprising a level planar cycle and five path augmentations extending to extreme levels. Since removing any edge from a minimal level non-planar graph makes it planar, there must be four non-crossing paths. This can be achieved only by having on each pillar either two diverging paths ending on opposite extreme levels or, parallel paths where one pair of paths starts out from a single corner vertex. In neither case is it possible to add a fifth path so that minimal level non-planarity holds.

3.3 Minimal Level Non-planar Subgraphs in Hierarchies

Having shown that our characterizations of trees, level non-planar cycles and path-augmented cycles are minimally level non-planar, it only remains for us now to show that this set is a complete characterization of minimal level non-planar subgraphs.

Theorem 11. *The set of MLNP patterns characterized in sections 3.1, 3.2, and 3.2 is complete for hierarchies.*

Proof. Every graph comprises either a tree, or one, or more, cycles. It remains to prove that there is no MLNP pattern containing more than one cycle. Suppose a graph is MLNP and it has more than one cycle. Then it must be a subcase of one of Di Battista and Nardelli's LNP patterns. Each of these, however, has at most one single cycle and the remainder of the patterns comprises chains. Then at least one of our cycles must be broken in order to match it to a chain, thus contradicting the hypothesis.

4 Minimal Level Non-planar Subgraphs in Level Graphs

We have given in the previous section a characterization of level planar hierarchies in terms of minimal forbidden subgraphs. It remains to show that the described patterns characterize level planarity for general level graphs as well.

Theorem 12. *Let $G = (V, E, \phi)$ be a level graph with $k > 1$ levels. Then G is not level planar if, and only if it contains one of the MLNP patterns as described in sections 3.1, 3.2, and 3.2.*

Proof. If a subgraph G_p of G corresponds to an MLNP pattern, then G must be non level planar.

It remains to proof the opposite direction. Suppose there exists a minimal pattern P of non level planarity that is not applicable for hierarchies. Let G be a level graph such that P is the only pattern of non level planarity in G.

Since G is not level planar, augmenting the graph by an incoming edge for every source preserving the leveling in the graph constructs a non level planar hierarchy $H = (V, E \cup E_H, \phi)$, where E_H is the set of all extra added edges. Let $\mathcal{P}$ be the set of all subgraphs of H corresponding to a MLNP pattern. By assumption, we have for any $G_p \in \mathcal{P}$, $G_p = (V_p, E_p, \phi)$ that there exists an edge $e_p \in E_p \cap E_H$. Removing for every $G_p \in \mathcal{P}$ the edge e_p from H, we construct a level planar graph H'. By construction, H' contains G as a subgraph. Since every subgraph of a level planar graph must be level planar itself, this contradicts G being a non level planar subgraph.

5   Conclusion

We have given a characterization of level planar graphs in terms of minimal forbidden subgraphs. This description of level planarity is a main contribution for solving the $\mathcal{NP}$-hard level planarization problem for practical instances. Based on the characterization of level planar graphs an integer linear programming formulation for the level planarization problem can be given, allowing to study the associated polytope in order to develop an efficient branch-and-cut approach.

References

[1] G. Di Battista and E. Nardelli. Hierarchies and planarity theory. *IEEE Transactions on Systems, Man, and Cybernetics*, 18(6):1035–1046, 1988.

[2] P. Eades, B. D. McKay, and N. C. Wormald. On an edge crossing problem. In *Proc. 9th Australian Computer Science Conference*, pages 327–334. Australian National University, 1986.

[3] P. Eades and S. Whitesides. Drawing graphs in two layers. *Theoretical Computer Science*, 131:361–374, 1994.

[4] F. Harary. *Graph Theory*. Addison-Wesley, 1969.

[5] M. Jünger, S. Leipert, and P. Mutzel. Level planarity testing in linear time. Technical Report 98.321, Institut für Informatik, Universität zu Köln, 1998.

[6] P. Mutzel. An alternative method to crossing minimization on hierarchical graphs. In Stephen North, editor, *Graph Drawing. Symposium on Graph Drawing, GD '96*, volume 1190 of *Lecture Notes in Computer Science*, pages 318–333. Springer-Verlag, 1996.

Rectangular Drawings of Plane Graphs Without Designated Corners

(Extended Abstract)

Md. Saidur Rahman[1], Shin-ichi Nakano[2], and Takao Nishizeki[3]

[1] Department of Computer Science and Engineering, Bangladesh University of
Engineering and Technology, Dhaka-100, Bangladesh.
saidur@cse.buet.edu
[2] Department of Computer Science, Gunma University, Kiryu 376-8515, Japan.
nakano@cs.gunma-u.ac.jp
[3] Graduate School of Information Sciences, Tohoku University, Aoba-yama 05,
Sendai 980-8579, Japan.
nishi@ecei.tohoku.ac.jp

Abstract. A rectangular drawing of a plane graph G is a drawing of
G such that each vertex is drawn as a point, each edge is drawn as a
horizontal or a vertical line segment, and the contour of each face is drawn
as a rectangle. A necessary and sufficient condition for the existence of
a rectangular drawing has been known only for the case where exactly
four vertices of degree 2 are designated as corners in a given plane graph
G. In this paper we establish a necessary and sufficient condition for the
existence of a rectangular drawing of G for the general case in which no
vertices are designated as corners. We also give a linear-time algorithm
to find a rectangular drawing of G if it exists.

Key words: Graph, Algorithm, Graph Drawing, Rectangular Drawing.

1 Introduction

Recently automatic drawings of graphs have created intense interest due to their
broad applications, and as a consequence, a number of drawing styles and corresponding drawing algorithms have come out [DETT99]. Among different drawing
styles a "rectangular drawing" has attracted much attention due to its applications in VLSI floorplanning and architectural floorplanning [BS88, H93, H95,
KH97, KK84, L90, RNN98]. A *rectangular drawing* of a plane graph G is a
drawing of G in which each vertex is drawn as a point, each edge is drawn as
a horizontal or a vertical line segment, and the contour of each face is drawn
as a rectangle. Note that in a rectangular drawing of a plane graph G the contour of the outer face of G is also drawn as a rectangle. Fig. 1(d) illustrates a
rectangular drawing of the graph in Fig. 1(c). Clearly not every plane graph has
a rectangular drawing. For example the graph in Fig. 1(e) has no rectangular
drawing.

D.-Z. Du et al. (Eds.): COCOON 2000, LNCS 1858, pp. 85–94, 2000.

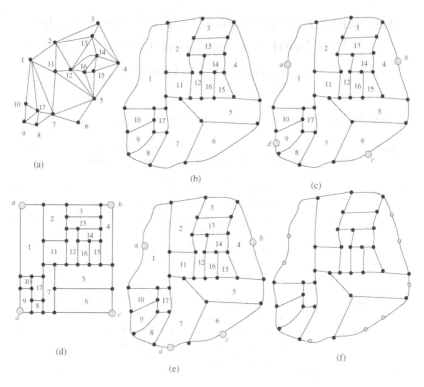

Fig. 1. (a) Interconnection graph, (b) dual-like graph, (c) good designation of corners, (d) rectangular drawing, (e) bad designation of corners, and (f) 2-3 plane graph.

In a VLSI floorplanning problem, the interconnection among modules is usually represented by a plane graph, every inner face of which is triangulated. Fig. 1(a) illustrates such an interconnection graph of 17 modules. The dual-like graph of an interconnection graph is a cubic graph in which every vertex has degree 3 [L90]. Fig. 1(b) illustrates a dual-like graph of the graph in Fig. 1(a), which has 17 inner faces. Inserting four vertices a, b, c and d of degree 2 in appropriate edges on the outer face contour of the dual-like graph as illustrated in Fig. 1(c), one wishes to find a rectangular drawing of the resulting graph as illustrated in Fig. 1(d). If there is a rectangular drawing, it yields a floorplan of modules. Each vertex of degree 2 is a corner of the rectangle corresponding to the outer rectangle. Thomassen [T84] gave a necessary and sufficient condition for such a plane graph with exactly four vertices of degree 2 to have a rectangular drawing. Linear-time algorithms are given in [BS88, H93, KH97, RNN98] to obtain a rectangular drawing of such a plane graph, if it exists. However, it has not been known how to appropriately insert four vertices of degree 2 as corners. If four vertices a, b, c and d of degree 2 are inserted in edges as in Fig. 1(e), then the resulting graph has no rectangular drawing.

In this paper we assume that four or more vertices of degree 2 have been inserted to all edges that may be drawn as edges incident to corners in a rectangular drawing; for example, insert one or two vertices of degree 2 to each edge on the outer contour. Thus we consider a plane connected graph G in which every vertex on the outer contour has degree 2 or 3, every vertex not on the outer contour has degree 3, and there are four or more vertices of degree 2 on the outer contour. We call such a graph a *2-3 plane graph*. (See Fig. 1(f).) We do not assume that four vertices of degree 2 are designated as corners in advance. We give a necessary and sufficient condition for such a 2-3 plane graph G to have a rectangular drawing, and give a linear-time algorithm to find appropriate four vertices of degree 2 and obtain a rectangular drawing of G with these vertices as corners.

The rest of the paper is organized as follows. Section 2 describes some definitions and presents preliminary results. Section 3 presents our main result on rectangular drawing. Finally, Section 4 concludes with an open problem.

2   Preliminaries

In this section we give some definitions and present preliminary results.

Let G be a connected simple graph. We denote the set of vertices of G by $V(G)$, and the set of edges of G by $E(G)$. The *degree* of a vertex v is the number of neighbors of v in G. We denote the maximum degree of a graph G by Δ.

A graph is *planar* if it can be embedded in the plane so that no two edges intersect geometrically except at a vertex to which they are both incident. A *plane* graph is a planar graph with a fixed embedding. A plane graph divides the plane into connected regions called *faces*. We regard the *contour* of a face as a clockwise cycle formed by the edges on the boundary of the face. We denote the contour of the outer face of graph G by $C_o(G)$. We call a vertex not on $C_o(G)$ an *inner vertex* of G.

For a simple cycle C in a plane graph G, we denote by $G(C)$ the plane subgraph of G inside C (including C). An edge of G which is incident to exactly one vertex of a simple cycle C and located outside C is called a *leg* of the cycle C. The vertex of C to which a leg is incident is called a *leg-vertex* of C. A simple cycle C in G is called a *k-legged cycle* of G if C has exactly k legs in G. A k-legged cycle C is a *minimal k-legged cycle* if $G(C)$ does not contain any other k-legged cycle of G. We say that cycles C and C' in a plane graph G are *independent* if $G(C)$ and $G(C')$ have no common vertex. A set S of cycles is independent if any pair of cycles in S are independent.

A *rectangular drawing* of a plane graph G is a drawing of G such that each edge is drawn as a horizontal or a vertical line segment, and each face is drawn as a rectangle. In a rectangular drawing of G, the contour $C_o(G)$ of the outer face is drawn as a rectangle and hence has four convex corners. Such a corner is called a *corner of the rectangular drawing*. Since a rectangular drawing D of G has no bend and each edge is drawn as a horizontal or a vertical line segment, only a vertex of degree 2 of G can be drawn as a corner of D. Therefore, a graph with less than four vertices of degree 2 on $C_o(G)$ has no rectangular drawing. Thus we consider a *2-3 plane graph* G in which four or more vertices on $C_o(G)$ have degree 2 and all other vertices have degree 3. We call a vertex of degree 2 in G that is drawn as a corner of D a *corner vertex*.

The following result on rectangular drawings is known [T84, RNN98].

Lemma 1. *Let G be a 2-3 plane graph. Assume that four vertices of degree 2 are designated as corners. Then G has a rectangular drawing if and only if G satisfies the following three conditions [T84]: (c1) G has no 1-legged cycles; (c2) every 2-legged cycle in G contains at least two designated vertices; and (c3) every 3-legged cycle in G contains at least one designated vertex. Furthermore one can check in linear time whether G satisfies the conditions above, and if G does then one can find a rectangular drawing of G in linear time [RNN98].* □

It is rather difficult to determine whether a 2-3 plane graph G has a rectangular drawing unless four vertices of degree 2 are designated as corners. Considering all combinations of four vertices of degree 2 as corners and applying the algorithm in Lemma 1 for each of the combinations, one can determine whether G has a rectangular drawing. Such a straightforward method requires time $O(n^5)$ since there are $O(n^4)$ combinations, and one can determine in linear time whether G has a rectangular drawing for each combination. In the next section we obtain a necessary and sufficient condition for G to have a rectangular drawing for the general case in which no vertices of degree 2 are designated as corners. The condition leads to a linear-time algorithm.

3   Rectangular Drawings Without Designated Corners

The following theorem is our main result on rectangular drawings.

Theorem 2. *A 2-3 plane graph G has a rectangular drawing if and only if G satisfies the following four conditions:*

(1) G has no 1-legged cycle;
(2) every 2-legged cycle in G contains at least two vertices of degree 2;
(3) every 3-legged cycle in G contains at least one vertex of degree 2; and
(4) if an independent set S of cycles in G consists of c_2 2-legged cycles and c_3 3-legged cycles, then $2c_2 + c_3 \leq 4$. □

Before proving Theorem 2, we observe the following fact.

Fact 3 *In any rectangular drawing D of G, every 2-legged cycle of G contains at least two corners, every 3-legged cycle of G contains at least one corner, and every cycle with four or more legs may contain no corner.* □

We now prove the necessity of Theorem 2.

Necessity of Theorem 2. Assume that G has a rectangular drawing D. Then G has no 1-legged cycle, since the face surrounding a 1-legged cycle cannot be drawn as a rectangle in D. By Fact 3 every 2-legged cycle in D contains at least two corners and every 3-legged cycle in D contains at least one corner. Since a corner is a vertex of degree 2, every 2-legged cycle in G must have at least two vertices of degree 2 and every 3-legged cycle must have at least one vertex of degree 2.

Let an independent set S consist of c_2 2-legged cycles and c_3 3-legged cycles in G. Then by Fact 3 each of the c_2 2-legged cycles in S contains at least two corners, and each of the c_3 3-legged cycle in S contains at least one corner. Since all cycles in S are independent, they are vertex-disjoint each other. Therefore there are at least $2c_2 + c_3$ corners in D. Since there are exactly four corners in D, we have $2c_2 + c_3 \leq 4$. □

In the rest of this section we give a constructive proof for the sufficiency of Theorem 2, and show that the proof leads to a linear-time algorithm for finding a rectangular drawing of G if it exists.

We first give some definitions. For a graph G and a set $V' \subseteq V(G)$, $G - V'$ denotes a graph obtained from G by deleting all vertices in V' together with all edges incident to them. For a plane graph G, we define a $C_o(G)$-*component* as follows. A subgraph F of G is a $C_o(G)$-component if F consists of a single edge which is not in $C_o(G)$ and both ends of which are in $C_o(G)$. The graph $G - V(C_o(G))$ may have a connected component. Add to such a component all edges of G, each joining a vertex in the component and a vertex in $C_o(G)$. The resulting subgraph F of G is a $C_o(G)$-component, too. All these subgraphs F of G and only these are the $C_o(G)$-components.

Let $\{v_1, v_2, \cdots, v_{p-1}, v_p\}$, $p \geq 3$, be a set of three or more consecutive vertices on $C_o(G)$ such that the degrees of the first vertex v_1 and the last vertex v_p are exactly three and the degrees of all intermediate vertices $v_2, v_3, \cdots, v_{p-1}$ are two. Then we call the set $\{v_2, v_3, \cdots, v_{p-1}\}$ a *chain* of G, and we call vertices v_1 and v_p the *ends* of the chain. Every vertex of degree 2 in G is contained in exactly one chain. A chain in G corresponds to an edge of the dual-like graph of an interconnection graph into which vertices of degree 2 have been inserted. Two chains are *consecutive* if they have a common end. We now have the following lemmas.

Lemma 4. *Let G be a 2-3 plane graph, and let k be the number of vertices of degree 3 on $C_o(G)$. If a $C_o(G)$-component F contains a cycle, then F contains a cycle with k or less legs.* □

Lemma 5. *Assume that a 2-3 plane graph G satisfies the four conditions in Theorem 2, and that G has at most three vertices of degree 3 on $C_o(G)$. Then G has a rectangular drawing.*

Proof. If G is a cycle, then clearly G has a rectangular drawing. Therefore we may assume that G has at least one vertex of degree 3 on $C_o(G)$. If G had exactly one vertex of degree 3 on $C_o(G)$, then G would have a 1-legged cycle or a vertex of degree 1, contrary to the assumption that G is a 2-3 plane graph satisfying the conditions in Theorem 2. Thus G has either two or three vertices of degree 3 on $C_o(G)$, and hence there are the following two cases to consider.

Case 1: G has exactly two vertices of degree 3 on $C_o(G)$.

In this case G has exactly one $C_o(G)$-component F. We now claim that F is a graph of a single edge. Otherwise, F has a cycle, and then by Lemma 4 F has a 2-legged cycle, but the 2-legged cycle contains no vertex of degree 2, contrary to the assumption that G satisfies Condition (2).

Clearly the ends of the edge in F are the two vertices of degree 3 on $C_o(G)$. Since G is a simple graph, the two vertices divide $C_o(G)$ into two chains of G. Furthermore, G has two 2-legged cycles C_1 and C_2 as indicated by dotted lines in Fig. 4(a). Since G satisfies Condition (2), each of C_1 and C_2 contains at least two vertices of degree 2. We arbitrarily choose two vertices of degree 2 on each of C_1 and C_2 as corner vertices. Then we can find a rectangular drawing of G as illustrated in Fig. 2(b).

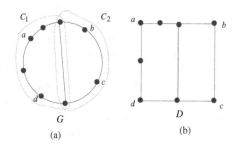

Fig. 2. Illustration for Case 1.

Case 2: G has exactly three vertices of degree 3 on $C_o(G)$.

In this case G has exactly one $C_o(G)$-component F. The $C_o(G)$-component F has no cycle; otherwise, by Lemma 4 G would have a 2-legged or 3-legged cycle containing no vertex of degree 2, contrary to the assumption that G satisfies Conditions (2) and (3). Therefore, F is a tree, and has exactly one vertex of degree 3 not on $C_o(G)$. Thus G is a "subdivision" of a complete graph K_4 of four vertices, and has three 3-legged cycles C_1, C_2 and C_3 as indicated by dotted lines in Fig. 3(a). Since G satisfies Condition (3), each 3-legged cycle contains at least one vertex of degree 2. We choose four vertices of degree 2; one vertex of degree 2 on each of the three 3-legged cycles, and one vertex of degree 2 arbitrarily. We then obtain a rectangular drawing D of G as illustrated in Fig. 3(b).   □

By Lemma 5 we may assume that G has four or more vertices of degree 3 on $C_o(G)$. We now have the following lemmas.

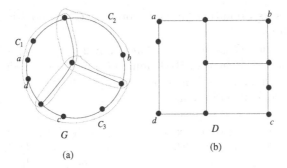

Fig. 3. Illustration for Case 2.

Lemma 6. *Assume that a 2-3 plane graph G satisfies the four conditions in Theorem 2, that G has four or more vertices of degree 3 on $C_o(G)$, and that there is exactly one $C_o(G)$-component. Then the following (a)-(d) hold:*

(a) for any 2-legged cycle C, at most one chain of G is not on C;
(b) one can choose four vertices of degree 2 in a way that at most two of them are chosen from any chain and at most three are chosen from any pair of consecutive chains;
(c) G has a 3-legged cycle; and
(d) if G has two or more independent 3-legged cycles, then the set of all minimal 3-legged cycles in G is independent. ☐

Lemma 7. *If a 2-3 plane graph G has two or more $C_o(G)$-components, then G has a pair of independent 2-legged cycle.* ☐

We are now ready to prove the following lemma.

Lemma 8. *Assume that a 2-3 plane graph G satisfies the four conditions in Theorem 2, and that $C_o(G)$ has four or more vertices of degree 3. Then G has a rectangular drawing.*

Proof. We shall consider the following two cases depending on the number of $C_o(G)$-components.
Case 1: G has exactly one $C_o(G)$-component.
By Lemma 6(c) G has a 3-legged cycle. We shall consider the following two subcases depending on whether G has a pair of independent 3-legged cycles or not.
Subcase 1a: G has no pair of independent 3-legged cycle.
By Lemma 6(b) we can designate four vertices of degree 2 as corners in a way that at most two of them are chosen from any chain and at most three are chosen from any pair of consecutive chains of G. By Lemma 1 it suffices to show that G satisfies the three conditions (c1)-(c3) regarding the four designated vertices.

Since G satisfies Condition (1) in Theorem 2, G has no 1-legged cycle and hence satisfies (c1).

We now claim that G satisfies (c2), that is, any 2-legged cycle C in G contains at least two designated vertices. By Lemma 6(a) at most one chain of G is not on C. Since any chain contains at most two of the four designated vertices, C must contain at least two of them.

We then claim that G satisfies (c3), that is, every 3-legged cycle C in G has at least one designated vertex. By Condition (3) C has a vertex of degree 2. Therefore exactly two of the three legs of C lie on $C_o(G)$. Let x and y be the leg-vertices of these two legs, respectively. Let P be the path on $C_o(G)$ starting at x and ending at y without passing through any edge on C. Then P contains at most two chains; otherwise, either G would have more than one $C_o(G)$-components or G would have a pair of independent 3-legged cycles, a contradiction. Furthermore, if P contains exactly one chain then it contains at most two designated vertices, and if P contains exactly two chains then they are consecutive and contain at most three designated vertices. Hence, in either case, C contains at least one of the four designated vertices.

Subcase 1b: G has a pair of independent 3-legged cycles.

Let $\mathcal{M}$ be the set of all minimal 3-legged cycles in G. By Lemma 6(d) $\mathcal{M}$ is independent. Let $k = |\mathcal{M}|$, then clearly $2 \le k$ and $k \le 4$ by Condition (4). For each 3-legged cycle C_m in $\mathcal{M}$, we arbitrarily choose a vertex of degree 2 on C_m. If $k < 4$, we arbitrarily choose $4 - k$ vertices of degree 2 on $C_o(G)$ which are not chosen so far in a way that at most two vertices are chosen from any chain. This can be done because G has four or more vertices of degree 2 and every 2-legged cycle has two or more vertices of degree 2. Thus we have chosen exactly four vertices of degree 2, and we regard them as the four designated corner vertices. In Fig. 4(a) four vertices a, b, c and d of degree 2 are chosen as the designated corner vertices; vertices a, c and d are chosen on three independent minimal 3-legged cycles indicated by dotted lines, whereas vertex b is chosen arbitrarily on $C_o(G)$. By Lemma 1 it suffices to show that G satisfies the conditions (c1)-(c3) regarding the four designated vertices. We can indeed prove that G satisfies Conditions (c1)-(c3) regarding the four designated vertices. The detail is omitted in this extended abstract.

Case 2: G has two or more $C_o(G)$-components.

In this case by Lemma 7 G has a pair of independent 2-legged cycles, C_1 and C_2. One may assume that both C_1 and C_2 are minimal 2-legged cycles. By Condition (4) at most two 2-legged cycles of G are independent. Therefore, for any other 2-legged cycle $C'(\neq C_1, C_2)$, $G(C')$ contains either C_1 or C_2.

Let k_i, $i = 1$ and 2, be the number of all minimal (not always independent) 3-legged cycles in $G(C_i)$. Then we claim that $k_i \le 2$, $i = 1$ and 2. First consider the case where C_i has exactly three vertices of degree 3 on $C_o(G)$. Then $G(C_i)$ has exactly two inner faces; otherwise, $G(C_i)$ would have a cycle which has 2 or 3 legs and has no vertex of degree 2, contrary to Condition (2) or (3). The contours of the two faces are minimal 3-legged cycles, and there is no other minimal 3-legged cycle in $G(C_i)$. Thus $k_i = 2$. We next consider the case where

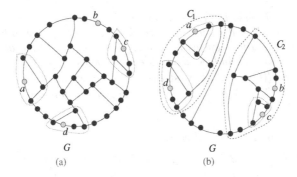

Fig. 4. (a)Illustration for Case 1, and (b) illustration for Case 2.

C_i has four or more vertices of degree 3 on $C_o(G)$. Then we can show, similarly as in the proof of Lemma 6(d), that the set of all minimal 3-legged cycles of G in $G(C_i)$ is independent. If $k_i \geq 3$, then Condition (4) would not hold for the independent set $\mathcal{S}$ of $k_i + 1$ cycles: the k_i 3-legged cycles in $G(C_i)$ and the 2-legged cycle C_j, $j = 1$ or 2 and $j \neq i$. Thus $k_i \leq 2$.

We choose two vertices of degree 2 on each C_i, $1 \leq i \leq 2$, as follows. For each of the k_i minimal 3-legged cycle in $G(C_i)$, we arbitrarily choose exactly one vertex of degree 2. If $k_i < 2$, then we arbitrarily choose $2 - k_i$ vertices of degree 2 in $V(C_i)$ which have not been chosen so far in a way that at most two vertices are chosen from any chain. This can be done because by Condition (2) C_i has at least two vertices of degree 2. Thus we have chosen four vertices of degree 2, and we regard them as the designated corner vertices for a rectangular drawing of G. In Fig. 4(b) G has a pair of independent 2-legged cycles C_1 and C_2, $k_1 = 2$, $k_2 = 1$, and four vertices a, b, c and d on $C_o(G)$ are chosen as the designated vertices. Vertices a and d are chosen from the vertices on C_1; each on a minimal 3-legged cycle in $G(C_1)$. Vertices b and c are chosen from the vertices on C_2; c is on the minimal 3-legged cycle in $G(C_2)$, and b is an arbitrary vertex on C_2 other than c.

By Lemma 1 it suffices to show that G satisfies the conditions (c1)-(c3) regarding the four designated vertices. We can indeed prove that G satisfies Conditions (c1)-(c3) regarding the four designated vertices. The detail is omitted in this extended abstract.   □

Lemmas 5 and 8 immediately imply that G has a rectangular drawing if G satisfies the conditions in Theorem 2. Thus we have constructively proved the sufficiency of Theorem 2.

We immediately have the following corollary from Theorem 2.

Corollary 9. *A 2-3 plane graph G has a rectangular drawing if and only if G satisfies the following six conditions:*

(1) G has no 1-legged cycle;
(2) every 2-legged cycle in G has at least two vertices of degree 2;

(3) every 3-legged cycle in G has a vertex of degree 2;

(4-1) at most two 2-legged cycles in G are independent of each other;

(4-2) at most four 3-legged cycles in G are independent of each other; and

(4-3) if G has a pair of independent 2-legged cycles C_1 and C_2, then each of $G(C_1)$ and $G(C_2)$ has at most two independent 3-legged cycles of G and the set $\{C_1, C_2, C_3\}$ of cycles is not independent for any 3-legged cycle C_3 in G.   □

We now have the following theorem.

Theorem 10. *Given a 2-3 plane graph G, one can determine in linear time whether G has a rectangular drawing or not, and if G has, one can find a rectangular drawing of G in linear time.*   □

4   Conclusions

In this paper we established a necessary and sufficient condition for a 2-3 plane graph G to have a rectangular drawing. We gave a linear-time algorithm to determine whether G has a rectangular drawing, and find a rectangular drawing of G if it exists. Thus, given a plane cubic graph G, we can determine in linear time whether one can insert four vertices of degree 2 into some of the edges on $C_o(G)$ so that the resulting plane graph G' has a rectangular drawing or not, and if G' has, we can find a rectangular drawing of G' in linear time. It is left as an open problem to obtain an efficient algorithm for finding rectangular drawings of plane graphs with the maximum degree $\Delta = 4$.

References

[BS88] J. Bhasker and S. Sahni, *A linear algorithm to find a rectangular dual of a planar triangulated graph*, Algorithmica, 3 (1988), pp. 247-278.

[DETT99] G. Di Battista, P. Eades, R. Tamassia and I. G. Tollis, *Graph Drawing*, Prentice Hall, Upper Saddle River, NJ, 1999.

[H93] X. He, *On finding the rectangular duals of planar triangulated graphs*, SIAM J. Comput., 22(6) (1993), pp. 1218-1226.

[H95] X. He, *An efficient parallel algorithm for finding rectangular duals of plane triangulated graphs*, Algorithmica 13 (1995), pp. 553-572.

[KH97] G. Kant and X. He, *Regular edge labeling of 4-connected plane graphs and its applications in graph drawing problems*, Theoretical Computer Science, 172 (1997), pp. 175-193.

[KK84] K. Kozminski and E. Kinnen, *An algorithm for finding a rectangular dual of a planar graph for use in area planning for VLSI integrated circuits*, Proc. 21st DAC, Albuquerque, June (1984), pp. 655-656.

[L90] T. Lengauer, *Combinatirial Algorithms for Integrated Circuit Layout*, John Wiley & Sons, Chichester, 1990.

[RNN98] M. S. Rahman, S. Nakano and T. Nishizeki, *Rectangular grid drawings of plane graphs*, Comp. Geom. Theo. Appl., 10(3) (1998), pp. 203-220.

[T84] C. Thomassen, *Plane representations of graphs*, (Eds.) J. A. Bondy and U. S. R. Murty, Progress in Graph Theory, Academic Press Canada, (1984), pp. 43-69.

Computing Optimal Embeddings for Planar Graphs

Petra Mutzel* and René Weiskircher

Technische Universität Wien, Karlsplatz 13, 1040 Wien
{mutzel, weiskircher}@apm.tuwien.ac.at

Abstract. We study the problem of optimizing over the set of all com-binatorial embeddings of a given planar graph. At IPCO' 99 we pre-sented a first characterization of the set of all possible embeddings of a given biconnected planar graph G by a system of linear inequalities. This system of linear inequalities can be constructed recursively using SPQR-trees and a new splitting operation. In general, this approach may not be practical in the presence of high degree vertices.

In this paper, we present an improvement of the characterization which allows us to deal efficiently with high degree vertices using a separation procedure. The new characterization exposes the connection with the asymmetric traveling salesman problem thus giving an easy proof that it is NP-hard to optimize arbitrary objective functions over the set of combinatorial embeddings.

Computational experiments on a set of over 11000 benchmark graphs show that we are able to solve the problem for graphs with 100 vertices in less than one second and that the necessary data structures for the optimization can be build in less than 12 seconds.

1 Introduction

A graph is called *planar* if it admits a drawing into the plane without edge-crossings (*planar drawing*). We call two planar drawings of the same graph *equivalent* when the circular sequence of the edges around each vertex is the same in both drawings. The equivalence classes of planar drawings are called *combinatorial embeddings*. A combinatorial embedding also defines the set of cycles in the graph that bound faces in a planar drawing.

The complexity of embedding planar graphs has been studied by various authors in the literature [5, 4, 6]. In this paper we deal with the following opti-mization problem concerning embeddings: Given a planar biconnected graph and a cost function on the cycles of the graph, find an embedding Π such that the sum of the cost of the cycles that appear as face cycles in Π is minimized. The objective function is chosen only to demonstrate the feasibility of the approach in computational experiments. However, our description of the set of combinatorial

---

* Partially supported by DFG-Grant Mu 1129/3-1, Forschungsschwerpunkt "Effiziente Algorithmen für diskrete Probleme und ihre Anwendungen"

D.-Z. Du et al. (Eds.): COCOON 2000, LNCS 1858, pp. 95–104, 2000.

embeddings of a planar graph as an Integer Linear Program (ILP) makes some important NP-hard problems arising in graph drawing accessible to ILP-based optimization methods.

One example is the minimization of the number of bends in an orthogonal planar drawing of a graph. This number highly depends on the chosen planar embedding. Whereas bend minimization of a planar graph is NP-hard ([9], a branch and bound algorithm for the problem is given in [3]), it can be solved in polynomial time for a fixed embedding ([13]). Figure 1 shows two different orthogonal drawings of the same graph that were generated using the bend minimization algorithm by Tamassia ([13]). The algorithm used different combinatorial embeddings as input. Drawing 1(a) has 13 bends while drawing 1(b) has only 7 bends.

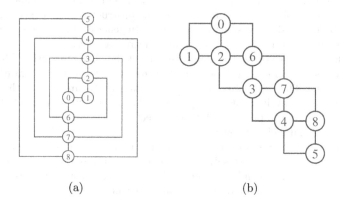

(a)   (b)

Fig. 1. The impact of the chosen planar embedding on the drawing

For a fixed embedding, the bend minimization problem can be formulated as a flow problem in the geometric dual graph. This is the point where we plan to use our description: Once we have characterized the set of all embeddings via an integer linear formulation on the variables associated with cycles of the graph, we can combine this formulation with the flow problem to solve the bend minimization problem over all embeddings.

In [12] we introduced an integer linear program (ILP) whose set of feasible solutions corresponds to the set of all possible combinatorial embeddings of a given biconnected planar graph. The program is constructed recursively with the advantage that we only introduce variables for those simple cycles in the graph that form the boundary of a face in at least one combinatorial embedding of the graph, thus reducing the number of variables tremendously. The constraints are derived using the structure of the graph. We use a data structure called SPQR-tree suggested by Di Battista and Tamassia ([2]) for the on-line maintenance of triconnected components. SPQR-trees can be used to code and enumerate all possible combinatorial embeddings of a biconnected planar graph.

The problem of our original formulation was that it may be inefficient for graphs with high-degree vertices, because it requires to compute the convex hull of a number of points that may be exponential in the degree of a vertex. When looking for a solution to this problem, we discovered a connection of the embedding problem with the asymmetric traveling salesman problem (ATSP) enabling us to apply the same machinery used for solving the ATSP-problem to the problem of finding an optimal embedding for a planar biconnected graph.

Our computational results on a set of more than 11000 benchmark graphs show that our new approach preserves the positive properties of our first approach while making it possible to deal efficiently with high degree vertices. As in our first version, the size of the integer linear system computed for the benchmark graphs grows only linearly with the size of the problem and the computed systems can be solved fast using a mixed integer program solver. There are only a few graphs with high degree vertices for which we need to apply our separation procedure and we never need more than three separation steps.

Section 2 gives a brief overview of SPQR-trees. In Section 3 we sketch the recursive construction of the linear constraint system using a splitting operation on the SPQR-tree. Section 4 introduces the connection of our problem with the ATSP problem and describes the separation procedure. Our computational results are described in Section 5.

2   SPQR-Trees

In this section, we give a brief overview of the SPQR-tree data structure for biconnected graphs. SPQR-trees have been suggested by Di Battista and Tamassia ([2]). They represent a decomposition of a biconnected graph into triconnected components. A connected graph is triconnected, if there is no pair of vertices in the graph whose removal splits the graph into two or more components.

An SPQR-tree has four types of nodes and with each node is associated a biconnected graph which is called the *skeleton* of that node. This graph can be seen as a simplified version of the original graph and its vertices are also contained in the original graph. The edges in a skeleton represent subgraphs of the original graph. The node types and their skeletons are as follows:

1. *Q*-node: The skeleton consists of two vertices that are connected by two edges. One of the edges represents an edge e of the original graph and the other one the rest of the graph.
2. *S*-node: The skeleton is a simple cycle with at least 3 vertices.
3. *P*-node: The skeleton consists of two vertices connected by at least three edge.
4. *R*-node: The skeleton is a triconnected graph with at least four vertices.

All leaves of the SPQR-tree are *Q*-nodes and all inner nodes *S*-,*P* or *R*-nodes. When we see the SPQR-tree as an unrooted tree, then it is unique for each biconnected planar graph. Another important property of these trees is that their size (including the skeletons) is linear in the size of the original graph and that they can be constructed in linear time [2].

As described in [2], SPQR-trees can be used to represent all combinatorial embeddings of a biconnected planar graph. This is done by choosing embeddings for the skeletons of the nodes in the tree. The skeletons of S- and Q-nodes are simple cycles, so they have only one embedding. Therefore, we only have to look at the skeletons of R- and P-nodes. The skeletons of R-nodes are triconnected graphs. Our definition of combinatorial embeddings distinguishes between two combinatorial embeddings of a triconnected graph, which are mirror-images of each other (the circular order of the edges around each vertex in clockwise order is reversed in the second embedding). The number of different embeddings of a P-node skeleton is $(k-1)!$ where k is the number of edges in the skeleton.

Every combinatorial embedding of the original graph defines a unique combinatorial embedding for each skeleton of a node in the SPQR-tree. Conversely, when we define an embedding for each skeleton of a node in the SPQR-tree, we define a unique embedding for the original graph. Thus, if the SPQR-tree of G has r R-nodes and the P-nodes P_1 to P_k where the skeleton of P_i has L_i edges, then the number of combinatorial embeddings of G is exactly

$$2^r \prod_{i=1}^{k} (L_i - 1)! \quad .$$

Because the embeddings of the R- and P-nodes determine the embedding of the graph, we call these nodes the *decision nodes* of the SPQR-tree.

3   Recursive Construction of the Integer Linear Program

3.1   The Variables of the Integer Linear Program

A *face cycle* in a combinatorial embedding of a planar graph is a directed cycle of the graph with the following property: In any planar drawing realizing the embedding, the left side of the cycle is empty. Note that the number of face cycles of a planar biconnected graph with m edges and n vertices is $m - n + 2$.

We construct an integer linear program (ILP) where the feasible solutions correspond to the combinatorial embeddings of a graph. The variables of the program are the same as in [12]. They are binary and represent directed cycles in the graph. In a feasible solution of the ILP, a variable x_c has value 1 if the associated cycle c is a face cycle in the represented embedding and 0 otherwise. To keep the number of variables as small as possible, we only introduce variables for those cycles that are indeed face cycles in a combinatorial embedding of the graph.

3.2   Splitting an SPQR-Tree

We construct the variables and constraints of the ILP recursively. Therefore, we need an operation that constructs a number of smaller problems from our original problem such that we can use the variables and constraints computed for the smaller problems to compute the ILP for the original problem. This is done by splitting the SPQR-tree at some decision-node v.

The splitting operation deletes all edges in the SPQR-tree incident to v whose other endpoint is not a Q-node, thus producing smaller trees. Then we attach new Q-nodes to all nodes that were incident to deleted edges to make sure that the trees we produce are again SPQR-trees. The new edges we use to attach the Q-nodes are called *split-edges* and the trees we produce *split-trees*. The graphs represented by the split-trees are called *split-graphs*. The new tree containing v is the *center split-tree* and the associated graph the *center split-graph*. The split-trees either have only one decision node (like the center split-tree) or at least one less than the original tree. The splitting process is depicted in Fig. 2.

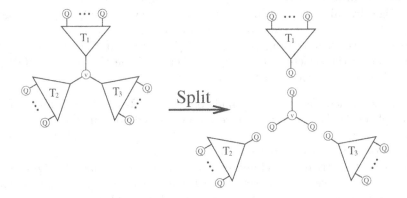

Fig. 2. Splitting an SPQR-tree at an inner node

3.3 Construction of the ILP for SPQR-Trees

Since this topic is treated in detail in [12] and [11], we only give a short sketch of our approach. Let T be the SPQR-tree of a biconnected planar graph G, v the node used for splitting the tree and $T_1, \ldots, T_k$ the split-trees of v. We assume that T_1 is the center split-tree and that the graph G_i is the split-graph belonging to split-tree T_i. We can distinguish two types of directed cycles in G:

1. *Local cycles* are contained in one of the graphs $G_1, \ldots, G_k$.
2. *Global cycles* of are not contained in any of the G_i.

We assume that we have already computed the ILP I_i for each T_i. The variables in I_i that represent local cycles will also be variables in the ILP for T. We compute the global cycles of G for which we need variables by combining cycles in the split-graphs that are represented by variables in the I_i.

The set C of all constraints of the ILP of T is given by $C = C_l \cup C_c \cup C_g$. C_l is the set of *lifted constraints*. For each constraint contained in I_i, we compute a constraint that is valid for T by replacing each variable x_c by the sum of the variables in $R(x_c)$ (The set of variables for T whose associated cycles have been constructed using c).

The set C_c is the set of *choice constraints*. They state that for each variable x_c computed for T_i, the sum of the variables in the set $R(x_c)$ can be at most one. This is true because all of the cycles in $R(x_c)$ either pass an edge or one of the split-graphs in the same direction and therefore at most one of the cycles can be a face cycle in any embedding. The proof is omitted but can be found in [11].

The only element of C_g is the *center graph constraint*, which states that the number of global face-cycles in any feasible solution plus the number of local face-cycles contained in G_1 is equal to the number of faces of G_1. This is true, because any drawing of G can be generated from a drawing of G_1 by replacing some edges by subgraphs. In this process, the face cycles of G_1 may be replaced by global cycles or are preserved (if they are local cycles of G).

We observe that a graph G whose SPQR-tree has only one inner node is isomorphic to the skeleton of this node. Therefore, the ILP for an SPQR-tree with only one inner node is defined as follows:

- S-node: When the only inner node of the SPQR-tree is an S-node, G is a simple cycle. Thus it has two directed cycles and both are face-cycles in the only combinatorial embedding of G. So the ILP consists of two variables, both of which must be equal to one.
- R-node: G is triconnected and according to our definition of combinatorial embeddings, every triconnected graph has exactly two embeddings, which are mirror-images of each other. When G has m edges and n vertices, we have $k = 2(m - n + 2)$ variables and two feasible solutions. The constraints are given by the convex hull of the two points in k-dimensional space that correspond to the solutions.
- P-node: The ILP for graphs whose only inner node in the SPQR-tree is a P-node is described in detail in Section 4 because this is where the connection to the asymmetric travelings salesman problem (ATSP) is used.

The proof of correctness differs from the proof given in [12] and more detailed in [11] only in the treatment of the skeletons of P-nodes, so we only look at the treatment of P-node skeletons in more detail in the next section.

4 P-Node Skeletons and the ATSP

A P-node skeleton P consists of two vertices v_a and v_b connected by $k \geq 3$ edges and has therefore $(k - 1)!$ different combinatorial embeddings. Every directed cycle in P is a face cycle in at least one embedding of P, so we need $k^2 - k$ variables in an ILP description of all combinatorial embeddings.

Let C be the set of variables in the ILP and $c : C \to \mathbb{R}$ a weight function on C. We consider the problem of finding the embedding of P that minimizes the sum of the weights of the cycles that are face cycles. We will show that this problem can be stated as the problem of finding a Hamiltonian cycle with minimum weight in the complete directed graph $B = (V, E)$ with k vertices or simply as the asymmetric traveling salesman problem (ATSP). The transformation we will show here also works in the opposite direction thus showing NP-hardness of optimizing over all embeddings. But we show here the reduction to ATSP because this is what our algorithm does when it computes the ILP.

The complete directed graph B has one vertex for every edge of P. Let e_1 and e_2 be two edges in P and v_1 and v_2 the corresponding vertices in B. Then the edge (v_1, v_2) corresponds to the cycle in P that traverses edge e_1 from v_a to v_b and edge e_2 from v_b to v_a. The edge (v_2, v_1) corresponds to the same cycle in the opposite direction. In this way, we define a bijection $b : C \rightarrow E$ from the cycles in P to the edges in B. This bijection defines a linear function $c' : E \rightarrow \mathbb{R}$ on the edges of B such that $c'(e) = c(b^{-1}(e))$.

Now it is not hard to see that each embedding of P corresponds to a Hamiltonian cycle in B and vice versa. If the Hamiltonian cycle is given by the sequence $(v_1, v_2, \ldots, v_k)$ of vertices, then the sequence of the edges in P in the corresponding embedding in counter-clockwise order around v_a is $(e_1, e_2, \ldots, e_k)$. Figure 3 shows an example of a P-node skeleton with four edges and the corresponding ATSP-graph. The embedding of the P-node skeleton on the left corresponds to the Hamiltonian cycle marked in the ATSP-graph. The marked edges correspond to the face cycles in the embedding of the P-node skeleton. The sequence of the edges in P in counter-clockwise order around vertex v_a corresponds to the sequence of the vertices in the Hamiltonian cycle of the ATSP-graph.

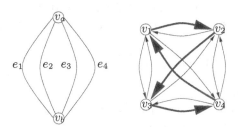

Fig. 3. A P-node skeleton and its corresponding ATSP-graph

The sum of the weights of all edges on the Hamiltonian cycle is equal to the sum of the weights of the cycles of P that are face cycles in this embedding. So finding an embedding of P that minimizes the sum of the weights of the face cycles is equivalent to finding a traveling salesman tour with minimum weight in B. Since we can easily construct a corresponding P-node embedding problem for any ATSP-problem, we have a simple proof that optimizing over all embeddings of a graph is in general NP-hard.

It also enables us to use the same ILP used for ATSP for the ILP that describes all combinatorial embeddings of a P-node skeleton. The formulation for the ATSP ILP found in [7] has two types of constraints.

1. The *degree constraints* state that each vertex must have exactly one incoming edge and one outgoing edge in every solution.
2. The *subtour elimination constraints* state that the number of edges with both endpoints in a nonempty subset S of the set of all vertices can be at most $|S| - 1$.

The number of degree constraints is linear in the number of edges in a P-node skeleton, while the number of subtour elimination constraints is exponential. Therefore, we define the ILP for a graph whose SPQR-tree has a P-node as the only inner node just as the set of degree constraints for the corresponding ATSP-problem.

To cope with the subtour elimination constraints, we store for each P-node skeleton the corresponding ATSP-graph. For each edge in the ATSP-graph we store the corresponding cycle in the P-node skeleton. During the recursive construction, we update the set of corresponding cycles for each edge in the ATSP-graph, so that we always know the list of cycles represented by an edge in the ATSP-graph. This is done in the same way as the construction of the lifted constraints in subsection 3.3.

When the construction of the recursive ILP is finished, we use a mixed integer programming solver to find an integer solution. Then we check if any subtour elimination constraint is violated. We do this by finding a minimum cut in each ATSP-graph. The weight of each edge in the ATSP graph is defined as the sum of the values of the variables representing the cycles associated with the edge. If the value of this minimum cut is smaller than one, we have found a violated subtour elimination constraint and add it to the ILP. As we will show in the next section, separation of the subtour elimination constraint is rarely necessary.

5   Computational Results

In our computational experiments, we used a benchmark set of 11491 graphs collected by the group around G. Di Battista in Rome ([1]). Since some of these graphs are not planar, we first computed a planar subgraph using the algorithm from [10]. Then we made the resulting graphs biconnected while preserving planarity using the algorithm from [8]. For the resulting graphs, we first computed the recursive part of our ILP and then used a branch and cut algorithm with CPLEX as integer program solver to optimize randomly chosen linear functions over the set of all combinatorial embeddings of the graph. After each optimization phase, we checked if there was a violated subtour elimination constraint and if this was the case, we added the found constraint and re-optimized the ILP. Our experiments ran on a Sun Enterprise 10000.

Figure 4(a) shows that the number of embeddings can vary greatly for graphs with similar size. One graph with 97 vertices and 135 edges had 2,359,296 combinatorial embeddings while another one with 100 vertices and 131 edges had only 64 embeddings (note that the y-axis in the figure is logarithmic).

Figure 4(b) shows that the number of constraints and variables of our formulation grows roughly linear with the size of the graphs which is surprising when

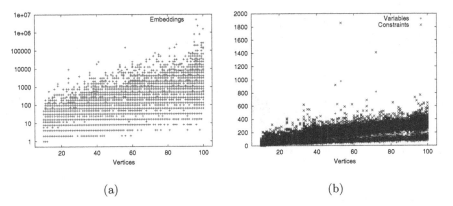

Fig. 4. The number of embeddings of the tested graphs and the number of variables and constraints

we consider the growth of the number of embeddings. Our ILP always has more constraints than variables. Figure 5(a) shows the time needed for building the recursive ILP and the time needed for optimization including separation. Optimization is very fast and the longest time we needed was 0.75 seconds. The time needed for building the ILP grows sub-exponential with the number of vertices and never exceeded 11 seconds.

Figure 5(b) shows that separation was rarely necessary and in the cases where we separated constraints, the number of the separation steps was small. The boxes show the number of graphs that needed 0, 1, 2 or 3 separation steps, e.g. 1, 2, 3 or 4 optimization rounds (note that the y-axis is logarithmic). We needed at most three separation steps and this was only the case for one of the 11,491 graphs. For 11,472 of the graphs, no re-optimization was necessary.

Our future goal will be to extend our formulation such that each solution will correspond to an orthogonal representation of the graph. This will enable us to find drawings with the minimum number of bends over all embeddings. Of course, this will make the solution of the ILP much more difficult.

Acknowledgments We thank the group of G. Di Battista in Rome for giving us the opportunity to use their implementation of SPQR-trees in *GDToolkit*, a software library that is part of the ESPRIT ALCOM-IT project (work package 1.2), and to use their graph generator.

References

[1] G. Di Battista, A. Garg, G. Liotta, R. Tamassia, E. Tassinari, and F. Vargiu. An experimental comparison of four graph drawing algorithms. *Comput. Geom. Theory Appl.*, 7:303–326, 1997.

[2] G. Di Battista and R. Tamassia. On-line planarity testing. *SIAM Journal on Computing*, 25(5):956–997, 1996.

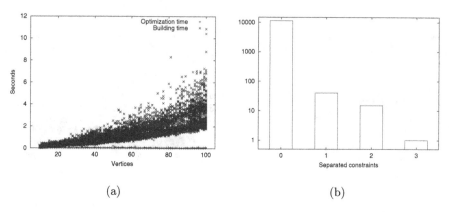

(a)   (b)

Fig. 5. The time needed for building the ILP and for optimization and the benchmark set divided up by the number of separation steps

[3] P. Bertolazzi, G. Di Battista, and W. Didimo. Computing orthogonal drawings with the minimum number of bends. *Lecture Notes in Computer Science*, 1272:331–344, 1998.

[4] D. Bienstock and C. L. Monma. Optimal enclosing regions in planar graphs. *Networks*, 19(1):79–94, 1989.

[5] D. Bienstock and C. L. Monma. On the complexity of embedding planar graphs to minimize certain distance measures. *Algorithmica*, 5(1):93–109, 1990.

[6] J. Cai. Counting embeddings of planar graphs using DFS trees. *SIAM Journal on Discrete Mathematics*, 6(3):335–352, 1993.

[7] G. Carpaneto, M. Dell'Amico, and P. Toth. Exact solution of large scale asymmetric travelling salesman problems. *ACM Transactions on Mathematical Software*, 21(4):394–409, 1995.

[8] S. Fialko and P. Mutzel. A new approximation algorithm for the planar augmentation problem. In *Proceedings of the Ninth Annual ACM-SIAM Symposium on Discrete Algorithms*, pages 260–269, San Francisco, California, 1998.

[9] A. Garg and R. Tamassia. On the computational complexity of upward and rectilinear planarity testing. *Lecture Notes in Computer Science*, 894:286–297, 1995.

[10] M. Jünger, S. Leipert, and P. Mutzel. A note on computing a maximal planar subgraph using PQ-trees. *IEEE Transactions on Computer-Aided Design*, 17(7):609–612, 1998.

[11] P. Mutzel and R. Weiskircher. Optimizing over all combinatorial embeddings of a planar graph. Technical report, Max-Planck-Institut für Informatik, Saarbrücken, 1998.

[12] P. Mutzel and R. Weiskircher. Optimizing over all combinatorial embeddings of a planar graph. In G. Cornuéjols, R. Burkard, and G. Wöginger, editors, *Proceedings of the Seventh Conference on Integer Programming and Combinatorial Optimization (IPCO)*, volume 1610 of *LNCS*, pages 361–376. Springer Verlag, 1999.

[13] R. Tamassia. On embedding a graph in the grid with the minimum number of bends. *SIAM Journal on Computing*, 16(3):421–444, 1987.

Approximation Algorithms for Independent Sets in Map Graphs

Zhi-Zhong Chen

Department of Mathematical Sciences, Tokyo Denki University,
Hatoyama, Saitama 350-0394, Japan.
chen@r.dendai.ac.jp

Abstract. This paper presents polynomial-time approximation algorithms for the problem of computing a maximum independent set in a given map graph G with or without weights on its vertices. If G is given together with a map, then a ratio of $1+\delta$ can be achieved in $O(n^2)$ time for any given constant $\delta > 0$, no matter whether each vertex of G is given a weight or not. In case G is given without a map, a ratio of 4 can be achieved in $O(n^7)$ time if no vertex is given a weight, while a ratio of $O(\log n)$ can be achieved in $O(n^7 \log n)$ time otherwise. Behind the design of our algorithms are several fundamental results for map graphs; these results can be used to design good approximation algorithms for coloring and vertex cover in map graphs, and may find applications to other problems on map graphs as well.

1 Introduction

An *independent set* in a graph $G = (V, E)$ is a subset I of V such that no two vertices of I are connected by an edge in E. If each vertex of G is associated with a weight, the *weight* of an independent set I in G is the total weight of vertices in I. Given a vertex-weighted graph $G = (V, E)$, the *maximum independent set* (MIS) problem requires the computation of an independent set of maximum weight in G. The special case where each vertex of G has weight 1, is called the *unweighted maximum independent set* (UMIS) problem. The MIS problem is a fundamental problem in many areas; unfortunately, even the UMIS problem is widely known to be NP-hard.

Since the MIS problem is NP-hard, we are interested in designing *approximation algorithms*, i.e. polynomial-time algorithms that find an independent set of large but not necessarily maximum weight in a given graph. An approximation algorithm $\mathcal{A}$ *achieves a ratio* of ρ if for every vertex-weighted graph G, the independent set I found by $\mathcal{A}$ on input G has weight at least $\frac{OPT(G)}{\rho}$, where $OPT(G)$ is the maximum weight of an independent set in G.

Håstad [6] gave strong evidence that no approximation algorithm for the UMIS problem achieves a ratio of $n^{1-\epsilon}$ for any $\epsilon > 0$, where n is the number of vertices in the input graph. Due to this, a lot of work has been devoted to designing approximation algorithms for the MIS problem restricted to certain

D.-Z. Du et al. (Eds.): COCOON 2000, LNCS 1858, pp. 105–114, 2000.

classes of graphs (such as planar graphs, bounded-degree graphs, etc). This paper considers the MIS problem restricted to *map graphs*, which were introduced by Chen et al. [5] recently.

Intuitively speaking, a *map* is a plane graph whose faces are classified into *nations* and *lakes*. Traditional planarity says that two nations on a map $\mathcal{M}$ are adjacent if and only if they share at least a borderline. Motivated by topological inference, Chen et al. [5] considered a modification of traditional planarity, which says that two nations on $\mathcal{M}$ are adjacent if and only if they share at least a border-point. Consider the simple graph G whose vertices are the nations on $\mathcal{M}$ and whose edges are all $\{f_1, f_2\}$ such that f_1 and f_2 are nations sharing at least a border-point. G is called a *map graph* [5,10]; if for some integer k, no point in $\mathcal{M}$ is shared by more than k nations, G is called a *k-map graph* [5]. The UMIS problem restricted to map graphs is NP-hard, because planar graphs are exactly 3-map graphs [5] and the UMIS problem restricted to planar graphs is NP-hard.

It is known [1,7] that the MIS problem restricted to planar graphs has a *polynomial-time approximation scheme* (PTAS). Recall that a PTAS for a maximization (respectively, minimization) problem Π is a polynomial-time algorithm $\mathcal{A}$ such that given an instance I of Π and an error parameter $\epsilon > 0$, $\mathcal{A}$ computes a solution S of I in time polynomial in the size of I such that the value of S is at least $(1 - \epsilon) \cdot OPT(I)$ (respectively, at most $(1 + \epsilon) \cdot OPT(I)$), where $OPT(I)$ is the optimal value of a solution of I. An interesting question is to ask whether the MIS problem restricted to map graphs also has a PTAS. Regarding this question, we show that if the input graph is given together with a map, then the MIS problem restricted to map graphs has a practical PTAS. Combining this result and the Thorup algorithm [10] for constructing a map from a given map graph, we obtain a PTAS for the MIS problem restricted to map graphs. Unfortunately, the Thorup algorithm is extremely complicated and the exponent of its polynomial time bound is about 120.

So, it is natural to assume that no map is given as part of the input and no map construction is allowed in approximation algorithms. Under this assumption, we first show that the MIS problem restricted to k-map graphs for any fixed integer k has a PTAS. This PTAS runs in linear time, but is impractical because it uses the impractical Bodlaender algorithm [2] for tree-decomposition. Using this PTAS, we then obtain an $O(n^7)$-time approximation algorithm for the UMIS problem restricted to general map graphs that achieves a ratio of 4. We also give two relatively more practical approximation algorithms. One of them is for the MIS problem restricted to general map graphs, runs in $O(n^7 \log n)$ time, and achieves a ratio of $O(\log n)$. The other is for the UMIS problem restricted to general map graphs, runs in $O(n^7)$ time, and achieves a ratio of 5.

Behind the design of our algorithms are several fundamental results for map graphs; these results can be used to obtain the following results. First, there is a PTAS for the minimum weighted vertex cover (MWVC) problem restricted to map graphs. Second, a map graph G with n vertices and m edges can be colored in $O(n^2 m)$ time using at most $2 \cdot \chi(G)$ colors, where $\chi(G)$ is the chromatic number of G. Third, for all optimization problems Π such that Π restricted to planar graphs has a PTAS as shown in [1], Π restricted to k-map graphs has a PTAS as well for any fixed integer k. These results do not involve map constructions. They may find applications to other problems on map graphs as well.

The rest of this paper is organized as follows. Section 2 gives precise definitions and states known results that are needed later on. Section 3 presents a PTAS for the MIS problem where the input graph is given together with a map. Section 4 describes two PTASs for the MIS problem restricted to k-map graphs. Section 5 presents nontrivial approximation algorithms for the MIS and UMIS problems restricted to map graphs. Section 6 asks several open questions.

See http://rnc2.r.dendai.ac.jp/~chen/papers/mapis.ps.gz, for a full version of this paper.

2   Preliminaries

Throughout this paper, a graph may have multiple edges but no loops, while a *simple* graph has neither multiple edges nor loops. Let $G = (V, E)$ be a graph. The subgraph of G induced by a subset U of V is denoted by $G[U]$. An *independent set* of G is a subset U of V such that $G[U]$ has no edge. The *independence number* of G, denoted by $\alpha(G)$, is the maximum number of vertices in an independent set of G. Let $v \in V$. The *degree* of v in G is the number of edges incident to v in G; v is an *isolated* vertex if its degree is 0. The *neighborhood* of v in G, denoted by $N_G(v)$, is the set of vertices adjacent to v in G. Note that $|N_G(v)|$ bounds the degree of v in G from below. For a subset U of V, let $N_G(U) = \cup_{v \in U} N_G(v)$.

Let $\mathcal{E}$ be a plane graph. Let f be a face of $\mathcal{E}$. The *boundary* of f is the graph (V_f, E_f), where V_f and E_f are the set of vertices and edges surrounding f, respectively. The *size* of f, denoted by $|f|$, is $|V_f|$. We call f a *cycle face* if its boundary is a simple cycle.

A *map* $\mathcal{M}$ is a pair $(\mathcal{E}, \varphi)$, where (1) $\mathcal{E}$ is a plane graph $(\mathcal{P}, E_{\mathcal{E}})$ and each connected component of $\mathcal{E}$ is biconnected, and (2) φ is a function that maps each inner face f of $\mathcal{E}$ to 0 or 1 in such a way that whenever $\varphi(f) = 1$, f is a cycle face. A face f of $\mathcal{E}$ is called a *nation* on $\mathcal{M}$ if f is an inner face with $\varphi(f) = 1$, while called a *lake* on $\mathcal{M}$ otherwise. For an integer k, a *k-point* in $\mathcal{M}$ is a $p \in \mathcal{P}$ that appears on the boundaries of exactly k nations on $\mathcal{M}$; a *k^+-point* in $\mathcal{M}$ is a $p \in \mathcal{P}$ that appears on the boundaries of more than k nations on $\mathcal{M}$. $\mathcal{M}$ is a *k-map* if there is no k^+-point in $\mathcal{M}$. Two nations are *adjacent* if their boundaries share a $p \in \mathcal{P}$. The *graph* of $\mathcal{M}$ is the simple graph $G = (V, E_G)$ where V consists of the nations on $\mathcal{M}$ and E_G consists of all $\{f_1, f_2\}$ such that f_1 and f_2 are adjacent. We call G a *map graph*, while call G a *k-map graph* if $\mathcal{M}$ is a k-map. For clarity, we call the elements of $\mathcal{P}$ *points*, while call the elements of V *vertices*. A nation f on $\mathcal{M}$ is *isolated* if f is an isolated vertex of G.

The *half square* of a bipartite graph $H = (X, Y; E_H)$ is the simple graph $G = (X, E)$ where $E = \{\{x_1, x_2\} \mid \text{there is a } y \in Y \text{ with } \{x_1, y\} \in E_H \text{ and } \{x_2, y\} \in E_H\}$.

Lemma 1. [5] *Every map graph* $G = (V, E)$ *is the half square of a bipartite planar graph* $H = (V, \mathcal{P}; E_H)$ *with* $|\mathcal{P}| \leq 3|V| - 6$.

For a map $\mathcal{M} = (\mathcal{E}, \varphi)$ and a face f in $\mathcal{E}$, we say that f *touches* a point p if p appears on the boundary of f.

3   A PTAS for the Case with a Given Map

Let $G = (V, E)$ be the input map graph given with a map $\mathcal{M} = ((\mathcal{P}, E_\mathcal{E}), \varphi)$ and a nonnegative weight to each nation on $\mathcal{M}$. Let ϵ be the given error parameter.

Let H be the bipartite planar graph $(V, \mathcal{P}; E_H)$ where $E_H = \{\{v, p\} \mid v \in V, p \in \mathcal{P},$ and p is a point on the boundary of nation $v\}$. Obviously, H can be constructed in linear time from $\mathcal{M}$. Moreover, G is the half square of H. Indeed, Lemma 1 is proved by further removing all *redundant points* from H where a point $p \in \mathcal{P}$ is redundant if for every pair $\{u, v\}$ of neighbors of p in H, there is a point $q \in \mathcal{P} - \{p\}$ with $\{u, v\} \subseteq N_H(q)$.

Our PTAS is a nontrivial application of the Baker approach. Let $h = \lceil \frac{1}{\epsilon} \rceil$. We may assume that G and hence H are connected. Starting at an arbitrary vertex $r \in V$, we perform a breadth-first search (BFS) on H to obtain a BFS tree T_H. For each vertex v in H, we define the level number $\ell(v)$ of v to be the number of edges on the path from r to v in T_H. Note that only r has level number 0. For each $i \in \{0, \ldots, h-1\}$, let H_i be the subgraph of H obtained from H by deleting all vertices v with $\ell(v) \equiv 2i \pmod{2h}$. Obviously, each $v \in V$ appears in exactly $h - 1$ of H_0 through H_{h-1}. Moreover, H_i is clearly a $(2h - 1)$-outerplanar graph.

For each $i \in \{0, \ldots, h - 1\}$, let G_i be the half square of H_i. The crux is that each G_i is an induced subgraph of G, because all $p \in \mathcal{P}$ appear in H_i. Our strategy is to compute an independent set I_i of maximum weight in G_i for each $i \in \{0, \ldots, h - 1\}$. Since each vertex of G appears in all but one of G_0 through G_{h-1}, the independent set of maximum weight among $I_0, \ldots, I_{h-1}$ has weight at least $(1 - \frac{1}{h}) \cdot OPT(G) \geq (1 - \epsilon) \cdot OPT(G)$. Recall that $OPT(G)$ is the maximum weight of an independent set in G.

So, it remains to show how to compute a maximum weight independent set I_i in G_i for each $i \in \{0, \ldots, h-1\}$. To do this, we first recall the notion of *treewidth*. A *tree-decomposition* of a graph $K = (V_K, E_K)$ is a pair $(\{X_j \mid j \in J\}, \mathcal{T})$, where $\{X_j \mid j \in J\}$ is a family of subsets of V_K and $\mathcal{T}$ is a tree with vertex-set J such that the following hold:

- $\cup_{j \in J} X_j = V_K$.
- For every edge $\{v, w\} \in E_K$, there is a $j \in J$ with $\{v, w\} \subseteq X_j$.
- For all $j_1, j_2, j_3 \in J$, if j_2 lies on the path from j_1 to j_3 in $\mathcal{T}$, then $X_{j_1} \cap X_{j_3} \subseteq X_{j_2}$.

Each X_j is called the *bag* of vertex j. The *treewidth* of a tree-decomposition $(\{X_j \mid j \in J\}, \mathcal{T})$ is $\max\{|X_j| - 1 \mid j \in J\}$. The *treewidth* of K is the minimum treewidth of a tree-decomposition of K, taken over all possible tree-decompositions of K.

It is widely known that the MIS problem restricted to graphs with bounded treewidth can be solved in linear time. However, the treewidths of our graphs $G_0, \ldots, G_{h-1}$ may not be bounded. Our strategy is to resort to $H_0, \ldots, H_{h-1}$ instead, and do a dynamic programming on them. The result is the following:

Theorem 1. *There is a PTAS for the special case of the MIS problem restricted to map graphs where the input graph is given together with a map $\mathcal{M} = (\mathcal{E}, \varphi)$. It runs in $O(h|\mathcal{E}|^2 + 2^{12h-8}h^2|\mathcal{E}|)$ time, where $h = \lceil \frac{1}{\epsilon} \rceil$ and ϵ is the given error parameter.*

4 PTASs for k-Map Graphs

We use an extension of the Baker approach, suggested in [4]. The only difference between the extension and the Baker approach is that instead of decomposing the given graph G into k'-outerplanar subgraphs, we decompose G into subgraphs of bounded treewidth such that each vertex of G appears in all but one of these subgraphs. The next lemma shows that such a decomposition is possible for a k-map graph.

Lemma 2. *Given a k-map graph G and an integer $h \geq 2$, it takes $O(h|G|)$ time to compute a list $\mathcal{L}$ of h induced subgraphs of G such that (1) each subgraph in $\mathcal{L}$ is of treewidth $\leq k(6h - 4)$ and (2) each vertex of G appears in exactly $h - 1$ of the subgraphs in $\mathcal{L}$.*

By the proof of Lemma 2, for all optimization problems Π such that Π restricted to planar graphs has a PTAS as shown in [1], we can emulate Baker [1] to show that Π restricted to k-map graphs has a PTAS as well for any fixed integer k. In particular, we have:

Theorem 2. *For any fixed integer k, there is a PTAS for the MIS problem restricted to k-map graphs. It runs in $O(2^{32(6kh-4k)^3}hn)$ time, where n is the size of the given k-map graph, $h = \lceil \frac{1}{\epsilon} \rceil$, and ϵ is the given error parameter.*

The PTAS in Theorem 2 is impractical. We next show a different PTAS for the special case of the UMIS problem where the input graph is given together with a k-map. This PTAS is based on the Lipton and Tarjan approach. Although this PTAS is not better than the one in Section 3, the tools used in its design seem to be fundamental. Indeed, the tools will be used in the next section, and we believe that they can be used in other applications.

A map $\mathcal{M} = (\mathcal{E}, \varphi)$ is *well-formed* if (1) the degree of each point in $\mathcal{E}$ is at least 3, (2) no edge of $\mathcal{E}$ is shared by two lakes, (3) no lake touches a point whose degree in $\mathcal{E}$ is 4 or more, and (4) the boundary of each lake contains at least 4 edges.

The next lemma shows that a map can easily be made well-formed.

Lemma 3. *Given a map $\mathcal{M} = (\mathcal{E}, \varphi)$ with at least three nations and without isolated nations, it takes $O(|\mathcal{E}|)$ time to obtain a well-formed map $\mathcal{M}'$ such that (1) the graph of $\mathcal{M}'$ is the same as that of $\mathcal{M}$ and (2) for every integer $k \geq 2$, $\mathcal{M}$ is a k-map if and only if so is $\mathcal{M}'$.*

Let $\mathcal{M} = (\mathcal{E}, \varphi)$ be a well-formed map. Then, each point on the boundary of a lake on $\mathcal{M}$ is a 2-point in $\mathcal{M}$ and has degree 3 in $\mathcal{E}$; moreover, for all integers $i \geq 3$, each i-point in $\mathcal{M}$ has degree i in $\mathcal{E}$. Using these facts, we can prove the following lemma which improves a result in [5] significantly.

Lemma 4. *Let $k \geq 3$ and G be a k-map graph with $n \geq 3$ vertices and m edges. Then, $m \leq kn - 2k$.*

From Lemma 4, we can prove the following corollary.

Corollary 1. *The following hold:*

1. *Given an integer k and a k-map graph G, it takes $O(|G|)$ time to color G with at most $2k$ colors. Consequently, given a k-map $\mathcal{M} = (\mathcal{E}, \varphi)$, it takes $O(k|\mathcal{E}|)$ time to color the graph of $\mathcal{M}$ with at most $2k$ colors.*
2. *Given a map graph $G = (V, E)$, we can color G in $O(|V|^2 \cdot |E|)$ time using at most $2 \cdot \chi(G)$ colors, where $\chi(G)$ is the chromatic number of G.*

Indeed it has been known for a long time that k-map graphs are $2k$-colorable [9], but Corollary 1 has the advantage of being algorithmic. It is also known that k-map graphs with $k \geq 8$ can be colored with at most $2k - 3$ colors [3], but the proof does not yield an efficient algorithm.

Let G be a connected graph with nonnegative weights on its vertices. The *weight* of a subgraph of G is the total weight of vertices in the subgraph. A *separator* of G is a subgraph H of G such that deleting the vertices of H from G leaves a graph whose connected components each have weight $\leq \frac{2W}{3}$, where W is the weight of G. The *size* of a separator H is the number of vertices in H. The following lemma shows that k-map graphs have small separators.

Lemma 5. *Given a k-map $\mathcal{M} = (\mathcal{E}, \varphi)$ on which each nation has an associated nonnegative weight and at least two nations exist, it takes $O(|\mathcal{E}|)$ time to find a separator H of the graph G of $\mathcal{M}$ such that the size of H is at most $2\sqrt{2k(n-1)}$.*

Using Lemma 5, we can emulate Lipton and Tarjan [8] to prove the following theorem:

Theorem 3. *For any fixed integer k, there is a PTAS for the special case of the UMIS problem restricted to k-map graphs where the input graph is given together with a k-map $\mathcal{M} = (\mathcal{E}, \varphi)$. It runs in $O(|\mathcal{E}| \log |\mathcal{E}| + 2^{288h^2 k^3} n)$ time, where n is the number of nations on $\mathcal{M}$, $h = \lceil \frac{1}{\epsilon} \rceil$, and ϵ is the given error parameter.*

5   Approximation Algorithms for Map Graphs

This section presents four approximation algorithms. The first three are for the UMIS problem restricted to map graphs. The fourth is for the MIS problem restricted to map graphs. The input to the first algorithm is a map instead of a map graph; this algorithm is given here for the purpose of helping the reader understand the other three algorithms. Also, the second algorithm is relatively more practical than the third.

For a well-formed map $\mathcal{M} = (\mathcal{E}, \varphi)$, let $\mathcal{G}(\mathcal{M})$ be the plane graph obtained from the dual plane graph $\mathcal{E}^*$ of $\mathcal{E}$ by deleting all vertices that are lakes on $\mathcal{M}$. Note that corresponding to each 2^+-point p in $\mathcal{E}$, there is a unique face $\mathcal{F}_p$ in $\mathcal{G}(\mathcal{M})$ such that the vertices on the boundary of $\mathcal{F}_p$ are exactly the nations on $\mathcal{M}$ that touch p; the boundary of $\mathcal{F}_p$ is a simple cycle. We call $\mathcal{F}_p$ a *point-face* of $\mathcal{G}(\mathcal{M})$. Those faces of $\mathcal{G}(\mathcal{M})$ that are not point-faces are called *lake-faces*.

Consider the following algorithm, called $MapIS_1$:

Input : A map $\mathcal{M} = (\mathcal{E}, \varphi)$.
1. If at most two nations are on $\mathcal{M}$, then return a maximum weight independent set of the graph of $\mathcal{M}$.
2. If there is an isolated nation on $\mathcal{M}$, then find such a nation f; otherwise, perform the following substeps:
    2.1. Modify $\mathcal{M}$ to a well-formed map, without altering the graph of $\mathcal{M}$.
    2.2. Construct $\mathcal{G}(\mathcal{M})$.
    2.3. Find a minimum-degree vertex f in $\mathcal{G}(\mathcal{M})$.
3. Construct a new map $\mathcal{M}' = (\mathcal{E}, \varphi')$, where for all faces f_1 of $\mathcal{E}$, if $f_1 = f$ or f_1 is a neighbor of f in the graph of $\mathcal{M}$, then $\varphi'(f_1) = 0$, while $\varphi'(f_1) = \varphi(f_1)$ otherwise.
4. Recursively call $MapIS_1$ on input $\mathcal{M}'$, to compute an independent set J in the graph of $\mathcal{M}'$.
5. Return $\{f\} \cup J$.

Theorem 4. *$MapIS_1$ achieves a ratio of 5 and runs in $O(|\mathcal{E}|^2)$ time.*

Proof. (Sketch) The number of edges in $\mathcal{G}(\mathcal{M})$ is at most $3n - 6$. So, the degree of f in $\mathcal{G}(\mathcal{M})$ is at most 5. We can then claim that the neighborhood of f in G can be divided into at most five cliques.

Consider the following algorithm, called $MapIS_2$:

Input : A map graph $G = (V, E)$.
1. If G has exactly at most two vertices, then return a maximum weight independent set of G.
2. Find a vertex f in G such that $\alpha(G_f) \leq 5$, where $G_f = G[N_G(f)]$.
3. Remove f and its neighbors from G.
4. Recursively call $MapIS_2$ on input G, to compute an independent set J.
5. Return $\{f\} \cup J$.

Theorem 5. *$MapIS_2$ achieves a ratio of 5, and runs in $O(|V| \cdot \sum_{f \in V} |N_G(f)|^6)$ time.*

Proof. By the proof of Theorem 4, Step 2 of $MapIS_2$ can always be done.

Consider the following algorithm, called $MapIS_3$:

Input : A map graph $G = (V, E)$, an integer $k \geq 4$, and a positive real number ϵ less than 1.

1. If G has exactly at most two vertices, then return a maximum weight independent set of G.
2. Find a vertex f in G such that $\alpha(G_f)$ is minimized among all vertices f of G, where $G_f = G[N_G(f)]$.
3. If $\alpha(G_f) \leq 4$, then perform the following substeps:
   - **3.1.** Remove f and its neighbors from G.
   - **3.2.** Recursively call $MapIS_3$ on input G to compute an independent set J.
   - **3.3.** Return $\{f\} \cup J$.
4. Repeat the following substeps until G has no clique of size $\geq k+1$ (and hence is a k-map graph):
   - **4.1.** Find a maximal clique C in G such that C contains at least $k + 1$ vertices.
   - **4.2.** Remove all vertices of C from G.
5. Color G with at most $2k$ colors (cf. Corollary 1); let J_1 be the maximum set of vertices with the same color.
6. Compute an independent set J_2 in G with $|J_2| \geq (1 - \epsilon)\alpha(G)$.
7. Return the bigger one among J_1 and J_2.

To analyze $MapIS_3$, we need the following lemma:

Lemma 6. [5] *Let G be the graph of a map $\mathcal{M} = (\mathcal{E}, \varphi)$. Let C be a clique of G containing at least five vertices. Then, at least one of the following holds:*

- *C is a pizza, i.e., there is a point p in $\mathcal{E}$ such that all $f \in C$ touch p.*
- *C is a pizza-with-crust, i.e., there are a point p in $\mathcal{E}$ and an $f_1 \in C$ such that (1) all $f \in C - \{f_1\}$ touch p and (2) for each $f \in C - \{f_1\}$, some point $q \neq p$ in $\mathcal{E}$ is touched by both f_1 and f.*
- *C is a hamantasch, i.e., there are three points in $\mathcal{E}$ such that each $f \in C$ touches at least two of the points and each point is touched by at least two nations.*

Theorem 6. *If we fix $k = 28$ and $\epsilon = 0.25$, then $MapIS_3$ achieves a ratio of 4 and runs in $O(|V| \cdot \sum_{f \in V} |N_G(f)|^6 + |V|^3 \cdot |E|)$ time.*

Proof. Let G_0 be the input graph. We may assume that G_0 is connected. By the proof of Theorem 5, it suffices to prove that if $\alpha(G_f) \geq 5$, then J_1 found in Step 5 or J_2 found in Step 6 has size $\geq \frac{\alpha(G_0)}{4}$. So, suppose that $\alpha(G_f) \geq 5$.

Let n be the number of vertices in G_0. Fix a well-formed map $\mathcal{M}$ of G_0. For each integer $i > k$, let h_i be the number of cliques C in G_0 such that (1) C contains exactly i vertices and (2) C is found in Substep 4.1 by the algorithm. We claim that the number x of edges in $\mathcal{G}(\mathcal{M})$ is bounded from above by $3n - 6 - \sum_{i=k+1}^{\infty} h_i(i - 4)$. To see this claim, let $\mathcal{F}_1, \ldots, \mathcal{F}_t$ be the point-faces in $\mathcal{G}(\mathcal{M})$ whose boundaries each contain at least three vertices. For each $j \in \{1, \ldots, t\}$, fix a vertex f_j on the boundary of $\mathcal{F}_j$. We can triangulate each face $\mathcal{F}_j$ by adding exactly $|\mathcal{F}_j| - 3$ new edges incident to f_j; call these new edges *imaginary edges* incident to f_j. There are exactly $|\mathcal{F}_j| - 3$ imaginary edges incident to f_j. Thus, $x \leq 3n - 6 - \sum_{j=1}^{t}(|\mathcal{F}_j| - 3)$. Let C be a clique of G_0 found in Substep 4.1. We say that an imaginary edge $\{f_j, f'\}$ with $j \in \{1, \ldots, t\}$ is *contributed* by C, if f' is contained in C. Let i be the number of vertices in C. To prove the claim, it remains to prove that the number of imaginary edges contributed by C is at least $i - 4$. By Lemma 6, only the following three cases can occur:

- *Case 1:* C is a pizza. Then, there is a point p in $\mathcal{M}$ such that all nations f' in C touch p on $\mathcal{M}$. Let $\mathcal{F}_j$ be the point-face in $\mathcal{G}(\mathcal{M})$ that corresponds to p. No matter whether f_j is in C or not, at least $i - 3$ imaginary edges incident to f_j are contributed by C.
- *Case 2:* C is a pizza-with-crust. Then, there is a point p in $\mathcal{M}$ such that $|C| - 1$ nations in C touch p on $\mathcal{M}$. Similarly to Case 1, at least $(i - 1) - 3$ imaginary edges are contributed by C.
- *Case 3:* C is a hamantasch. Then, there are three points p_1 through p_3 in $\mathcal{M}$ such that each nation $f' \in C$ touches at least two of the three points on $\mathcal{M}$ and each point is touched by at least two nations in C. Since $i \geq k+1 \geq 5$, at most one of p_1 through p_3 is a 2-point in $\mathcal{M}$. Moreover, if one of p_1 through p_3 is a 2-point, then C is actually a pizza-with-crust and Case 2 applies. So, we may assume that p_1 through p_3 are 2^+-points in $\mathcal{M}$. Let $\mathcal{F}_{j_1}$ be the point-face in $\mathcal{G}(\mathcal{M})$ that corresponds to p_1, and a_1 be the number of nations in C that touch p_1 on $\mathcal{M}$. Define $\mathcal{F}_{j_2}$, a_2, $\mathcal{F}_{j_3}$, and a_3 similarly. Then, the number of imaginary edges incident to f_{j_1} (respectively, f_{j_2}, or f_{j_3}) contributed by C is at least $a_1 - 3$ (respectively, $a_2 - 3$, or $a_3 - 3$). Thus, the number of imaginary edges contributed by C is at least $(a_1+a_2+a_3)-9 \geq 2i-9 \geq i-4$.

By the claim, $x \leq 3n - 6 - \sum_{i=k+1}^{\infty} h_i(i - 4)$. The degree of f in $\mathcal{G}(\mathcal{M})$ is at most $\lfloor \frac{2x}{n} \rfloor$. On the other hand, since $\alpha(G_f) \geq 5$, the degree of f in $\mathcal{G}(\mathcal{M})$ is at least 5. Thus, $2\sum_{i=k+1}^{\infty} h_i(i - 4) \leq n$. In turn, $\sum_{i=k+1}^{\infty} h_i i \leq \frac{n(k+1)}{2(k-3)}$ and $\sum_{i=k+1}^{\infty} h_i \leq \frac{n}{2(k-3)}$.

Let J be the bigger one among J_1 and J_2. By Step 6, $|J| \geq (1 - \epsilon)(\alpha(G_0) - \sum_{i=k+1}^{\infty} h_i)$. By Step 5 and Corollary 1, $|J| \geq \frac{1}{2k} \cdot (n - \sum_{i=k+1}^{\infty} i h_i)$. Thus, $\frac{|J|}{\alpha(G_0)} \geq (1 - \epsilon)(1 - \frac{1}{2(k-3)} \cdot \frac{n}{\alpha(G_0)})$ and $\frac{|J|}{\alpha(G_0)} \geq \frac{k-7}{4k(k-3)} \cdot \frac{n}{\alpha(G_0)}$. As the value of $\frac{n}{\alpha(G_0)}$ decreases, the first lower bound on $\frac{|J|}{\alpha(G_0)}$ increases while the second lower bound on $\frac{|J|}{\alpha(G_0)}$ decreases. The worst case happens when $\frac{n}{\alpha(G_0)} = \frac{4k(k-3)(1-\epsilon)}{3k-2k\epsilon-7}$. Therefore, $\frac{|J|}{\alpha(G_0)} \geq (1 - \epsilon) \cdot \frac{k-7}{3k-2k\epsilon-7}$. By fixing $\epsilon = \frac{1}{4}$ and $k = 28$, we have $\frac{|J|}{\alpha(G_0)} \geq \frac{1}{4}$.

Finally, the proof of Theorem 4 leads to an approximation algorithm for the MIS problem restricted to map graphs that achieves a logarithmic ratio.

Theorem 7. *Given a vertex-weighted map graph $G = (V, E)$, we can find an independent set of G whose weight is at least $\frac{\log \frac{7}{6}}{6 \log |V|} \cdot OPT(G)$, in $O((\sum_{f \in V} |N_G(f)|^7 + |G|) \cdot \log |V|)$ time.*

6   Concluding Remarks

The impracticalness of the PTAS in Theorem 2 and hence that of $MapIS_3$ mainly result from the impracticalness of the Bodlaender algorithm [2] for tree-decomposition. A more practical algorithm for tree-decomposition can make our algorithms more practical.

Is there a practical PTAS for the MIS or UMIS problem restricted to map graphs? It would also be nice if one can design approximation algorithms for the problems that achieve better ratios than ours. Can $MapIS_1$ and $MapIS_2$ be implemented to run faster than as claimed?

Finally, we point out that Lemma 2 can be used to design a PTAS for the *minimum weighted vertex cover* (MWVC) problem restricted to map graphs.

Theorem 8. *There is a PTAS for the MWVC problem restricted to map graphs. It runs in $O(n^2)$ time.*

Acknowledgments: I thank Professors Seinosuke Toda and Xin He for helpful discussions, and a referee for helpful comments on presentation.

References

1. B.S. Baker. Approximation algorithms for NP-complete problems on planar graphs. *J. ACM* **41** (1994) 153-180.
2. H.L. Bodlaender. A linear time algorithm for finding tree-decompositions of small treewidth. *SIAM J. Comput.* **25** (1996) 1305-1317.
3. O.V. Borodin. Cyclic coloring of plane graphs. *Discrete Math.* **100** (1992) 281-289.
4. Z.-Z. Chen. Efficient approximation schemes for maximization problems on $K_{3,3}$-free or K_5-free graphs. *J. Algorithms* **26** (1998) 166-187.
5. Z.-Z. Chen, M. Grigni, and C.H. Papadimitriou. Planar map graphs. *STOC'98*, 514-523. See http://rnc2.r.dendai.ac.jp/~chen/papers/mg2.ps.gz for a full version.
6. J. Håstad. Clique is hard to approximate within $n^{1-\epsilon}$. *STOC'96*, 627-636.
7. R.J. Lipton and R.E. Tarjan. A separator theorem for planar graphs. *SIAM J. Appl. Math.* **36** (1979) 177-189.
8. R.J. Lipton and R.E. Tarjan. Applications of a planar separator theorem. *SIAM J. Computing* **9** (1980) 615-627.
9. O. Ore and M.D. Plummer. Cyclic coloration of plane graphs. *Recent Progress in Combinatorics*, pp. 287-293, Academic Press, 1969.
10. M. Thorup. Map graphs in polynomial time. *FOCS'98*, 396-405.

Hierarchical Topological Inference on Planar Disc Maps

Zhi-Zhong Chen[1] and Xin He[2]

[1] Department of Mathematical Sciences, Tokyo Denki University,
Hatoyama, Saitama 350-0394, Japan.
chen@r.dendai.ac.jp
[2] Dept. of Comput. Sci. and Engin., State Univ. of New York at Buffalo,
Buffalo, NY 14260, U.S.A.
xinhe@cse.buffalo.edu

Abstract. Given a set V and three relations $\bowtie_d$, $\bowtie_m$ and $\bowtie_i$, we wish to ask whether it is possible to draw the elements $v \in V$ each as a closed disc homeomorph $\mathcal{D}_v$ in the plane in such a way that (1) $\mathcal{D}_v$ and $\mathcal{D}_w$ are disjoint for every $(v, w) \in \bowtie_d$, (2) $\mathcal{D}_v$ and $\mathcal{D}_w$ have disjoint interiors but share a point of their boundaries for every $(v, w) \in \bowtie_m$, and (3) $\mathcal{D}_v$ includes $\mathcal{D}_w$ as a sub-region for every $(v, w) \in \bowtie_i$. This problem arises from the study in geographic information systems. The problem is in NP but not known to be NP-hard or polynomial-time solvable. This paper shows that a nontrivial special case of the problem can be solved in almost linear time.

1 Introduction

1.1 Topological Inference Problems

An instance of the *planar topological inference* (PTI) problem is a triple $(V, \bowtie_d, \bowtie_m)$, where V is a finite set, $\bowtie_d$ (subscript d stands for *disjoint*) and $\bowtie_m$ (subscript m stands for *meet*) are two irreflexive symmetric relations on V with $\bowtie_d \cap \bowtie_m = \emptyset$. The problem is to determine whether we can draw the elements v of V in the plane each as a closed disc homeomorph $\mathcal{D}_v$ in such a way that (1) $\mathcal{D}_v$ and $\mathcal{D}_w$ are disjoint for every $(v, w) \in \bowtie_d$, and (2) $\mathcal{D}_v$ and $\mathcal{D}_w$ have disjoint interiors but share a point of their boundaries for every $(v, w) \in \bowtie_m$. The PTI problem and its extensions have been studied for geographic information systems [2,5,6,7,8,9]. In the *fully-conjunctive* case of the PTI problem, it holds that for every pair of distinct $v, w \in V$, either $\{(v, w), (w, v)\} \subseteq \bowtie_d$ or $\{(v, w), (w, v)\} \subseteq \bowtie_m$. The fully-conjunctive PTI problem was first studied by Chen et al. [1] and subsequently by Thorup [10]; the latter gave a complicated inefficient but polynomial-time algorithm for the problem. Motivated by the maps of contemporary countries in the world, Chen et al. [1] restricts the fully-conjunctive PTI problem by requiring that no point of the plane is shared by more than k closed disc homeomorphs $\mathcal{D}_v$ with $v \in V$. Call this restriction the fully-conjunctive k-PTI problem. The fully-conjunctive 3-PTI problem is just the

D.-Z. Du et al. (Eds.): COCOON 2000, LNCS 1858, pp. 115–125, 2000.
© Springer-Verlag Berlin Heidelberg 2000

problem of deciding whether a given graph is planar. However, even the fully-conjunctive 4-PTI problem is difficult to solve; a very complicated $O(n^3)$-time algorithm for it was given in [1].

As has been suggested previously [1,5,6,7,9], a generalization of the PTI problem is obtained by adding another irreflexive, antisymmetric and transitive relation $\bowtie_i$ (subscript i stands for *inclusion*) on V with $\bowtie_i \cap (\bowtie_d \cup \bowtie_m) = \emptyset$ and requiring that $\mathcal{D}_v$ includes $\mathcal{D}_w$ as a sub-region for all $(v, w) \in \bowtie_i$. Unfortunately, no meaningful result has been obtained for this generalization. A natural restriction on $\bowtie_i$ is to require that each $\mathcal{D}_v$ with $v \in V$ and $\{(v, w) \in \bowtie_i \mid w \in V\} \neq \emptyset$ is the union of all $\mathcal{D}_w$ with $(v, w) \in \bowtie_i$. We call this generalization the *hierarchical topological inference* (HTI) problem. In the *fully-conjunctive* case of the HTI problem, for every pair of distinct $v, w \in V$, exactly one of the following (1) through (3) holds: (1) $\{(v, w), (w, v)\} \subseteq \bowtie_d$; (2) $\{(v, w), (w, v)\} \subseteq \bowtie_m$; (3) $\{(v, w), (w, v)\} \cap (\bowtie_d \cup \bowtie_m) = \emptyset$ and $|\{(v, w), (w, v)\} \cap \bowtie_i| = 1$.

Since no efficient algorithm for the fully-conjunctive HTI problem is in sight, we seek to restrict it in a meaningful way so that an efficient algorithm can be designed. As Chen et al. [1] restricted the fully-conjunctive PTI problem, we restrict the fully-conjunctive HTI problem by requiring that no point of the plane is shared by more than k *minimal* closed disc homeomorphs $\mathcal{D}_v$, i.e., those $\mathcal{D}_v$ with $v \in V$ such that $\{w \in V \mid (v, w) \in \bowtie_i\} = \emptyset$. Call this restriction the fully-conjunctive k-HTI problem. Unlike the fully-conjunctive 3-PTI problem, the fully-conjunctive 3-HTI problem is nontrivial.

1.2   A Graph-Theoretic Formulation

Throughout this paper, a graph may have multiple edges but no loops, while a *simple* graph has neither multiple edges nor loops. Also, a cycle in a graph H always means a vertex-simple cycle in H.

A *hierarchical map* $\mathcal{M}$ is a pair $(\mathcal{E}, \mathcal{F})$, where (1) $\mathcal{E}$ is a plane graph whose connected components are biconnected, and (2) $\mathcal{F}$ is a rooted forest whose leaves are distinct faces of $\mathcal{E}$. The faces of $\mathcal{E}$ that are not leaves of $\mathcal{F}$ are called the *lakes* on $\mathcal{M}$, while the rest faces of $\mathcal{E}$ are called the *leaf districts* on $\mathcal{M}$. For each non-leaf vertex α of $\mathcal{F}$, the union of the faces that are leaves in the subtree of $\mathcal{F}$ rooted at α is called a *non-leaf district* on $\mathcal{M}$. $\mathcal{M}$ is a *planar* map if each vertex of $\mathcal{E}$ is incident to at most three edges of $\mathcal{E}$. $\mathcal{M}$ is a *disc* map if (1) the boundary of each leaf district on $\mathcal{M}$ is a cycle and (2) for every non-leaf district $\mathcal{D}$ on $\mathcal{M}$, there is a cycle C such that $\mathcal{D}$ is the union of the faces in the interior of C [1]. The *map graph* G of $\mathcal{M}$ is the simple graph whose vertices are the leaf districts on $\mathcal{M}$ and whose edges are those $\{f_1, f_2\}$ such that the boundaries of f_1 and f_2 intersect. Note that two districts on a map $\mathcal{M}$ may share two or more disjoint borderlines (e.g. China and Russia share two disjoint borderlines with Mongolia in between). However, the map graph G of $\mathcal{M}$ is always simple, meaning that there is at most one edge between each pair of vertices. Note that G is planar when $\mathcal{M}$ is a planar map. We call $(G, \mathcal{F})$ the *abstract* of $\mathcal{M}$.

---

[1] In topology, $\mathcal{D}$ is a closed disc homeomorph.

A *graph-forest* pair is a pair of a simple graph G and a rooted forest $\mathcal{F}$ whose leaves are exactly the vertices of G. In the study of geographic information systems, we are given G and $\mathcal{F}$ and need to decide whether the graph-forest pair $(G, \mathcal{F})$ is the abstract of a disc map. This problem is equivalent to the fully-conjunctive HTI problem. Its special case where each tree in $\mathcal{F}$ is a single vertex is equivalent to the fully-conjunctive PTI problem. Moreover, its special case where we ask whether a given graph-forest pair $(G, \mathcal{F})$ is the abstract of a planar disc map is equivalent to the fully-conjunctive 3-HTI problem.

1.3   Our Result and Its Significance

Our result is an $O(|\mathcal{F}| + |G| \log |G|)$-time algorithm for deciding whether a given graph-forest pair $(G, \mathcal{F})$ is the abstract of a planar disc map $\mathcal{M}$, where $|G|$ denotes the total number of vertices and the edges in G. The difficulty of this problem comes from the lack of information in G about whether there are lakes on $\mathcal{M}$ and how many borderlines two leaf districts on $\mathcal{M}$ can share.

Since Chen et al.'s algorithm [1] for the fully-conjunctive 4-PTI problem is a reduction to the fully-conjunctive 3-PTI problem, we suspect that our result may serve as a basis for tackling the fully-conjunctive 4-HTI problem or even more general cases.

The rest of the paper is organized as follows. § 2 gives definitions needed in later sections. Theorem 1 reduces the problem to the special case where the input planar graph G is biconnected. For this special case, if we know that two vertices z_1 and z_2 of the input planar graph G can be embedded on the outer face, then the algorithm in § 3 works. In § 4, we discuss how to identify two such vertices z_1, z_2.

See http://rnc2.r.dendai.ac.jp/~chen/papers/3map.ps.gz for a full version.

2   Definitions

Let G be a graph. $V(G)$ and $E(G)$ denotes the vertex-set and the edge-set of G, respectively. $|G|$ denotes the total number of vertices and edges in G. Let $E' \subseteq E(G)$ and $U \subseteq V(G)$. The *subgraph of G spanned by E'*, denoted by $G[E']$, is the graph (V', E') where V' consists of the endpoints of the edges of E'. $G - E'$ denotes the graph $(V(G), E(G) - E')$. The *subgraph of G induced by U* is denoted by $G[U]$. $G - U$ denotes $G[V(G) - U]$. If G is a *plane* graph, then each of $G[E']$, $G - E'$, $G[U]$, and $G - U$ *inherits* an embedding from G in the obvious way.

A *cut vertex* of G is a $v \in V(G)$ such that $G - \{v\}$ has more connected components than G. G is *biconnected* if G is connected and has no cut vertex. A *biconnected component* of G is a maximal biconnected subgraph of G. We simply call connected components *1-components*, and biconnected components *2-components*.

Let G be a plane graph. A *k-face* of G is a face of G whose boundary is a cycle consisting of exactly k edges. A *superface* of G is $G[U]$ for some $U \subseteq V(G)$ such that $G[U]$ is connected and each inner face of $G[U]$ is an inner 2- or 3-face of G. The *interior* of a superface $G[U]$ of G is the open set consisting of all points p in the plane that lie neither on the boundary nor in the interior of the outer face of $G[U]$.

A graph-forest pair $(G, \mathcal{F})$ is *planar* if G is planar. A *graph-forest-set* (GFS) triple is a triple $(G, \mathcal{F}, S)$ such that $(G, \mathcal{F})$ is a planar graph-forest pair and S is a subset of $V(G)$. For a rooted forest $\mathcal{F}$ and a non-leaf vertex f of $\mathcal{F}$, let $L_{\mathcal{F}}(f)$ denote the set of all leaves of $\mathcal{F}$ that are descendants of f. An *edge-duplicated supergraph* of a simple graph G is a graph $\mathcal{G}$ such that (1) the vertices of $\mathcal{G}$ are those of G and (2) for each pair of adjacent vertices in $\mathcal{G}$, G has an edge between the two vertices. A plane graph $\mathcal{G}$ *satisfies* a GFS triple $(G, \mathcal{F}, S)$ if (1) $\mathcal{G}$ is an edge-duplicated supergraph of G and (2) for every non-leaf vertex f of $\mathcal{F}$, $\mathcal{G}[L_{\mathcal{F}}(f)]$ is a superface of $\mathcal{G}$ whose interior contains no vertex of S. A GFS triple is *satisfiable* if some plane graph satisfies it. A planar graph-forest pair $(G, \mathcal{F})$ is *satisfiable* if the GFS triple $(G, \mathcal{F}, \emptyset)$ is satisfiable.

We can assume that a GFS triple $(G, \mathcal{F}, S)$ (resp., graph-forest pair $(G, \mathcal{F})$) always satisfies that no vertex of $\mathcal{F}$ has exactly one child. By this, $|\mathcal{F}| = O(|G|)$.

Theorem 1. *The problem of deciding whether a given planar graph-forest pair $(G, \mathcal{F})$ is the abstract of a planar disc map can be reduced in $O(|G| \log |G|)$ time, to the problem of deciding whether a given GFS triple $(G, \mathcal{F}, S)$ with G biconnected is satisfiable.*

3   Biconnected Case

Fix a GFS triple $(G, \mathcal{F}, S)$ such that G is biconnected. We want to check if $(G, \mathcal{F}, S)$ is satisfiable. To this end, we first check whether $G[L_{\mathcal{F}}(f)]$ is connected for all vertices f of $\mathcal{F}$. This checking takes $O(|G| \log |G|)$ time. If $(G, \mathcal{F})$ fails the checking, then we stop immediately. So, suppose that $(G, \mathcal{F})$ passes the checking.

3.1   SPQR Decompositions

A *two-terminal graph* [2] (TTG) $\mathcal{G} = (V, E)$ is an acyclic plane digraph with a pair (s, t) of specified vertices on the outer face such that s is the only *source* (i.e. with no entering edge) and t is the only *sink* (i.e. with no leaving edge). We denote s and t by $s(\mathcal{G})$ and $t(\mathcal{G})$, respectively. We also call s and t the *poles* of $\mathcal{G}$ and the rest vertices the *nonpoles* of $\mathcal{G}$. A *two-terminal subgraph* of $\mathcal{G}$ is a subgraph H of $\mathcal{G}$ such that H is a TTG and no edge of $\mathcal{G}$ is incident both to a nonpole of H and to a vertex outside H. Although $\mathcal{G}$ is a digraph, the directions on its edges are used only in the definition of its two-terminal subgraphs but have no meaning anywhere else. With this in mind, we will view $\mathcal{G}$ as an undirected graph from now on.

A *split pair* of $\mathcal{G}$ is either an unordered pair of adjacent vertices, or an unordered pair of vertices whose removal disconnects the graph $(V(\mathcal{G}), E(\mathcal{G}) \cup \{\{s(\mathcal{G}), t(\mathcal{G})\}\})$. A *split component* of a split pair $\{u, v\}$ is either an edge $\{u, v\}$

---

[2] Previously called a *planar st-graph*.

or a maximal two-terminal subgraph H (of $\mathcal{G}$) with poles u and v such that $\{u,v\}$ is not a split pair of H. A split pair $\{u,v\}$ of $\mathcal{G}$ is *maximal* if there is no other split pair $\{w_1, w_2\}$ in $\mathcal{G}$ such that $\{u,v\}$ is contained in a split component of $\{w_1, w_2\}$.

The *decomposition tree* T of $\mathcal{G}$ describes a recursive decomposition of $\mathcal{G}$ with respect to its split pairs [3,4]. For clarity, we call the elements of $V(T)$ *nodes*. T is a rooted ordered tree whose nodes are of four types: S, P, Q, and R. Each node μ of T has an associated TTG, called the *skeleton* of μ and denoted by $\mathrm{Skl}(\mu)$. Also, it is associated with an edge in the skeleton of the parent χ of μ, called the *virtual edge* of μ in $\mathrm{Skl}(\chi)$. T is recursively defined as follows.

Trivial case: $\mathcal{G}$ consists of a single edge $\{s(\mathcal{G}), t(\mathcal{G})\}$. Then, T consists of a single Q-node whose skeleton is $\mathcal{G}$ itself.

Series case: $\mathcal{G}$ is not biconnected. Let $c_1, \ldots, c_{k-1}$ $(k \geq 2)$ be the cut vertices of $\mathcal{G}$. Since $\mathcal{G}$ is a TTG, each $c_i \in \{c_1, \ldots, c_{k-1}\}$ is contained in exactly two 2-components $\mathcal{G}_i$ and $\mathcal{G}_{i+1}$ of $\mathcal{G}$ such that $s(\mathcal{G})$ is in $\mathcal{G}_1$ and $t(\mathcal{G})$ is in $\mathcal{G}_k$. The root of T is an S-node μ. $\mathrm{Skl}(\mu)$ is a path $s(\mathcal{G}), e_1, c_1, e_2, \ldots, c_{k-1}, e_k, t(\mathcal{G})$.

Parallel case: $s(\mathcal{G})$ and $t(\mathcal{G})$ constitute a split pair of $\mathcal{G}$ with split components $\mathcal{G}_1, \ldots, \mathcal{G}_k$ $(k \geq 2)$. Then, the root of T is a P-node μ. $\mathrm{Skl}(\mu)$ consists of k parallel edges between $s(\mathcal{G})$ and $t(\mathcal{G})$, denoted by $e_1, \ldots, e_k$.

Rigid case: none of the above cases applies. Let $\{s_1, t_1\}, \ldots, \{s_k, t_k\}$ be the maximal split pairs of $\mathcal{G}$. For $i = 1, \ldots, k$, let $\mathcal{G}_i$ be the union of all the split components of $\{s_i, t_i\}$. The root of T is an R-node μ. $\mathrm{Skl}(\mu)$ is obtained from $\mathcal{G}$ by replacing each $\mathcal{G}_i \in \{\mathcal{G}_1, \ldots, \mathcal{G}_k\}$ with an edge $e_i = \{s_i, t_i\}$.

In the last three cases, μ has children $\nu_1, \ldots, \nu_k$ (in this order), such that ν_i is the root of the decomposition tree of graph $\mathcal{G}_i$ for all $i \in \{1, \ldots, k\}$. The virtual edge of node ν_i is the edge e_i in $\mathrm{Skl}(\mu)$. $\mathcal{G}_i$ is called the *pertinent graph* of ν_i and is denoted by $\mathrm{Prt}(\nu_i)$.

By the definition of T, no child of an S-node is an S-node and no child of a P-node is a P-node. Also note that, for an R-node μ, the two poles of $\mathrm{Skl}(\mu)$ are not adjacent in $\mathrm{Skl}(\mu)$. A *block* of $\mathcal{G}$ is the pertinent graph of a node in T. For a non-leaf node μ in T and each child ν of μ, we call $\mathrm{Prt}(\nu)$ a *child block* of $\mathrm{Prt}(\mu)$.

Lemma 1. *Let z_1 and z_2 be two vertices of G such that adding a new edge between z_1 and z_2 does not distroy the planarity of G. Then, it takes $O(|G|)$ time to convert G to a TTG $\mathcal{G}$ with poles z_1 and z_2 and to construct the corresponding decomposition tree T of $\mathcal{G}$. Moreover, every TTG $\mathcal{G}'$ with poles z_1 and z_2 such that $\mathcal{G}'$ satisfies $(G, \mathcal{F}, S)$, can be obtained by performing a sequence of $O(|G|)$ following operations:*

1. *Flip the skeleton of an R-node of T around the poles of the skeleton.*
2. *Permute the children of a P-node of T.*
3. *Add a new Q-node child to a P-node μ of T such that μ originally has a Q-node child; further add a virtual edge to the skeleton of μ.*

A *desired pair* for $(G, \mathcal{F}, S)$ is an unordered pair $\{z_1, z_2\}$ of two vertices of G such that if $(G, \mathcal{F}, S)$ is satisfiable, then it is satisfied by a plane graph in which both z_1 and z_2 appear on the boundary of the outer face.

Our algorithm is based on Lemma 1. In order to save time, we want to first find a desired pair $\{z_1, z_2\}$ for $(G, \mathcal{F}, S)$. If $\mathcal{F}$ is not a tree, then we can let $\{z_1, z_2\}$ be an edge such that there are two roots f_1 and f_2 in $\mathcal{F}$ with $z_1 \in L_{\mathcal{F}}(f_1)$ and $z_2 \in L_{\mathcal{F}}(f_2)$. If $\mathcal{F}$ is a tree but $|S| \geq 2$, then we can let z_1 and z_2 be two arbitrary vertices in S. In case $\mathcal{F}$ is a tree and $|S| \leq 1$, it is not easy to find out z_1 and z_2; §4 is devoted to this case.

3.2   Basic Ideas

Suppose that we have found a desired pair $\{z_1, z_2\}$ for $(G, \mathcal{F}, S)$ and have converted G to a TTG $\mathcal{G}$ with poles z_1 and z_2. Throughout §3.2, 3.3, and 3.4, when we say that $(G, \mathcal{F}, S)$ is not satisfiable, we mean that $(G, \mathcal{F}, S)$ is satisfied by no plane graph in which z_1 and z_2 are on the boundary of the outer face.

Let T be the decomposition tree of $\mathcal{G}$. Our algorithm processes the P- or R-nodes of T in post-order. Processing a node μ is to perform Operations 1 through 3 in Lemma 1 on the subtree of T rooted at μ, and to modify $\mathrm{Prt}(\mu)$ and hence $\mathcal{G}$ accordingly.

Before proceeding to the details of processing nodes, we need several definitions. For a block B of $\mathcal{G}$, $\mathcal{G} - B$ denotes the plane subgraph of $\mathcal{G}$ obtained from $\mathcal{G}$ by deleting all vertices in $V(B) - \{s(B), t(B)\}$ and all edges in $E(B)$. For a plane subgraph H of $\mathcal{G}$ and a vertex f of $\mathcal{F}$, $H \cap L_{\mathcal{F}}(f)$ denotes the plane subgraph of H induced by $V(H) \cap L_{\mathcal{F}}(f)$.

Let μ be a node of T. Let $B = \mathrm{Prt}(\mu)$. A vertex f in $\mathcal{F}$ is μ-through if $B \cap L_{\mathcal{F}}(f)$ contains a path between $s(B)$ and $t(B)$; is μ-out-through if $(\mathcal{G} - B) \cap L_{\mathcal{F}}(f)$ contains a path between $s(B)$ and $t(B)$; is μ-critical if it is both μ-through and μ-out-through. Note that if a vertex f of $\mathcal{F}$ is μ-critical (resp., μ-through, or μ-out-through), then so is every ancestor of f in $\mathcal{F}$. A vertex f of $\mathcal{F}$ is extreme μ-critical if f is μ-critical but no child of f in $\mathcal{F}$ is. Define extreme μ-through vertices and extreme μ-out-through vertices similarly.

Let $\nu_1, \ldots, \nu_k$ be the children of μ in T. For each $i \in \{1, \ldots, k\}$, let e_i be the virtual edge of ν_i in $\mathrm{Skl}(\mu)$, and $B_i = \mathrm{Prt}(\nu_i)$. For a vertex f of $\mathcal{F}$, $E_{\mu, f}$ denotes the set $\{e_i \mid f \text{ is } \nu_i\text{-through}\}$. We say f is μ-cyclic, if the plane subgraph of $\mathrm{Skl}(\mu)$ spanned by $E_{\mu, f}$ contains a cycle. Note that if f is μ-cyclic, then so is every ancestor of f in $\mathcal{F}$. We say f is extreme μ-cyclic, if f is μ-cyclic but no child of f in $\mathcal{F}$ is.

Lemma 2. *For each node μ of T, there is at most one extreme μ-critical (resp., μ-through, or μ-out-through) vertex in $\mathcal{F}$.*

For each node μ in T, if there is a μ-through (resp., μ-out-through, or μ-critical) vertex in $\mathcal{F}$, then let x-thr(μ) (resp., x-out-thr(μ), or x-crt(μ)) denote the extreme μ-through (resp., μ-out-through, or μ-critical) vertex in $\mathcal{F}$; otherwise, let x-thr$(\mu) = \bot$ (resp., x-out-thr$(\mu) = \bot$, or x-crt$(\mu) = \bot$).

Let μ be a node of T. An *embedding* of $\mathrm{Prt}(\mu)$ is a TTG that is an edge-duplicated supergraph of $\mathrm{Prt}(\mu)$ and has the same poles as $\mathrm{Prt}(\mu)$. In an embedding B of $\mathrm{Prt}(\mu)$, the poles of B divide the boundary of the outer face of

B into two paths, which are called the *left side* and the *right side* of B, respectively. An embedding B of $\mathrm{Prt}(\mu)$ is *potential* if for every vertex f in $\mathcal{F}$ with $V(B) \cap L_{\mathcal{F}}(f) \neq \emptyset$, each connected component of $B \cap L_{\mathcal{F}}(f)$ is a superface of $B \cap L_{\mathcal{F}}(f)$ whose interior contains no vertex of S. When μ is a P-node (resp., R-node), a potential embedding B_1 of $\mathrm{Prt}(\mu)$ is *better* than another potential embedding B_2 of $\mathrm{Prt}(\mu)$ if B_1 has more sides consisting of exactly one edge (resp., two edges) than B_2. An embedding of $\mathrm{Prt}(\mu)$ is *optimal* if it is potential and no potential embedding of $\mathrm{Prt}(\mu)$ is better than it.

Lemma 3. *Suppose that an embedding $\mathcal{G}'$ of G satisfies $(G, \mathcal{F}, S)$. Let μ be a node of T. Let B_μ be the subgraph of $\mathcal{G}'$ that is an embedding of $\mathrm{Prt}(\mu)$. Let O_μ be an optimal embedding of $\mathrm{Prt}(\mu)$. Then, the embedding of G obtained from $\mathcal{G}'$ by replacing B_μ with O_μ still satisfies $(G, \mathcal{F}, S)$.*

Right before the processing of a P- or R-node μ, the following invariant holds for all ν that is a non-S-child of μ or a child of an S-child of μ in T:

– An optimal embedding B_ν of $\mathrm{Prt}(\nu)$ has been fixed, except for a possible flipping around its poles.

In the description of our algorithm, for convenience, we will denote the fixed embedding of $\mathrm{Prt}(\nu)$ still by $\mathrm{Prt}(\nu)$ for all nodes ν of T that have been processed.

3.3   Processing R-Nodes

Let μ be an R-node of T. Let $B = \mathrm{Prt}(\mu)$ and $K = \mathrm{Skl}(\mu)$. Let $\nu_1, \ldots, \nu_k$ be the children of μ in T. For each $i \in \{1, \ldots, k\}$, let e_i be the virtual edge of ν_i in K, and $B_i = \mathrm{Prt}(\nu_i)$.

Since μ is an R-node, each side of B contains at least two edges. If x-crt$(\mu) = f \neq \bot$, at least one side of B must be a part of the boundary of an inner 3-face of the superface induced by $L_{\mathcal{F}}(f)$. In this case, if $\{s(B), t(B)\} \notin E(G)$, then $(G, \mathcal{F}, S)$ is not satisfiable because no side of B can be part of the boundary of a 3-face in an embedding of $\mathcal{G}$. So, we assume that $\{s(B), t(B)\} \in E(G)$ when x-crt$(\mu) \neq \bot$. Processing μ is accomplished by performing the following steps:

R-Procedure:

1. Compute the rooted forest $\mathcal{F}'$ obtained from $\mathcal{F}$ by deleting all vertices that are not μ-cyclic.
2. For all vertices f in $\mathcal{F}'$, perform the following substeps:
   (a) For every inner face F of $K[E_{\mu,f}]$, check if F is also an inner 3-face of K; if not, stop immediately because $(G, \mathcal{F}, S)$ is not satisfiable.
   (b) For every edge $e_i \in E_{\mu,f}$ that is not on the outer face of $K[E_{\mu,f}]$, check if $V(B_i) \subseteq L_{\mathcal{F}}(f)$; if $V(B_i) \nsubseteq L_{\mathcal{F}}(f)$, then stop immediately because $(G, \mathcal{F}, S)$ is not satisfiable.

3. For all roots f in $\mathcal{F}'$, perform the following substeps:

   (a) For every edge $e_i \in E_{\mu,f}$ that is not on the outer face of $K[E_{\mu,f}]$, check if (1) $S \cap (V(B_i) - \{s(B), t(B)\}) = \emptyset$ and (2) both sides of B_i consist of a single edge; if (1) or (2) does not hold, then stop immediately because $(G, \mathcal{F}, S)$ is not satisfiable.

   (b) For every vertex v of K that is not on the boundary of the outer face of $K[E_{\mu,f}]$, check if $v \notin S$; if $v \in S$, then stop immediately because $(G, \mathcal{F}, S)$ is not satisfiable.

   (c) For every edge e_i shared by the boundaries of the outer face of $K[E_{\mu,f}]$ and some inner face F of $K[E_{\mu,f}]$, check if some side of B_i consists of a single edge and has not been labeled with the index of a face other than F; if no such side exists, then stop immediately because $(G, \mathcal{F}, S)$ is not satisfiable, while label one such side with the index of F otherwise.

4. For each side P of K consisting of exactly two edges, and for each edge e_i on P, if some side P_i of B_i consists of a single edge and has not been labeled, then mark the side P_i of B_i "outer".

5. For all child blocks B_j of B, if at least one of the following holds, then modify B by flipping B_j around the poles of B_j:

   − A side of B_j is marked "outer" but is not on the boundary of the outer face of B.

   − A side of B_j is labeled with the index of some face F of K but that side is not on the boundary of the corresponding face of B.

6. If x-crt$(\mu) = f_1 \neq \perp$, check whether some side of B consists of exactly two edges and the common endpoint of the two edges is in $L_{\mathcal{F}}(f_1)$; if no such side of B exists, then stop immediately because $(G, \mathcal{F}, S)$ is not satisfiable.

After μ is processed without failure, the invariant in §3.2 hold for μ.

3.4   Processing of P-Nodes

Let μ be a P-node of T. Let $B = \text{Prt}(\mu)$. Let $\nu_1, \ldots, \nu_k$ be the children of μ in T. For each $i \in \{1, \ldots, k\}$, let e_i be the virtual edge of ν_i in $\text{Skl}(\mu)$, and $B_i = \text{Prt}(\nu_i)$.

If there is neither μ-cyclic vertex nor μ-critical vertex in $\mathcal{F}$, then the processing of μ is finished by (1) checking if $\{s(B), t(B)\} \in E(G)$ and (2) if so, modifying B by permuting its child blocks so that edge $\{s(B), t(B)\}$ forms one side of B. So, assume that there is a μ-cyclic or μ-critical vertex in $\mathcal{F}$. Then, $(G, \mathcal{F}, S)$ is not satisfiable, if $\{s(B), t(B)\} \notin E(G)$. So, further assume that $\{s(B), t(B)\} \in E(G)$. Since $\{s(B), t(B)\} \subseteq L_{\mathcal{F}}(f)$ for all μ-through vertices f in $\mathcal{F}$, every μ-cyclic or μ-critical vertex must appear on the path P from $f_0 = \text{x-thr}(\mu)$ to the root f_r of the tree containing f_0 in $\mathcal{F}$. If f_r is a μ-critical vertex, we process μ by calling P-Procedure1 below; otherwise, we process μ by calling P-Procedure2 below.

P-Procedure1:

1. Let $E_{\mu,f_r} = \{e_{i_1}, \ldots, e_{i_h}\}$ where ν_{i_1} is a Q-node.
2. For all $j \in \{2, \ldots, h\}$, mark e_{i_j} if (1) some side of B_{i_j} does not consist of exactly two edges, or (2) $S \cap (V(B_{i_j}) - \{s(B), t(B)\}) \neq \emptyset$, or (3) $V(B_{i_j}) \not\subseteq L_{\mathcal{F}}(f_0)$.
3. Depending on the number m of marked edges, distinguish three cases:
   - Case 1: $m = 0$. Then, perform the following substeps:
     (a) Modify B by permuting its child blocks so that (1) the edge B_{i_1} is on the left side of B and (2) $B_{i_1}, \ldots, B_{i_h}$ appear consecutively.
     (b) For each $j \in \{2, \ldots, h\}$, add an edge $\{s(B), t(B)\}$ right after B_{i_j}.
   - Case 2: $m = 1$. Let the marked edge be e_{i_ℓ}, and perform the following:
     (a) If ν_{i_ℓ} is an R-node, perform the following substeps:
        i. Check whether some side P' of B_{i_ℓ} satisfies that (1) there are exactly two edges e_1' and e_2' on P' and (2) the common endpoint of e_1' and e_2' is not in S but is in $L_{\mathcal{F}}(f)$ where $f = $ x-thr(ν_{i_ℓ}). (Comment: f is on P, i.e., f is an ancestor of f_0 in $\mathcal{F}$.)
        ii. If P' does not exist, then stop immediately because $(G, \mathcal{F}, S)$ is not satisfiable; otherwise, flip B_{i_ℓ} around its poles if P' is not the left side of B_{i_ℓ}.
     (b) If ν_{i_ℓ} is an S-node, then perform the following substeps:
        i. If ν_{i_ℓ} has more than two children, then stop immediately because $(G, \mathcal{F}, S)$ is not satisfiable.
        ii. Let the children of ν_{i_ℓ} be χ_1 and χ_2. For each $j \in \{1, 2\}$, if no side of Prt(χ_j) consists of a single edge, then stop immediately for failure; otherwise, flip Prt(χ_j) around its poles when the left side of Prt(χ_j) does not consist of a single edge.
     (c) Modify B by permuting its child blocks so that (1) the edge B_{i_1} is on the left side of B and (2) $B_{i_1}, \ldots, B_{i_\ell-1}, B_{i_\ell+1}, \ldots, B_{i_h}, B_{i_\ell}$ appear consecutively in this order.
     (d) For all $j \in \{2, \ldots, h\} - \{\ell\}$, add an edge $\{s(B), t(B)\}$ right after B_{i_j}.
   - Case 3: $m \geq 2$. Then, stop because $(G, \mathcal{F}, S)$ is not satisfiable.

P-Procedure2:

1. Perform Steps 1 and 2 of P-Procedure1.
2. Depending on the number m of marked edges, distinguish three cases:
   - Case 1: $m = 0$. Same as Case 1 in P-Procedure1.
   - Case 2: $m = 1$. Same as Case 2 in P-Procedure1.
   - Case 3: $m = 2$. Let e_{i_ℓ} and $e_{i_{\ell'}}$ be the two marked edges where $i_\ell < i_{\ell'}$. Then, perform the following substeps:
     (a) If ν_{i_ℓ} is an R-node, then perform Substeps 3(a)i and 3(a)ii of P-Procedure1; otherwise, perform Substeps 3(b)i and 3(b)ii of P-Procedure1.

(b) If $\nu_{i_{\ell'}}$ is an R-node, perform the modification of Substeps 3(a)i and 3(a)ii of P-Procedure1 obtained by (1) changing i_ℓ to $i_{\ell'}$ and (2) changing the word "left" to "right"; otherwise, perform the modification of Substeps 3(b)i and 3(b)ii of P-Procedure1 obtained by doing (1) and (2).

(c) Modify B by permuting its child blocks so that $B_{i_{\ell'}}, B_{i_1}, \ldots, B_{i_\ell-1}$, $B_{i_\ell+1}, \ldots, B_{i_{\ell'}-1}, B_{i_{\ell'}+1}, \ldots, B_{i_h}, B_{i_\ell}$ appear consecutively in this order.

(d) For all $j \in \{2, \ldots, h\} - \{\ell, \ell'\}$, add an edge $\{s(B), t(B)\}$ right after B_{i_j}.

– Case 4: $m \geq 3$. Then, stop because $(G, \mathcal{F}, S)$ is not satisfiable.

After μ is processed without failure, the invariant in §3.2 hold for μ.

Theorem 2. *The overall algorithm can be implemented in $O(|G| \log |G|)$ time.*

4 The Single-Tree Case

Assume that $\mathcal{F}$ is a tree and $|S| \leq 1$. A *candidate edge* is an edge $\{z_1, z_2\}$ of G such that (1) the lowest common ancestor of z_1 and z_2 in $\mathcal{F}$ is the root of $\mathcal{F}$, and (2) if $|S| = 1$, then $\{z_1, z_2\} \cap S \neq \emptyset$.

A plane graph $\mathcal{G}$ *witnesses* $(G, \mathcal{F}, S)$ if (1) $\mathcal{G}$ is an edge-duplicated supergraph of G, (2) for every non-leaf and *non-root* vertex f of $\mathcal{F}$, $\mathcal{G}[L_{\mathcal{F}}(f)]$ is a superface of $\mathcal{G}$ whose interior contains no vertex of S, (3) at most one face of $\mathcal{G}$ is a k-face with $k \geq 4$, and (4) if $|S| = 1$ and a k-face F with $k \geq 4$ exists in $\mathcal{G}$, then the unique vertex of S appears on the boundary of F.

Lemma 4. *Let $\{z_1, z_2\}$ be a candidate edge of G. Then, $(G, \mathcal{F}, S)$ is satisfiable if and only if there is a plane graph H such that (1) H witnesses $(G, \mathcal{F}, S)$ and (2) both z_1 and z_2 appear on the boundary of the outer face in H.*

Using Lemma 4 and modifying the algorithm in §3, we can prove:

Theorem 3. *Given a graph-forest pair $(G, \mathcal{F})$, it takes $O(|\mathcal{F}| + |G| \log |G|)$ time to decide whether $(G, \mathcal{F})$ is the abstract of a planar disc map.*

References

1. Z.-Z. Chen, M. Grigni, and C.H. Papadimitriou. Planar map graphs. *Proc. 30th STOC* (1998) 514–523. See http://rnc2.r.dendai.ac.jp/~chen/papers/mg2.ps.gz for a full version.
2. Z.-Z. Chen, X. He, and M.-Y. Kao. Nonplanar topological inference and political-map graphs. *Proc. 10th SODA* (1999) 195–204.
3. G. Di Battista and R. Tamassia. On-line maintenance of triconnected components with *spqr*-trees. *Algorithmica* **15** (1996) 302–318.
4. G. Di Battista and R. Tamassia. On-line planarity testing. *SIAM J. Comput.* **25** (1996) 956–997.

5. M.J. Egenhofer. Reasoning about binary topological relations. In: O. Gunther and H.J. Schek (eds.): Proc. Advances in Spatial Database (1991) 143–160.
6. M.J. Egenhofer and J. Sharma. Assessing the consistency of complete and incomplete topological information. *Geographical Systems* **1** (1993) 47–68.
7. M. Grigni, D. Papadias, and C.H. Papadimitriou. Topological inference. *Proc. 14th IJCAI* (1995) 901–906.
8. D. Papadias and T. Sellis. The qualitative representation of spatial knowledge in two-dimensional space. *Very Large Data Bases Journal* **4** (1994) 479–516.
9. T.R. Smith and K.K. Park. Algebraic approach to spatial reasoning. *Int. J. on Geographical Information Systems* **6** (1992) 177–192.
10. M. Thorup. Map graphs in polynomial time. *Proc. 39th FOCS* (1998) 396–405.

Efficient Algorithms for the Minimum Connected Domination on Trapezoid Graphs

Yaw-Ling Lin*, Fang Rong Hsu**, and Yin-Te Tsai***

Providence University,
200 Chung Chi Road, Sa-Lu, Taichung Shang, Taiwan 433, R.O.C.
{yllin,frhsu,yttsai}@pu.edu.tw

Abstract. Given the trapezoid diagram, the problem of finding the minimum cardinality connected dominating set in trapezoid graphs was solved in $O(m + n)$ time [7]; the results is recently improved to be $O(n)$ time by Kohler [5]. For the (vertex) weighted case, finding the minimum weighted connected dominating set in trapezoid graphs can be solved in $O(m + n \log n)$ time [11]. Here n (m) denotes the number of vertices (edges) of the trapezoid graph.

In this paper, we show a different approach for finding the minimum cardinality connected dominating set in trapezoid graphs using $O(n)$ time. For finding the minimum weighted connected dominating set, we show the problem can be efficiently solved in $O(n \log \log n)$ time.

1   Introduction

The intersection graph of a collection of trapezoids with corner points lying on two parallel lines is called the *trapezoid graph*. Trapezoid graphs is first proposed by Dagan *et al.* [3] showing that the channel routing problem is equivalent to the coloring problems on trapezoid graphs. Throughout the paper, we use n (m) to represent the number of vertices (edges) in the given graph. Ma and Spinrad [10] show that trapezoid graphs can be recognized in $O(n^2)$ time. Note that trapezoid graphs are cocomparability graphs, and properly contain both interval graphs and permutation graphs [1]. Note that interval graphs are the intersection graph of intervals on a real line; permutation graphs are the intersection graph of line segments whose endpoints lying on two parallel lines.

A *dominating set* of a graph $G = (V, E)$ is a subset D of V such that every vertex not in D is adjacent to at least one vertex in D. Each vertex $v \in V$ can be associated with a (non-negative) real weight, denoted by $w(v)$. The *weighted domination problem* is to find a dominating set, D, such that its weight $w(D) = \sum_{v \in D} w(v)$ is minimized. A dominating set D is *independent, connected* or *total*

---

* Corresponding author. Department of Computer Science and Information Management. The work was partly supported by the National Science Council, Taiwan, R.O.C, grant NSC-88-2213-E-126-005.

** Department of Accounting.

*** Department of Computer Science and Information Management.

D.-Z. Du et al. (Eds.): COCOON 2000, LNCS 1858, pp. 126–136, 2000.
© Springer-Verlag Berlin Heidelberg 2000

if the subgraph induced by D has no edge, is connected, or has no isolated vertex, respectively. Dominating set problem and its variants have many applications in areas like bus routing, communication network, radio broadcast, code-word design, and social network [4].

The decision version of the weighted domination problem is NP-complete even for cocomparability graphs [2]. For trapezoid graphs, Liang [8] shows that the minimum weighted domination and the total domination problem can be solved in $O(mn)$ time. Lin [9] show that the minimum weighted independent domination in trapezoid graphs can be found in $O(n \log n)$ time. Srinivasan *et al.* [11] show that the minimum weighted connected domination problem in trapezoid graphs can be solved in $O(m + n \log n)$ time. For the unweighted case, the $O(n \log n)$ factor is improved by Liang [7], who show that the minimum-cardinality connected domination problem in trapezoid graphs can be solved in $O(m+n)$ time. However, since the number of edges, m, can be as large as $O(n^2)$, the potential time-complexity of the algorithm is still $O(n^2)$. In this paper, we show that the minimum cardinality connected domination problem in trapezoid graphs can be solved optimally in $O(n)$ time. Further, for the weighted case, we show the problem can be efficiently solved in $O(n \log \log n)$ time.

This paper is organized as follows. Section 2 establishes basic notations and some interesting properties of trapezoid graphs concerning the connected domination problem. Section 3 gives our $O(n)$-time algorithm of finding the minimum-sized connected dominating set in trapezoid graphs. Section 4 shows that finding the minimum weighted connected domination can be done in $O(n \log \log n)$ time. Finally, we conclude our results in Section 5.

2   Basic Notations and Properties

A graph $G = (V, E)$ is a *trapezoid graph* if there is a trapezoid diagram T such that each vertex v_i in V corresponds to a trapezoid t_i in T and $(v_i, v_j) \in E$ if and only if t_i and t_j intersects in the trapezoid diagram. Here the *trapezoid diagram* consists of two parallel lines L_1 (the upper line) and L_2 (the lower line). Further, a trapezoid t_i is defined by four corner points a_i, b_i, c_i, d_i such that $[a_i..b_i]$ ($[c_i..d_i]$) is the top (bottom) interval of trapezoid t_i on L_1 (L_2); note that $a_i < b_i$ and $c_i < d_i$.

It can be shown that any trapezoid diagram can be transformed into another trapezoid diagram with all corner points are distinct while the two diagrams still correspond to the same trapezoid graph. Thus, we assume that the two parallel lines, L_1 and L_2, are labeled with consecutive integer values $1, 2, 3, \ldots, 2n$ from left to right; that is, $\cup_{i=1}^{n}\{a_i, b_i\} = \cup_{i=1}^{n}\{c_i, d_i\} = [1..2n]$. We assume that these trapezoids are labeled in increasing order of their corner points b_i's. We define two trapezoids $t_i \prec t_j$ if $b_i < a_j$ and $d_i < c_j$; that is, the top (bottom) interval of t_i totally lies on the left of the top (bottom) interval of t_j. Given a trapezoid diagram, the corresponded partial ordered set is a *trapezoid order*. Note that two trapezoids t_k, t_l intersect each other (denoted by $t_k \sim t_l$) if and only if neither

$t_k \prec t_l$ nor $t_l \prec t_k$; that is, they are incomparable in the trapezoid (partial) order.

Throughout the paper, we use the words "trapezoid" and "vertex" interchangeably for convenience. A vertex v_i (v_j) is called a *left (right) dominator* if and only if there are no vertices lie on the left (right) of v_i (v_j) in the trapezoid diagram. That is, $(\forall v \in V)(v \nprec v_i)$ and $(\forall v \in V)(v_j \nprec v)$. Let D be a minimum cardinality connected dominating set of the trapezoid graph G. Clearly, $|D| = 1$ if and only if there is a vertex $v \in G$ that is a left and right dominator at the same time.

Two adjacent vertices u, v form a *source pair* (*sink pair*) if and only if there are no vertices lie on the left (right) of u and v at the same time. A vertex v is a source (sink) vertex if there is another vertex u such that $\{u, v\}$ form a source (sink) pair.

Proposition 1. *Given the trapezoid diagram, v is a source (sink) vertex if and only if there is a left (right) dominator $d \in V$ such that $v \sim d$.* It is easily seen that $|D| = 2$ if and only if $|D| \neq 1$ and there is a pair $\{u, v\} \in G$ that is a source pair and a sink pair at the same time. For $|D| \geq 3$, we have the following lemma [6] for cocomparability graphs; it thus holds for trapezoid graphs as well.

Lemma 1. *For a connected cocomparability graph G, there exists a minimum cardinality connected dominating set that induces a simple path in G. Furthermore, let $P = \langle p_1, \ldots, p_k \rangle$ be such a path with $k \geq 3$, then $p_i \prec p_{i+2}$ for $1 \leq i \leq k - 2$, and every vertex before $\min\{p_1, p_2\}$ is covered by $\{p_1, p_2\}$ and every vertex after $\max\{p_{k-1}, p_k\}$ is covered by $\{p_{k-1}, p_k\}$.*

Now consider the case that $|D| \geq 3$. Within all left dominators, let b (d) be the vertex having the rightmost top (bottom) corner; it is possible that b, d are the same vertex. By Lemma 1 and Proposition 1, it follows that the minimum cardinality connected dominating set for $|D| \geq 3$ can be found by finding a shortest-length path, from b or d, until one of the leftmost left dominators $\{a, c\}$ is reached. That is, let $P_{ba}, P_{bc}, P_{da}, P_{dc}$ denote these four shortest paths. Among these four paths, the one with the smallest cardinality is the desired minimum cardinality connected dominating set.

3   Minimum Cardinality Connected Domination

Immediately following the previous discussions, the strategy of the algorithm can be described as following: for each vertex v, decide whether v is a left (right) dominator, and check for the special cases for $1 \leq |D| \leq 2$. If it turns out that $|D| \geq 3$, we can find the shortest-length path from the rightmost corners of left dominators until reaching the first right dominator.

We first show that all the dominators can be found in $O(n)$ time. Consider the *right (left) tangent edge* of a trapezoid t_i, the vertical line segment $\overline{b_i d_i}$ $(\overline{a_i c_i})$ connecting L_1 and L_2, is called a *right (left) stick*, denoted by $r_i = \overline{b_i d_i}$ $(l_i = \overline{a_i c_i})$. Treating a stick as a degenerated case of trapezoid, we say sticks

$r_i \prec r_j$ if and only if $b_i < b_j$ and $d_i < d_j$. A group of sticks, S, is called a *pencil* if every two sticks of S intersect each other. Consider the set of all right sticks $S = \{r_1, r_2, \ldots, r_n\}$. A stick in S is *eliminated* if there is another stick lies on the left of it; the resulting sticks form a *left-pencil*, denoted by $L = \langle r_{i_1}, r_{i_2}, \ldots, r_{i_k} \rangle$. Note that $b_{i_1} < b_{i_2} < \cdots < b_{i_k}$, but $d_{i_1} > d_{i_2} > \cdots > d_{i_k}$. It is readily seen that vertices $\{v_{i_1}, v_{i_2}, \ldots, v_{i_k}\}$ are left dominators. The following discussions concern properties of the left-pencil; similar properties hold for the right-pencil as well by symmetry .

Proposition 2. *Let L be the left-pencil of the trapezoid diagram. A vertex v is not a left dominator if and only if $(\exists r \in L)(r \prec v)$. Further, the left-pencil L can be found in $O(n)$ time.* The left-pencil structure provides an efficient way of checking whether a given vertex is a left dominator.

Lemma 2. *All left (right) dominators of the trapezoid graph can be identified in $O(n)$ time.*

After obtaining all left dominators V_L and right dominators V_R in $O(n)$ time, we compute the rightmost top (bottom) corners among all left dominators; let $b = \max\{b_i \mid v_i \in V_L\}, d = \max\{d_i \mid v_i \in V_L\}$. The computation of corners b and d clearly takes $O(n)$ time. Symmetrically, the leftmost top (bottom) corners among all right dominators, $a = \min\{a_i \mid v_i \in V_R\}, c = \min\{c_i \mid v_i \in V_R\}$ is computed in $O(n)$ time. For brevity, when the context is clear, we also use a, c (b, d) to denote the vertices v_i's whose a_i's, c_i's (b_i's, d_i's) values are minimized (maximized).

Let D denote the minimum cardinality connected dominating set. Recall that, $|D| = 1$ if and only if there is a vertex v being a left and right dominator at the same time; it can be easily checked within $O(n)$ time. Now consider the case that $|D| = 2$. In which case, we say that $D = \{u, v\}$ is a *dominating pair*. It is possible that both vertices of D are dominators, one left dominator and one right dominator. In which case, we must have $a < b$ or $c < d$; it follows that $D = \{a, b\}$ or $\{c, d\}$. However, there is another possibility for $|D| = 2$:

Proposition 3. *Let $D = \{u, v\}$ be a dominating pair of a trapezoid diagram T. If none of the left dominators of T intersects with any of the right dominators, i.e., $b < a$ and $d < c$, then neither u nor v is a dominator.*

Now consider the left-pencil $L = \langle r_{i_1}, r_{i_2}, \ldots, r_{i_g} \rangle$ and the right-pencil $R = \langle \ell_{j_1}, \ell_{j_2}, \ldots, \ell_{j_h} \rangle$. By Proposition 3 and 2, neither u nor v fully dominates L or R although $\{u, v\}$ collectively dominate both L and R. Thus, we have two types of coverage of L and R in terms of the shapes of intersection of trapezoids u and v.

Case 1: the *X-shaped* intersection. That is, we have $u \sim \{r_{i_{g'+1}}, \ldots, r_{i_g}\} \cup \{\ell_{j_{h'+1}}, \ldots, \ell_{j_h}\}$ and $v \sim \{r_{i_1}, \ldots, r_{i_{g'}}\} \cup \{\ell_{j_1}, \ldots, \ell_{j_h}\}$.

Case 2: the *8-shaped* intersection. That is, we have $u \sim \{r_{i_{g'+1}}, \ldots, r_{i_g}\} \cup \{\ell_{j_1}, \ldots, \ell_{j_{h'}}\}$ and $v \sim \{r_{i_1}, \ldots, r_{i_{g'}}\} \cup \{\ell_{j_{h'+1}}, \ldots, \ell_{j_h}\}$.

We can show that

Lemma 3. *The X-shaped (8-shaped) dominating pair of a trapezoid diagram can be found in $O(n)$ time.*

Theorem 1. *The minimum cardinality connected dominating set of a trapezoid diagram can be found in $O(n)$ time.*

ALGORITHM **MCDS**(T)

Input: A trapezoid diagram T with n trapezoids. Each trapezoid v_i is defined by four corner points a_i, b_i, c_i, d_i, $1 \leq i \leq n$. They are labeled with consecutive integer values $1, 2, 3, \ldots, 2n$; that is, $\cup_{i=1}^{n}\{a_i, b_i\} = \cup_{i=1}^{n}\{c_i, d_i\} = [1..2n]$.

Output: A minimum cardinality connected dominating set D.

Step 1: Identify all left dominators V_L and right dominators V_R.

Step 2: $|D| = 1$ if $V_L \cap V_R \neq \emptyset$; output any vertex $v \in V_L \cap V_R$ as the singleton dominating set and stop.

Step 3: Compute $b = \max\{b_i \mid v_i \in V_L\}$; $d = \max\{d_i \mid v_i \in V_L\}$; $a = \min\{a_i \mid v_i \in V_R\}$; $c = \min\{c_i \mid v_i \in V_R\}$.

Step 4: (* Deal with $|D| = 2$. *) Stop and output $D = \{a, b\}$ if $a < b$; stop output $\{c, d\}$ if $c < d$. Further, stop and output D if there is a dominating pair with X-shaped or 8-shaped intersection.

Step 5: (* Now, $|D| \geq 3$. *) Let $P_{ba}, P_{bc}, P_{da}, P_{dc}$ be the four shortest paths from $\{b, d\}$ to $\{a, c\}$. Among these four paths, output the one with the smallest cardinality.

END OF **MCDS**

Fig. 1. Finding the minimum cardinality connected dominating set in trapezoid graphs.

Proof. The algorithm is shown in Figure 1. Note that Step 1 of the algorithm takes $O(n)$ time by Lemma 2. Clearly Step 2 and Step 3 can be done in $O(n)$ time. Step 4 is the most complicated case of the algorithm; it has been shown in Lemma 3 that the dominating set can be correctly found in $O(n)$ time.

The rest of the proof appears in the complete paper. □

4   Minimum Weighted Connected Domination

A *minimal* connected dominating set is a connected dominating set such that the removal of any vertex leaves the resulting subset being no longer a connected dominating set. It is easily seen that a minimum weighted connected dominating set is minimal since the assigned weights are non-negative. It can be shown [11] that a minimal connected dominating set of a trapezoid graph is consisted of three parts: $S, P,$ and T; here S denotes the set of a *dominating source* (a left dominator or a source pair), T denotes the set of a *dominating target* (a right dominator or a sink pair), and P denotes a (lightest) chordless path from S to T. Note that the dominating source (target) can be a singleton or a pair of vertices.

We associate with each vertex v an *aggregated weight*, $w'(v)$, to be the sum of weights of the minimum connected dominating vertices from the source set to v, containing the vertex v as the dominating set. In other words,

$$w'(v) = \min\{\sum_{u \in P} w(u) \mid P : \text{a path from a left dominator (source pair) to } v\}$$

We first show that the source vertices can be initialized efficiently. The aggregated weight for a left dominator u is, by definition, $w'(u) = w(u)$. For a non-dominator source vertex v, it joins with another vertex x to form a minimum weighted source pair containing v. It is possible that the joint partner x is a dominator; the aggregated weight of v can be thus deduced from the dominator. The general scheme of the algorithm will deal with the situation properly. Otherwise, x is another non-dominator but together $\{v, x\}$ constitute a minimum weighted source pair. We call such kind of a source pair a *non-dominator source pair*.

Lemma 4. *The aggregated weights of all non-dominator source pairs of the trapezoid graph can be found in $O(n)$ time.*

Proof. The proof appears in the complete paper. □

Let two adjacent vertices $\{u, v\}$ form a sink pair. If neither u nor v is a right dominator, we call such kind of a sink pair a *non-dominator sink pair*. Note that the target vertices can either be a singleton right dominator or two vertices forming a non-dominator sink pair. For each non-dominator sink vertex v, there is a (possibly none existed) *lightest* non-dominator vertex x so that together $\{v, x\}$ form a non-dominator sink pair. We assign such a non-dominator sink vertex v a *modified weight* of v with a value $w(v) + w(x)$. We can show that

Lemma 5. *The modified weights of all non-dominator sink pairs of the trapezoid graph can be found in $O(n)$ time.*

Proof. The proof appears in the complete paper. □

We now present our method of calculating the aggregated weights of every vertex. The idea is to set up two priority queues, say Q_1 and Q_2, along with the two parallel lines, L_1 and L_2, on the trapezoid diagram. The (so far calculated) aggregated weights of the vertices are stored upon the priority queues. The algorithm will consider *both* of the priority queues and sweep along left (top or bottom) corners of trapezoids in a zigzag fashion. Note that we record the aggregated weights upon the *positions of right corners* of the trapezoids considered so far; the aggregated weights are kept according to the increasing order along the parallel lines. Since the input diagram of the trapezoid graph is given in a finite range of integers, we can use the data structure of van Emde Boas[12] to maintain a priority queue such that each operation of inserting, searching, and deleting an element in the priority queue can be done in $O(\log \log n)$ time. Thus, by carefully maintaining the priority queues and balancing the order of the visited vertices, we can show that

Lemma 6. *The aggregated weights of all vertices of the trapezoid graph can be found in $O(n \log \log n)$ time.*

Proof. We begin by inserting the aggregated weights of all the left dominators and non-dominator source pairs into the priority queues Q_1 and Q_2. Note that the top right corners of source vertices are inserted into Q_1, and the bottom right corners are inserted into Q_2. The aggregated weights are kept on both priority in increasing (weighted) order. Let the currently maintained list on $Q_1 = \langle v_{i_1}, v_{i_2}, \ldots, v_{i_k} \rangle$. It follows that $b_{i_1} < b_{i_2} < \cdots < b_{i_k}$ and $w'(v_{i_1}) < w'(v_{i_2}) < \cdots < w'(v_{i_k})$. Symmetrically, let $Q_2 = \langle v_{j_1}, v_{j_2}, \ldots, v_{j_k} \rangle$; we have $d_{j_1} < d_{j_2} < \cdots < d_{j_k}$ and $w'(v_{j_1}) < w'(v_{j_2}) < \cdots < w'(v_{j_k})$. In particular, assume that a vertex v with position $b(v) \in [v_{i_g}..v_{i_{g+1}}]$ is inserted into Q_1. In the following, we describe how to *join* a vertex v into Q_1: The vertex v is discarded if $w'(v) > w'(v_{i_{g+1}})$. Otherwise, we insert the vertex v into Q_1 and remove all vertices lied on the left of $b(v)$ with $w'(\cdot)$ larger than $w'(v)$. The operation of joining $d(v)$'s into Q_2 follows symmetrically.

After the source vertices have been properly joined into Q_1 and Q_2, let $Q_1 = \langle v_{i_1}, v_{i_2}, \ldots \rangle$ and $Q_2 = \langle v_{j_1}, v_{j_2}, \ldots \rangle$. The algorithm starts by checking the smaller value of $\{w'(v_{i_1}), w'(v_{j_1})\}$. Assume that $w'(v_{i_1}) \leq w'(v_{j_1})$; it follows that vertices v_i's with their top left corners satisfying $a_i < b_{i_1}$ now have their aggregated values decided. That is, $w'(v_i) = w(v_i) + w'(v_{i_1})$. The situation is illustrated in Figure 2. Note that the top right corner of the vertex v_i will then

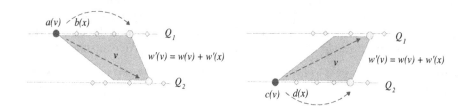

Fig. 2. Finding the aggregated weights of all vertices.

be joined into Q_1, and the bottom right corner of the vertex v_i will be joined into Q_2. We then discard the vertex v_{i_1} from Q_1. The case that $w'(v_{j_1}) < w'(v_{i_1})$ can be handle symmetrically.

It is readily verified that the algorithm described above gives the correct aggregated weights of all vertices, and the total efforts spent is at most $O(n \log \log n)$ time. The rest of the proof appears in the complete paper.   □

Let two adjacent vertices $\{u, v\}$ form a dominating pair of the trapezoid graph. There are three kinds of dominating pairs. The first case of a dominating pair is when one of the vertices is a left dominator and the other vertex is a right dominator. It is easily seen that the general scheme of the algorithm deal with the situation correctly. The second case of a dominating pair is when one of the vertices is a dominator, but the other vertex is *not* a dominator. Here we show how to find the minimum weighted connected dominating pair of such kind.

Lemma 7. *The minimum weighted connected dominating pair, forming by a dominator and a non-dominator, can be found in $O(n)$ time given the trapezoid diagram.*

Proof. The proof appears in the complete paper.   □

The third case of a dominating pair is when none of the vertices is a dominator. That is, we have a X-shaped or a 8-shaped dominating pair. The case considered here is actually the most complicated situation of the algorithm. We can also show that

Lemma 8. *The minimum weighted connected, X-shaped or 8-shaped, dominating pair of a trapezoid diagram can be found in $O(n \log \log n)$ time.*

Proof. First we show that the minimum weighted dominating pair with X-shaped intersection can be found in $O(n \log \log n)$ time. Let the left-pencil $L = \langle r_{i_1}, r_{i_2}, \ldots, r_{i_g} \rangle$ and the right-pencil $R = \langle \ell_{j_1}, \ell_{j_2}, \ldots, \ell_{j_h} \rangle$. Further, let the minimum weighted dominating pair be the set $\{v, x\}$. Without loss of generality we may assume that vertex v dominates the lower part of L and the upper part of R, while x dominates the upper part of L and the lower part of R. The idea here is that, while the bottom left corner of v located at interval $[d_{i_{k+1}}..d_{i_k}]$, the top left corner of x must be located before the position b_{i_k}; i.e., $a(x) \in [b_{i_1}..b_{i_k}]$. So, by pairing these $\{v, x\}$'s pair and checking their domination of the right-pencil, we can find the minimum, X-shaped, dominating pair.

Maintain two priority queues: Q_1 for the upper intervals of right-pencil and Q_2 for the lower intervals of the right-pencil. Note that Q_1 is initialized by inserting all upper positions of the right-pencil R. Scan each vertex x within $[b_{i_1}..b_{i_2}]$, the first upper interval of L. Recall the *join* operation we defined in Lemma 6. We will join the bottom right corners of x (as well as their corresponded weights $w(x)$'s) into the queue Q_2. Then, for each vertex v within the last lower interval of L, $[d_{i_1}..d_{i_2}]$, we *probe* the top right corners of v to seek its position in Q_1, say $b(v) \in [a_{j_k}..a_{j_{k+1}}]$. It follows that those x's whose bottom right corners greater than $c_{j_{k+1}}$ can be paired with v to become a dominating pair. To get the lightest joint vertex, we can *probe* $c_{j_{k+1}}$ into Q_2, *seeking to the right*, and obtain the desired minimum joint partner x. We maintain the sum of weights of the lightest dominating pair obtained so far; each time a joint partner is found, the sum of their weights is compared with it. The minimum weighted dominating pair can thus be found by this manner. The situation is illustrated in Figure 3.

We obtain the final minimum weighted dominating pair by successively joining the vertices cornered at $[b_{i_2}..b_{i_3}]$ into Q_2 and probing vertices cornered at $[d_{i_2}..d_{i_3}]$ into Q_1, and so on, until the final intervals $[b_{i_{g-1}}..b_{i_g}]$ and $[d_{i_{g-1}}..d_{i_g}]$ have been reached. Note that positions $d(x)$'s are *accumulatively* joined (with side effects) into Q_2 while positions $b(v)$'s are probed into Q_1 interleavedly with joining of $d(x)$'s. Since each operation to the priority queue takes $O(\log \log n)$

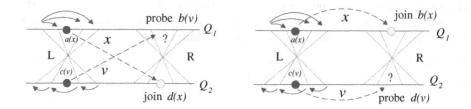

Fig. 3. Finding a X-shaped (left), or a 8-shaped (right) minimum weighted dominating pair.

time, and we spend only a constant number of queue operations for each vertex, it is not hard to see that the total efforts spent is at most $O(n \log \log n)$ time.

The case of 8-shaped dominating pair is similar to the case discussed above. The rest of proof appears in the complete paper.   □

We are now ready to show that

Theorem 2. *The minimum weighted connected dominating set of a trapezoid diagram can be found in* $O(n \log \log n)$ *time.*

ALGORITHM **WCDS**(T)
Input: A trapezoid diagram T with n trapezoids.
Output: A minimum weighted connected dominating set D.
Step 1: Find the left-pencil (right-pencil) and all left (right) dominators.
Step 2: Initialize the aggregated weights of left-dominators and non-dominator source pair.
Step 3: Find the non-dominator sink pairs, set their modified weights, and mark these right dominators and non-dominator sink pairs as the target vertices.
Step 4: Calculate the aggregated weights $w'(\cdot)$ of all the vertices along the two parallel lines. Scan through each right dominator and the modified sink pairs, and find the one with the minimum aggregated weight. Let the resulting aggregated weight of the dominating set be w_1.
Step 5: Compute the minimum dominating pair, forming by a dominator and a non-dominator. Let the resulting weight of the pair be w_2.
Step 6: Compute the minimum dominating pair with X-shaped or 8-shaped intersection. Let the resulting weight of the minimum dominating pair be w_3.
Step 7: The smallest value within the set $\{w_1, w_2, w_3\}$ is the desired weight of the minimum weighted connected dominating set. Output the corresponded dominating set D by tracing back the predecessor links.
END OF **WCDS**

Fig. 4. Finding the minimum weighted connected dominating set in trapezoid graphs.

Proof. The algorithm is shown in Figure 4. Note that Step 1 to Step 3 takes $O(n)$ time by Lemma 2, 4, and 5. Step 4, by Lemma 6, can be done in $O(n \log \log n)$ time. Step 5 can be done within $O(n)$ time by Lemma 7. Step 6 is the most complicated case of the algorithm; Lemma 8 shows that it takes $O(n \log \log n)$ time. The rest of the proof appears in the complete paper.   □

5   Concluding Remarks

Note that previous results [7, 11] solved the problems by checking all possible source and sink pairs; thus they had to spend $O(m+n)$ time for the unweighted case and $O(m + n \log n)$ time for the weighted case even when the trapezoid diagram was given. Note that the size of the trapezoid diagram is just $O(n)$, but these algorithms still spend potentially $\Omega(n^2)$ time when the size of edges is large.

Recently, Kohler [5] presents an $O(n)$ time algorithm for finding the minimum cardinality connected dominating set in trapezoid graphs. Using different approach, our algorithm exploits the *pencil* structure within the trapezoid diagram to find the minimum-sized connected dominating set in $O(n)$ time. For the weighted problem, our algorithm again uses the pencil structure as well as priority queues of finite integers [12] to find the minimum weighted connected dominating set in $O(n \log \log n)$ time.

References

[1] Andreas Brandstädt, Van Bang Le, and Jeremy P. Spinrad. *Graph Classes: A Survey*. SIAM monographs on discrete mathematics and applications, Philadelphia, P. A., 1999.

[2] M.-S. Chang. Weighted domination of cocomparability graphs. *Discr. Applied Math.*, 80:135–148, 1997.

[3] I. Dagan, M.C. Golumbic, and R.Y. Pinter. Trapezoid graphs and their coloring. *Discr. Applied Math.*, 21:35–46, 1988.

[4] T.W. Haynes, S.T. Hedetniemi, and P.J. Slater. *Fundamentals of Domination in Graphs*. Marcel Dekker, Inc., N. Y., 1998.

[5] E. Kohler. Connected domination and dominating clique in trapezoid graphs. *Discr. Applied Math.*, 99:91–110, 2000.

[6] D. Kratsch and L. K. Stewart. Domination on cocomparability graphs. *SIAM J. Discrete Math.*, 6(3):400–417, 1993.

[7] Y. D. Liang. Steiner set and connected domination in trapezoid graphs. *Information Processing Letters*, 56(2):101–108, 1995.

[8] Y. Daniel Liang. Dominations in trapezoid graphs. *Information Processing Letters*, 52(6):309–315, December 1994.

[9] Yaw-Ling Lin. Fast algorithms for independent domination and efficient domination in trapezoid graphs. In *ISAAC'98*, LNCS 1533, pages 267–276, Taejon, Korea, December 1998. Springer-Verlag.

[10] T.-H. Ma and J.P. Spinrad. On the 2-chain subgraph cover and related problems. *J. Algorithms*, 17:251–268, 1994.

[11] Anand Srinivasan, M.S. Chang, K. Madhukar, and C. Pandu Rangan. Efficient algorithms for the weighted domination problems on trapezoid graphs. Manuscript, 1996.

[12] P. van Emde Boas. Preserving order in a forest in less than logarithmic time and linear space. *Information Processing Letters*, 6:80–82, 1977.

Parameterized Complexity of Finding Subgraphs with Hereditary Properties[*]

Subhash Khot[1] and Venkatesh Raman[2]

[1] Department of Computer Science, Princeton University, New Jersey, USA
khot@cs.princeton.edu
[2] The Institute of Mathematical Sciences, Chennai-600113, INDIA
vraman@imsc.ernet.in

Abstract. We consider the parameterized complexity of the following problem under the framework introduced by Downey and Fellows[4]: Given a graph G, an integer parameter k and a non-trivial hereditary property Π, are there k vertices of G that induce a subgraph with property Π? This problem has been proved NP-hard by Lewis and Yannakakis[9]. We show that if Π includes all independent sets but not all cliques or vice versa, then the problem is hard for the parameterized class $W[1]$ and is fixed parameter tractable otherwise. In the former case, if the forbidden set of the property is finite, we show, in fact, that the problem is $W[1]$-complete (see [4] for definitions). Our proofs, both of the tractability as well as the hardness ones, involve clever use of Ramsey numbers.

1 Introduction

Many computational problems typically involve two parameters n and k, e.g. finding a vertex cover or a clique of size k in a graph G on n vertices. The parameter k contributes to the complexity of the problem in two qualitatively different ways. The parameterized versions of VERTEX COVER and UNDIRECTED FEEDBACK VERTEX SET problems can be solved in $O(f(k)n^\alpha)$ time where n is the input size, α is a constant independent of k and f is an arbitrary function of k (against a naive $\Theta(n^{ck})$ algorithm for some constant c). This "good behavior", which is extremely useful in practice for small values of k, is termed *fixed parameter tractability* in the theory introduced by Downey and Fellows[2,3,4].

On the other hand, for problems like CLIQUE and DOMINATING SET, the best known algorithms for the parameterized versions have complexity $\Theta(n^{ck})$ for some constant c. These problems are known to be hard for the parameterized complexity classes $W[1]$ and $W[2]$ respectively and are considered unlikely to be fixed parameter tractable (denoted by FPT) (see [4] for the definitions and more on the parameterized complexity theory). In this paper, we investigate the parameterized complexity of finding induced subgraphs of any non-trivial hereditary property in a given graph.

---

[*] Part of this work was done while the first author was at IIT Bombay and visited the Institute of Mathematical Sciences, Chennai

D.-Z. Du et al. (Eds.): COCOON 2000, LNCS 1858, pp. 137–147, 2000.
© Springer-Verlag Berlin Heidelberg 2000

A graph property Π is a collection of graphs. A graph property Π is non-trivial if it holds for at least one graph and does not include all graphs. A non-trivial graph property is said to be *hereditary* if a graph G is in property Π implies that every *induced subgraph* of G is also in Π. A graph property is said to be *interesting* [9] if the property is true (as well as false) for infinite families of graphs. Lewis and Yannakakis[9] (see also [6]) showed that if Π is a non-trivial and interesting hereditary property, then it is NP-hard to decide whether in a given graph, k vertices can be deleted to obtain a graph which satisfies Π.

For a hereditary property Π, let $\mathcal{F}$ be the family of graphs not having the property. The set of minimal members (minimal with respect to the operation of taking induced subgraphs) of $\mathcal{F}$ is called the forbidden set for the property Π. For example, the collection of all bipartite graphs is a hereditary property whose forbidden set consists of all odd cycles. Conversely, given any family $\mathcal{F}$ of graphs, we can define a hereditary property by declaring its forbidden set to be the set of all minimal members of $\mathcal{F}$.

Cai[1] has shown that the graph modification problem for a non-trivial hereditary property Π with a finite forbidden set is fixed parameter tractable (FPT). This problem includes the node deletion problem addressed by Lewis and Yannakakis (mentioned above). While the parameterized complexity of the question, when Π is a hereditary property with an infinite forbidden set, is open, we address the parametric dual problem in this paper. Given any property Π, let $P(G, k, \Pi)$ be the problem defined below.

Given: A simple undirected graph $G(V, E)$

Parameter: An integer $k \leq |V|$

Question: Is there a subset $V' \subseteq V$ with $|V'| = k$ such that the subgraph of G induced by V' has property Π ?

This problem is the same as '$|V| - k$' node deletion problem (i.e. can we remove all but k vertices of G to get a graph with property Π) and hence NP-hard. However the parameterized complexity of this problem doesn't follow from Cai's result. We prove that if Π includes all independent sets, but not all cliques or vice versa, then the problem $P(G, k, \Pi)$ is $W[1]$-complete when the forbidden set of Π is finite and $W[1]$-hard when the forbidden set is infinite. The proof is by a parametric reduction from the INDEPENDENT SET problem. If Π includes all independent sets and all cliques, or excludes some independent sets and some cliques then we show that the problem is fixed parameter tractable.

Note, from our and Cai's result, that the parameterized dual problems dealt with, have complimentary parameterized complexity. This phenomenon has been observed in a few other parameterized problems as well. In a graph $G(V, E)$, finding a vertex cover of size k is FPT whereas finding an independent set of size k (or a vertex cover of size $|V| - k$) is $W[1]$-complete; In a given boolean 3-CNF formula with m clauses, finding an assignment to the boolean variables that satisfies at least k clauses is FPT whereas finding an assignment that satisfies at least $(m - k)$ clauses (i.e. all but at most k clauses) is known to be $W[P]$-hard [2] (k is the parameter in both these problems). The k-IRREDUNDANT SET problem is $W[1]$-hard whereas CO-IRREDUNDANT set or $(n - k)$ IRREDUNDANT

SET problem is FPT [5]. Our result adds one other (general) problem to this list.

Throughout the paper, by a graph we mean an undirected graph with no loops or multiple edges. By a non-trivial graph, we mean a graph with at least one edge. Given a graph G and $A \subseteq V(G)$, by $I_G(A)$ we mean the subgraph of G induced by vertices in A. For two graphs H and G, we use the notation $H \subseteq G$ to mean that H is isomorphic to an induced subgraph of G. For the graph properties Π we will be concerned with in this paper, we assume that Π is recursive; i.e. given a graph G on n vertices, one can decide whether or not G has property Π in $f(n)$ time for some function of n.

We had already defined the notion of fixed parameter tractable problems. To understand the hardness result, we give below some definitions. See [4] for more details.

A parameterized language L is a subset of $\Sigma^* \times N$ where Σ is some finite alphabet and N is the set of all natural numbers. For $(x, k) \in L$, k is the parameter. We say that a parameterized problem A reduces to a parameterized problem B, if there is an algorithm Φ which transforms (x, k) into $(x', g(k))$ in time $f(k)|x|^\alpha$ where $f, g : N \to N$ are arbitrary functions and α is a constant independent of k, so that $(x, k) \in A$ if and only if $(x', g(k)) \in B$. The essential property of parametric reductions is that if A reduces to B and if B is FPT, then so is A.

Let F be a family of boolean circuits with *and*, *or* and *not* gates; We allow that F may have many different circuits with a given number of inputs. Let the weight of a boolean vector be the number of 1's in the vector. To F we associate the parameterized circuit problem $L_F = \{(C, k) : C \text{ accepts an input vector of weight } k\}$. Let the weft of a circuit be the maximum number of gates with fan-in more than two, on an input-output path in the circuit.

A parameterized problem L belongs to $W[t]$ if L reduces to the parameterized circuit problem $L_{F(t,h)}$ for the family $F(t, h)$ of boolean circuits with the weft of the circuits in the family bounded by t, and the depth of the circuits in the family bounded by a constant h. This naturally leads to a completeness program based on a hierarchy of parameterized problem classes:

$$FPT \subseteq W[1] \subseteq W[2] \subseteq \cdots$$

The parameterized analog of NP is $W[1]$, and $W[1]$-hardness is the basic evidence that a parameterized problem is unlikely to be FPT.

The next section deals with the hereditary properties for which the problem is fixed parameter tractable, and Section 3 proves the $W[1]$-hardness result for the remaining hereditary properties. Section 4 concludes with some remarks and open problems.

2   Hereditary Properties That Are FPT to Find

Lemma 1. *If a hereditary property Π includes all independent sets and all cliques, or excludes some independent sets as well as some cliques, then the problem $P(G, k, \Pi)$ is fixed parameter tractable.*

Proof. For any positive integers p and q, there exists a minimum number $R(p, q)$ (the Ramsey number) such that any graph on at least $R(p, q)$ vertices contains either a clique of size p or an independent set of size q. It is well-known that $R(p, q) \leq \binom{p+q-2}{q-1}$ [8].

Assume that Π includes all cliques and independent sets. For any graph G with $|V(G)| \geq R(k, k)$, G contains either a clique of size k or an independent set of size k. Since all independent sets and all cliques have property Π, the answer to the problem $P(G, k, \Pi)$ in this case is "yes".

When $|V(G)| \leq R(k, k)$, we can use brute force by picking all k-elements subsets of $V(G)$ and checking whether the induced subgraph on the subset has property Π. This will take time $\binom{R(k,k)}{k} f(k)$ where $f(k)$ is the time to decide whether a given graph on k vertices has property Π. Thus the problem $P(G, k, \Pi)$ is fixed parameter tractable.

If Π excludes some cliques and some independent sets, let s and t respectively be the sizes of the smallest clique and independent set which do not have property Π. Since any graph with at least $R(s, t)$ vertices has either a clique of size s or an independent set of size t, no graph with at least $R(s, t)$ vertices can have property Π (since Π is hereditary). Hence any graph in Π has at most $R(s, t)$ vertices and hence Π contains only finitely many graphs. So if $k > R(s, t)$, then the answer to the $P(G, k, \Pi)$ problem is NO for any graph G. If $k \leq R(s, t)$, then check, for each k subset of the given vertex set, whether the induced subgraph on the subset has property Π. This will take time $\binom{n}{k} f(k) \leq Cn^{R(s,t)}$ for an n vertex graph, where C is time taken to check whether a graph of size at most $R(s, t)$ has property Π. Since s and t depend only on the property Π, and not on k or n, and $k \leq R(s, t)$, the problem $P(G, k, \Pi)$ is fixed parameter tractable in this case also.   □

We list below a number of hereditary properties Π (dealt with in [11]) each of which includes all independent sets and cliques, and hence for which the problem $P(G, k, \Pi)$ is fixed parameter tractable.

Corollary 1. *Given any simple undirected graph G, and an integer k, it is fixed parameter tractable to decide whether there is a set of k vertices in G that induces (a) a perfect graph, (b) an interval graph (c) a chordal graph, (d) a split graph, (e) an asteroidal triple free (AT-free) graph, (f) a comparability graph, or (g) a permutation graph. (See [11] or [7] for the definitions of these graphs.)*

3   Hereditary Properties That Are W-Hard to Find

In this section, we show that the problem $P(G, k, \Pi)$ is $W[1]$-hard if Π includes all independent sets but not all cliques or vice versa.

For a graph G, let $\overline{G}$ denote the edge complement of G. For a property Π, let $\overline{\Pi} = \{\overline{G} \mid G \text{ has property } \Pi\}$. We note that Π is hereditary if and only if $\overline{\Pi}$ is hereditary, and Π includes all independent sets but not all cliques if and only if $\overline{\Pi}$ includes all cliques but not all independent sets. Thus it suffices to prove $W[1]$-hardness when Π includes all independent sets, but not all cliques.

First we will show that, the problem is in $W[1]$ if the forbidden set for Π is finite.

Lemma 2. *Let Π be a non-trivial hereditary property having a finite forbidden set $\mathcal{F} = \{H_1, H_2, \ldots, H_s\}$. Then the problem $P(G, k, \Pi)$ is in $W[1]$.*

Proof. Let $\nu = max(|H_i|)$. Let $A_1, \ldots, A_q$ be all the subsets of $V(G)$ such that for every $1 \leq j \leq q$, $I_G(A_j)$ is isomorphic to some H_i. The sets A_j's can be determined in $O(f(\nu)n^\nu)$ time by trying every subset of $V(G)$ of size at most ν. Here f is some function of ν alone.

Consider the boolean formula

$$\bigwedge_{j=1}^{q} \left(\bigvee_{u \in A_j} \overline{x_u} \right)$$

If this formula has a satisfying assignment with weight (the number of true variables in the assignment) k, then the subset X of $V(G)$ defined by $X = \{u \in V \mid x_u = 1\}$ is a subset of $V(G)$ of cardinality k such that $A_j \not\subseteq X$ for any j. This implies that G has an induced subgraph of size k with property Π.

Conversely if G has an induced subgraph of size k with property Π, then setting $x_u = 1$ for vertices u in the induced subgraph gives a satisfying assignment with weight k. Since $q \leq n^{\nu+1}$ and $|A_i| \leq \nu$ (a constant) for all i, it follows from the definitions that the problem is in $W[1]$. □

We now show that the problem is $W[1]$-hard when one of the graphs in the forbidden set of Π is a complete bipartite graph.

Lemma 3. *Let Π be a hereditary property that includes all independent sets but not all cliques, having a finite forbidden set $\mathcal{F} = \{H_1, H_2, \ldots, H_s\}$. Assume that some H_i, say H_1 is a complete bipartite graph. Then the problem $P(G, k, \Pi)$ is $W[1]$-complete.*

Proof. In Lemma 2 we have shown that the problem is in $W[1]$.

Let Π be as specified in the Lemma. Let $t = max(|V_1|, |V_2|)$ where $V_1 \bigcup V_2$ is bipartition of H_1. If $t = 1$, $H_1 = K_2$, and the given problem P is identical to the k-independent set problem, hence $W[1]$-hard. So assume $t \geq 2$. Note that $H_1 \subseteq K_{t,t}$. Let H_s be the clique of smallest size that is not in Π, hence in the forbidden set $\mathcal{F}$.

Now we will show that the problem is $W[1]$-hard by a reduction from the Independent Set Problem. Let G_1 be a graph in which we are interested in finding an independent set of size k_1. For every vertex $u \in G_1$ we take r independent vertices (r to be specified later) $u^1, \ldots, u^r$ in G. If (u, v) is an edge in G_1, we add all r^2 edges (u^i, v^j) in G. G has no other edges.

We claim that that G_1 has an independent set of size k_1 if and only if G has rk_1 vertices that induce a subgraph with property Π.

Suppose G_1 has an independent set $\{u_i | 1 \leq i \leq k_1\}$ of size k_1. Then the set of rk_1 vertices $\{u_i^j \mid 1 \leq i \leq k_1, 1 \leq j \leq r\}$ is an independent set in G and hence has property Π.

Conversely let S be a set of rk_1 vertices in G which induces a subgraph with property Π. This means $I_G(S)$ does not contain any H_i, in particular it does not contain H_1. Group the rk_1 vertices according to whether they correspond to the same vertex in G_1 or not. Let $X_1, \ldots, X_h, Y_1, \ldots, Y_p$ be the groups and $u_1, \ldots, u_h, v_1, \ldots, v_p$ be the corresponding vertices in G_1 such that $|X_i| \geq t \, \forall \, i$ and $|Y_j| < t \, \forall \, j$. Observe that $\{u_1, \ldots, u_h\}$ must be independent in G_1 because if we have an edge (u_i, u_j), $H_1 \subseteq K_{t,t} \subseteq I_G(X_i, X_j) \subseteq I_G(S)$, a contradiction. If $h \geq k_1$ we have found an independent set of size at least k_1 in G_1. Therefore assume that $h \leq k_1 - 1$. Then $\sum_{i=1}^h |X_i| \leq r(k_1 - 1)$ which implies that $\sum_{j=1}^p |Y_j| \geq r$ or $p \geq r/(t-1)$. Since vertices in distinct groups (one vertex per group) in G and the corresponding vertices in G_1 induce isomorphic subgraphs, the vertices $v_1, \ldots, v_p$ induce a subgraph of G_1 with property Π (since Π is hereditary). Since this subgraph has property Π, it does not contain H_s as an induced subgraph. We choose r large enough so that any graph on $r/(t-1)$ vertices that does not contain a clique of size $|H_s|$ has an independent set of size k_1. With this choice of r, it follows that G_1 does contain an independent set of size k_1. The number r depends only on $|H_s|$ and the parameter k_1 and not on $n_1 = |V(G_1)|$. So the reduction is achieved in $O(f(k_1)n_1^\alpha)$ time where f is some function of k_1 and α is some fixed constant independent of k_1. □

Next, we will show that the problem is $W[1]$-hard even if none of the graphs in the forbidden set is complete-bipartite.

Theorem 1. *Let Π be a hereditary property that includes all independent sets but not all cliques, having a finite forbidden set $\mathcal{F} = \{H_1, H_2, \ldots, H_s\}$. Then the problem $P(G, k, \Pi)$ is $W[1]$-complete.*

Proof. The fact that the problem is in $W[1]$ has already been proved in Lemma 2. Assume that none of the graphs H_i in the forbidden set of Π is complete-bipartite. Let H_s be the clique of smallest size that is not in Π, hence in the forbidden set $\mathcal{F}$.

For a graph H_i in $\mathcal{F}$, select (if possible) a subset of vertices Z such that the vertices in Z are independent and every vertex in Z is connected to every vertex in $H_i \setminus Z$. Let $\{H_{ij} | 1 \leq j \leq s_i\}$ be the set of graphs obtained from H_i by removing such a set Z for every possible choice of Z. Since H_i is not complete-bipartite, every H_{ij} is a non-trivial graph. Let $\mathcal{F}_1 = \mathcal{F} \bigcup \{H_{ij} | 1 \leq i \leq s, 1 \leq j \leq s_i\}$. Note that $\mathcal{F}_1$ contains a clique of size $|H_s| - 1$ because a set Z, consisting of a single vertex, can be removed from the clique H_s. Let Π_1 be the hereditary property defined by the forbidden set $\mathcal{F}_1$. Observe that Π_1 also includes all independent sets but not all cliques. Let P_1 be the problem $P(G_1, k_1, \Pi_1)$.

We will prove that P_1 is $W[1]$-hard later. Now, we will reduce P_1 to the problem $P(G, k, \Pi)$ at hand.

Given G_1, we construct a graph G as follows. Let $V(G) = V(G_1) \bigcup D$ where D is a set of r independent vertices (r to be specified later). Every vertex in $V(G_1)$ is connected to every vertex in D. Let $\nu = max_i(|H_i|)$.

We claim that G_1 has an induced subgraph of size k_1 with property Π_1 if and only if G has $k_1 + r$ vertices that induce a subgraph with property Π.

Let A be a subset of $V(G_1)$, $|A| = k_1$ such that $I_{G_1}(A) \in \Pi_1$. Let $S = A \bigcup D$. Suppose on the contrary that $I_G(S)$ contains some H_i as a subgraph. If this H_i contains some vertices from D, we throw away these independent vertices. The remaining portion of H_i, which is some $H_{ij}, 1 \leq j \leq s_i$, must lie in $I_G(A)$. But this is a contradiction because $I_G(A) = I_{G_1}(A)$ and by hypothesis, $I_{G_1}(A)$ has property Π_1 and it cannot contain any H_{ij}. Similarly H_i cannot lie entirely in $I_G(A)$ because $\mathcal{F} \subseteq \mathcal{F}_1$, so $I_G(A)$ does not contain any H_i as induced subgraph. Therefore $I_G(S)$ does not contain any H_i, hence it has property Π and $|S| = k_1 + r$.

Conversely, suppose we can choose a set S, $|S| = k_1 + r$ such that $I_G(S)$ does not contain any H_i. Since $|D| = r$ we must choose at least k_1 vertices from $V(G_1)$. Let $A \subseteq S \bigcap V(G_1)$ with cardinality k_1. If $I_G(A)$ does not contain any H_{ij}, we are through. Otherwise let $H_{i_0 j_0} \subseteq I_G(A)$ for some i_0, j_0. Now $H_{i_0 j_0}$ is obtained from H_{i_0} by deleting an independent set of size at most ν. Hence S can contain at most $\nu - 1$ vertices from D, otherwise we could add sufficient number of vertices from D to the graph $H_{i_0 j_0}$ to get a copy of H_{i_0} which is not possible. Hence $|S \bigcap D| < \nu$ which implies that $|S \bigcap V(G_1)| > k_1 + r - \nu$. Thus $I_{G_1}(S \bigcap V(G_1))$ is an induced subgraph of G_1 of size at least $k_1 + r - \nu$ that does not contain any H_i, in particular it does not contain H_s which is a clique of size say μ. We can select r (as before, by Ramsey Theorem) such that any graph on $k_1 + r - \nu$ vertices that does not contain a μ-clique has an independent set of size k_1. Hence G_1 has an independent set of size k_1 which has property Π_1. The number r depends only on the family $\mathcal{F}$ and parameter k_1 and not on $n_1 = |V(G_1)|$. So the reduction is achieved in $O(g(k_1)n_1{}^\beta)$ time where g is some function of k_1 and β is a constant. Also $|V(G)| = |V(G_1)| + r$, so the size of the input problem increases only by a constant.

We will be through provided the problem P_1 is $W[1]$−hard. If any of the H_{ij} is complete-bipartite, then it follows from Lemma 3. Otherwise, we repeatedly apply the construction, given at the beginning of the proof, of removing set Z of vertices from each graph in the forbidden set, to get families $\mathcal{F}_2, \mathcal{F}_3, \ldots$ and corresponding problems $P_2, P_3, \ldots$ such that there is a parametric reduction from P_{m+1} to to P_m. Since $\mathcal{F}_{m+1}$ contains a smaller clique than a clique in $\mathcal{F}_m$, eventually some family $\mathcal{F}_{m_0}$ contains a clique of size 2 (the graph K_2) or a complete-bipartite graph. In the former case, the problem P_{m_0} is same as parameterized independent set problem, so $W[1]$-hard. In the latter case P_{m_0} is $W[1]$-hard by Lemma 3. $\square$

We can extend Theorem 1 to the case when the forbidden set is infinite. However we could prove the problem only $W[1]$-hard; we don't know the precise class in the W-hierarchy the problem belongs to.

Theorem 2. *Let Π be a hereditary property that includes all independent sets but not all cliques (or vice versa). Then the problem $P(G, k, \Pi)$ is $W[1]$-hard.*

Proof. Every hereditary property is defined by a (possibly infinite) forbidden set [9] and so let the forbidden family for Π be $\mathcal{F} = \{H_1, H_2, \ldots\}$. The proof is almost the same as in Theorem 1. Note that Lemma 3 does not depend on finiteness of the forbidden family. Also the only point where the finiteness of $\mathcal{F}$ is used in Theorem 1 is in the argument that if $I_G(A)$ does contain some $H_{i_0 j_0}$ then S can contain at most $\nu - 1$ vertices from the set D. This argument can be modified as follows. Since $I_G(A)$ contains some $H_{i_0 j_0}$, $|V(H_{i_0 j_0})| \leq |A| = k_1$. Also $H_{i_0 j_0}$ is obtained from some H_i by removing an independent set adjacent to all other vertices of H_i. (If there are more than one such H_is from which $H_{i_0 j_0}$ is obtained, we choose an arbitrary H_i.) Let $\nu_1 = max \; (|V(H_i)| - |V(H_{ij})|)$ where the maximum is taken over all H_{ij} such that $|V(H_{ij})| \leq k_1$. Hence if $I_G(A)$ does contain some $H_{i_0 j_0}$, we can add at most ν_1 vertices from D to get H_{i_0}. So S must contain less than ν_1 vertices from D. The choice of r will have to be modified accordingly. $\square$

Corollary 2 follows from Theorem 2 since the collection of forests is a hereditary property with the forbidden set as the set of all cycles. This collection includes all independent sets and does not include any clique of size ≥ 3.

Corollary 2. *The following problem is $W[1]$-hard:*
Given (G, k), does G have k vertices that induce a forest?

This problem is the parametric dual of the UNDIRECTED FEEDBACK VERTEX SET problem which is known to be fixed parameter tractable [4].

Corollary 3. *Following problem is $W[1]$-complete:*
Given (G, k), does there exist an induced subgraph of G with k vertices that is bipartite ?

Proof. Hardness follows from Theorem 2 since all independent sets are bipartite and no clique of size at least 3 is bipartite.

To show that the problem is in $W[1]$, given the graph G, consider the boolean formula

$$\bigwedge_{u \in V(G)} (\overline{x_u} \vee \overline{y_u}) \; \bigwedge_{(u,v) \in E(G)} ((\overline{x_u} \vee \overline{x_v}) \bigwedge (\overline{y_u} \vee \overline{y_v}))$$

We claim that G has an induced bipartite subgraph of size k if and only if the above formula has a satisfying assignment with weight k. Suppose G has an induced bipartite subgraph with k vertices with partition V_1 and V_2. Now for each vertex in V_1 assign $x_u = 1, y_u = 0$, for each vertex in V_2 assign $x_u = 0, y_u = 1$ and assign $x_u = y_u = 0$ for the remaining vertices. It is easy to see that this assignment is a weight k satisfying assignment for the above formula.

Conversely, if the above formula has a weight k satisfying assignment, the vertices u such that $x_u = 1, y_u = 0$ or $x_u = 0, y_u = 1$ induce a bipartite subgraph of G with k vertices.

The corollary follows as the above formula can be simulated by a $W[1]$-circuit (i.e. a circuit with bounded depth, and weft 1). $\square$

Corollary 4 can be proved along similar lines of Corollary 3.

Corollary 4. *Following problem is $W[1]$-complete: Given (G, k) and a constant l, does there exist an l-colorable induced subgraph of size k?*

Finally we address the parametric dual of the problem addressed in Corollary 3. Given a graph G, and an integer k, are there k vertices in G whose removal makes the graph bipartite? We will call this problem '$n - k$ bipartite'.

The precise parameterized complexity of this problem is unknown since though bipartiteness is a hereditary property, it has an infinite forbidden set, and so the problem is not covered by Cai's result [1].

The 'edge' counterpart of the problem, given a graph G with m edges, and an integer k, are there k edges whose removal makes the graph bipartite, is the same as asking for a cut in the graph of size $m - k$. It is known[10] that there exists a parameterized reduction from this problem to the following problem, which we call 'all but k 2-SAT'.

Given: A Boolean 2 CNF formula F
Parameter: An integer k
Question: Is there an assignment to the variables of F that satisfies all but at most k clauses of F?

We show that there is also a parameterized reduction from the '$n-k$ bipartite problem' to the 'all but k 2-SAT' problem.

Theorem 3. *There is a parameterized reduction from the '$n - k$ bipartite problem' to the 'all but k 2-SAT' problem.*

Proof. Given a graph G, for every vertex, we set two variables (x_u, y_u) and construct clauses in the same manner as in the proof of Corollary 3. The clauses are as follows:

Set 1:

$$\overline{x_u} \vee \overline{y_u} \quad \forall \, u \in V(G)$$
$$\overline{x_u} \vee \overline{x_v} \; ; \; \overline{y_u} \vee \overline{y_v} \quad \forall (u, v) \in E(G)$$

Each clause in Set 1 is repeated $k + 1$ times.

Set 2: $x_u \vee y_u \quad \forall \, u \in V(G)$.

We show that it is possible to remove k vertices to make the given graph bipartite if and only if there is an assignment to the variables in the above formula that makes all but at most k clauses true.

If there is an assignment that makes all but at most k clauses true, then the clauses in Set 1 must be true because each of them occurs $k + 1$ times. This ensures that the variables x_u, y_u corresponding to the vertices are assigned respectively 0,0 or 0,1 or 1,0 and each edge $e = (u, v)$ has $x_u = x_v = y_u = y_v = 0$ or $x_u = 0, y_u = 1$ and $x_v = 1, y_u = 0$ or vice versa. The vertices s for which

$x_s = y_s = 0$ are removed to get a bipartite graph. At most k clauses in Set 2 are false. This ensures that at most k vertices are removed.

Conversely if there exist k vertices whose removal results in a bipartite graph with partition $V_1 \bigcup V_2$, consider the assignment corresponding to each vertex u in the graph, $x_u = y_u = 0$ if the vertex u is removed, $x_u = 1, y_u = 0$ if $u \in V_1$ and $x_u = 0, y_u = 1$ if $u \in V_2$.

It is easy to see that this assignment makes all but at most k clauses of the formula true.

Note that the reduction is actually a polynomial time reduction.   □

4   Concluding Remarks

We have characterized the hereditary properties for which finding an induced subgraph with k vertices having the property in a given graph is $W[1]$- hard. In particular, using Ramsey Theorem, we have shown that if the property includes all independent sets and all cliques or if it excludes some independent sets as well as cliques, then the problem is fixed parameter tractable. However, for some of these specific properties, we believe that a more efficient fixed parameter algorithms (not based on Ramsey numbers) is possible.

It remains an open problem to determine the parameterized complexity of both the problems stated in Theorem 3 (the '$n - k$ bipartite problem' and the 'all but k 2-SAT' problem). More generally, the parameterized complexity of the node-deletion problem for a hereditary property with an infinite forbidden set is open.

Finally we remark that our results prove that the parametric dual of the problem considered by Cai[1] (and was proved FPT) is $W[1]$-hard. This observation adds weight to the conjecture (first made in [10]) that typically parametric dual problems have complimentary parameterized complexity. It would be interesting to explore this in a more general setting.

References

1. L. Cai, Fixed Parameter Tractability of Graph Modification Problem for Hereditary Properties, *Information Processing Letters*, **58** (1996), 171-176.
2. R.G. Downey, M.R. Fellows, Fixed Parameter Tractability and Completeness   I: Basic theory, *SIAM Journal on Computing* **24** (1995) 873-921.
3. R.G. Downey, M.R. Fellows, Fixed Parameter Tractability and Completeness   II: Completeness for W[1], *Theoretical Computer Science* **141** (1995) 109-131.
4. R.G. Downey, M.R. Fellows, *Parameterized Complexity*, Springer Verlag New York, 1999.
5. R.G. Downey, M.R. Fellows, V. Raman, The Complexity of Irredundant sets Parameterized by Size, *Discrete Applied Mathematics* **100** (2000) 155-167.
6. M. R. Garey and D. S. Johnson, Problem [GT21], *Computers and Intractability, A Guide to The Theory of NP-Completeness*, Freeman and Company, New York, 1979.
7. M. C. Golumbic, *Algorithmic Graph Theory and Perfect Graphs*. Academic Press, New york, 1980.

8. Harary, *Graph Theory*, Addison-Wesley Publishing Company, 1969.
9. J. M. Lewis and M. Yannakakis, The Node-Deletion Problem for Hereditary Properties is NP-complete, *Journal of Computer and System Sciences* **20(2)** (1980) 219-230.
10. M. Mahajan and V. Raman, Parameterizing above Guaranteed Values: MaxSat and MaxCut, *Journal of Algorithms* **31** (1999) 335-354.
11. A. Natanzon, R. Shamir and R. Sharan, Complexity Classification of Some Edge Modification Problems, in *Proceedings of the 25th Workshop on Graph-Theoretic Concepts in Computer Science* (WG'99), Ascona, Switzerland (1999); Springer Verlag Lecture Notes in Computer Science.

Some Results on Tries with Adaptive Branching

Yuriy A. Reznik[1]

RealNetworks, Inc., 2601 Elliott Avenue, Seattle, WA 98121

Yreznik@real.com

Abstract. We study a modification of *digital trees* (or *tries*) with *adaptive multi-digit branching*. Such *tries* can dynamically adjust degrees of their nodes by choosing the number of digits to be processed per each lookup. While we do not specify any particular method for selecting the degrees of nodes, we assume that such selection can be accomplished by examining the number of strings remaining in each sub-tree, and/or estimating parameters of the input distribution. We call this class of digital trees *adaptive multi-digit tries* (or *AMD-tries*) and provide a preliminary analysis of their expected behavior in a memoryless model. We establish the following results: 1) there exist *AMD-tries* attaining a constant (O(1)) expected time of a successful search; 2) there exist *AMD-tries* consuming a linear (*O(n)*, *n* is the number of strings inserted) amount of space; 3) both constant search time and linear space usage can be attained if the (memoryless) source is symmetric. We accompany our analysis with a brief survey of several known types of adaptive trie structures, and show how our analysis extends (and/or complements) the previous results.

1   Introduction

Digital trees (also known as *radix search trees*, or *tries*) represent a convenient way of organizing alphanumeric sequences (strings) of variable lengths that facilitates their fast retrieving, searching, and sorting (cf. [11,17]). If we designate a set of n distinct strings as $S = \{s_1,...,s_n\}$, and assume that each string is a sequence of symbols from a finite alphabet $\Sigma = \{\alpha_1,...,\alpha_v\}$, $|\Sigma| = v$, then a *trie* $T(S)$ over S can be constructed recursively as follows. If $n = 0$ ($n = |S|$), the *trie* is an empty external node. If $n = 1$ (i.e. S has only one string), the *trie* is an external node containing a pointer to this single string in S. If $n > 1$, the *trie* is an internal node containing v pointers to the child *tries*: $T(S_1),...,T(S_v)$, where each set S_i ($1 \le i \le v$) contains suffixes of all strings from S that begin with a corresponding first symbol. For example, if a string $s_j = u_j w_j$ (u_j is a first symbol, and w_j is a string containing the remaining symbols from s_j), and $u_j = \alpha_i$, then the string w_j will go into S_i. Thus, after all child *tries* $T(S_1),...,T(S_v)$ are recursively processed, we arrive at a tree-like data structure, where the original strings $S = \{s_1,...,s_n\}$ can be uniquely identified by the *paths* from the root node to non-empty external nodes (cf. Fig 1.a).

---

[1] The author is also on leave from the Institute of Mathematical Machines and Systems of National Academy of Sciences of Ukraine.

D.-Z. Du et al. (Eds.): COCOON 2000, LNCS 1858, pp. 148-158, 2000.
© Springer-Verlag Berlin Heidelberg 2000

A simple extension of the above structure is obtained by allowing more than one symbol of input alphabet to be used for branching. Thus, if a *trie* uses some constant number of symbols $r \geq 1$ per lookup, then the effective branching coefficient is v^r, and we essentially deal with a *trie* built over an alphabet Σ^r (cf. Fig.1.c). To underscore the fact that branching is implemented using multiple input symbols (digits) at a time, such *tries* are sometimes called *multi-digit tries* (cf. [2]).

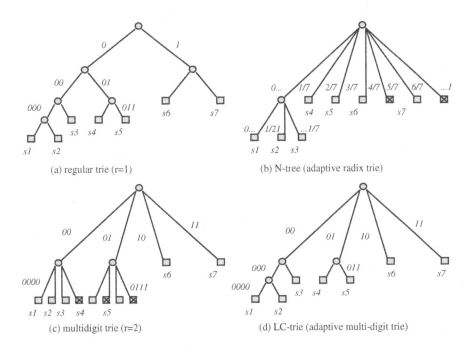

Fig. 1. Examples of tries built from 7 binary strings: $s1=0000...$, $s2=0001...$, $s3=0010...$, $s4=0100...$, $s5=0110...$, $s6=10...$, $s7=11...$.

The behavior of regular *tries* is thoroughly analyzed (cf. [5, 9, 10, 11, 16, 19]). For example, it has been shown (cf. [16]), that the expected number of nodes examined during a successful search in a v-ary *trie* is asymptotically $O(\log n/h)$, where h is the entropy of a process used to produce n input strings. The expected size of such *trie* is asymptotically $O(nv/h)$. These estimates are known to be correct for a rather large class of stochastic processes, including *memoryless*, *Markovian*, and ψ-*mixed* models (cf. [16, 2]). In many special cases (e.g. when a source is *memoryless*, or *Markovian*) the complete characterizations (expectation, variance and higher moments) of these parameters have been obtained, and their precise (up to $O(1)$ term) asymptotic expansions have been deducted (cf. [19, 10]).

Much less known are modifications of *tries* that use *adaptive branching*. That is, instead of using nodes of some fixed degree (e.g., matching the cardinality of an input alphabet), *adaptive tries* select branching factor dynamically, from one node to another. Perhaps the best-known example of this idea is *sorting by distributive partitioning*, due to Dobosiewicz (cf. [6]). This algorithm (also known as an *N-tree* (cf. Ehrlich [8]) selects the degrees of nodes to be equal exactly the number of strings inserted in the *sub-tries* they originate. For example, the *N-tree* displayed on Fig.1.b, contains seven strings overall, and therefore its root node has seven branches. In turn, its first branch receives three strings to be inserted, and thus the *N-tree* creates a node with three additional branches, and so on.

While *N-trees* have extremely appealing theoretical properties (thus, according to Tamminen [20], they attain a constant ($O(1)$) expected search time, and use a linear $O(n)$ amount of memory), there are several important factors that limit their practical usage. The main problem is that the *N-tree* is not a dynamic data structure. It is more or less suitable for a multi-pass construction when all n strings are given, but an addition or removal of a string in an existing structure is rather problematic. In the worst case, such an operation may involve the reconstruction of the entire tree, making the cost of its maintenance extremely high. Somewhat more flexible is a *B-b* parametrized version of the *distributive partitioning* proposed in [7]. This algorithm selects the branching factors to be equal n/b ($b \geq 1$), and split child nodes only when the number of strings there becomes larger than B. When both B and b equal one, we have a normal *N-tree*, however, when they are large, the complexity of updates can be substantially reduced.

Unfortunately, this does not help with another problem in adaptive *tries*. Since the degrees of nodes are being selected dynamically, *N-trees* cannot use simple per-symbol lookups. Instead, they must implement dynamic conversions from one size alphabet to another, or even treat the input strings as real numbers in $[0,1)$ (cf. [6, 8]).

In either scenario, the transition between levels in *N-trees* (and their *B-b* variants) is much slower than one in regular *tries*, and combined with the complexity of the maintenance, it creates serious constraints for the usage of *N-trees* in practice.

In this paper, we focus on another implementation of adaptive branching that promises to be (at least partially) free from the above mentioned shortcomings. We are trying to create an *adaptive* version of *multi-digit tries* by allowing the number of digits processed by their nodes (parameter r) to be selected dynamically. Due to the luck of the standard name, we will call these structures *adaptive multi-digit tries* (or *AMD-tries*). To cover a wide range of possible implementations of such *tries*, we (at least initially) do not specify any particular algorithm for selecting the depths (numbers of digits) of their nodes. Instead, we assume that such selection can be accomplished by examining the number of strings remaining in each sub-tree, and estimating parameters of the input distribution, and attempt to study the resulting *class* of data structures. The goal of this research is to find performance bounds (in both search time and space domains) attainable by *AMD-tries*, and deduct several particular implementations, that can be of interest in practice.

Somewhat surprising, there were only few attempts to explore the potential of the *adaptive multi-digit* branching in the past. Perhaps the only studied implementation in this class is a *level-compressed trie* (or *LC- trie*) by Andersson and Nilsson (cf. [2]). The node depth selection heuristic in a *LC-trie* is actually very simple; it combines the (v-ary) levels of the corresponding regular *trie* until it reaches the first external node (cf. Fig 1.c). It has been shown (cf. [2, 15]), that in a *memoryless model* this algorithm produces nodes with depths $\bar{r} = Er = O(\log k - \log\log k)$, where k is the number of strings in a sub-trie originated by the node. When source is symmetric, the expected search time in a *LC-trie* is only $O(\log^* n)$, however, it grows as $O(\log\log n)$ in the asymmetric case (cf. [15]).

In our analysis, we extend the above results, and in particular, we show that there exist implementations of *AMD-tries* attaining the constant ($O(1)$) complexity of a successful search in both symmetric and asymmetric cases. Moreover, if source distribution is symmetric, such *tries* can be implemented in linear ($O(n)$) amount of space. Compared to an *N-tree*, an equally fast *AMD-trie* appears to have a larger memory usage, however, it is a much more suitable scheme for dynamic implementation, and combined with the benefits of fast multi-digit processing, it promises to be a structure of choice for many practical situations.

This paper is organized as follows. In the next section, we will provide some basic definitions and present our main results. The proofs are delayed until Section 3, where we also derive few recurrent expressions necessary for the analysis of these algorithms. Finally, in our concluding remarks, we emphasize the limitations of our present analysis and show several possible directions for future research.

2   Definitions and Main Results

In our analysis, we only consider *AMD-tries* built over strings from a *binary* alphabet $\Sigma = \{0,1\}$, but the extension to any finite alphabet is straightforward. We also limit our study to a situation when n strings to be inserted in a *trie* are generated by a *memoryless* (or *Bernoulli*) source (cf. [3]). In this model, symbols of the alphabet Σ occur independently of one another, so that if x_j - is the j-th symbol produced by this source, then for any j: $\Pr\{x_j = 0\} = p$, and $\Pr\{x_j = 1\} = q = 1 - p$. If $p = q = 1/2$, such source is called *symmetric*, otherwise it is *asymmetric* (or *biased*). In our analysis, we will also use the following additional parameters of memoryless sources:

$$h = -p\log p - q\log q; \qquad \eta_\infty = -\log p_{\min}; \qquad (1)$$
$$h_2 = p\log^2 p + q\log^2 q; \qquad \eta_{-\infty} = -\log p_{\max};$$

where: h is the (Shannon's) entropy of the source (cf. [3]), η_∞ and $\eta_{-\infty}$ are special cases of the Rényi's k-order entropy (cf. [18]): $\eta_k = -\frac{1}{k}\log\left(p^{k+1} + q^{k+1}\right)$, $p_{\max} = \max\{p,q\}$, and $p_{\min} = \min\{p,q\}$. Observe that η_∞, $\eta_{-\infty}$, and h have the following relationship:

$$\eta_{-\infty} \leq h \leq \eta_\infty \qquad (2)$$

(the equality is attained when the source is *symmetric*, however in the *asymmetric* case, these bounds are rather weak).

In this paper, we will evaluate two major the performance characteristics of the *AMD-tries*: the expected *time of a successful search* and the expected *amount of memory* used by a *trie* built over n strings. To estimate the search time we can use the expected *depth* D_n or the *external path length* (i.e. the combined length of paths from root to all non-empty external nodes) C_n in a *trie* $D_n = C_n/n$. To estimate the size of a *trie* we will use the expected number of its branches B_n. Note that the last metric is slightly different from one used for the regular *tries* (cf. [11, 9]). In that case, it was sufficient to find the number of internal nodes A_n in a trie. However, since internal nodes in *adaptive tries* have different sizes by themselves, we need to use another parameter (B_n) to take into account these differences as well.

As we have mentioned earlier, we allow *AMD-tries* to use an arbitrary (but not random) mechanism for selecting the depths of their nodes. However, in a Bernoulli model, there are only two parameters that can affect the expected outcome of such selection applied to all *sub-tries* in a *trie*: a) the number of strings inserted in a *sub-trie*, and b) the parameters of source distribution. Thus, without significant loss of generality or precision, we can assume that node-depth selection logic can be presented as a (integer-valued) function:

$$r_n := r(n, p), \tag{3}$$

where n indicates the number of strings inserted in the corresponding *sub-trie*, and p is a probability of 0 in the Bernoulli model.

We are now ready to present our main results regarding the expected behavior of *AMD-tries*. The following theorem establishes the existence of *AMD-tries* attaining the constant expected time of a successful search.

THEOREM 1. There exist *AMD-tries* such that:

$$1 < \xi_1 \le C_n/n \le \xi_2 < \infty , \tag{4}$$

where ξ_1, ξ_2 are some positive constants. The upper bound holds if:

$$r_n \ge \tfrac{-1}{\eta_{-\infty}} \log\left(1 - \exp\left(\tfrac{-\log \xi_2}{n-1}\right)\right) = \tfrac{1}{\eta_{-\infty}}\left(\log n - \log\log \xi_2\right) + O\left(\tfrac{1}{n}\right), \tag{5}$$

and the lower bound holds if:

$$r_n \le \tfrac{-1}{\eta_{\infty}} \log\left(1 - \exp\left(\tfrac{-\log \xi_1}{n-1}\right)\right) = \tfrac{1}{\eta_{\infty}}\left(\log n - \log\log \xi_1\right) + O\left(\tfrac{1}{n}\right). \tag{6}$$

Notice that the above conditions are based on parameters η_{∞} and $\eta_{-\infty}$, which, in turn, can be considered as bounds for the entropy h of the source (2). This may suggest that there should be a more accurate way to estimate the depth of nodes needed to attain a certain performance (and vice versa), expressed in terms of the entropy h. The following theorem answers this conjecture in affirmative.

THEOREM 2. Let

$$r_n^* = \log n / h , \tag{7}$$

and assume that the actual depths of nodes r_n in an *AMD-trie* are selected to be sufficiently close to r_n^*:

$$\left|r_n - r_n^*\right| = O\left(\sqrt{r_n^*}\right). \tag{8}$$

Then the complexity of a successful search in such trie is (with $n \to \infty$):

$$C_n/n \quad \Phi^{-1}\left(\left(r_n - r_n^*\right)/\left(\sigma\sqrt{\log n}\right)\right)\left(1 + O\left(1/\sqrt{\log n}\right)\right), \tag{9}$$

where:

$$\Phi(x) = \tfrac{1}{\sqrt{2\pi}} \int_{-\infty}^{x} e^{-\tfrac{t^2}{2}} dt , \tag{10}$$

is the distribution function of the standard normal distribution (cf.[1]), and $\sigma = \tfrac{h_2 - h^2}{h^3}$.

Now, we will try to evaluate *AMD-tries* from the memory usage perspective. The next theorem establishes bounds for *AMD-tries* that attain a linear (with the number of strings inserted) size.

THEOREM 3. There exist *AMD-tries* such that:
$$1 < \xi_1 \le B_n/n \le \xi_2 < \infty, \tag{11}$$
where ξ_1, ξ_2 are some positive constants. The upper bound holds if:
$$
\begin{aligned}
r_n &\ge \tfrac{1}{\log 2}\left(\log n + \log \xi_2 + \tfrac{1}{\kappa_2} W_{-1}\left(-\kappa_2 \xi_2^{-\kappa_2} n^{1-\kappa_2} \right) \right) + O\left(\tfrac{1}{n} \right) \\
&= \tfrac{1}{\eta_\infty}\left(\log n - \Delta(n, \kappa_2, \xi_2) - \log\log \xi_2 \right) + O\left(\tfrac{\log\log n}{\log n} \right),
\end{aligned} \tag{12}
$$
and the lower bound holds if:
$$
\begin{aligned}
r_n &\le \tfrac{1}{\log 2}\left(\log n + \log \xi_1 + \tfrac{1}{\kappa_1} W_{-1}\left(-\kappa_1 \xi_1^{-\kappa_1} n^{1-\kappa_1} \right) \right) + O\left(\tfrac{1}{n} \right) \\
&= \tfrac{1}{\eta_\infty}\left(\log n - \Delta(n, \kappa_1, \xi_1) - \log\log \xi_1 \right) + O\left(\tfrac{\log\log n}{\log n} \right),
\end{aligned} \tag{13}
$$
where: $\kappa_1 = \eta_\infty/\log 2$, $\kappa_2 = \eta_{-\infty}/\log 2$, $W_{-1}(x)$ is a branch $W(x) \le -1$ of the Lambert W function: $W(x)e^{W(x)} = x$ (cf. [4]), and
$$\Delta(n, \kappa, \xi) = \log\left(1 + \left(\kappa\log n - \log(n\kappa)\right)/\left(\kappa\log\xi\right)\right). \tag{14}$$

Observe that in the symmetric case ($p = q = 1/2$), we have $\kappa_1 = \kappa_2 = 1$, and thus, the term (14) becomes zero: $\Delta(n, 1, \xi) = 0$. The resulting bounds for r_n (12) and (13) will be identical to one we have obtained in the Theorem 1 (5), and (6), and thus, we have the following conclusion.

COROLLARY 4. In the symmetric Bernoulli model, an *AMD-trie* can attain a constant expected depth $D_n = C_n/n = O(1)$ and have a linear size $B_n = O(n)$ at the same time.

Unfortunately, in the asymmetric case the situation is not the same. Thus, if $\kappa \ne 1$, the term $\Delta(n, \kappa, \xi)$ can be as large as $O(\log\log n)$, and the conditions for the constant time search (5), (6) may not hold. Actually, from the analysis of *LC-tries* (cf.. [15]), we know that the use of depths $r = O(\log n - \log\log n)$ leads to the expected search time of $O(\log\log n)$, and it is not yet clear if this bound can be improved considering the other linear in space implementations of the *AMD-tries*.

3   Analysis

In this section, we will introduce a set of recurrent expressions for various parameters in *AMD-tries*, derive some necessary asymptotic expansions, and will sketch the proofs of our main theorems.

For the purposes of compact presentation of our intermediate results, we will introduce the following two variables. A sequence $\{x_n\}$ will represent an abstract property of *AMD-tries* containing n strings, and a sequence $\{y_n\}$ will represent a property of root nodes of these *tries* that contribute to $\{x_n\}$. The following mapping to the standard *trie* parameters is obvious:

$$\begin{matrix} x_n \\ y_n \end{matrix} \begin{vmatrix} A_n & B_n & C_n \\ 1 & 2^{r_n} & n \end{vmatrix},$$

(15)

where: A_n is the average number of internal nodes, B_n is the average number of branches, C_n is the external path length in an *AMD-trie* containing n strings.

Using the above notation, we can now formulate the following recurrent relationship between these parameters of *AMD-tries*.

LEMMA 1. The properties x_n and y_n of an *AMD-trie* in a Bernoulli model satisfy:

$$x_n = y_n + \sum_{k=2}^{n} \binom{n}{k} \sum_{s=0}^{r_n} \binom{r_n}{s} \left(p^s q^{r_n-s} \right)^k \left(1 - p^s q^{r_n-s} \right)^{n-k} x_k; \ x_0 = x_1 = y_0 = y_1 = 0;$$

(16)

where: n is the total number of strings inserted in an *AMD-trie*, and the rest of parameters are as defined in (3) and (16).

Proof. Consider a root node of depth r_n in an *AMD-trie*. If $n < 2$ the result is 0 by definition. In $n \geq 2$, the property of trie x_n should include a property of a root node y_n plus the sum of properties of all the child *tries*. It remains to enumerate child tries and estimate probabilities of them having $2 \leq k \leq n$ strings inserted. This could be done using the following approach.

Assume that each of the 2^{r_n} possible bit-patterns of length r_n has a probability of occurrence $\pi_i = \Pr\{Bin_{r_n}(s_j) = i\}$ ($Bin_{r_n}(s_j)$ means first r_n bits of a string s_j), $0 \leq i < 2^{r_n}$, $\sum \pi_i = 1$, and that this is a memoryless scheme (π_i is the same for all strings). Then, the probability that exactly k strings match the i-th pattern is $\binom{N}{k} \pi_i^k (1 - \pi_i)^{N-k}$. The i-th pattern corresponds to a child *trie* only if there are at least two strings that match it. Therefore, the contribution of the i-th pattern to the property x_n is $\sum_{k \geq 2} \binom{n}{k} \pi_i^k (1 - \pi_i)^{n-k} x_k$, and it remains to scan all 2^{r_n} patterns to obtain the complete expression for x_n.

Recall now, that the actual strings we are inserting in the *trie* are produced by a binary memoryless source with probability of *0* equal p. Therefore, the probabilities of the r_n-bit sequences produced by this source will only depend on the number of *0*'s (*s*) they contain: $\pi(s) = p^s q^{r_n-s}$. Also, given s zeros, the total number of patterns yielding this probability is $\binom{r_n}{s}$. Combining all these formulas, we arrive at

$$\sum_{0 \leq s \leq r_n} \binom{r_n}{s} \sum_{k \geq 2} \binom{n}{k} \left(p^s q^{r_n-s} \right)^k \left(1 - p^s q^{r_n-s} \right)^{n-k} x_k,$$

(17)

which, after the addition of y_n yields the claimed expression (16).

It should be stressed that due to dependency upon the unknown parameter r_n the rigorous analysis of a recurrent expression (16) appears to be a very difficult task. It is not clear for example, if it is possible to convert (16) to a closed form (for any of the parameters involved). Moreover, the attempts to use some particular formulas (or algorithms) for r_n can actually make the situation even more complicated.

The analysis of (16) that we provide in this paper is based on a very simple approach, which nevertheless is sufficient to evaluate few corner cases in the behavior of these algorithms. Thus, most of our theorems claim the existence of a solution in linear form, and we use (16) to find conditions for r_n such that the original claim holds. We perform the first step in this process using the following lemma.

LEMMA 2. Let ξ_1 and ξ_n be two positive constants ($1 < \xi_1 \le \xi_2 < \infty$), such that

$$\xi_1 n \le x_n \le \xi_2 n . \tag{18}$$

Then, the recurrent expression (16) holds when:

$$\xi_1 \le y_n / f(n, p, r_n) \le \xi_2 , \tag{19}$$

where:

$$f(n, p, r_n) = n \sum_{s=0}^{r_n} \binom{r_n}{s} p^s q^{r_n - s} \left(1 - p^s q^{r_n - s}\right)^{n-1} . \tag{20}$$

Proof. Consider the upper bound in (18) first, and substitute x_k with $k\xi_2$ in the right side of (16). This yields:

$$x_n \le y_n + \sum_{s=0}^{r_n} \binom{r_n}{s} \sum_{k=2}^{n} \binom{n}{k} \left(p^s q^{r_n - s}\right)^k \left(1 - p^s q^{r_n - s}\right)^{n-k} \xi_2 k =$$
$$y_n + \xi_2 n \sum_{s=0}^{r_n} \binom{r_n}{s} p^s q^{r_n - s} \left(1 - \left(1 - p^s q^{r_n - s}\right)^{n-1}\right) = y_n + \xi_2 n - \xi_2 f(n, p, r_n). \tag{21}$$

Now, according to (18), $x_n \le \xi_2 n$, and combined with (21), the upper bound holds only when $y_n - \xi_2 f(n, p, r_n) \le 0$. Hence $\xi_2 \ge y_n / f(n, p, r_n)$, and repeating this procedure for the lower bound ($x_n \ge \xi_1 n$), we arrive at formula (19), claimed by the lemma.

The next step in our analysis is to find bounds for the sum (20) that would allow us to separate r_n. The following lemma summarizes few such results.

LEMMA 3. Consider a function $f(n, p, r_n)$ defined in (20). The following holds:

$$f\left(n, \tfrac{1}{2}, r_n\right) = n\left(1 - 2^{-r_n}\right)^{n-1} , \tag{22.a}$$

$$f(n, p, r_n) \le n\left(1 - e^{-r_n h_\infty}\right)^{n-1} \le n \exp\left(-(n-1)e^{-r_n h_\infty}\right), \tag{22.b}$$

$$f(n, p, r_n) \ge n\left(1 - e^{-r_n h_{-\infty}}\right)^{n-1} \ge n \exp\left(-(n-1)e^{-r_n h_{-\infty}} / \left(1 - e^{-r_n h_{-\infty}}\right)\right), \tag{22.c}$$

In addition, if $n, r_n \to \infty$, and $|r_n - \log n / h| = O\left(\sqrt{\log n / h}\right)$, then asymptotically:

$$f(r_n, n, p) = n\Phi\left((r_n - \log n / h) / \left(\sigma \sqrt{\log n}\right)\right)\left(1 + O\left(1 / \sqrt{\log n}\right)\right), \tag{22.d}$$

where $\Phi(x)$ and σ are as defined in the Theorem 2 (7-10).

Proof. The equality for the symmetric case (22.a) follows from by direct substitution. Two other bounds (22.b) and (22.c) can be obtained with the help of the following estimate:

$$e^{-r_n \eta_\infty} = p_{min}^{r_n} \le p^s q^{r_n - s} \le p_{max}^{r_n} = e^{-r_n \eta_{-\infty}},$$

where p_{max}, p_{min}, η_∞, and $\eta_{-\infty}$ are as defined in (1). We apply these bounds to the last component in the sum (20):

$$\left(1 - e^{-r_n \eta_\infty}\right)^{n-1} \le \left(1 - p^s q^{r_n - s}\right)^{n-1} \le \left(1 - e^{-r_n \eta_{-\infty}}\right)^{n-1},$$

which allows the first portion of sum (20) to converge: $\sum \binom{r_n}{s} p^s q^{r_n - s} = 1$.

The additional (right-side) inequalities in (22.b) and (22.c) are due to: $e^{-x/(1-x)} \le 1 - x \le e^{-x}, 0 \le x < 1$, (cf. [1], 4.2.29).

The derivation of an asymptotic expression (22.d) is a complicated task, and here we will rather refer to a paper of Louchard and Szpankowski (cf. [14]) which discusses it in detail. A somewhat different approach of solving it has also been provided in [12] and [13].

Now using the results of the above lemmas, we can sketch the proofs of our main theorems. Consider the problem of evaluating the behavior of *AMD-tries* from the expected complexity of a successful search perspective first.

We can use a recurrent expression (16) where, according to (15), we can substitute x_n with a parameter of the external path length C_n, and y_n with n. Observe that the bounds for C_n (4) claimed by the Theorem 1 and condition (18) in Lemma 2 are equivalent, and the combination of the result of this lemma (18) with (22.b) and (22.c) yields the following:

$$\xi_1 \le \left(1 - e^{-r_n \eta_{-\infty}}\right)^{1-n} \le n / f(n, p, r_n) \le \left(1 - e^{-r_n \eta_{-\infty}}\right)^{1-n} \le \xi_2 . \tag{23}$$

Here, it remains to find an expression for r_n such that $1 < \xi_1 \le \xi_2 < \infty$ regardless of n. Considering the right part of 18 (upper bound), the last requirement is obviously satisfied when $e^{-r_n \eta_{-\infty}} = \frac{1}{n-1}$. More precisely, for any given constant $1 < \xi_2 < \infty$ we need:

$$r_n \ge \frac{-1}{\eta_{-\infty}} \log\left(1 - \exp\left(\frac{-\log \xi_2}{n-1}\right)\right) = \frac{1}{\eta_{-\infty}}\left(\log n - \log\log \xi_2\right) + O\left(\frac{1}{n}\right),$$

which is exactly the condition (5) claimed by the Theorem (1). The expression for the lower bound (6) is obtained in a similar way.

The result of the Theorem 2 follows directly from (19), (20) and asymptotic expression (22.d).

To evaluate the expected size an *AMD-trie* in memory we will use a very similar technique. Consider a recurrent expression (16) with x_n substituted by the expected number of branches B_n, and y_n with 2^{r_n}. Following the Theorem 3 and Lemma 2, we will assume that B_n is bounded (11), and using (18) combined with (22.b) and (22.c) we arrive at:

$$\xi_1 \le 2^{r_n} n^{-1}\left(1 - e^{-r_n \eta_{-\infty}}\right)^{1-n} \le 2^{r_n} / f(n, p, r_n) \le 2^{r_n} n^{-1}\left(1 - e^{-r_n \eta_{-\infty}}\right)^{1-n} \le \xi_2 . \tag{24}$$

Compared to (23) this appears to be a slightly more complicated expression, which does not yield an exact solution for r_n (unless the source is symmetric). To find an asymptotic (for large n) solution of (24), we will write:

$$r_n \ge \frac{-1}{\eta_{-\infty}} \log\left(1 - \exp\left(\frac{-\log\left(\xi_2 n 2^{-r_n}\right)}{n-1}\right)\right) = \frac{1}{\eta_{-\infty}}\left(\log n - \log\log\left(\xi_2 n 2^{-r_n}\right)\right) + O\left(\frac{1}{n}\right),$$

and after some manipulations we arrive at:

$$r_n \ge \frac{1}{\log 2}\left(\log n + \log \xi_2 + \frac{1}{\kappa_2} W_{-1}\left(-\kappa_2 \xi_2^{-\kappa_2} n^{1-\kappa_2}\right)\right) + O\left(\frac{1}{n}\right), \tag{25}$$

where according to (12) $\kappa_2 = \eta_{-\infty} / \log 2$, and $W_{-1}(x)$ is a branch $W(x) \le -1$ of the Lambert W function $W(x) e^{W(x)} = x$ (cf. [4]). To perform further simplifications in (25), we can use the following asymptotic expansion of $W_{-1}(x)$ (cf. [4]):

$$W_{-1}(x) = \log(-x) - \log(-\log(-x)) + O\left(\frac{\log(-\log(-x))}{\log(-x)}\right), \quad -e^{-1} < x < 0, \tag{26}$$

which, after some algebra yields the second part of inequality (12) claimed by the Theorem 3. The expression for the lower bound (13) is obtained in essentially the same way.

4   Concluding Remarks

We absolutely believe that *AMD-tries* have great potential that should be explored in practice. To that end, our conclusion that they attain $O(1)$ expected search time, while preserving all the benefits of the regular *tries*, should be a good starting point.

At the same time, *AMD-tries* pose several interesting theoretical problems, and from this perspective, our results just scratch the surface. Our analysis was only sufficient to produce (rather coarse) bounds for the expected characteristics of *AMD-tries*, and the derivation of their exact expressions (including lower-magnitude terms and oscillating components) remains an open (and difficult) problem.

Another interesting problem is to find parameters of an *AMD-trie* that uses the minimum possible amount of memory. Thus, analyzing the set of expressions in the proof of the Theorem 3 we can conjecture that such trie exists, but finding its actual parameters requires a more solid analytical framework.

Acknowledgment

The author wishes to express his gratitude to Prof. W. Szpankowski from Purdue University, for many useful discussions and encouragement in starting this research.

References

1. M. Abramowitz, and I. Stegun, Handbook of Mathematical Functions, Dover, NY (1972)
2. Andersson. and S. Nilsson, Improved Behaviour of Tries by Adaptive Branching, Information Processing Letters, 46 (1993) 295-300.
3. T. M. Cover and J. M. Thomas, Elements of Information Theory, John Wiley & Sons, New York (1991)
4. R. M. Corless, G. H. Gonnet, D. E. G. Hare, D. J. Jeffrey, and D. E. Knuth, On the Lambert W Function, Advances in Computational Mathematics, 5 (1996) 329-359
5. L. Devroye, A Note on the Average Depths in Tries, Computing, 28 (1982) 367-371
6. W. Dobosiewitz, Sorting by Distributive Partitioning, Information Processing Letters, 7, 1, (1978) 1-6
7. W. Dobosiewitz, The Practical Significance of DP Sort Revisited, Information Processing Letters, 8, 4 (1979) 170-172
8. G. Ehrlich, Searching and Sorting Real Numbers, J. Algorithms, 2 (1981) 1-14
9. P. Flajolet and R. Sedgewick, Digital Search Trees Revisited, SIAM J. Computing, 15, (1986) 748-767
10. P. Jacquet and W. Szpankowski, Analysis of Digital Trees with Markovian Dependency, IEEE Trans. Information Theory, 37 (1991) 1470-1475
11. D. Knuth, The Art of Computer Programming. Sorting and Searching. Vol. 3., Addison-Wesley (1973)
12. G. Louchard, The Brownian Motion: A Neglected Tool for the Complexity Analysis of Sorted Tables Manipulations, RAIRO Theoretical Informatics, 17 (1983) 365-385
13. G. Louchard, Digital Search Trees Revisited, Cahiers du CERO, 36 (1995) 259-27

14. G. Louchard and W. Szpankowski, An Exercise in Asymptotic Analysis, reprint (1995)
15. S. Nilsson, Radix Sorting and Searching, Ph.D. thesis, Department of Computer Science, Lund University (1996)
16. Pittel, Paths in a Random Digital Tree: Limiting Distributions. Advances in Applied Probability, 18 (1986) 139-155
17. R. Sedgewick, and P. Flajolet, An Introduction to the Analysis of Algorithms, Addison-Wesley, Reading, MA (1996)
18. W. Szpankowski, Techniques for the Average Case Analysis of Algorithms on Words, John Wiley & Sons, to be published
19. W. Szpankowski, Some results on V-ary asymmetric tries, Journal of Algorithms, 9 (1988) 224-244
20. M. Tamminen, Analysis of N-Trees, Information Processing Letters, 16, 3 (1983) 131-137

Optimal Coding with One Asymmetric Error: Below the Sphere Packing Bound

Ferdinando Cicalese[1] and Daniele Mundici[2]

[1] Dipartimento di Informatica ed Applicazioni, University of Salerno,
84081 Baronissi (SA), Italy
cicalese@dia.unisa.it,
http://www.dia.unisa.it/∼cicalese
[2] Dipartimento Scienze Informazione, University of Milan,
Via Comelico 39-41, 20135 Milan, Italy
mundici@mailserver.unimi.it

Abstract. Ulam and Rényi asked what is the minimum number of yes-no questions needed to find an unknown m-bit number x, if up to ℓ of the answers may be erroneous/mendacious. For each ℓ it is known that, up to only finitely many exceptional m, one can find x asking *Berlekamp's minimum number* $q_\ell(m)$ of questions, i.e., the smallest integer q satisfying the sphere packing bound for error-correcting codes. The Ulam-Rényi problem amounts to finding optimal error-correcting codes for the binary *symmetric* channel with noiseless feedback, first considered by Berlekamp. In such concrete situations as optical transmission, error patterns are highly asymmetric—in that only one of the two bits can be distorted. Optimal error-correcting codes for these asymmetric channels with feedback are the solutions of the *half-lie* variant of the Ulam-Rényi problem, asking for the minimum number of yes-no questions needed to find an unknown m-bit number x, *if up to ℓ of the negative answers may be erroneous/mendacious*. Focusing attention on the case $\ell = 1$, in this self-contained paper we shall give tight upper and lower bounds for the half-lie problem. For infinitely many m's our bounds turn out to be matching, and the optimal solution is explicitly given, thus strengthening previous estimates by Rivest, Meyer et al.

1 Introduction

In his autobiographical book "Adventures of a Mathematician" [11, p.281], Ulam posed the problem of optimal binary search in the presence of faulty tests. Independently, the very same problem had been formulated by Rényi [8, p.47]. In fact, the search of an unknown m-bit number by asking the minimum number of "yes-no" questions, when up to ℓ of the answers may be erroneous/mendacious, is the same as the problem of finding shortest ℓ-error-correcting codes for Berlekamp's noiseless delayless feedback channel [1].

As a typical example, let us consider a Transmitter sending binary messages (e.g., m-bit numbers) on a noisy channel. Assume that the received bits can

D.-Z. Du et al. (Eds.): COCOON 2000, LNCS 1858, pp. 159–169, 2000.

be sent back to the Transmitter on a noiseless channel. Thus Transmitter-to-Receiver bits are interleaved with Receiver-to-Transmitter bits. In other words, before sending the $(i+1)$th bit, the Transmitter does know which ones among the previously sent i bits b have been distorted and received as $1-b$ because of the noise effect. The main problem here is to minimize the number of bits sent by the Transmitter, while still guaranteeing that the m-bit number can be fully recovered by the Receiver.

Assuming the 1 and 0 bits to be equally subject to distortion, it is known that an optimal solution to this problem is given by sending $q_\ell(m)$ bits, where $q_\ell(m)$ is the smallest integer q satisfying the Sphere Packing Bound $\sum_{j=0}^{\ell} \binom{q}{j} \le 2^{q-m}$. Trivially, no solution exists using less than $q_\ell(m)$ bits.

In other concrete situations the distribution of errors/distortions is highly asymmetric: for instance, in optical communication systems photons may fail to be detected ($1 \rightarrow 0$), but the creation of spurious photons ($0 \rightarrow 1$) is impossible [7]. Similarly, in most LSI memory protection schemes one can safely assume that only one of the two possible error patterns ($1 \rightarrow 0$) and ($0 \rightarrow 1$) can occur [4].

Optimal error-correcting codes for these asymmetric channels with feedback are the same as shortest binary strategies to find an m-bit number, when only negative tests can be erroneous.

To the best of our knowledge, the first paper dealing with this "half-lie" variant of the Ulam-Rényi problem was [9]. The authors proved an asymptotic result to the effect that the half-lie game has the same complexity as the original Ulam problem—henceforth referred to as the "full-lie" game. They also proved that the half-lie variant of the Ulam-Rényi game *with comparison questions* corresponds to the problem of finding the minimum root of a set of unimodal functions.

Focusing on the case $\ell = 1$, in this paper we improve the result of [9], by giving very tight and non-asymptotic, lower and upper bounds for the half-lie game. More precisely, we prove that

- for each m, $q_1(m) - 1$ questions suffice;
- for each m, $q_1(m) - 3$ questions do not suffice;
- for infinitely many m, $q_1(m) - 2$ questions do not suffice, and we give an optimal strategy using exactly $q_1(m) - 1$ questions;
- for infinitely many m, $q_1(m) - 2$ questions suffice, and we give a (necessarily optimal) strategy using exactly $q_1(m) - 2$ questions.

2   Preliminaries

We shall consider a game between two players, Paul (the Questioner), and Carole (the Responder). The rules are as follows: Paul and Carole first fix an integer m, and a *search space E* coinciding with the set of m-bit integers $E = \{0, 1, \ldots, 2^m - 1\}$. Now Carole chooses a number x in E, and Paul must find x by asking questions, to which Carole can only answer yes or no. Carole can lie at most once, but only by answering "no" to a question whose correct (sincere) answer is "yes". Thus, only Carole's negative answers are dubious for Paul. This is the the half-lie game over the search space E.

Paul's *state* of knowledge is completely represented by a pair (A_0, A_1) of disjoint subsets of E, where A_0 (the *truth-set*) is the set of those elements of E satisfying all of Carole's answers, and A_1 (the *lie-set*) is the set of elements of E falsifying exactly one (necessarily negative) answer. By a *question* we understand any arbitrary subset $T \subseteq E$. Thus T represents Paul's question "Does the secret number x belong to T?".

In particular, Paul's initial state of knowledge, before any question is asked, is represented by the pair $(E, \emptyset)$. At any stage of the game, suppose that Paul's state of knowledge is given by $\sigma = (A_0, A_1)$, and his next question is T. If Carole's answer is "yes" (whence, by hypothesis, this answer is correct) then the resulting state of knowledge of Paul is given by

$$\sigma^{yes} = (A_0 \cap T, A_1 \cap T).$$

On the other hand, if Carole's answer is "no", Paul's new state of knowledge is given by

$$\sigma^{no} = (A_0 \setminus T, (A_0 \cap T) \cup (A_1 \setminus T)).$$

This is so because the lie-set of σ^{no} has two kinds of elements: (i) members of the truth-set of σ that falsify Carole's answer, and (ii) members in the lie-set of σ that satisfy Carole's answer.

The game is over when Paul's state $\sigma = (A_0, A_1)$ satisfies $|A_0 \cup A_1| \leq 1$, where $|\cdot|$ denotes cardinality. As a matter of fact, if $|A_0 \cup A_1| = 1$ then the only element of $A_0 \cup A_1$ must coincide with Carole's secret number. On the other hand, if $|A_0 \cup A_1| = 0$ then every element $x \in E$ falsifies at least two of Carole's answers—against the rules of the game.

Our main concern here is to estimate the minimum *number* of questions needed to find x, rather than actually exhibiting such questions. Accordingly, only the number of elements of A_0 and A_1 is relevant here, and we can safely represent Paul's state (A_0, A_1), by the pair of integers $(|A_0|, |A_1|)$. Similarly, any question $T \subseteq E$ will be conveniently represented by the pair of integers $[t_0, t_1]$, where $t_i = |A_i \cap T|$.

Using this notation, for any state $\sigma = (x_0, x_1)$ and question $[t_0, t_1]$ the two possible states resulting from Carole's answer are given by

$$\sigma^{yes} = (t_0, t_1) \qquad \text{and} \qquad \sigma^{no} = (x_0 - t_0, \ t_0 + x_1 - t_1) \qquad (1)$$

Definition 1. *A* final state *is a state* $\sigma = (x_0, x_1)$ *such that* $x_0 + x_1 \leq 1$.

Starting with the state σ, suppose that questions $T_1, \ldots, T_t$ have been asked, and a corresponding t-tuple of answers $\boldsymbol{b} = b_1, \ldots, b_t$, has been received. Iterated application of the above formulas yields a sequence of states $\sigma_0, \sigma_1, \ldots, \sigma_t$, where

$$\sigma_0 = \sigma, \quad \sigma_1 = \sigma_0^{b_1}, \quad \sigma_2 = \sigma_1^{b_2}, \quad \ldots, \quad \sigma_t = \sigma_{t-1}^{b_t}. \qquad (2)$$

By a *strategy* $\mathcal{S}$ *with* q *questions* we mean the full binary tree of depth q, where each node ν is mapped into a question T_ν, and the two edges $\eta_{left}, \eta_{right}$

below ν are respectively labeled *yes* and *no*. Let $\boldsymbol{\eta} = \eta_1, \ldots, \eta_q$ be a path in $\mathcal{S}$, from the root to a leaf, with respective labels $b_1, \ldots, b_q$, generating nodes $\nu_1, \ldots, \nu_q$ and associated questions $T_{\nu_1}, \ldots, T_{\nu_q}$. Fixing an arbitrary state σ, iterated application of (1) naturally transforms σ into σ^η (where the dependence on the b_j's and T_{ν_j}'s is understood). We say that strategy $\mathcal{S}$ is *winning* for σ iff for every path $\boldsymbol{\eta}$ from the root to a leaf, the state σ^η is final.

Definition 2. *We say that $\sigma = (x_0, x_1)$ is a* winning q-state *iff there exists a strategy with q questions which is winning for σ.*

Trivially, σ is a winning q-state iff there exists a question T such that both resulting states σ^{yes} and σ^{no} are winning $(q-1)$-states.

Definition 3. *We say that $\sigma' = (y_0, y_1)$ is a* substate *of a state $\sigma = (x_0, x_1)$ iff $y_j \leq x_j$, for each $j = 0, 1$.*

Lemma 1. *Let $\sigma = (x_0, x_1)$ be a winning q-state. Then every substate of σ is a winning q-state.*

To help the reader, we shall recall here two basic definitions concerning the full-lie game [5,6,10].

Definition 4. *For every integer $m \geq 1$, Berlekamp's minimum number of questions $q(m)$ is defined by $q(m) = \min\{q \mid q + 1 \leq 2^{q-m}\}$.*

Definition 5. *For every integer $s \geq 0$ the sth critical index m_s is defined by $m_s = \max\{m \mid m + s + 1 \leq 2^s\}$.*

It turns out [6] that $q(m)$ is the minimum number of questions that are not only necessary, but also sufficient to solve the full-lie game with one lie over the search space of cardinality 2^m. The critical index m_s is the largest integer m such that $m + s$ questions are necessary and sufficient to solve the full-lie game with one lie over a search space of cardinality 2^m.

3   Lower Bounds

Definition 6. *Fix two integers $m = 1, 2, 3, \ldots$ and $q > m$. Then the integer $\kappa(m, q)$ is the smallest k such that $\sum_{i=0}^{k} \binom{q}{i} > 2^m$. The integer $\rho = \rho(m, q)$ is defined by $\rho = 2^m - \sum_{i=0}^{\kappa(m,q)-1} \binom{q}{i}$.*

Trivially, $1 \leq \kappa(m, q) \leq q$ and $0 \leq \rho < \binom{q}{\kappa(m,q)}$.

Theorem 1. *Fix an integer $m = 1, 2, 3, \ldots$. Let $q^\ddagger$ be the smallest integer q such that*

$$2^q \geq 2^m + \sum_{i=0}^{\kappa(m,q)-1} i\binom{q}{i} + \kappa(m, q)\rho(m, q).$$

Note that $1 \leq q^{\ddagger} \leq q(m)$. Assume there exists a winning strategy with q^* questions to find an unknown integer in the set $E = \{0, 1, \ldots 2^m - 1\}$ in the half-lie game, i.e., when up to one of the "no" answers may be mendacious and all "yes" answers are true. Then $q^* \geq q^{\ddagger}$.

Proof. Let $\mathcal{S}$ be such strategy, where we can safely assume that the number q^* of questions in $\mathcal{S}$ is the smallest possible. By a path in $\mathcal{S}$ we shall understand a path from the root to a leaf of $\mathcal{S}$. For each $x \in E$ there exists precisely one path μ_x in $\mathcal{S}$ leading to the final state $(\{x\}, \emptyset)$. This final state is obtained if Carole chooses x as the secret number and then decides to give true answers to all q^* questions. Let $southwest_x$ be the number of left branches (i.e., branches whose label is "yes") in this path. The $southwest_x$ many branches of μ_x are a record of Carole's "yes" answers, once she decides to choose x as the secret number, and to always tell the truth. If, on the other hand, Carole decides to give one mendacious answer, she can deviate from this path in $southwest_x$ many ways, replacing a "yes" true answer by a mendacious "no" answer. [1] Since $\mathcal{S}$ is a winning strategy, there exist in $\mathcal{S}$ precisely $southwest_x$ many paths leading to the final state $(\emptyset, \{x\})$. Furthermore, whenever $x \neq y \in E$ the state $(\emptyset, \{y\})$ will be obtained by a different path. Now, each path in $\mathcal{S}$ has q^* branches, and paths are in one-one correspondence with subsets of $\{1, \ldots, q^*\}$ (giving the depths of their left branches). To obtain a set of 2^m paths having the smallest possible total number of left branches, with reference to Definition 6, one must proceed as follows: first take the only path in $\mathcal{S}$ with no left branch, then the q^* paths with 1 left branch, the $\binom{q^*}{2}$ paths with 2 left branches, $\ldots$, the $\binom{q^*}{\kappa(m,q^*)-1}$ paths with $\kappa(m, q^*) - 1$ left branches, and finally, $\rho(m, q^*)$ many paths with $\kappa(m, q^*)$ left branches. Then the total number N_{once} of possibilities for Carole to lie precisely once in $\mathcal{S}$ will satisfy the inequality

$$N_{\text{once}} \geq \sum_{i=0}^{\kappa(m,q^*)-1} i \binom{q^*}{i} + \kappa(m, q^*)\rho(m, q^*).$$

On the other hand, if Carole decides not to lie, she can still choose $x \in E$, and the corresponding path leading to the leaf $(\{x\}, \emptyset)$, in 2^m ways. Summing up, the 2^{q^*} many paths of the winning strategy $\mathcal{S}$ cannot be less than the total number of Carole's answering strategies, whence

$$2^{q^*} \geq 2^m + N_{\text{once}} \geq \sum_{i=0}^{\kappa(m,q^*)-1} i \binom{q^*}{i} + \kappa(m, q^*)\rho(m, q^*).$$

We shall now prove that, for all integers $m \geq 1$, at least $q(m) - 2$ questions are necessary to find a number in the half-lie game over the search space $\{0, 1, \ldots, 2^m - 1\}$. We prepare the following

---

[1] Note the effect of the assumed asymmetry in Carole's lies: in the classical Ulam game Carole can choose to lie in q^* many ways.

Lemma 2. *Let $m = 2, 3, \ldots$. Let the integer q have the property that there exists a winning strategy S for the state $\sigma = (2^m, 0)$ with q questions. Then there exists a winning strategy for the state $\sigma' = (2^{m-1}, 0)$ with $q - 1$ questions.*

Proof. Let $T = [t_0, 0]$ be the first question in S. The resulting states are $\sigma^{yes} = (t_0, 0)$ and $\sigma^{no} = (2^m - t_0, t_0)$, both having a winning $(q-1)$-strategy. According as $t_0 \geq 2^{m-1}$ or $t_0 < 2^{m-1}$, the state $(2^{m-1}, 0)$ is either a substate of σ^{yes} or a substate of σ^{no}, respectively. In either case, $(2^{m-1}, 0)$ has a winning $(q-1)$-strategy.

Notation: For each real number t we let $\lfloor t \rfloor$ (resp., $\lceil t \rceil$) denote the largest integer $\leq t$ (resp., the smallest integer $\geq t$). For each integer $r > 0$, we let $powerset_r$ denote the powerset of the set $\{1, 2, \ldots, r\}$. For any family Θ of subsets of $\{1, 2, \ldots, r\}$ we let

$$\overline{card}(\Theta)$$

denote the average cardinality of the sets in Θ. For all real numbers $0 \leq a \leq b \leq r$ we let

$$powerset_r[a, b] = powerset_r[\lceil a \rceil, \lfloor b \rfloor] = \{G \subseteq \{1, 2, \ldots, r\} \mid a \leq |G| \leq b\}.$$

Given an integer $0 \leq k \leq 2^r$, the *initial k-segment*

$$powerset_r \upharpoonright k$$

of $powerset_r$ is the family of the *first k smallest* subsets of $\{1, 2, \ldots, r\}$, i.e., the 0-element subset, the r many 1-element subsets, the $\binom{r}{2}$ many 2-element subsets, $\ldots$, until k subsets have been collected having the smallest possible cardinality.

Lemma 3. *The following inequalities hold:*

(i) For all $q \geq 57$, $|powerset_q[0, q/3]| < \frac{2^q}{2q}$;

(ii) For all $q \geq 171$, $|powerset_q[0, 2q/5]| < \frac{2^q}{1.1q}$;

(iii) For all $q \geq 12$, $\overline{card}(powerset_q, [0, q/3]) > \frac{q}{4}$;

(iv) For all $q \geq 23$, $\overline{card}(powerset_q, [0, 2q/5]) > \frac{q}{3}$.

Proof. A routine inspection, using the properties of the binomial coefficients.

Theorem 2. *For each $m \geq 1$, let $q(m)$ be Berlekamp's minimum number of questions. Then at least $q(m) - 2$ questions are necessary to win the half-lie game (with one lie) over the search space $\{0, 1, \ldots 2^m - 1\}$.*

Proof. First we consider the case $1 \leq m \leq 50$. Let $q_{\text{half}-\text{lie}}(m)$ denote the length of the shortest winning strategy for the half-lie game (with one lie) over the search space $\{0, 1, \ldots 2^m - 1\}$. By Definition 5 we have $m_0 = 0$ and $m_6 = 57$. Hence for each $1 \leq m \leq 50$ we have $m_0 + 1 \leq m \leq m_6$. For all $0 \leq s \leq 5$, a direct inspection using Theorem 1 yields $q_{\text{half}-\text{lie}}(m_s + 1) \geq q(m_s + 1) - 2$. By Lemma 2, the desired result now follows for all $m_s + 1 \leq m \leq m_{s+1}$.

We now consider the case $m \geq 51$. It follows that $q(m) \geq 57$. Writing q instead of $q(m)$, we have

$$2^{q-1} < q2^m, \tag{3}$$

whence, a fortiori,

$$2^{q-3} < 2^m(1 + q/4). \tag{4}$$

By Lemma 3(iii) we get

$$2^{q-3} < 2^m(1 + \overline{card}(powerset_q[0, q/3]). \tag{5}$$

From Lemma 3(i) it follows that

$$\sum_{j=0}^{j=\lfloor q/3 \rfloor} \binom{q}{j} < \frac{2^q}{2q}. \tag{6}$$

Since by (3) $\frac{2^q}{2q} < 2^m$, the monotonicity properties of the binomial coefficients, together with inequality (6) imply

$$\overline{card}(powerset_q[0, q/3]) = \overline{card}(powerset_q[0, \lfloor q/3 \rfloor]) < \overline{card}(powerset_q \lceil 2^m).$$

Recalling (5) we can write

$$2^{q-3} < 2^m(1 + \overline{card}(powerset_q \lceil 2^m)). \tag{7}$$

A moment's reflection shows that

$$\overline{card}(powerset_q \lceil 2^m) < \overline{card}(powerset_{q-3} \lceil 2^m). \tag{8}$$

Let $\mathcal{S}$ be a strategy with $q - 3$ of questions (absurdum hypothesis). As in the proof of Theorem 1, let N_{once} be the total number of possibilities for Carole to lie precisely once in $\mathcal{S}$: this is the same as the number of left branches in the totality of paths of depth $q - 3$ leading to a final state $(\{x\}, \emptyset)$, for all possible $x \in E$. In the light of (7)-(8), and recalling the estimates in the proof of Theorem 1, we can now write in a self-explanatory notation

$$2^{q-3} < 2^m(1 + \overline{card}(powerset_{q-3} \lceil 2^m)) = 2^m + 2^m \frac{\sum_{powerset_{q-3} \lceil 2^m} j \binom{q-3}{j}}{\sum_{powerset_{q-3} \lceil 2^m} \binom{q-3}{j}}$$

$$= 2^m + \sum_{powerset_{q-3} \lceil 2^m} j \binom{q-3}{j} \leq 2^m + N_{once} \quad (m = 51, 52, 53, \ldots),$$

which is impossible.

Theorem 3. *With reference to Definitions 4 and 5, fix an integer $s \geq 1$, and let $m = m_s$ and $q = q(m_s)$. Then at least $q - 1$ questions are necessary to solve the half-lie game with one lie, over the search space of cardinality 2^m.*

Proof. By definition, $m = m_s = 2^s - s - 1$, and $q = q(m_s) = m + s = 2^s - 1$. For $1 \leq s \leq 8$, direct verification shows that the integer $q^{\ddagger}(m_s)$ defined in Theorem 1 coincides with $q(m_s) - 1$, and hence the thesis is a direct consequence of Theorem 1.

Let us now consider the case $s \geq 9$. Suppose that there exists a winning strategy for Paul using $q - 2$ questions (absurdum hypothesis). In the light of Theorem 1, in order to reach a contradiction, it is sufficient to show

$$2^s < 4(1 + \overline{card}(powerset_{q-2} \upharpoonright 2^m)). \tag{9}$$

Claim: $\overline{card}(powerset_{q-2} \upharpoonright 2^m)) > m/3$.

As a matter of fact, for all $s \geq 9$, by Lemma 3(iv) and (ii) we have $m/3 < q/3 < \overline{card}(powerset_q[0, 2q/5]) = \overline{card}(powerset_q[0, \lfloor 2q/5 \rfloor]) < \overline{card}(powerset_q \upharpoonright \frac{2q}{1.1q})$. It follows that $m/3 < \overline{card}(powerset_{q-2} \upharpoonright \frac{2q}{1.1q})$. Since for all $s \geq 4$, $\frac{2q}{1.1q} = \frac{2^{2^s}-1}{1.1(2^s-1)} < \frac{2^{2^s}-1}{2^s} = 2^{2^s - s - 1} = 2^m$, we then get $m/3 < \overline{card}(powerset_{q-2} \upharpoonright 2^m)$ and the claim is proved.

From equation (9) we now see that, to complete the proof of the theorem it is sufficient to settle the inequality $2^s < 4(1 + m/3)$. Since $m = 2^s - s - 1$, this inequality trivially holds for all integers $s \geq 1$.

4   Upper Bounds

For all $m \geq 1$, we shall construct a winning strategy to solve the half-lie game over the search space $\{0, 1, \dots, 2^m - 1\}$ using no more than $q(m) - 1$ questions. Moreover, we shall show that for infinitely many m this strategy uses no more than $q(m) - 2$ questions.

Theorem 4. *For every integer $m \geq 1$, let $\sigma = (2^m, 0)$. Then there exists a winning strategy for σ in the half-lie game, using exactly $m - 1 + \lceil \log(m + 3) \rceil$ questions.*

Proof. The proof proceeds through several claims:

Claim 1: For each integer $r \geq 1$, the state $\sigma = (1, 2^r - 1)$ has a winning r-strategy.

By induction on r. For $r = 1$, we have $\sigma = (1, 1)$. Upon asking the question $[0, 1]$, we obtain the two final states $(0, 1)$ and $(1, 0)$. For the induction step, suppose the claim is true for $r \geq 1$, with the intent of proving it for the state $(1, 2^{r+1} - 1)$. Let $\sigma = (1, 2^{r+1} - 1)$ and ask the question $[0, 2^r]$. We then obtain the two states $\sigma^{yes} = (0, 2^r)$ and $\sigma^{no} = (1, 2^r - 1)$. It is easy to see that σ^{yes} is a winning r-state. Moreover, σ^{no} is a winning r-state by induction hypothesis. Therefore, σ is a winning $(r + 1)$-state, and the claim is settled.

Claim 2: For each integer $r \geq 1$, the state $\sigma = (2, 2^r - 3)$ is a winning r-state.

Upon asking question $[1, 2^{r-1} - 1]$, the two resulting states are $\sigma^{yes} = \sigma^{no} = (1, 2^{r-1} - 1)$. By Claim 1, both states have a winning strategy with $r - 1$ questions. A fortiori, there exists a winning strategy with r questions for the state $(2, 2^r - 3)$.

Claim 3: For every integer $m \geq 1$, the state $\sigma = (2, m)$ has a winning strategy with $\lceil \log(m + 3) \rceil$ questions.

As a matter of fact, let $r = \lceil \log(m + 3) \rceil$. Then $r = \min\{i \mid 2^{i-1} - 3 < m \leq 2^i - 3\}$. Therefore, σ is a substate of $(2, 2^r - 3)$ and the desired result directly follows, by Claim 2 and Lemma 1.

Claim 4: For any two integers $m \geq 1$ and $k = 1, 2, \ldots, m$, there exists a winning strategy for the state $\sigma = \left(2^k, (m - k + 1)2^{k-1}\right)$, using exactly $k - 1 + \lceil \log(m + 3) \rceil$ questions.

We proceed by induction on k. The basis is a direct consequence of Claim 3. For the induction step, let $k = i + 1$. Then $\sigma = \left(2^{i+1}, (m - i)2^i\right)$. By asking the question $\left[2^i, (m - i + 1)2^{i-1}\right]$, the two resulting states coincide—specifically, $\sigma^{yes} = \sigma^{no} = \left(2^i, (m - i + 1)2^{i-1}\right)$. By induction one immediately sees that $\sigma^{yes} = \sigma^{no}$ is a winning $(i - 1 + \lceil \log(m + 3) \rceil)$-state. Therefore, σ is a winning $(k - 1 + \lceil \log(m + 3) \rceil)$-state, as required to settle the claim.

End of the Proof of Theorem 4. Let $\sigma = (2^m, 0)$. Then σ is a substate of $(2^m, 2^{m-1})$ and we have the desired result by Claim 4 and Lemma 1.

In the following two lemmas, upper bounds on the number of questions in the shortest winning strategy for the half-lie game over a search space of cardinality 2^m, are given in terms of Berlekamp's minimum number $q(m)$.

Lemma 4. *For every integer $m \geq 1$, let $E = \{0, 1, \ldots, 2^m - 1\}$. Then there exists a winning strategy to solve the half-lie game over E, using $\leq q(m) - 1$ questions.*

Proof. By Theorem 4 there exists a winning strategy using $m - 1 + \lceil \log(m + 3) \rceil$ questions. Then we need only show that for all $m \geq 1$,

$$m - 1 + \lceil \log(m + 3) \rceil \leq q(m) - 1.$$

Let $k = m - 1 + \lceil \log(m + 3) \rceil$. By definition of Berlekamp's number $q(m)$, it is sufficient to settle the inequality $k + 1 > 2^{k-m}$, that is,

$$m + \lceil \log(m + 3) \rceil > 2^{\lceil \log(m+3) \rceil - 1}. \tag{10}$$

To this purpose, let $s \geq 0$ be the smallest integer such that $m = 2^s - t$ for some $0 \leq t < 2^{s-1}$. Then we argue by cases:

Case 1: $t \geq 3$. Then (10) becomes $2^s - t + s > 2^{s-1}$. This latter inequality holds true, because $2^{s-1} > t \geq 3$ implies $s > 2$, whence $2^{s-1} + s > t$.

Case 2: $0 \leq t \leq 2$. Then (10) becomes $2^s - t + s + 1 > 2^s$, or equivalently,

$$s + 1 > t. \tag{11}$$

Since $2^{s-1} > t$, the only possible cases to be considered are

$$t = 0, \ s \geq 0; \qquad t = 1, \ s \geq 2; \qquad t = 2, \ s \geq 3.$$

In every case, inequality (11) is satisfied and the proof is complete.

Lemma 5. *Fix integers $n \geq 1$ and $3 \leq t \leq n$. Let $m = 2^n - t$. Then there exists a winning strategy for the state $\sigma = (2^m, 0)$ using $\leq q(m) - 2$ questions.*

Proof. Let $k = m - 1 + \lceil \log(m + 3) \rceil$. By Theorem 4 there exists a winning strategy for σ using k questions. Then we need only prove that $k < q(m) - 1$. Recalling the definition of $q(m)$, it is enough to settle the inequality $k + 2 > 2^{k+1-m}$, that is, $m + 1 + \lceil \log(m + 3) \rceil > 2^{\lceil \log(m+3) \rceil}$. From our hypotheses $m = 2^n - t \leq 2^n - 3$ and $t \leq n \leq 2^{n-1}$, we obtain $\lceil \log(m + 3) \rceil = n$. It follows that $m + 1 + \lceil \log(m + 3) \rceil = 2^n - t + 1 + n > 2^n = 2^{\lceil \log(m+3) \rceil}$, as required to conclude the proof.

5   Conclusion: Main Results

Theorem 5. *For each $m = 1, 2, \ldots$, let $q(m)$ be the (Berlekamp) minimum integer q such that $q + 1 \leq 2^{q-m}$. Let $q_{\text{half-lie}}(m)$ denote the number of questions of the shortest winning strategy for the half-lie game over the search space $E = \{0, 1, \ldots, 2^m - 1\}$. Let the sets of integers $\mathcal{C}$ and $\mathcal{E}_s$ ($s = 1, 2, 3, \ldots$) be defined by $\mathcal{C} = \{m \geq 0 \mid q(m) = 2^{q(m)-m} - 1\}$ and $\mathcal{E}_s = \{m \geq 0 \mid m_s + 1 \leq m \leq m_s + s - 2\}$. Then*

(A) $q(m) - 2 \leq q_{\text{half-lie}}(m) \leq q(m) - 1$, for each $m = 1, 2, \ldots$.

(B) $q_{\text{half-lie}}(m) = q(m) - 1$, for all $m \in \mathcal{C}$.

(C) $q_{\text{half-lie}}(m) = q(m) - 2$, for each $s = 1, 2, \ldots$ and $m \in \mathcal{E}_s$.

Proof. (A) By Lemma 4 and Theorem 2. (B) By Lemma 4 and Theorem 3. (C) By Theorem 2 and Lemma 5.

Acknowledgment

We thank Andrzej Pelc for drawing our attention to the half-lie variant of the Ulam-Rényi game.

References

1. E. R. Berlekamp, *Block coding for the binary symmetric channel with noiseless, delayless feedback*, In: Error-correcting Codes, H.B. Mann (Editor), Wiley, New York (1968) pp. 61-88.
2. F. Cicalese, D. Mundici, *Optimal binary search with two unreliable tests and minimum adaptiveness*, In: Proc. European Symposium on Algorithms, ESA '99, J. Nesetril, Ed., *Lecture Notes in Computer Science* 1643 (1999) pp.257-266.
3. F. Cicalese, U. Vaccaro, *Optimal strategies against a liar*, Theoretical Computer Science, **230** (1999) pp. 167-193.
4. S.D. Constantin, T.R.N. Rao, *On the Theory of Binary Asymmetric Error Correcting Codes*, Information and Control **40** (1979) pp. 20-26.
5. R. Hill, *Searching with lies*, In: Surveys in Combinatorics, Rowlinson, P. (Editor), Cambridge University Press (1995) pp. 41-70.

6. A. Pelc, *Solution of Ulam's problem on searching with a lie*, J. Combin. Theory, Ser. A, **44** (1987) pp. 129-142.

7. J.R. Pierce, *Optical Channels: Practical limits with photon counting*, IEEE Trans. Comm. COM-26 (1978) pp. 1819-1821.

8. A. Rényi, *Napló az információelméletről*, Gondolat, Budapest, 1976. (English translation: *A Diary on Information Theory*, J.Wiley and Sons, New York, 1984).

9. R. L. Rivest, A. R. Meyer, D. J. Kleitman, K. Winklmann, J. Spencer, *Coping with errors in binary search procedures*, Journal of Computer and System Sciences, **20** (1980) pp. 396-404.

10. J. Spencer, *Ulam's searching game with a fixed number of lies*, Theoretical Computer Science, **95** (1992) pp. 307-321.

11. S.M. Ulam, *Adventures of a Mathematician*, Scribner's, New York, 1976.

Closure Properties of Real Number Classes under Limits and Computable Operators

Xizhong Zheng

Theoretische Informatik, FernUniversität Hagen,
58084-Hagen, Germany
xizhong.zheng@fernuni-hagen.de

Abstract. In effective analysis, various classes of real numbers are discussed. For example, the classes of computable, semi-computable, weakly computable, recursively approximable real numbers, etc. All these classes correspond to some kind of (weak) computability of the real numbers. In this paper we discuss mathematical closure properties of these classes under the limit, effective limit and computable function. Among others, we show that the class of weakly computable real numbers is not closed under effective limit and partial computable functions while the class of recursively approximable real numbers is closed under effective limit and partial computable functions.

1   Introduction

In computable analysis, a real number x is called *computable* if there is a computable sequence $(x_n)_{n \in \mathbb{N}}$ of rational numbers which converges to x effectively. That is, the sequence satisfies the condition that $|x_n - x| < 2^{-n}$, for any $n \in \mathbb{N}$. In this case, the real number x is not only approximable by some effective procedure, there is also an effective error-estimation in this approximation. In practice, it happens very often that some real values can be effectively approximated, but an effective error-estimation is not always available. To characterize this kind of real numbers, the concept of recursively approximable real numbers is introduced. Namely, a real number x is *recursively approximable* (r.a., in short) if there is a computable sequence $(x_n)_{n \in \mathbb{N}}$ of rational numbers which converges to x. It is first noted by Ernst Specker in [15] that there is a recursively approximable real number which is not computable by encoding the halting problem into the binary expansion of a recursively approximable real numbers.

The class $\mathbf{C}_e$ of computable real numbers and the class $\mathbf{C}_{ra}$ of recursively approximable real numbers shares a lot of mathematical properties. For example, both $\mathbf{C}_e$ and $\mathbf{C}_{ra}$ are closed under the arithmetical operations and hence they are algebraic fields. Furthermore, these two classes are closed under the computable real functions, namely, if x is computable (r.a.) real number and f is a computable real function in the sense of, say, Grzegorczyk [6], then $f(x)$ is also computable (resp. r.a.).

The classes of real numbers between $\mathbf{C}_e$ and $\mathbf{C}_{ra}$ are also widely discussed (see e.g. [12,13,4,2,18]). Among others, the class of, so called, recursively enumerable real numbers might be the first widely discussed such class. A real

D.-Z. Du et al. (Eds.): COCOON 2000, LNCS 1858, pp. 170–179, 2000.
© Springer-Verlag Berlin Heidelberg 2000

number x is called *recursive enumerable* if its left Dedekind cut is an r.e. set of rational numbers, or equivalently, there is an increasing computable sequence $(x_n)_{n \in \mathbb{N}}$ of rational numbers which converges to x. We prefer to call such real numbers *left computable* because it is very naturally related to the left topology $\tau_< := \{(a; \infty) : a \in \mathbb{R}\}$ of the real numbers by the admissible representation of Weihrauch [16]. Similarly, a real number x is called *right computable* if it is a limit of some decreasing computable sequence of rational numbers. Left and right computable real numbers are called *semi-computable*. Robert Soare [12,13] discusses widely the recursion-theoretical properties of the left Dedekind cuts of the left computable real numbers. G. S. Ceĭtin [4] shows that there is an r.a. real number which is not semi-computable. Another very interesting result, shown by a series works of Chaitin [5], Solovay [14], Calude et al. [2] and Slaman [10], says that a real number x is r.e. random if and only if it is an Ω-number of Chaitin which is the halting probability of an universal self-delimiting Turing machine. We omit the details about these notions here and refer the interested readers to a nice survey paper of Calude [3].

Although the class of left computable real numbers has a lot of nice properties, it is not symmetrical in the sense that the real number $-x$ is right computable but usually not left computable for a left computable real number x. Furthermore, even the class of semi-computable real numbers is also not closed under the arithmetical operations as shown by Weihrauch and Zheng [18]. Namely, there are left computable real numbers y and z such that $y - z$ is neither left nor right computable. As the arithmetical closure of semi-computable real numbers, Weihrauch and Zheng [18] introduces the class of weakly computable real numbers. That is, a real number x is *weakly computable* if there are two left computable real numbers y and z such that $x = y - z$. It is shown in [18] that x is weakly computable if and only if there is a computable sequence $(x_n)_{n \in \mathbb{N}}$ which converges to x weakly effectively, i.e. $\lim_{n \to \infty} x_n = x$ and $\sum_{n=0}^{\infty} |x_n - x_{n+1}|$ is finite. By this characterization, it is also shown in [18] that the class of weakly computable real numbers is an algebraic field and is strictly between the classes of semi-computable and r.a. real numbers. In this paper we will discuss other closure properties of weakly computable real numbers for limits, effective limits and computable real functions. We show that weakly computable real numbers are not closed under the effective limits and partial computable real functions. For other classes mentioned above, we carry out also a similar discussion.

At the end of this section, let us explain some notions at first. For any set $A \subseteq \mathbb{N}$, denote by $x_A := \sum_{n \in A} 2^{-n}$ the real number whose binary expansion corresponds to set A. For any $k \in \mathbb{N}$, we define $kA := \{kn : n \in A\}$. For any function $f : \mathbb{N} \to \mathbb{N}$, a set A is called f-r.e. if there is a computable sequence $(A_n)_{n \in \mathbb{N}}$ of finite subsets of $\mathbb{N}$ such that $A = \cup_{i \in \mathbb{N}} \cap_{j \geq i} A_j$ and $|\{s : n \in A_s \Delta A_{s+1}\}| < f(n)$ for all $n \in \mathbb{N}$, where $A \Delta B := (A \backslash B) \cup (B \backslash A)$. If $f(n) := k$ is a constant function, then f-r.e. sets are also called k-r.e. A is called ω-r.e. iff there is a recursive function f such that A is f-r.e.

2   Computability of Real Numbers

In this section we give at first the formal definition of various versions of computability of real numbers and then recall some important properties about these notions. We assume that the reader familiar the computability about subsets of the natural nmbers $\mathbb{N}$ and number-theoretical functions. A sequence $(x_n)_{n\in\mathbb{N}}$ of rational numbers is computable iff there are recursive functions $a, b, c : \mathbb{N} \to \mathbb{N}$ such that $x_n = (a(n) - b(n))/(c(n)+1)$. We summarize the computability notions for real numbers as follows.

Definition 1. For any real number $x \in \mathbb{R}$,

1. x is *computable* iff there is a computable sequence $(x_n)_{n\in\mathbb{N}}$ of rational numbers such that $x = \lim_{n\to\infty} x_n$ and $\forall n\, (|x_n - x_{n+1}| < 2^{-n})$. In this case, the sequence $(x_n)_{n\in\mathbb{N}}$ is called *fast convergent* and it converges to x *effectively*.

2. x is *left (right) computable* iff there is an increasing (decreasing) computable sequence $(x_n)_{n\in\mathbb{N}}$ of rational numbers such that $x = \lim_{n\to\infty} x_n$. Left and right computable real numbers are all called *semi-computable*.

3. x is *weakly computable* (w.c. in short) iff there is a computable sequence $(x_n)_{n\in\mathbb{N}}$ of rational numbers such that $x = \lim_{n\to\infty} x_n$ and $\sum_{n=0}^{\infty} |x_n - x_{n+1}|$ is finite. $(x_n)_{n\in\mathbb{N}}$ is called converging to x *weakly effectively*.

4. x is *recursively approximable* (r.a., in short) iff there is a computable sequence $(x_n)_{n\in\mathbb{N}}$ of rational numbers such that $x = \lim_{n\to\infty} x_n$.

The class of computable, left computable, right computable, semi-computable, w.c., r.a. real numbers is denoted by $\mathbf{C}_e, \mathbf{C}_{lc}, \mathbf{C}_{rc}, \mathbf{C}_{sc}, \mathbf{C}_{wc}, \mathbf{C}_{ra}$, respectively.

As shown in [18], the relationship among these classes looks like the following

$$\mathbf{C}_e = \mathbf{C}_{lc} \cap \mathbf{C}_{rc} \subsetneq \begin{matrix} \mathbf{C}_{lc} \\ \mathbf{C}_{rc} \end{matrix} \subsetneq \mathbf{C}_{sc} = \mathbf{C}_{lc} \cup \mathbf{C}_{rc} \subsetneq \mathbf{C}_{wc} \subsetneq \mathbf{C}_{ra}.$$

Note that in above definition, we define various versions of computability of real numbers in a similar way. Namely, a real number x is of some version of computability iff there is a computable sequence of rational numbers which satisfies some special property and converges to x. For example, if $P_{lc}[(x_n)]$ means that $(x_n)_{n\in\mathbb{N}}$ is increasing, then $x \in \mathbf{C}_{lc}$ iff there is a computable sequence $(x_n)_{n\in\mathbb{N}}$ of rational numbers such that $P_{lc}[(x_n)]$ and $\lim_{n\to\infty} x_n = x$. In general, for any reasonable property on sequences, we can define a corresponding class of real numbers which have some kind of (weaker) computability. This can even be extended to the case of sequences of real numbers as in the following definition.

Definition 2. Let P be any property about the sequences of real numbers. Then

1. A real number x is called *P-computable* if there is a computable sequence $(x_n)_{n\in\mathbb{N}}$ of rational numbres which satisfies property P and converges to x. The class of all *P-computable* real numbers is denoted by $\mathbf{C}_P$

2. A sequence $(x_n)_{n\in\mathbb{N}}$ of real numbers is called *P-computable*, or it is a *computable sequence of* $\mathbf{C}_P$ iff there is a computable double sequence $(r_{nm})_{nm\in\mathbb{N}}$ of rational numbers such that $(r_{nm})_{m\in\mathbb{N}}$ satisfies P and $\lim_{m\to\infty} r_{nm} = x_n$ for all $n \in \mathbb{N}$.

3. The class $\mathbf{C}_P$ is called *"closed under limits"*, iff for any computable sequences $(x_n)_{n\in\mathbb{N}}$ of $\mathbf{C}_P$, the limits $x := \lim_{n\to\infty} x_n$ is also in $\mathbf{C}_P$ whenever $(x_n)_{n\in\mathbb{N}}$ satisfies P and converges.

4. The class $\mathbf{C}_P$ defined in 2. is called *"closed under effective limits"*, iff for any fast convergent computable sequences $(x_n)_{n\in\mathbb{N}}$ of $\mathbf{C}_P$, the limits $x := \lim_{n\to\infty} x_n$ is also in $\mathbf{C}_P$.

Now we remind the notion of computable real function. There are a lot of approaches to define the computability of real functions. Here we use Grzegorczyk-Ko-Weihrauch's approach and define computable real function in terms of "Type-two Turing Machine" (TTM, in short) of Weihrauch.

Let Σ be any alphabet. Σ^* and Σ^∞ are sets of all finite strings and infinite sequences on Σ, respectively. Roughly, TTM M extends the classical Turing machine in such a way that it can be inputed and also can output infinite sequences as well as finite strings. For any $p \in \Sigma^* \cup \Sigma^\infty$, $M(p)$ outputs a finite string q, if $M(p)$ writes q in output tape and halt in finite steps similar to the case of classical Turing machine. $M(p)$ outputs an infinite sequence q means that $M(p)$ will compute forever and keep writing q on the output tape. We omit the formal details about TTM here and refer the interested readers to [16,17]. We will omit also the details about the encoding of rational numbers by Σ^* and take directly the sequences of rational numbers as inputs and outputs to TTM's.

Definition 3. A real function $f :\subseteq \mathbb{R} \to \mathbb{R}$ is *computable* if there is a TTM M such that, for any $x \in \mathrm{dom}(f)$ and any sequence $(u_n)_{n\in\mathbb{N}}$ of rational numbers which converges effectively to x, $M((u_n)_{n\in\mathbb{N}})$ outputs a sequence $(v_n)_{n\in\mathbb{N}}$ of rational numbers which converges to $f(x)$ effectively.

Note that, in this definition we do not add any restriction on the domain of computable real function. Hence a computable real function can have any type of domain, because $f \restriction A$ is always computable whenever f is computable and $A \subseteq \mathrm{dom}(f)$. Furthermore, for a total function $f : [0,1] \to \mathbb{R}$, f is computable iff f is sequentially computable and effectively uniformly continuous (see [9]).

Definition 4. For any subset $\mathbf{C} \subseteq \mathbb{R}$,

1. $\mathbf{C}$ is *closed under computable operators*, iff $f(x) \in \mathbf{C}$ for any $x \in \mathbf{C}$ and any total computable real function $f : \mathbb{R} \to \mathbb{R}$.

2. $\mathbf{C}$ is *closed under partial computable operators*, iff $f(x) \in \mathbf{C}$, for any $x \in \mathbf{C}$ and any partial computable real function $f :\subseteq \mathbb{R} \to \mathbb{R}$ with $x \in \mathrm{dom}(f)$.

Following proposition follows immediately from the definition. Remember that, $A \subseteq \mathbb{N}$ is Δ_2^0 iff A is Turing reducible to the halting problem $\emptyset'$.

Proposition 1. *1.* $x_A \in \mathbf{C}_e \iff A$ *is recursive.*

2. $x_A \in \mathbf{C}_{ra} \iff A$ *is a* Δ_2^0*-set, or equivalently,* $A \leq_T \emptyset'$.

3. $\mathbf{C}_e$ *and* $\mathbf{C}_{ra}$ *are closed under arithmetical operations* $+, -, \times$ *and* $\div$*, hence they are algebraic fields.*

4. $\mathbf{C}_e$ *are closed under limits and computable real functions.*

5. $\mathbf{C}_{lc}$ *and* $\mathbf{C}_{rc}$ *are closed under addition.*

Some other non-trivial closure properties are shown in [18] and [19].

Theorem 1 (Weihrauch and Zheng). *1.* $\mathbf{C}_{sc}$ *is not closed under addition, i.e. there are left computable* y *and right computable* z *such that* $y + z$ *is neither left nor right computable.*

2. $\mathbf{C}_{wc}$ *is closed under arithmetical operations. In fact* $\mathbf{C}_{wc}$ *is just the closure of* $\mathbf{C}_{sc}$ *under the arithmetical operations.*

It is not very surprising that the classes $\mathbf{C}_{lc}$ and $\mathbf{C}_{rc}$ are not closed under "subtraction" and, in general, under computable real functions, because they are not symmetrical. On the other hand, the class $\mathbf{C}_{wc}$ is symmetrical and closed under arithmetical operations. So it is quite natural to ask whether it is also closed under limits and computable real functions. In the following we will give the negative answers to both questions. To this end we need the following observations about weakly computable real numbers.

Theorem 2 (Ambos-Spies, [1]). *1. If* $A, B \subseteq \mathbb{N}$ *are incomparable under Turing reduction, then* $x_{A \oplus \overline{C}}$ *is not semi-computable.*

2. For any set $A \subseteq \mathbb{N}$, *if* x_{2A} *is weakly computable, then* A *is* f-*r.e. for* $f(n) := 2^{3n}$, *hence* A *is* ω-*r.e.*

Theorem 3 (Zheng [20]). *There is a non-*ω-*r.e.* Δ_2^0-*set* A *such that* x_A *is weakly computable.*

3   Closure Property under Limits

In this section, we will discuss the closure properties of several classes of real numbers under limits. We first consider the classes of left and right computable real numbers. The following result is quite straightforward.

Theorem 4. *The classes of left and right computable real numbers are closed under limits, respectively.*

For semi-computable real numbers, the situation is different.

Theorem 5. *The class* $\mathbf{C}_{sc}$ *is not closed under limits.*

Proof. Define, for any $n, s \in \mathbb{N}$, at first the following sets:

$$A := \{e \in \mathbb{N} : \varphi_e \text{ is total}\}$$
$$A_n := \{e \in \mathbb{N} : (\forall x \leq n)\varphi_e(x) \downarrow\}$$
$$A_{n,s} := \{e \in \mathbb{N} : (\forall x \leq n)\varphi_{e,s}(x) \downarrow\}$$

Since $A_{n,s} \subseteq A_{e,s+1}$, $(x_{A_{n,s}})_{n,s \in \mathbb{N}}$ is obviously a computable sequence of rational numbers such that, for and $n \in \mathbb{N}$, $(x_{A_{n,s}})_{s \in \mathbb{N}}$ is nondecreasing and converges to x_{A_n}. That is, $(x_{A_n})_{n \in \mathbb{N}}$ is a computable sequence of $\mathbf{C}_{lc}$, hence it is a computable sequence of $\mathbf{C}_{sc}$. But its limit x_A is not semi-computable. In fact x_A is even not r.a. by Proposition 1, since A is not a Δ_2^0-set. $\qquad\square$

Note that in above proof, as a computable sequence of $\mathbf{C}_{lc}$, $(x_{A_n})_{n\in\mathbb{N}}$ is also a computable sequence of $\mathbf{C}_{wc}$ and $\mathbf{C}_{ra}$. Then the folowing corollary follows immediately.

Corollary 1. *The classes $\mathbf{C}_{wc}$ and $\mathbf{C}_{ra}$ are not closed under the limit.*

Now we discuss the closure property under the effective limits. We will show that the class of semi-computable real numbers is closed under effective limits and the class of weakly computable real numbers, hence also the class of r.a. real numbers, is not closed under effective limits.

Theorem 6. *The class $\mathbf{C}_{sc}$ is closed under the effective limits.*

Proof. Let $(x_n)_{n\in\mathbb{N}}$ be a computable sequence of $\mathbf{C}_{sc}$ which satisfies the condition that $\forall n(|x_n - x_{n+1}| < 2^{-(n+1)})$ and converges to x. We shall show that $x \in \mathbf{C}_{sc}$.

By Definition 2, there is a computable sequence $(r_{ij})_{i,j\in\mathbb{N}}$ of rational numbers such that, for any $n \in \mathbb{N}$, $(r_{nj})_{n\in\mathbb{N}}$ is monotonic and converges to x_n. For any n, we can effectively determine whether x_n is left or right computable by comparing, say, r_{n0} and r_{n1}. Therefore, the sequence $(x_n)_{n\in\mathbb{N}}$ can be split into two computable subsequences $(x_{n_i})_{i\in\mathbb{N}}$ and $(x_{m_i})_{i\in\mathbb{N}}$ of left and right computable real numbers, respectively. At least one of them is infinite. Suppose w.l.o.g. that $(x_{n_i})_{i\in\mathbb{N}}$ is an infinite sequence. Obviously it is also a fast convergent computable sequence, i.e., $|x_{n_i} - x_{n_{i+1}}| < 2^{-i}$, since $(x_n)_{n\in\mathbb{N}}$ converges fast. Define a new sequence $(y_n)_{n\in\mathbb{N}}$ by $y_i := x_{n_i} - 2^{-(i-1)}$. Since $y_{i+1} = x_{n_{i+1}} - 2^{-i} = (x_{n_{i+1}} - x_{n_i} + 2^{-i} + (x_{n_i} - 2^{-(i-1)}) \geq x_{n_i} - 2^{-(i-1)} = y_i$ $(y_i)_{i\in\mathbb{N}}$ is an increasing sequence. Let $r'_{ij} := r_{ij} - 2^{-(i-1)}$. Then $(r'_{ij})_{i,j\in\mathbb{N}}$ is a computable sequence of rational numbers such that, for any i, $(r'_{ij})_{j\in\mathbb{N}}$ is increasing and converges to y_i. Namely, $(y_i)_{i\in\mathbb{N}}$ is an increasing computable sequence of $\mathbf{C}_{lc}$. By Theorem 4, its limit $\lim_{i\to\infty} y_i = \lim_{i\to\infty} x_{n_i} = \lim_{i\to\infty} x_i = x$ is also left computable, i.e., $x \in \mathbf{C}_{lc} \subseteq \mathbf{C}_{sc}$. □

Theorem 7. *The class $\mathbf{C}_{wc}$ is not closed under effective limits.*

Proof. Suppose by Theorem 3 that A is a non-ω-r.e. Δ_2^0-set such that x_A is weakly computable. Then x_{2A} is not weakly computable by Theorem 2. Let $(A_s)_{s\in\mathbb{N}}$ be a recursive approximation of A such that $(x_{A_s})_{s\in\mathbb{N}}$ converges to x_A weakly effectively, i.e. $\sum_{s=0}^{\infty} |x_{A_s} - x_{A_{s+1}}| \leq C$ for some $C \in \mathbb{N}$. Define, for $n, s \in \mathbb{N}$,

$$B_{n,s} := 2(A_s \upharpoonright (n+1)) \cup (A_s \downharpoonright 2n)$$
$$B_n := 2(A \upharpoonright (n+1)) \cup (A \downharpoonright 2n)$$

It is easy to see that $(B_{n,s})_{n,s\in\mathbb{N}}$ is a computable sequence of finite subsets of $\mathbb{N}$, hence $(x_{B_{n,s}})_{n,s\in\mathbb{N}}$ is a computable sequence of rational numbers.

Since $\lim_{s\to\infty} A_s = A$, there is an $N(n)$, for any $n \in \mathbb{N}$ such that, for any $s \geq N(n)$, $A_s \upharpoonright (n+1) = A \upharpoonright (n+1)$. Let $C_1 = \sum_{s=0}^{N(n)} |x_{B_{n,s}} - x_{B_{n,s+1}}|$. Then

$\sum_{s=0}^{\infty} |x_{B_{n,s}} - x_{B_{n,s+1}}| = \sum_{s=0}^{N(n)} |x_{B_{n,s}} - x_{B_{n,s+1}}| + \sum_{s>N(n)}^{\infty} |x_{B_{n,s}} - x_{B_{n,s+1}}|$
$= C_1 + \sum_{s>N(n)}^{\infty} |x_{A_{n,s}} - x_{A_{n,s+1}}| < C_1 + C$. On the other hand, it is easy to
see that $\lim_{s\to\infty} x_{B_{n,s}} = x_{B_n}$. Therefore, the sequence $(x_{B_{n,s}})_{n,s\in\mathbb{N}}$ converges
to x_{B_n} weakly effectively. Hence $(x_{B_n})_{n\in\mathbb{N}}$ is a weakly computable sequence of
real numbers. By the definition of B_n, $B_n \Delta 2A \subseteq \{2n+1, 2n+2, \ldots\}$. It follows
that $|x_{B_n} - x_{2A}| \leq 2^{-2n} \leq 2^{-n}$ This means that $(x_{B_n})_{n\in\mathbb{N}}$ converges to x_{2A}
effectively and this ends the proof of the theorem. $\square$

Theorem 8. *The class $\mathbf{C}_{ra}$ is closed under effective limits.*

Proof. Let $(x_n)_{n\in\mathbb{N}}$ be any computable sequence of $\mathbf{C}_{ra}$ which converges effectively to x. Assume w.l.o.g. that it satisfies, for all $n \in \mathbb{N}$, the condition
$|x_n - x_{n+1}| < 2^{-(n+1)}$. By Definition 2, there is a computable sequence $(r_{ij})_{i,j\in\mathbb{N}}$
of rational numbers such that, for any $n \in \mathbb{N}$, $\lim_{s\to\infty} r_{ns} = x_n$. We shall show
that $x \in \mathbf{C}_{ra}$.

It suffices to construct a computable sequence $(u_s)_{s\in\mathbb{N}}$ of rational numbers
such that $\lim_{s\to\infty} u_s = x$. This sequence will be constructed from $(r_{ij})_{i,j\in\mathbb{N}}$ in
following stages:

The construction of sequence $(u_s)_{s\in\mathbb{N}}$:

Stage $s = 0$: Define $u_0 := r_{00}$, $t(0,0) := 0$ and $i(0) := 0$.

Stage $s+1$: Given $i(s)$, $u_0, \ldots, u_{i(s)}$ and $t(j,s)$ for all $j \leq s$. If there is $j \leq s$
satisfying $|u_{t(j-1,s)} - r_{js}| < 2^{-(j-1)}$ such that either $t(j,s) \neq -1$ & $|u_{t(j,s)} - r_{js}| \geq
2^{-(j+1)}$ or $t(j-1,s) \neq -1$ & $t(j,s) = -1$, then choose j_0 as minimal such j and
define

$$(*) \quad \begin{cases} i(s+1) & := i(s) + 1 \\ u_{i(s+1)} & := r_{j_0,s} \\ t(j,s+1) := \begin{cases} t(j,s) & \text{if } 0 \leq j < j_0 \\ i(s)+1 & \text{if } j = j_0 \\ -1 & \text{if } j_0 < j \leq s+1 \end{cases} \end{cases}$$

Otherwise, if no such j exists, then define, $i(s+1) := i(s)$, $t(s+1, s+1) := -1$
and $t(j, s+1) := t(j,s)$ for all $j \leq s$.

To show this construction succeeds, we need only to prove the following
claims.

1. For any $j \in \mathbb{N}$, the limit $t(j) := \lim_{s\to\infty} t(j,s)$ exists and satisfies the condi-
   tions that $t(j) \neq -1$ and $|u_{t(j)} - x_j| \leq 2^{-(j+1)}$.
2. $\lim_{s\to\infty} i(s) = +\infty$.
3. For any $j \in \mathbb{N}$, $(\forall s \leq t(j))(|u_{t(j)} - u_s| \leq 2^{-j})$.

Now it is clear that the sequence $(u_s)_{s\in\mathbb{N}}$ constructed above is a computable
infinite sequence of rational numbers. Furthermore, this sequence converges to
x. This completes the proof of Theorem. $\square$

4   Closure Property under Computable Operators

In this section we will discuss the closure property under computable operators. The following result about left and right computable real numbers is immediate by the fact that the real function f defined by $f(x) = -x$ is computable.

Proposition 2. *The classes $\mathbf{C}_{lc}$ and $\mathbf{C}_{rc}$ are not closed under the computable operators, hence is also not closed under partial computable operators.*

To discuss the closure property under partial computable operators for other classes, we will apply the following observation of Ko [7].

Theorem 9 (Ker-I Ko [7]). *For any sets $A, B \subseteq \mathbb{N}$, $A \leq_T B$ iff there is a (partial) computable real function $f :\subseteq \mathbb{R} \to \mathbb{R}$ such that $f(x_B) = x_A$.*

¿From this result, it is easy to show that a lot of classes of real numbers are not closed under the partial computable operators.

Theorem 10. *The classes $\mathbf{C}_{sc}$ and $\mathbf{C}_{wc}$ are not closed under the partial computable operators. The class $\mathbf{C}_{ra}$ is closed under partial computable operators.*

Proof. 1. For class $\mathbf{C}_{sc}$. By Muchnik-Friedberg Theorem (see [11]), there are two r.e. sets A and B such that they are incomparable under Turing reduction. Then $x_{A \oplus \overline{B}}$ is not semi-computable by Theorem 2. On the other hand, $x_{A \oplus B}$ is left computable since $A \oplus B$ is r.e. Obviously, we have the reduction that $A \oplus \overline{B} \leq_T A \oplus B$. By Theorem 9, there is a computable real function f such that $f(x_{A \oplus B}) = x_{A \oplus \overline{B}}$. Therefore, $\mathbf{C}_{sc}$ is not closed under partial computable operators.

2. For class $\mathbf{C}_{wc}$. By Theorem 3, there is a non-ω-r.e. set A such that x_A is weakly computable. On the other hand, x_{2A} is not weakly computable by Theorem 2 since $2A$ is obviously not ω-r.e. Because $2A \leq_T A$, by Theorem 9, there is a computable real function f such that $f(x_A) = x_{2A}$. That is, $\mathbf{C}_{wc}$ is not closed under the partial computable operators.

3. For class $\mathbf{C}_{ra}$, it follows immediately from the fact that a real number x_A is r.a iff A is a Δ_2^0-set and the class of all Δ_2^0-sets is closed under the Turing reduction, i.e. if $A \leq_T B$ and B is Δ_2^0-set, then A is also Δ_2^0-set. $\square$

It is shown in Theorem 1 that the class $\mathbf{C}_{sc}$ is not closed under addition. Hence it is not closed under the polynomial functions with several arguments. Namely, if $p(x, \cdots, x_n)$, $n \geq 2$, is a polynomial with rational coefficients and $a_1, \cdots, a_n$ are semi-computable real numbers, then $p(a_1, \cdots, a_n)$ is not necessary semi-computable. But for the case of $n = 1$, it is not clear. Furthermore it is also not known whether the class of semi-computable real numbers is closed under the total computable real functions.

Similarly, it remains still open whether the class $\mathbf{C}_{wc}$ is closed under (total) computable operators. We guess it is not. One possible approach is to define a computable real function which maps some weakly computable x_A for a non-ω-r.e. set A to a not weakly computable real number x_{2A}. Using the idea in the proof of Theorem 9, it is not difficult to show that there is a computable real function $f :\subseteq \mathbb{R} \to \mathbb{R}$ such that $f(x_A) = x_{2A}$ for any irrational x_A. Unfortunately, such function cannot be extended to a total computable real function as shown by next result.

Theorem 11. *1. Let $f :\subseteq \mathbb{N} \to \mathbb{N}$ be a function such that $f(x_A) = x_{2A}$ for any irrational x_A. If x_A is a rational number, then there is a sequence $(x_n)_{n \in \mathbb{N}}$ of irrational numbers such that $\lim_{n \to \infty} x_n = x_{2A}$ and $\lim_{n \to \infty} f(x_n) = x_{2A}$.*
2. The function $f : [0; 1] \to \mathbb{R}$ defined by $f(x_A) := x_{2A}$ for any $A \subseteq \mathbb{N}$ is not continuous at any rational points, hence it is not computable.

Proof. 1. Suppose that function $f :\subseteq \mathbb{N} \to \mathbb{N}$ satisfies $f(x_A) = x_{2A}$ for any irrational x_A. Let x_A be rational, hence A is a finite set. We define a sequence $(x_n)_{n \in \mathbb{N}}$ of irrational numbers by $x_n := x_A + \sqrt{2} \cdot 2^{-(n+1)}$. Let n_0 be the maximal element of A. Define a set A_n by $x_{A_n} = \sqrt{2} \cdot 2^{-(n+1)}$ for any $n \in \mathbb{N}$. Then for any $n > n_0$, $x_{A_n} < 2^{-n} \le 2^{-n_0}$. This implies that A_n contains only the elements which are bigger than n_0. Therefore, for any $n \ge n_0$, $A \cap A_n = \emptyset$ and $f(x_n) = f(x_A + \sqrt{2} \cdot 2^{-(n+1)}) = f(x_A + x_{A_n}) = f(x_{A \cup A_n}) = x_{2(A \cup A_n)} = x_{(2A) \cup (2A_n)} = x_{2A} + x_{2A_n}$. Since $\lim_{n \to \infty} x_{A_n} = 0$, it is easy to see that $\lim_{n \to \infty} x_{2A_n} = 0$ too. So we conclude that $\lim_{n \to \infty} f(x_n) = x_{2A}$.

2. Suppose that $f : [0; 2] \to \mathbb{R}$ satisfies $f(x_A) = x_{2A}$ for any $A \subseteq \mathbb{N}$. For any rational x_A, A is finite. Let n_0 be the maximal element of A and $A' := A \backslash \{n_0\}$ and define, for all $n \in \mathbb{N}$, a finite set A_n by $A_n := A' \cup \{n_0 + 1, n_0 + 2, \cdots, n_0 + n\}$. Then it is easy to see that $\lim_{n \to \infty} x_{A_n} = x_A$. On the other hand we have:

$$f(x_{A_n}) = x_{2A_n} = x_{2A'} + \sum_{i=1}^{n} 2^{-2(n_0 + i)}$$
$$= x_{2A} - 2^{-2n_0} + 2^{-2n_0} \cdot (1 - 2^{-2n})/3$$

This implies that $\lim_{n \to \infty} f(x_{A_n}) = x_{2A} - 2^{-2n_0 + 1}/3 \ne x_{2A}$. $\square$

In summary, the closure properties of several classes of real numbers under arithmetic operations ($+$, $-$, $\times$ and $\div$), limits, effective limits, partial computable operators and computable operators are listed in the following table:

	arithmetic operations	limits	effective limits	computable operators	partial computable operators
C_e	Yes	Yes	Yes	Yes	Yes
C_{lc}	No	Yes	Yes	No	No
C_{rc}	No	Yes	Yes	No	No
C_{sc}	No	No	Yes	?	No
C_{wc}	Yes	No	No	?	No
C_{ra}	Yes	No	Yes	Yes	Yes

References

1. K. Ambos-Spies A note on recursively approximable real numbers, Research Report on Mathematical Logic, University of Heidelberg, No. 38, September 1998.
2. C. Calude, P. Hertling, B. Khoussainov, and Y. Wang, Recursive enumerable reals and Chaintin's Ω-number, in *STACS'98*, pp596–606.
3. C. Calude. A characterization of c.e. random reals. CDMTCS Research Report Series 095, March 1999.
4. G. S. Ceĭtin A pseudofundamental sequence that is not equivalent to a monotone one. (Russian) *Zap. Naučn. Sem. Leningrad. Otdel. Mat. Inst. Steklov.* (LOMI) 20 1971 263–271, 290.
5. G. J. Chaitin A theory of program size formally identical to information theory, *J. of ACM.*, 22(1975), 329–340.
6. A. Grzegorczyk. On the definitions of recursive real continuous functions, *Fund. Math.* 44(1957), 61–71.
7. Ker-I Ko Reducibilities of real numbers, *Theoret. Comp. Sci.* 31(1984) 101–123.
8. Ker-I Ko *Complexity Theory of Real Functions*, Birkhäuser, Berlin, 1991.
9. M. Pour-El & J. Richards *Computability in Analysis and Physics*. Springer-Verlag, Berlin, Heidelberg, 1989.
10. T. A. Slaman Randomness and recursive enumerability, preprint, 1999.
11. R. Soare *Recursively Enumerable Sets and Degrees*, Springer-Verlag, Berlin, Heidelberg, 1987.
12. R. Soare Recursion theory and Dedekind cuts, *Trans, Amer. Math. Soc.* 140(1969), 271–294.
13. R. Soare Cohesive sets and recursively enumerable Dedekind cuts, *Pacific J. of Math.* 31(1969), no.1, 215–231.
14. R. Solovay. Draft of a paper (or series of papers) on Chaitin's work … done for the most part during the period of Sept. –Dec. 1975, unpublished manuscript, IBM Thomas J. Watson Research Center, Yorktoen Heights, New York, May 1975.
15. E. Specter Nicht konstruktive beweisbare Sätze der Analysis, *J. Symbolic Logic* 14(1949), 145–158
16. K. Weihrauch *Computability*. EATCS Monographs on Theoretical Computer Science Vol. 9, Springer-Verlag, Berlin, Heidelberg, 1987.
17. K. Weihrauch. *An Introduction to Computable Analysis*. Springer-Verlag, 2000. (to appear).
18. K. Weihrauch & X. Zheng A finite hierarchy of the recursively enumerable real numbers, *MFCS'98* Brno, Czech Republic, August 1998, pp798–806.
19. K. Weihrauch & X. Zheng Arithmetical hierarchy of ral numbers. in *MFCS'99*, Szklarska Poreba, Poland, September 1999, pp 23–33.
20. X. Zheng. Binary enumerability of real numbers. in *Computing and Combinatorics, Proc. of COCOON'99*, Tokyo, Japan. July 26–28, 1999, pp300–309.

A Characterization of Graphs
with Vertex Cover Six

Michael J. Dinneen and Liu Xiong

Department of Computer Science, University of Auckland,
Private Bag 92019, Auckland, New Zealand
(mjd@cs.auckland.ac.nz & liuxiong@microsoft.com)

Abstract. The complete list of forbidden minors (obstructions) for the family of graphs with vertex cover 6 have been found. This paper shows how one can limit the search space of graphs and how to simplify the process for deciding whether a graph is an obstruction for k–Vertex Cover. The upper bounds $2k + 1$ $(2k + 2)$ on the maximum number of vertices for connected (disconnected) obstructions are shown to be sharp for all $k > 0$.

1 Introduction

The main contribution of this paper is the characterization of graphs with vertex cover at most 6 by its obstruction set (forbidden minors). The general problem of vertex cover (which is $\mathcal{N}P$-complete; see [GJ79]) asks whether a graph has a set of vertices of size at most k that covers all edges (a more formal definition is given below). Earlier Cattell and Dinneen in [CD94] classified the families of graphs with vertex cover at most 5 by using the computational machinery now described in [CDD+00]. Our current results are based on a more family-specific approach where we limit the search space of graphs. In this paper, as our primary limiting factor, we prove an exact upper bound on the number of vertices for an obstruction to any k–Vertex Cover family.

The numbers of obstructions for 1–Vertex Cover to 5–Vertex Cover, along with our new result for 6–Vertex Cover, are listed below in Table 1.

We had known that the set of obstructions for 6–Vertex Cover is finite by the now-famous Graph Minor Theorem (GMT) of Robertson and Seymour [RS85]. They proved Wagner's conjecture which states that there are a finite number of obstructions for any graph family closed under the minor order. Unfortunately the proof of the GMT does not indicate how to find these obstructions. The set of planar graphs are the best known example of a family with "forbidden graphs", where Kuratowski's characterization provides us with K_5 and $K_{3,3}$ as the only obstructions to planarity. A lot of work has recently been done concerning the development of general methods for computing minor-order obstructions, such as mentioned in the papers [FL89, APS91, LA91, Pro93].

In this paper we use standard graph theory definitions. A graph is a pair (V, E), where V is a finite set of vertices and E is a set of undirected edges

D.-Z. Du et al. (Eds.): COCOON 2000, LNCS 1858, pp. 180–192, 2000.

Table 1. Numbers of obstructions for k–VERTEX COVER, $1 \leq k \leq 6$.

k	Connected obstructions	Disconnected obstructions	Total obstructions
1	1	1	2
2	2	2	4
3	3	5	8
4	8	10	18
5	31	25	56
6	**188**	**72**	**260**

connecting two vertices of V. An edge between vertices x and y of V will be denoted by xy.

A graph H is a *minor* of a graph G, denoted $H \leq_m G$, if a graph isomorphic to H can be obtained from G by a sequence of a operations chosen from: (1) delete a vertex, (2) delete an edge, or (3) contract an edge (removing any multiple edges or loops that form). The *minor order* is the set of finite graphs ordered by $\leq_m$ and is easily seen to be a partial order.

A family $\mathcal{F}$ of graphs is a *(minor-order) lower ideal*, if whenever a graph $G \in \mathcal{F}$ implies that $H \in \mathcal{F}$ for any $H \leq_m G$. An *obstruction* (often called *forbidden minor*) O for a lower ideal $\mathcal{F}$ is a minor-order minimal graph not in $\mathcal{F}$. Thus, for example, K_5 and $K_{3,3}$ are the 'smallest' non-planar graphs (under the minor order). Recall that by the GMT, a complete set of obstructions provides a *finite characterization* for any (minor-order) lower ideal $\mathcal{F}$.

The graph families of interest in this paper are based on the following problem [GJ79].

Problem. Vertex Cover
Input: Graph $G = (V, E)$ and a positive integer $k \leq |V|$.
Question: Is there a subset $V' \subseteq V$ with $|V'| \leq k$ such that V' contains at least one vertex from every edge in E?

A set V' in the above problem is called a *vertex cover* for the graph G. The family of graphs that have a vertex cover of size at most k will be denoted by k–VERTEX COVER, which is easily seen to be a lower ideal (see [CD94]). For a given graph G, let $VC(G)$ denote the least k such that G has vertex cover of cardinality k. Figure 1 shows an example of a graph G with $VC(G) = 4$.

We now describe the organization of our paper. In the next section we first explain our computational model and give some general results relating to how to compute obstructions for any k–VERTEX COVER lower ideal. We then state some specific results regarding 6–VERTEX COVER, namely vertex and edge bounds. After a short conclusion and references, we compactly list all of the connected obstructions for 6–VERTEX COVER in Figures 4–5. Complete proofs for some of the results and larger drawings of the obstructions may be found in [DX00].

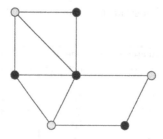

Fig. 1. An example of a graph of vertex cover 4. The black vertices denote one possible vertex cover.

2   Computing Minor-Order Obstructions

Our basic computational method is simply to generate and check all graphs that are potential obstructions. In practice, this search method can be used for an arbitrary lower ideal if one can (1) bound the search space of graphs to a reasonable size and (2) easily decide whether an arbitrary graph is an obstruction. With respect to k–VERTEX COVER, we can do both of these tasks efficiently.

In earlier work, Cattell and Dinneen in [CD94] bounded the search space to graphs of pathwidth at most $k + 1$ when they computed the k–VERTEX COVER obstructions for $1 \leq k \leq 5$. Their generation process was self-terminating but, unfortunately, many isomorphic graphs were created during the generation of the search space. These superfluous graphs had to be either caught by an isomorphism checking program or eliminated by other means. For $k = 6$ using this approach did not seem feasible. For 6–VERTEX COVER, we utilized McKay's graph generation program geng (part of his Gtools/Nauty package [McK90]) and obtained some tight new upper bounds on the structure of vertex cover obstructions.

2.1   Directly Checking Non-isomorphic Graphs

We now discuss a natural method for finding a set of obstructions for a lower ideal from an available set of non-isomorphic graphs. We need to generate a complete set of non-isomorphic (connected) graphs, which is large enough to contain the set of obstructions sought.

To decide if a graph is an obstruction it helps to have an efficient membership algorithm GA for the targeted lower ideal $\mathcal{F}$. For the 6–VERTEX COVER family of graphs we used an implementation of the Balasubramanian *et. al.* linear-time vertex cover algorithm [BFR98]. By using GA for $\mathcal{F}$, an algorithm to decide if a graph is an obstruction is almost trivial. If a given graph G is an obstruction for $\mathcal{F}$, then G is a minimal graph such that $G \notin \mathcal{F}$. For a lower ideal $\mathcal{F}$ we only need to check that each 'one-step' minor of G is in $\mathcal{F}$. A general algorithm to decide if G is an obstruction for $\mathcal{F}$ is presented below.

Procedure IsObstruction(GraphMembershipAlgorithm GA, Graph G)
    if $GA(G) =$ true **then return** false
    for each edge e **in** G **do**
        $G' =$ the resulting graph after deleting e in G
        if $GA(G') =$ false **then return** false
\+ $G'' =$ the resulting graph after contracting e in G
\+   **if** $GA(G'') =$ false **then return** false
    endfor
    return true
end

The above algorithm `IsObstruction` will work for any minor-order lower ideal. For a given input finite graph G, if the graph family membership algorithm GA operates in polynomial time, then this algorithm `IsObstruction` has polynomial-time complexity. However, for some particular lower ideals, the algorithm can be simplified (e.g., remove lines marked with $+$'s).

Lemma 1. *A graph $G = (V, E)$ is an obstruction for k–VERTEX COVER if and only if $VC(G) = k + 1$ and $VC(G \backslash uv) = k$ for all $uv \in E$.*

Proof. Let the graph G be an obstruction for k–VERTEX COVER. This implies that if any edge in G is deleted then the vertex cover decreases, by the definition of an obstruction. The size of the vertex cover decreases by one to exactly k, otherwise $VC(G) < k + 1$.

Now we prove the other direction. Let $G = (V, E)$ be a graph, and $\mathcal{F}$ denote a fixed k–VERTEX COVER lower ideal such that $G \notin \mathcal{F}$. Suppose if any edge in G is deleted, then the resulting graph $G' \in \mathcal{F}$. Thus for each edge uv in G, a set of vertices V' of cardinality k can be found which covers all edges in G except edge uv.

Let u be the reserved vertex and v be the deleted vertex after uv is contracted. Since V' covers all edges in G except uv, V' covers each edge wv where $w \neq u$. We have $w \in V'$. After contraction of uv, for each edge wv, a new edge wu is made in G'. Since $w \in V'$, V' covers wu. Thus all new edges are covered by V'. Hence after doing any edge contraction for any edge uv in G, a vertex cover V' where $|V'| = k$ covers all edges in the resulting graph G'. That is $G' \in \mathcal{F}$. Therefore if each edge deletion causes $G' \in \mathcal{F}$ and $G \notin \mathcal{F}$, then G is an obstruction.   □

Thus according to the above lemma, for k–VERTEX COVER, our obstruction deciding algorithm is simplified. This means it does not have to check edge contraction minors (i.e., create the graphs G'' in the `IsObstruction` procedure). This greatly reduces the overall computation time for deciding if a graph is an obstruction since doing any one edge contraction, with most graph data structures, is not relatively efficient (as compared with the actual time needed to delete any one edge).

2.2   Properties of Connected k–VERTEX COVER Obstructions

The number of connected non-isomorphic graphs of order n increases exponentially. For $n = 11$, the number of connected graphs is 1006700565. If we could process one graph per microsecond we would still need over 11 days of running time. Since there exist obstructions larger than this for 6–VERTEX COVER we clearly need a more restricted search space. Thus we will filter the input by exploiting other properties of the k–VERTEX COVER obstructions.

A few important properties for the k–VERTEX COVER obstructions are now systematically presented.

Lemma 2. *Any connected obstruction for the k–VERTEX COVER lower ideal is a biconnected graph.*

Proof. Let v be a cut-vertex in a connected obstruction O for k–VERTEX COVER. Also let $C_1, C_2, \ldots, C_{m \geq 2}$ be the connected components of $O \backslash \{v\}$ and $C'_i = O[V(C_i) \cup \{v\}]$, where $O[X]$ denotes the subgraph induced by vertices X. Each C'_i denotes the part of the graph containing the component C_i, the vertex v, and the edges between v and C_i. Since O is an obstruction to k–VERTEX COVER. We have $\sum_{i=1}^{m} VC(C_i) = k$. Any vertex cover for $\bigcup_{i=1}^{m} C'_i$ is also a vertex cover for O, where vertex v may be repeated in several C'_i. Thus, $\sum_{i=1}^{m} VC(C'_i) \geq VC(O) = k + 1$. This implies that there exists an i such that $VC(C'_i) = VC(C_i) + 1$. Now $O' = (\bigcup_{j \neq i} C_j) \cup C'_i$ is a proper subgraph of O. But $VC(O') = k + 1$ contradicts O being an obstruction. So O does not have any cut-vertices. Therefore, O is a biconnected graph.   □

A nice filter for our graph generator is the following result.

Lemma 3. *A vertex in an obstruction for k–VERTEX COVER has degree at most $k + 1$.*

Proof. Suppose u is a vertex with degree at least $k + 2$ in an obstruction $O = (V, E)$ for k–VERTEX COVER. Let O' be the resulting graph by deleting any edge uv of O. Since O is an obstruction for k–VERTEX COVER, we have $VC(O') = k$. Hence, in G there is a set of vertices $V' \subseteq V$ which covers all edges in G except uv and $|V'| = k$. Since V' does not cover edge uv, $u \notin V'$ and $v \notin V'$.

Thus V' must contain all the neighbors of u except v. Hence V' contains at least $k + 2 - 1$ vertices in the neighborhood of u. Thus we have $|V'| \geq k + 1 > k$, contradicting V' being a witness vertex cover to O'.

Hence the degree of u is at most $k + 1$. Therefore a vertex in an obstruction for k–VERTEX COVER has maximum degree $k + 1$.   □

Using the above result, along with Lemma 1, we can easily derive the following.

Lemma 4. *There is only one connected obstruction for k–VERTEX COVER with $k + 2$ vertices, which is the complete graph K_{k+2}. Furthermore, no other obstruction for k–VERTEX COVER has fewer vertices.*

The previous lemma shows that K_{k+2} is the smallest connected obstruction for k–VERTEX COVER. We can also show that this is the unique obstruction with maximum degree.

Lemma 5. *If a vertex in a connected obstruction for k–VERTEX COVER has degree $k + 1$, then this obstruction is the complete graph with $k + 2$ vertices (K_{k+2}).*

Proof. Suppose $O = (V, E)$ is a connected obstruction for k–VERTEX COVER and u is a vertex in O with degree $k + 1$ and let $N(u)$ be the neighborhood of u. Thus $|N(u)| = k + 1$.

Let uv be any edge incident to u. Let $O' = O\backslash uv$; we know $VC(O') = k$. This is illustrated in Figure 2. Let V' be a minimum vertex cover of O'. We have $u \notin V'$ and $v \notin V'$ (otherwise O' is not an obstruction).

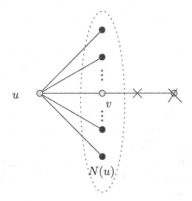

Fig. 2. A vertex u in O with degree $k + 1$. The black vertices must be in the vertex cover whenever uv is deleted.

Since V' covers all edges incident to u except uv, we have $V' = N(u)\backslash\{v\}$. Since $N(u)\backslash\{v\}$ is the vertex cover for O', v is not adjacent to any other vertex w where $w \notin N(u)$, otherwise, w must also be in V'. Thus $N(v) \subseteq N(u) \cup \{u\}$.

Since v is any vertex in $N(u)$, any vertex in $N(u)$ is connected to u or vertices in $N(u)$. Thus $\{u\} \cup N(u) = V$. Hence O has $k + 2$ vertices. By Lemma 4, a connected obstruction with $k + 2$ vertices for k–VERTEX COVER is the complete graph K_{k+2}, thus $O = K_{k+2}$ in this case.   □

This lemma shows that for any obstruction $O = (V, E)$ for k–VERTEX COVER, if Δ is the maximum degree of O and $|V| > k + 2$ then $\Delta < k + 1$.

2.3 Vertex Bounds for k–VERTEX COVER

We now present two important results that yield sharp upper bounds on the number of vertices for connected and disconnected k–VERTEX COVER obstructions. Later we will give some edge bounds that may be generalized (with some effort) to k–VERTEX COVER.

Theorem 1. *A connected obstruction for k–VERTEX COVER has at most $2k+1$ vertices.*

Proof. Assume $O = (V, E)$ is a connected obstruction for k–VERTEX COVER, where $|V| = 2k + 2$. We prove O is not a connected obstruction for k–VERTEX COVER by contradictions. The same argument also holds for graphs with more vertices.

If O is a connected obstruction of k–VERTEX COVER, then $VC(O) = k + 1$. Hence V can be split into two subsets V_1 and V_2 such that V_1 is a $k + 1$ vertex cover and $V_2 = V\backslash V_1$. Thus $|V_2| = 2k + 2 - (k + 1) = k + 1$. Obviously, no edge exists between any pair of vertices in V_2, otherwise V_1 is not a vertex cover. Each vertex in V_1 has at least one vertex in V_2 as a neighbor, otherwise it can be moved from V_1 to V_2. (i.e., the vertex is not needed in this minimal vertex cover).

Thus the neighborhood of V_2, $N(V_2)$ is V_1. We now prove that no subset S of V_2 has $|N(S)| < |S|$. [This result, in fact, will immediately exclude the case $|V| > 2k + 2$.]

By way of contradiction, assume $V_3 = S$ is a minimal subset in V_2 such that $|N(V_3)| < |V_3|$. We say V_3 is minimal whenever if T is any subset in V_3 then $|N(T)| \geq |T|$. Note V_3 will contain at least 3 vertices since every vertex has degree at least 2 in biconnected graphs (see Lemma 2).

Let $V_4 = N(V_3)$. Thus all edges adjacent to V_3 are covered by V_4. Thus $V_4 \subseteq V_1$ and $|V_4| < |V_3|$. Let $V_5 = V_2\backslash V_3$ and $V_6 = V_1\backslash V_4$. The graph O can be split as indicated in Figure 3.

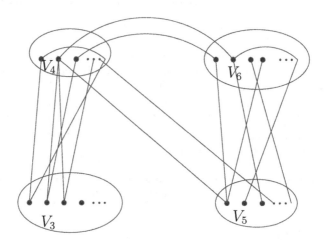

Fig. 3. Splitting the vertex set of O into four subsets.

We now prove that the existence of such a set V_3 forces O to be disconnected. Since V_3 is assumed minimal, no subset T in V_3 has $|N(T)| < |T|$. Thus if we delete any vertex in V_3, leaving V_3', then any subset T in V_3' has $|N(T)| \geq |T|$.

Recall that a matching in a bipartite graph is a set of independent edges (with no common end points). Recall Hall's Marriage Theorem [Hal35]:

Hall's Marriage Theorem:
A bipartite graph $B = (X_1, X_2, E)$ has a matching of cardinality $|X_1|$ if and only if for each subset $A \subseteq X_1$, $|N(A)| \geq |A|$.

Thus there is a matching of cardinality $|V_3'|$ in the bipartite induced subgraph $O' = (V_3', N(V_3))$ in O. Thus there are $|V_3'| = |V_3| - 1$ independent edges between the set V_3 and the set V_4.

Let us now delete all edges between V_4 and V_5 and all edges between V_4 and V_6. Let $C_1 = O[V_3 \cup V_4]$, $C_2 = O[V_5 \cup V_6]$ be these disconnected components in the resulting graph. (The induced graph C_1 is connected, while the induced graph C_2 may or may not be.)

As discussed above, there exists $|V_3| - 1$ independent edges in C_1. Thus to cover these edges, $VC(C_1) \geq |V_3| - 1$. We know $|V_4| < |V_3|$ and all edges incident to V_3 are covered by V_4, thus $VC(C_1) = |V_4|$.

Now consider the graph C_2. Suppose, $VC(C_2) < |V_6|$. Since all deleted edges are also covered by V_4, $V_4 \cup VC(C_2)$ must cover all edges in O. Thus $VC(O) = |V_4| + VC(C_2) < |V_4| + |V_6| = k + 1$. This contradicts that O is an obstruction for k–VERTEX COVER. Thus the assumption that $VC(C_2) < |V_6|$ is not correct. Hence $VC(C_2) = |V_6|$ even though edges between C_1 and C_2 were deleted.

Thus $VC(C_1 \cup C_2) = |V_4| + |V_6| = k + 1$. Therefore O can not be a connected obstruction for k–VERTEX COVER since the resulting graph still requires a $k + 1$ vertex cover whenever all edges between V_4 and V_5 and all edges between V_6 are deleted.

Therefore the assumption that there exists a minimal subset V_3 in V_2 is not correct. Hence any subset S in V_2 has $|N(S)| \geq |S|$.

Once again, by applying Hall's Marriage Theorem, there is a matching of cardinality $k + 1$ in the induced bipartite subgraph $O' = (V_2, V_1)$ of O. To cover these $k + 1$ independent edges, a vertex cover of size $k + 1$ is necessary. We know that if O is a connected graph, there must exist other edges in O except these $k + 1$ independent edges. If those edges are deleted, the resulting graph still has vertex cover $k + 1$. Thus O can not be a connected obstruction for k–VERTEX COVER if it has more than $2k + 1$ vertices.   □

By extending the above result, we can prove the following corollary by induction.

Corollary 1. *Any obstruction for k–VERTEX COVER has order at most $2k + 2$.*

We now mention some observations about the k–VERTEX COVER obstructions. The known obstructions for the small cases (see [CD94] for $k \leq 5$, in addition to this paper's $k = 6$) indicate a very interesting feature: *the more vertices, the fewer edges.* As proven above, a complete graph K_{k+2} is the smallest obstruction (and only one), but it seems to have the largest number of edges. The cycle obstruction C_{2k+1} appears to be the only connected obstruction with the largest number of vertices (and appears to have the smallest number of edges).

2.4   Computing All Obstructions for 6–VERTEX COVER

We divide our computational task for finding all the connected obstructions for 6–VERTEX COVER into two steps.

(1) We find all connected obstructions with order at most 11 for 6–VERTEX COVER. This step is very straight forward. Our algorithm IsObstruction is applied to all non-isomorphic biconnected graphs with a number of vertices between 9 and 11, of maximum degree 6, and of maximum number of edges 33.

(2) To find all connected obstructions of order 12 and 13 for 6–VERTEX COVER, new degree and edge bounds were found and used, as indicated below.

By the Lemma 5, we know that if a connected obstruction O has a vertex of degree 7 for 6–VERTEX COVER, then this obstruction must be K_8. Thus we only have to consider graphs with maximum degree 6. Furthermore, if the degree for O is 6, we can easily prove the following statement by cases.

Statement 1. *If a connected obstruction $O = (V, E)$ for 6–VERTEX COVER has a vertex of degree 6, then $|V| \le 10$. Furthermore, if $|V| = 10$ then $|E| \le 24$, and if $|V| = 9$ then $|E| \le 25$.*

A consequence of the proof for Statement 1 (see [DX00]) is the following: if a connected obstruction O for 6–VERTEX COVER has 11 or more vertices then the degree of every vertex is less than or equal to 5. With another series of case studies of graphs with vertices of degree 5 we have the following.

Statement 2. *If a connected obstruction $O = (V, E)$ for 6–VERTEX COVER has 12 vertices then $|E| \le 24$. Further, if O has 13 vertices then $|E| \le 26$.*

The proof of this statement (see [DX00]) also gives a degree bound of 4 for any obstruction with 13 vertices. The search space for finding all obstructions with orders 12 and 13 for 6–VERTEX COVER has been extremely reduced.

3   Final Remarks

Our obstruction checking procedure for k–VERTEX COVER (primarily based on Lemmata 1 and 5) was efficient enough to process all biconnected graphs of order at most 11 for $k = 6$. To find all of the obstructions with 12 vertices for 6–VERTEX COVER, we only needed to test all non-isomorphic (biconnected) graphs with maximum degree 5 and at most 24 edges. For finding all obstructions with 13 vertices for 6–VERTEX COVER, we only need to check all non-isomorphic graphs with maximum degree 4 and at most 26 edges. This search space is very manageable; it requires about two months of computation time! In Figures 4–5, we display all 188 connected obstructions for 6–VERTEX COVER.

We now mention a couple of areas left open by our research. It would be nice to have a proof that C_{2k+1} is the only connected obstruction for k–VERTEX COVER, since we now known that $2k + 1$ is an upper bound on the number of vertices. There are some interesting open questions regarding edge bounds for k–VERTEX COVER. As we pointed out earlier, the obstructions start having fewer edges as the number of vertices increases. More theoretical results that generalizes our specific bounds for $k = 6$ seem possible.

References

[APS91]   S. Arnborg, A. Proskurowski, and D. Seese. Monadic second order logic, tree automata and forbidden minors. In *Proceedings of the 4th Workshop on Computer Science Logic, CSL'90*, volume 533 of *Lecture Notes on Computer Science*, pages 1–16. Springer-Verlag, 1991.

[BFR98]   R. Balasubramanian, M. R. Fellows, and V. Raman. An improved fixed-parameter algorithm for vertex-cover. *Information Processing Letters*, 65(3):163–168, 1998.

[CD94]   K. Cattell and M. J. Dinneen. A characterization of graphs with vertex cover up to five. In V. Bouchitte and M. Morvan, editors, *Orders, Algorithms and Applications, ORDAL'94*, volume 831 of *Lecture Notes on Computer Science*, pages 86–99. Springer-Verlag, July 1994.

[CDD$^+$00]   K. Cattell, M. J. Dinneen, R. G. Downey, M. R. Fellows, and M. A. Langston. On computing graph minor obstruction sets. *Theoretical Computer Science*, 233(1-2):107–127, February 2000.

[DX00]   M. J. Dinneen and L. Xiong. The minor-order obstructions for the graphs of vertex cover six. Research Report CDMTCS-118, Centre for Discrete Mathematics and Theoretical Computer Science, University of Auckland, Auckland, New Zealand, 29 pp. January 2000. (http://www.cs.auckland.ac.nz/CDMTCS/researchreports/118vc6.pdf)

[FL89]   M. R. Fellows and M. A. Langston. An analogue of the Myhill-Nerode Theorem and its use in computing finite-basis characterizations. In *IEEE Symposium on Foundations of Computer Science Proceedings*, volume 30, pages 520–525, 1989.

[GJ79]   M. R. Garey and D. S. Johnson. *Computers and Intractability: A Guide to the Theory of NP-completeness*. W. H. Freeman and Company, 1979.

[Hal35]   P. Hall. On representation of subsets. *J. London Math. Soc.*, 10:26–30, 1935.

[LA91]   J. Lagergren and S. Arnborg. Finding minimal forbidden minors using a finite congruence. In *Proceedings of the International Colloquium on Automata, Languages and Programming*, volume 510 of *Lecture Notes on Computer Science*, pages 533–543. Springer-Verlag, 1991. 18th ICALP.

[McK90]   B. D. McKay. nauty User's Guide (version 2.0). Tech Report TR-CS-90-02, Australian National University, Department of Computer Science, 1990.

[Pro93]   A. Proskurowski. Graph reductions, and techniques for finding minimal forbidden minors. In N. Robertson and P. D. Seymour, editors, *Graph*

Fig. 4. All connected obstructions for 6–VERTEX COVER.

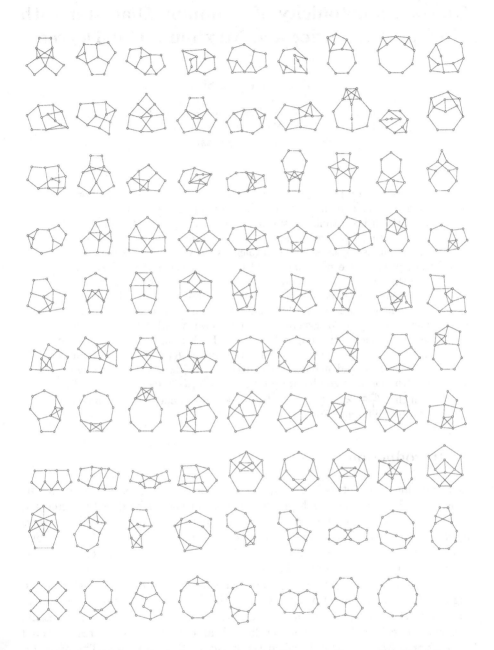

Fig. 5. All connected obstructions for 6–VERTEX COVER (continued).

On the Monotonicity of Minimum Diameter with Respect to Order and Maximum Out-Degree

Mirka Miller and Slamin

Department of Computer Science and Software Engineering,
The University of Newcastle, NSW 2308, Australia
{mirka, slamin}@cs.newcastle.edu.au

Abstract. Considering the three parameters of a directed graph: order, diameter and maximum out-degree, there are three optimal problems that arise if we optimise in turn each one of the parameters while holding the other two parameters fixed. These three problems are related but as far as we know not equivalent. One way to prove the equivalence of the three problems would be to prove that the optimal value of each parameter is monotonic in each of the other two parameters. It is known that maximum order is monotonic in both diameter and maximum out-degree and that minimum diameter is monotonic in maximum out-degree. In general, it is not known whether the other three possible monotonicity implications hold. In this paper, we consider the problem of determining the smallest diameter $K(n, d)$ of a digraph G given order n and maximum out-degree d. Using a new technique for construction of digraphs, we prove that $K(n, d)$ is monotonic for all n such that $\frac{d^k - d}{d - 1} < n \leq d^k + d^{k-1}$, thus solving an open problem posed in 1988 by Miller and Fris.

1 Introduction

In communication network design, there are several factors which should be considered. For example, each processing element should be directly connected to a limited number of other processing elements in such a way that there always exists a connection route from one processing element to another. Furthermore, in order to minimize the communication delay between processing elements, the directed connection route must be as short as possible.

A communication network can be modelled as a *digraph*, where each processing element is represented by a *vertex* and the directed connection between two processing elements is represented by an *arc*. The number of vertices (processing elements) is called the *order* of the digraph and the number of arcs (directed connections) from a vertex is called the *out-degree* of the vertex. The *diameter* is defined to be the largest of the shortest paths (directed connection routes) between any two vertices.

In graph-theoretical terms, the problems in the communication network designs can be modelled as optimal digraph problems. The example described above corresponds to the so-called *order/degree problem*: Construct digraphs with the

D.-Z. Du et al. (Eds.): COCOON 2000, LNCS 1858, pp. 193–201, 2000.

smallest possible diameter $K(n, d)$ for a given order n and maximum out-degree d. Before we discuss in more detail optimal digraph problems, we give more formal definitions of the terms in graph theory that will be used.

A *directed graph* or *digraph* G is a pair of sets (V, A) where V is a finite nonempty set of distinct elements called *vertices*; and A is a set of ordered pair (u, v) of distinct vertices $u, v \in V$ called *arcs*.

The *order* n of a digraph G is the number of vertices in G, that is, $n = |V|$. An *in-neighbour* (respectively *out-neighbour*) of a vertex v in G is a vertex u (respectively w) such that $(u, v) \in A$ (respectively $(v, w) \in A$). The set of all in-neighbour (respectively out-neighbour) of a vertex v is denoted by $N^-(v)$ (respectively $N^+(v)$). The *in-degree* (respectively *out-degree*) of a vertex v is the number of its in-neighbour (respectively out-neighbour). If the in-degree equals the out-degree $(= d)$ for every vertex in G, then G is called a *diregular* digraph of degree d.

A *walk* of length l in a digraph G is an alternating sequence $v_0 a_1 v_1 a_2 ... a_l v_l$ of vertices and arcs in G such that $a_i = (v_{i-1}, v_i)$ for each i, $1 \leq i \leq l$. A walk is closed if $v_0 = v_l$. If all the vertices of a $v_0 - v_l$ walk are distinct, then such a walk is called a *path*. A *cycle* is a closed path.

The *distance* from vertex u to v, denoted by $\delta(u, v)$, is the length of the shortest path from vertex u to vertex v. Note that in general $\delta(u, v)$ is not necessarily equal to $\delta(v, u)$. The *diameter* of digraph G is the longest distance between any two vertices in G.

Let v be a vertex of a digraph G with maximum out-degree d and diameter k. Let n_i, for $0 \leq i \leq k$, be the number of vertices at distance i from v. Then $n_i \leq d^i$, for $0 \leq i \leq k$, and so

$$n = \sum_{i=0}^{k} n_i \leq 1 + d + d^2 + \ldots + d^k = \frac{d^{k+1} - 1}{d - 1} \tag{1}$$

The right-hand side of (1) is called the *Moore bound* and denoted by $M_{d,k}$. If the equality sign holds in (1) then the digraph is called *Moore digraph*. It is well known that Moore digraphs exist only in the trivial cases when $d = 1$ (directed cycles of length $k+1$, C_{k+1}, for any $k \geq 1$) or $k = 1$ (complete digraphs of order $d + 1$, K_{d+1}, for any $d \geq 1$) [10,3]. Thus for $d \geq 2$ and $k \geq 2$, inequality (1) becomes

$$n \leq d + d^2 + \ldots + d^k = \frac{d^{k+1} - d}{d - 1} \tag{2}$$

Let $\mathcal{G}(n, d, k)$ denotes the set of all digraphs G, not necessarily diregular, of order n, maximum out-degree d, and diameter k. Given these three parameters, there are three problems for digraphs $G \in \mathcal{G}(n, d, k)$, namely,

1. The *degree/diameter problem*: determine the largest order $N(d, k)$ of a digraph given maximum out-degree d and diameter k.
2. The *order/degree problem*: determine the smallest diameter $K(n, d)$ of a digraph given order n and maximum out-degree d.

3. The *order/diameter problem*: determine the smallest degree $D(n, k)$ of a digraph given order n and diameter k.

Miller [6] pointed out the following implications regarding to the three problems above.

a. $d_1 < d_2 \Rightarrow N(d_1, k) < N(d_2, k)$;
b. $k_1 < k_2 \Rightarrow N(d, k_1) < N(d, k_2)$;
c. $d_1 < d_2 \Rightarrow K(n, d_1) \geq K(n, d_2)$.

It is not known whether the following three implications hold or not.

1. $k_1 < k_2 \Rightarrow D(n, k_1) \geq D(n, k_2)$, for $(k_1, k_2 \leq n - 1)$;
2. $n_1 < n_2 \Rightarrow K(n_1, d) \leq K(n_2, d)$;
3. $n_1 < n_2 \Rightarrow D(n_1, k) \leq D(n_2, k)$, for $(n_1, n_2 \geq k + 1)$.

Miller and Fris [7] posed the problem of whether or not $K(n, d)$ is monotonic on restricted intervals of n, for example, for all n such that $\frac{d^k - d}{d - 1} < n \leq d^k + d^{k-1}$. In this paper, we solve the open problem using a new technique for the construction of digraphs.

2   Construction

In this section we introduce a new technique for the construction of digraphs, as described in the following theorems. This construction will be used to prove the main results presented in the next section.

Theorem 1 *If $G \in \mathcal{G}(n, d, k)$ and $N^+(u) = N^+(v)$ for any vertex $u, v \in G$, then there exists $G_1 \in \mathcal{G}(n - 1, d, k')$, $k' \leq k$.*

Proof. Suppose that $N^+(u) = N^+(v)$ for any vertex $u, v \in G$. Let G_1 be a digraph deduced from G by deleting vertex u together with its outgoing arcs and reconnecting the incoming arcs of u to the vertex v. Obviously, the new digraph G_1 has maximum out-degree the same as the maximum out-degree of G. To prove that the diameter of G_1 is at most k, we only need to consider the distance from vertices of $N^-(u)$ to other vertices in G reached through u. Just deleting u might cause a vertex in $N^-(u)$ not to reach some vertex within distance at most k. However, this is overcome when we reconnect the vertices of $N^-(u)$ to v. Thus we obtain a digraph G_1 with order $n - 1$, maximum out-degree d and diameter k', where $k' \leq k$.   □

Figure 1(a) shows an example of digraph $G \in \mathcal{G}(6, 2, 2)$ with the property $N^+(1) = N^+(5)$, $N^+(2) = N^+(6)$, and $N^+(3) = N^+(4)$. Applying Theorem 1 to the digraph by deleting vertex 6 together with its outgoing arcs and then reconnecting its incoming arcs to vertex 2 (since $N^+(2) = N^+(6)$), we obtain a new digraph $G_1 \in \mathcal{G}(5, 2, 2)$ as shown in Figure 1(b).

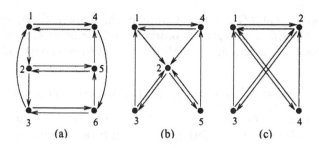

Fig. 1. The digraphs $G \in \mathcal{G}(6,2,2)$, $G_1 \in \mathcal{G}(5,2,2)$, and $G_2 \in \mathcal{G}(4,2,2)$

Digraphs which are obtained from *line digraph* construction contain some pairs of vertices with the same out-neighbourhoods. The *line digraph* $L(G)$ of a digraph G is the digraph where each vertex of $L(G)$ corresponds to arc of G; and a vertex uv of $L(G)$ is adjacent to a vertex wx if and only if $v = w$, that is, if the arc (u, v) is adjacent to the arc (w, x) in G.

The order of the line digraph $L(G)$ is equal to the number of arcs in the digraph G. For a diregular digraph G of out-degree $d \geq 2$, the sequence of line digraph iterations

$$L(G), L^2(G) = L(L(G)), ..., L^i(G) = L(L^{i-1}(G)), ...$$

is an infinite sequence of diregular digraphs of degree d. For more detail explanations about line digraph iterations, see [4]. The following facts about line digraph are obvious.

Fact 1 *If a digraph G is a diregular of degree $d \geq 2$, diameter k and order n then the line digraph $L(G)$ is also diregular of degree d, order dn and diameter $k + 1$.*

Fact 2 *If $L(G)$ is a line digraph of diregular digraph G with degree $d \geq 2$, diameter k and order n then there are n tuples of d vertices in $L(G)$ with identical out-neighbourhood. In other words, if $u_i \in N^-(v)$ and $w_j \in N^+(v)$ where $v, u_i, w_j \in V(G)$ for every $i = 1, 2, ..., d$ and $j = 1, 2, ..., d$ then*

$$N^+(u_1 v) = N^+(u_2 v) = ... = N^+(u_d v) = \{vw_1, vw_2, ..., vw_d\}$$

where $u_i v, vw_j \in V(LG)$.

Applying Theorem 1 to line digraphs, we obtain

Theorem 2 *If $LG \in \mathcal{G}(dn, d, k)$ is a line digraph of a diregular digraph $G \in \mathcal{G}(n, d, k-1)$, then there exists digraph $LG_r \in \mathcal{G}(dn - r, d, k')$, $k' \leq k$, for every $1 \leq r \leq (d-1)n - 1$.*

Proof. Suppose that LG is a line digraph of the digraph G. By Fact 1, LG is also diregular digraph of degree d, order dn and diameter k. From Fact 2, we have n tuples of d vertices in LG with identical out-neighbourhood. One of the tuples is

$$N^+(u_1v) = N^+(u_2v) = ... = N^+(u_dv) = \{vw_1, vw_2, ..., vw_d\}$$

where $u_iv, vw_j \in V(LG)$ for every $i = 1, 2, ..., d$ and $j = 1, 2, ..., d$.

We apply Theorem 1 by deleting vertex u_1v together with its outgoing arcs and reconnecting its incoming arcs to the vertex u_iv for any $i = 2, 3, ..., d$, we obtain the digraph with order one less than the order of LG. Repeating this notion can be done $d - 1$ times for each tuple. Since there are n tuples of d vertices with identical out-neighbourhood in LG, we can delete the vertices as many as $(d - 1)n - 1$ vertices to obtain the digraph with diameter at most k. Thus the digraphs $LG_r \in \mathcal{G}(dn - r, d, k')$, $k' \leq k$, for every $1 \leq r \leq (d-1)n - 1$ do exist.   □

Every time we delete a vertex u and reconnect its incoming arcs to other vertex v with the same out-neighbourhood as u leads to the number of incoming arcs of v increases. This implies that the resulting digraphs is not regular with respect to the in-degree, but still regular in out-degree. Therefore the resulting digraphs LG_r are non-diregular.

This situation becomes more interesting, when we consider *Kautz digraphs*, $Ka \in \mathcal{G}(d^k + d^{k-1}, d, k)$ for $d \geq 3$ and $k \geq 2$. Since Kautz digraphs can be obtained by $(k - 1)$-fold iterations of the line digraph construction applied to complete digraph K_{d+1}, Theorem 2 immediately implies that

Corollary 1 *For $d^{k-1} + d^{k-2} < n \leq d^k + d^{k-1}$, there exist digraphs $G \in \mathcal{G}(n, d, k')$, $k' \leq k$.*   □

Figure 1(a) shows an example of the Kautz digraph $G \in \mathcal{G}(6, 2, 2)$ which is a line digraph of K_3 and the non-diregular digraphs $G_1 \in \mathcal{G}(5, 2, 2)$ and $G_2 \in \mathcal{G}(4, 2, 2)$ (Figure 1(b) and Figure 1(c)) deduced from digraph G. In this example, to obtain $G_1 \in \mathcal{G}(5, 2, 2)$ we delete the vertex 6 from G together with its outgoing arcs and reconnect its incoming arcs to the vertex 2. Furthermore, to obtain $G_2 \in \mathcal{G}(4, 2, 2)$, we delete the vertex 5 from G_1 together with its outgoing arcs and reconnect its incoming arc to the vertex 1.

3   The Diameter

Kautz digraphs $Ka \in \mathcal{G}(d^k + d^{k-1}, d, k)$ are the largest known digraphs of maximum out-degree $d \geq 3$ and diameter $k \geq 2$. Kautz digraph Ka can be regarded as a line digraph $L^{k-1}(K_{d+1})$ of the complete directed graph K_{d+1}. There are exactly $d + 1$ different out-neighbourhoods, each shared by d^{k-1} vertices. Therefore Kautz digraphs can be used as base digraphs in our construction (Theorem 2) to produce other digraphs with smaller order while keeping the diameter no larger than Ka and the maximum out-degree the same. Consequently, we have

Theorem 3 *If $\frac{d^k-d}{d-1} < n \le d^k + d^{k-1}$ for $d \ge 2$ and $k \ge 2$, then $K(n,d) = k$.*

Proof. If the diameter of the digraph is $k-1$, the inequality (2) becomes

$$N(d,k) \le \frac{d^k - d}{d - 1}. \tag{3}$$

Obviously, if $n > \frac{d^k-d}{d-1}$, then $K(n,d) \ge k$. By Corollary 1, there exist digraphs of order n, for $d^{k-1} + d^{k-2} < n \le d^k + d^{k-1}$, out-degree d and diameter at most k. Since $d^{k-1} + d^{k-2} < \frac{d^k-d}{d-1}$, it follows that for digraphs with maximum out-degree d, order n, where $\frac{d^k-d}{d-1} < n \le d^k + d^{k-1}$, the minimum diameter is k, i.e., $K(n,d) = k$. $\qquad\square$

Theorem 3 can be restated as

Theorem 3′ *$K(n,d)$ is monotonic for all n such that $\frac{d^k-d}{d-1} < n \le d^k + d^{k-1}$.* $\square$

The interval of order n can be tightened if we consider particular cases of out-degree d and if we utilise some additional known results, as described below.

Let us first consider the case of maximum out-degree 3. Baskoro et al [1] proved that there is no digraph of maximum out-degree 3 and diameter $k \ge 3$ with order one less than the Moore bound. Therefore, for $d = 3$ and $k \ge 3$, $n \le \frac{3^{k+1}-5}{2}$. For diameter $k-1$, $n \le \frac{3^k-5}{2}$.

Consequently, if $n > \frac{3^k-5}{2}$, then $K(n,3) \ge k$. Combining this result and Corollary 1, we have

Theorem 4 *For every $k > 3$ and $\frac{3^k-5}{2} < n \le 4 \times 3^{k-1}$, $K(n,3) = k$.* $\qquad\square$

Another special case we can consider is digraphs of maximum out-degree 2. Alegre constructed diregular digraph of (constant) out-degree 2, diameter 4 and order 25, that is, $Al \in \mathcal{G}(25,2,4)$. This digraph is larger than Kautz digraph of the same maximum out-degree and diameter. Applying line digraph iterations and Theorem 2 to Al, we obtain

Lemma 1 *For every $k \ge 5$ and $25 \times 2^{k-5} < n \le 25 \times 2^{k-4}$, $\mathcal{G}(n,d,k)$ is not empty.* $\qquad\square$

Furthermore, Miller and Širáň [9] proved that

Theorem 5 *If $d = 2$ and $k > 2$, then $n \le 2^{k+1} - 4$.* $\qquad\square$

This result implies that for $n > 2^k - 3$, $K(n,2) \ge k$. Combining this result and Lemma 1, we have

Theorem 6 *For every $k \ge 5$ and $2^k - 3 \le n \le 25 \times 2^{k-4}$, $K(n,2) = k$.* $\qquad\square$

We conclude this paper with our current knowledge of the $K(n,2)$ problem for $n \le 200$ (Table 1). New results as displayed in bold-type, for example $G \in \mathcal{G}(49,2,5)$ which is obtained by applying Theorem 2 to the line digraph of Alegre digraph (see Figure 2).

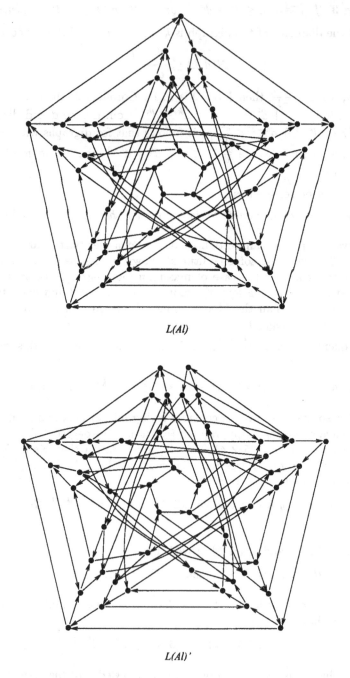

L(Al)

L(Al)'

Fig. 2. The line digraph of Alegre digraph $L(Al)$ and non-diregular digraph of order 49 $L(Al)'$

n	$K(n,2)$	n	$K(n,2)$	n	$K(n,2)$	n	$K(n,2)$	n	$K(n,2)$
1	0	41	5	81	6	121	6 or 7	161	7
2	1	42	5	82	6	122	6 or 7	162	7
3	1	43	5	83	6	123	6 or 7	163	7
4	2	44	5	84	6	124	6 or 7	164	7
5	2	45	5	85	6	125	6 or 7	165	7
6	2	46	5	86	6	126	7	166	7
7	3	47	5	87	6	127	7	167	7
8	3	48	5	88	6	128	7	168	7
9	3	49	5	89	6	129	7	169	7
10	3	50	5	90	6	130	7	170	7
11	3	51	5 or 6	91	6	131	7	171	7
12	3	52	5 or 6	92	6	132	7	172	7
13	4	53	5 or 6	93	6	133	7	173	7
14	4	54	5 or 6	94	6	134	7	174	7
15	4	55	5 or 6	95	6	135	7	175	7
16	4	56	5 or 6	96	6	136	7	176	7
17	4	57	5 or 6	97	6	137	7	177	7
18	4	58	5 or 6	98	6	138	7	178	7
19	4	59	5 or 6	99	6	139	7	179	7
20	4	60	5 or 6	100	6	140	7	180	7
21	4	61	6	101	6 or 7	141	7	181	7
22	4	62	6	102	6 or 7	142	7	182	7
23	4	63	6	103	6 or 7	143	7	183	7
24	4	64	6	104	6 or 7	144	7	184	7
25	4	65	6	105	6 or 7	145	7	185	7
26	4 or 5	66	6	106	6 or 7	146	7	186	7
27	4 or 5	67	6	107	6 or 7	147	7	187	7
28	4 or 5	68	6	108	6 or 7	148	7	188	7
29	5	69	6	109	6 or 7	149	7	189	7
30	5	70	6	110	6 or 7	150	7	190	7
31	5	71	6	111	6 or 7	151	7	191	7
32	5	72	6	112	6 or 7	152	7	192	7
33	5	73	6	113	6 or 7	153	7	193	7
34	5	74	6	114	6 or 7	154	7	194	7
35	5	75	6	115	6 or 7	155	7	195	7
36	5	76	6	116	6 or 7	156	7	196	7
37	5	77	6	117	6 or 7	157	7	197	7
38	5	78	6	118	6 or 7	158	7	198	7
39	5	79	6	119	6 or 7	159	7	199	7
40	5	80	6	120	6 or 7	160	7	200	7

Table 1. The values of $K(n,d)$ for $d = 2$ and $n \leq 200$.

References

1. E.T. Baskoro, Mirka Miller, J. Širáň and M. Sutton, Complete characterisation of almost Moore digraphs of degree 3, Discrete Mathematics, to appear.
2. E.T. Baskoro, Mirka Miller, J. Plesník and Š. Znám, Digraphs of degree 3 and order close to the Moore bound, Journal of Graph Theory **20** (1995) 339–349
3. W.G. Bridges and S. Toueg, On the impossibility of directed Moore graphs, J. Combinatorial Theory Series **B29** (1980) 339–341
4. M.A. Fiol, J.L.A. Yebra and I. Alegre, Line digraph iterations and the (d, k) digraph problem, IEEE Transactions on Computers **C-33** (1984) 400–403
5. M. Imase and M. Itoh, A design for directed graphs with minimum diameter, IEEE Trans. on Computers **C-32** (1983) 782–784
6. Mirka Miller, M. A. Thesis, Dept. of Math, Stats and Comp. Sci., U NE, Armidale (1986)
7. Mirka Miller and I. Fris, Minimum diameter of diregular digraphs of degree 2, Computer Journal **31** (1988) 71–75
8. Mirka Miller and I. Fris, Maximum order digraphs for diameter 2 or degree 2, Pullman volume of Graphs and Matrices, Lecture Notes in Pure and Applied Mathematics **139** (1992) 269–278
9. Mirka Miller and J. Širáň, Digraphs of degree two and defect two, submitted.
10. J. Plesník and Š. Znám, Strongly geodetic directed graphs, Acta F. R. N. Univ. Comen. - Mathematica XXIX (1974) 29–34

Online Independent Sets

Magnús M. Halldórsson*, Kazuo Iwama**, Shuichi Miyazaki, and
Shiro Taketomi

School of Informatics, Kyoto University, Kyoto 606-8501, Japan
{mmh, iwama, shuichi, taketomi}@kuis.kyoto-u.ac.jp

Abstract. At each step of the online independent set problem, we are
given a vertex v and its edges to the previously given vertices. We are
to decide whether or not to select v as a member of an independent
set. Our goal is to maximize the size of the independent set. It is not
difficult to see that no online algorithm can attain a performance ratio
better than $n - 1$, where n denotes the total number of vertices. Given
this extreme difficulty of the problem, we study here relaxations where
the algorithm can hedge his bets by maintaining multiple alternative
solutions simultaneously.

We introduce two models. In the first, the algorithm can maintain a
polynomial number of solutions (independent sets) and choose the largest
one as the final solution. We show that $\theta(\frac{n}{\log n})$ is the best competitive
ratio for this model. In the second more powerful model, the algorithm
can copy intermediate solutions and grow the copied solutions in different
ways. We obtain an upper bound of $O(n/(k \log n))$, and a lower bound of
$n/(e^{k+1} \log^3 n)$, when the algorithm can make n^k operations per vertex.

1 Introduction

An *independent set* (IS) in a graph is a set of vertices that are mutually non-
adjacent. In the *online* independent set problem, a graph is given one vertex at
a time along with edges to previous vertices. The algorithm is to maintain a
proper solution, and must at each step decide irrevocably whether to keep the
presented vertex as a part of the solution. Its goal is to find as large a set as
possible, relative to the size of the largest independent set in the graph.

Online computation has received considerable attention as a natural model
of some properties of the real world such as the irreversibility of time and the
unpredictability of the future. Aside from modeling real-time computation, it
also has been found to be important in contexts where access to data is limited,
e.g. because of unfamiliar terrain, space constraints, or other factors determining
the order of the input.

The online IS problem occurs naturally in resource scheduling. Requests for
resources or sets of resources arrive online, and two requests can only be serviced

---

* Current address: Science Institute, University of Iceland (mmh@hi.is)

** Supported in part by Scientific Research Grant, Ministry of Japan, 10558044,
09480055 and 10205215.

D.-Z. Du et al. (Eds.): COCOON 2000, LNCS 1858, pp. 202–209, 2000.

simultaneously if they do not involve the same resource. When the objective is to maximize the throughput of a server, we have a (possibly weighted) online IS problem. Previously, this has primarily been studied in the context of scheduling intervals [7].

The usual measure of the quality of an online algorithm A is its *competitive ratio* or *performance ratio*. For the online IS problem, it is defined to be $\rho(n) = \max_{I \in I_n} opt(I)/A(I)$, where I_n is the set of all the sequences of vertices and corresponding edges given as n-vertex graphs, $opt(I)$ is the cardinality of the maximum independent set of I, and $A(I)$ is the size of the independent set found by A on I.

One reason why the online IS problem has not received a great deal of attention is that any algorithm is bound to have a poor competitive ratio. It is not hard to see that in the basic model, once the algorithm decides to add a vertex to its set, the adversary can ensure that no further vertices can be added, even though all the other $n - 1$ vertices form an independent set. Thus, the worst possible performance ratio of $n - 1$ holds for any algorithm, even when the input graph is a tree. As a result, we are interested in relaxations of the basic model and their effects on the possible performance ratios.

In the *coloring model* studied recently in [3], the online algorithm constructs a coloring online. The IS solution output will be the largest color class. This circumvents the strait-jacket that the single current solution imposes on the algorithm, allowing him to hedge his bets. This model relates nicely to online coloring that has been studied for decades. In particular, positive results known for some special classes of graphs carry over to this IS problem: $O(\log n)$ for trees and bipartite graphs [1,2,8], and $n^{1/k!}$ for k-colorable graphs [5]. Also, 2-competitive ratio is immediate for line graphs, as well as ratio k for $k + 1$-claw free graphs. On the other hand, for general graphs, no algorithm has a competitive ratio better than $n/4$ [3], a ratio that is achieved by the FirstFit coloring algorithm.

Our results. We propose the *multi-solutions* model, where the algorithm can maintain a collection of independent sets. At each step, the current vertex can be added to up to $r(n)$ different sets. The coloring model is equivalent to the special case $r(n) = 1$. We are particularly interested in polynomially bounded computation, thus focusing on the case when $r(n) = n^k$, for some constant k. For this case, we derive an upper and a lower bound of $\theta(n/\log n)$ on the best possible competitive ratio. For $r(n)$ constant, we can argue a stronger $\Omega(n)$ lower bound. These results are given in Section 2.

A still more powerful model proposed here is the *inheritance model*. Again each vertex can participate in limited number of sets. However, the way these sets are formed is different. At each step, $r(n)$ different sets can be copied and the current vertex is added to one copy of each of the sets. Thus, adding a vertex to a set leaves the original set still intact. This model can be likened to forming a tree, or a forest, of solutions: at each step, $r(n)$ branches can be added to trees in the forest, all labeled by the current vertex. Each path from a root to a node (leaf or internal) corresponds to one current solution.

The inheritance model is the most challenging one in terms of proving lower bounds. This derives from the fact that all the lower bounds used the other models, as well as for online graph coloring [4,3], have a property called *transparency*: once the algorithm makes his choice, the adversary immediately reveals her classification of that vertex to the algorithm. This transparency property, however, trivializes the problem in the inheritance model. Still we are able to prove lower bounds in the same ballpark, or $\Omega(n/(e^k \log^3 n))$ when $r(n) = n^k$. These results are given in Section 3.

Notation. Throughout this paper, n denotes the total number of given vertices, and we write vertices $v_1, v_2, \cdots, v_n$, where vertices are given in this order. Namely, at the ith step, the vertex v_i is given and an online algorithm decides whether it selects v_i or not.

2 Multi-solutions Model

In the original online IS problem, the algorithm can have only one bin, one candidate solution. In the multi-solutions model, the algorithm can maintain a collection of independent sets, choosing the largest one as the final solution. The model is parameterized by a function $r(n)$; each vertex can be placed online into up to $r(n)$ sets. When $r(n) = 1$, this corresponds to forming a coloring of the nodes. When $r(n) = 2^n$, an online algorithm can maintain all possible independent sets and the problem becomes trivial. We are primarily interested in cases when $r(n)$ is polynomially bounded.

We find that the best possible performance ratio in this model is $\theta(n/\log r(n))$. In particular, when $r(n)$ is constant, the ratio is linear, while if $r(n)$ is polynomial, the ratio is $\theta(n/\log n)$.

Remark. In online problems, the algorithm does not know the length n of the input sequence. Thus, in the multi-solutions model, the algorithm can assign the i-th node to $r(i)$ different sets, rather than $r(n)$. However, it is well known that not informing of the value of n has very limited effect on the performance ratio, affecting at most the constant factor. Thus, in order to simplify the arguments, we may actually allow the algorithm $r(n)$ assignments for every vertex.

Theorem 1. *There is an online algorithm with competitive ratio at most* $O(n/\log r(n))$ *in the multi-solutions model.*

Proof. Let $t = \lceil \log r(n) - 1 \rceil$. View the input as sequence of blocks, each with t vertices. For each block, form a bin corresponding to each non-empty subset of the t vertices. Greedily assign each vertex as possible to the bins corresponding to sets containing that vertex. This completes the specification of the algorithm.

Each node is assigned to at most $r(n)$ bins. Each independent set in each block will be represented in some bin. Thus, the algorithm finds an optimal solution within each block. It follows that the performance ratio is at most n/t, the number of blocks. $\qquad\square$

Lower Mounds We first give a linear lower bound for the case of $r(n)$ being constant.

Theorem 2. *Any online IS algorithm has a competitive ratio at least $\frac{n}{2(r(n)+1)}$ in the multi-solutions model.*

Proof. We describe the action of our adversary. Vertices have one of two states: *good* and *bad*. When a vertex is presented to an online algorithm, its state is yet undecided. Immediately after an online algorithm decides its action for that vertex, its state is determined. If a vertex v becomes good, no subsequent vertex has an edge to v. If v becomes bad, all subsequent vertices have an edge to v. Note that all good vertices constitute an independent set, in fact the optimal one. Also, note that once a bad vertex is added to a bin, no succeeding vertex can be put into that bin.

The adversary determines the state of each vertex as follows: If an online algorithm put the vertex v into a bin with $r(n)$ vertices or more, then v becomes a bad vertex; otherwise it's good. Thus, no bin contains more than $r(n) + 1$ vertex. This upper bounds the algorithm's solution.

Each bad node becomes the last node in a bin with $r(n)$ copies of other good nodes. Since each node appears in at most $r(n)$ bins, we have that the number of good nodes is at least the number of bad nodes, or at least $n/2$. This lower bounds the size of the optimal solution.   □

Theorem 3. *Any online IS algorithm has a competitive ratio at least $\frac{n}{2\log(n^2 \cdot r(n))}$ in the multi-solutions model.*

Proof. In the lower bound argument here, we consider a randomized adversary. As before, we use good vertices and bad vertices. This time, our adversary makes each vertex good or bad with equal probability. We analyze the average performance of an arbitrary online algorithm. The probability that a fixed bin contains more than $C = \log(n^2 \cdot r(n))$ vertices is at most $(\frac{1}{2})^C = \frac{1}{n^2 \cdot r(n)}$ since all but the last vertex must be good. The online algorithm can maintain at most $n \cdot r(n)$ bins. Thus, the probability that there is a bin with more than C vertices is at most $nr(n) \times \frac{1}{n^2 r(n)} = \frac{1}{n}$. Namely, with high probability the online algorithm outputs an independent set of size at most C. On the other hand, the expected value of the optimal cost, namely the expected number of good vertices, is $\frac{1}{2} \times n = \frac{n}{2}$. Hence, the competitive ratio is at least $n/2(\log(n^2 \cdot r(n)))$.   □

Remark. Observe that the graphs constructed have a very specific structure. For one, they are *split graphs*, as the vertex set can be partitioned into an independent set (the good nodes) and a clique (the bad nodes). The graphs also belong to another subclass of chordal graphs: interval graphs. Vertex i can be seen to correspond to an interval on the real line with a starting point at i. If the node becomes a bad node, then the interval reaches far to the right with an endpoint at n. If it becomes a good node, then the right endpoint is set at

$i + 1/2$. It follows that the problem of scheduling intervals online for maximizing throughput is hard, unless some information about the intervals' endpoints are given. This contrasts with the case of scheduling with respect to makespan, which corresponds to online coloring interval graphs, for which a 3-competitive algorithm is known [6].

Also, observe that the latter lower bound (as well as the construction in the following section) is oblivious (in that it does not depend on the actions of the algorithm), and hold equally well against randomized algorithms.

3   Inheritance Model

The model we shall introduce in this section is much more flexible than the previous one. At each step, an algorithm can copy up to $r(n)$ bins and put the current vertex into those copies or a new empty bin. (We restrict the number of copies to $r(n) - 1$ when an online algorithm uses an empty bin, because it is natural to restrict the number of bins each vertex can be put into to $r(n)$ considering the consistency with multi-solutions model.) Note that this model is at least as powerful as the multi-solutions model with the same $r(n)$.

The upper bound follows from Theorem 1.

Corollary 4. *There is an online algorithm with competitive ratio at most $O(n/\log r(n))$ in inheritance model.*

3.1   A Lower Bound

As mentioned before, a lower bound argument is a challenging one. Before showing our result, we discuss some natural ideas and see why those arguments only give us poor bounds.

Discussion 1   Consider the *transparent* adversary strategy of the proof of Theorem 3. There is an online algorithm that against this adversary always outputs an optimal solution for split graphs. The algorithm uses only two bins at each step. At the first step, it maintains the empty bin and the bin that contains v_1. At the ith step ($i \geq 2$), the algorithm knows that the vertex v_{i-1} is good or bad by observing whether there is an edge (v_{i-1}, v_i) or not. If v_{i-1} is good, the algorithm copies the bin containing v_{i-1}, adds v_i to the copy, and forgets about the bin not containing v_{i-1}. Otherwise, if v_{i-1} is bad, exchange the operations for the two bins. Observe that this algorithm outputs a solution consisting of all good vertices.

What then is the difference between our new model and the old one? Note that, in the inheritance model, an algorithm can copy bins it currently holds. To simulate this operation in the multi-solutions model, an algorithm has to construct in advance two bins whose contents are the same. Since the above algorithm in the inheritance model copies at each step, an algorithm in the multi-solution model must prepare exponential number of bins in the worst case to simulate it. Hence, for the current adversary, the inheritance model is exponentially more powerful than the multi-solutions model. Thus, to argue a lower bound in the inheritance model, we need a more powerful adversary.

Discussion 2   In the discussion of the previous section, what supported the online algorithm in the new model is the information that the vertex is good or bad given *immediately after* the algorithm processed that vertex. Hence the algorithm knows whether it is good or bad at the next step. We now consider the possibility of postponing this decision of the adversary.

Divide the vertex sequence into rounds. At the end of each round, the adversary determines the states of vertices it gave in that round. A vertex then acts as a good vertex while its state is undetermined. Until the end of the round, it is impossible for an online algorithm to make a decision depending on states of vertices given in that round. If, at the end of a round, an online algorithm maintains a bin that contains many vertices of that round, we "kill" that bin by making one of those vertices bad. This can be easily done by setting an appropriate probability that a vertex becomes bad.

This approach still allows for an algorithm with a competitive ratio of $\sqrt{n}$. The algorithm maintains an independent set I containing one vertex from each round that contains a good vertex. It also puts all the vertices of each round i into an independent set I_i. It is easy to see that the largest of I and the I_i's is of size at least $\sqrt{opt}$, when the optimal solution contains opt vertices.

A Better Lower Bound   The reason why the above argument did not give a satisfactory lower bound is that the adversary strategy was still weak: Once a decision is made, the algorithm knows that a good vertex will never become bad. Here, we consider an adversary that makes a probabilistic decision more than once (approximately $\log n$ times) to each vertex.

Theorem 5. *Any online IS algorithm has a competitive ratio at least* $\frac{n}{e^{k+1}(\log n)^3}$ *in the inheritance model with* $r(n) = n^k$ *where* e *is the base of the natural logarithm.*

Proof. We first describe informally the action of the adversary. He presents $l_0 = \log^2 n$ new vertices in each round. Between rounds he makes a decision to make vertices bad, not only for vertices given in the previous round, but also for some other vertices already given.

The goal of the adversary is to stunt the growth of bins containing many vertices by making some of them bad. At the same time, it needs to ensure that the fraction of vertices that are good in the end is large. The adversary repeats the following steps n/l_0 times:

(1) The adversary presents a block of l_0 new vertices. All those vertices are good and each of them has edges to all bad vertices previously given.

(2) As long as there are two blocks b_1, b_2 of the same size, replace them by the union of the two. At the same time, each vertex of the combined block is changed into a bad vertex *independently* with probability $\frac{k+1}{\log n}$.

Let $t = \log(n/l_0)$. Note that each block is of size $2^i l_0$, $0 \leq i < t$, and the number of blocks is at most $t + 1$.

Each time blocks are merged, the adversary is said to *attack* the respective vertices by probabilistically changing them to bad. The probability of a vertex surviving an attack is $1 - \frac{k+1}{\log n}$. Each vertex receives at most t attacks. Hence, the probability that a vertex is good at the end of the game is at least $(1 - \frac{k+1}{\log n})^t \geq \frac{1}{e^{k+1}}$. The expected number of good vertices, which is the measure of the optimal solution, is then at least n/e^{k+1}.

We now evaluate the performance of the online algorithm. It then appears that a smaller l_0 is harder for the algorithm to keep large bins. Unfortunately, a smaller l_0 also makes proofs harder. The following setting, $l_0 = (\log n)^2$, is somewhat conservative:

Lemma 6. *Suppose that $l_0 = (\log n)^2$. Then any online IS algorithm in the inheritance model with $r(n) = n^k$ cannot produce independent sets whose size is $(\log n)^3$ or more in the worst case.*

Proof. We maintain the following invariant $(*)$:

> There is no bin B containing only good vertices that has
> more than $(\log n)^2$ vertices from a single block. $(*)$

When the first $l_0 = (\log n)^2$ vertices are given, $(*)$ is obviously true. Suppose now that $(*)$ is true at some moment. It then also remains true when the adversary introduces in step (1) a new block of l_0 vertices. Consider now the case when two blocks b_1 and b_2 are merged. Let S be the set of good vertices in $b_1 \cup b_2$ contained in a particular bin B. If $|S| \leq l_0$ the invariance remains true, so suppose otherwise. Recall that the adversary attacks each vertex in S with probability $\frac{k+1}{\log n}$. Therefore, the probability that all vertices in S survive the attacks is at most

$$(1 - \frac{k+1}{\log n})^{(\log n)^2} = \frac{1}{n^{(k+1)\log e}}.$$

The probability that there is some bin that contains more than l_0 vertices from $b_1 \cup b_2$ is at most

$$\frac{n^{k+1}}{n^{(k+1)\log e}} = n^{(k+1)(1-\log e)} < 1.$$

That is, it holds with some positive probability that no bin includes $(\log n)^2$ or more good vertices in this combined block. This implies that there exists an adversary strategy (a choice of assignments to at most $(k+1)/\log n$-fraction of the vertices of $b_1 \cup b_2$) that maintains the invariant $(*)$.

It follows that no bin contains more than $t \cdot l_0 < \log^3 n$ vertices. $\square$

References

1. D. Bean, "Effective coloration," *J. Symbolic Logic*, Vol. 41, pp. 469–480, 1976.
2. A. Gyárfás and J. Lehel, "On-line and first fit colorings of graphs," *J. Graph Theory*, Vol. 12(2), pp. 217–227, 1988.
3. M. Halldórsson "Online coloring known graphs," *Electronic J. Combinatorics*, Vol. 7, R6, 2000. www.combinatorics.org.
4. M. Halldórsson and M. Szegedy, "Lower bounds for on-line graph coloring," *Theoretical Computer Science*, Vol. 130, pp. 163–174, 1994.
5. H. Kierstead, "On-line coloring k-colorable graphs," *Israel J. of Math*, Vol. 105, pp. 93–104, 1998.
6. H. A. Kierstead and W. T. Trotter, "An extremal problem in recursive combinatorics," *Congressus Numerantium* Vol. 33, pp. 143–153, 1981.
7. R. J. Lipton and A. Tomkins, "Online interval scheduling," *Proc. SODA '94*, pp. 302–311, 1994.
8. L. Lovász, M. Saks, and W. T. Trotter, "An online graph coloring algorithm with sublinear performance ratio," *Discrete Math.*, Vol. 75, pp. 319–325, 1989.

Two-Dimensional On-Line Bin Packing Problem with Rotatable Items

Satoshi Fujita and Takeshi Hada

Faculty of Engineering, Hiroshima University
Higashi-Hiroshima, 739-8527, JAPAN
{fujita,hada}@se.hiroshima-u.ac.jp

Abstract. In this paper, we consider a two-dimensional version of the on-line bin packing problem, in which each rectangular item that should be packed into unit square bins is "rotatable" by 90 degrees. An on-line algorithm that uses an unbounded number of active bins is proposed. The worst case ratio of the algorithm is at least 2.25 and at most 2.565.

1 Introduction

The bin packing problem is one of the basic problems in the fields of theoretical computer science and combinatorial optimization. It has many important real-world applications, such as memory allocation and job scheduling, and is well-known to be NP-hard [4]; that is a main motivation of the study and development of approximation algorithms for solving the problem.

The classical (one-dimensional) on-line bin packing problem is the problem of, given a list L of items $\langle x_1, x_2, \ldots, x_n \rangle$ where $x_i \in (0, 1]$, packing all items in L into a minimum number of unit-capacity bins. Note that term "on-line" implies that items in L are consecutively input, and the packing of an item must be determined before the arrival of the next item. In the literature, it is known that the worst case ratio[1] of an optimal on-line bin packing algorithm OPT, denoted by R_{OPT}^{∞}, is at least 1.5401 [7] and at most 1.588 [6] if the number of active bins[2] used by the algorithm is unbounded; and is exactly 1.69103 [5] if it is bounded by some constant k.

In this paper, we consider a two-dimensional version of the problem, in which each rectangular item that should be packed into unit square bins is "rotatable" by 90 degrees. The goal of the problem we consider is to pack all items in L into a minimum number of unit square bins in such a way that 1) each item is entirely contained inside its bin with all sides parallel to the sides of the bin, and 2) no two items in a bin overlap. It should be worth noting that in the "normal" setting of higher dimensional bin packing problems, it is further requested that 3) the orientation of any item is the same as the orientation of the bin, i.e., each

---

[1] A formal definition will be given in Section 2.

[2] We consider a class of algorithms in which each bin becomes *active* when it receives its first item; and once a bin is declared to be *inactive* (or *closed*), it can never become active again.

D.-Z. Du et al. (Eds.): COCOON 2000, LNCS 1858, pp. 210–220, 2000.
© Springer-Verlag Berlin Heidelberg 2000

item must *not* be rotated; and under such a setting, it is known that an optimal on-line algorithm OPT fulfills $1.907 \le R_{OPT}^{\infty} \le 2.85958 \ (= 1.69103^2)$ [1,3].

The two-dimensional on-line bin packing problem with rotatable items has several interesting applications, e.g., job scheduling on two-dimensional mesh computers and the floor planning of VLSI layouts, while the rotation of items is prohibited in some of other applications, such as the assignment of newspaper articles to newspaper pages. A naive conjecture is that the worst case ratio could be improved by removing the condition on the rotatability since there is an instance in which the removal of the condition reduces an optimal number of bins to a half (e.g., consider a sequence of items of height $\frac{3}{8}$ and width $\frac{5}{8}$ each; by rotating a half of items, we can pack four such items into a bin, while each bin can accommodate at most two items without rotation). Another conflicting conjecture is that it is not very easy to improve the ratio, since the lower bound can also be improved by allowing rotations. In the following, we propose two algorithms A_1 and A_2 such that $2.5624 \le R_{A_1}^{\infty} \le 2.6112$ and $2.25 \le R_{A_2}^{\infty} \le 2.565$, where the second algorithm is an extension of the first algorithm.

The remainder of this paper is organized as follows. Section 2 introduces some basic definitions. The classifications of items and strips used in the proposed algorithms will also be given. Sections 3 and 4 describe the proposed algorithms. Section 5 concludes the paper with some future directions of research.

2   Preliminaries

2.1   Basic Definitions

Let A be an on-line bin packing algorithm, and $A(L)$ the number of bins used by algorithm A for input sequence L. The *asymptotic worst case ratio* (or simply, *worst case ratio*) of algorithm A, denoted as R_A^{∞}, is defined as follows:

$$R_A^{\infty} \overset{\text{def}}{=} \limsup_{n \to \infty} R_A^n$$

$$\text{where } R_A^n \overset{\text{def}}{=} \max \left\{ \frac{A(L)}{OPT(L)} \mid OPT(L) = n \right\}.$$

where $OPT(L)$ denotes the number of bins used by an optimal bin packing problem provided that the input is L. An on-line bin packing problem is said to use k-bounded-space (or, simply *bounded*) if, for each item, that is given in an on-line manner, the choice of bins into which it may be packed is restricted to a set of k or fewer active bins. If there is no such restriction, we say that the problem uses unbounded space (or, simply *unbounded*).

2.2   Classification of Items

Let T be the set of all rectangular items, denoted as (x, y), such that $0 < x \le y \le 1$,[3] where x and y represent the height and width of the rectangle,

---

[3]   Note that we may assume $x \le y$ without loss of generality, since each item is rotatable by 90 degrees.

respectively. We first classify items in T into four subsets by the value of the x-coordinate, as follows: $T_0 = \{(x,y)|\ \frac{2}{3} < x \le 1\ \}$, $T_1 = \{(x,y)|\ \frac{1}{2} < x \le \frac{2}{3}\ \}$, $T_2 = \{(x,y)|\ \frac{1}{3} < x \le \frac{1}{2}\ \}$, and $T_3 = \{(x,y)|\ 0 < x \le \frac{1}{3}\ \}$. See Figure 1 (a) for illustration. Subsets T_2 and T_3 are further partitioned into subsets by the value of the y-coordinate, as follows.

Definition 1 (Partition of T_3). Subset T_3 is partitioned into four subsets $T_3^0, T_3^1, T_3^2, T_3^3$ by the value of the y-coordinate, as follows: $T_3^0 = \{(x,y) \in T_3|\ \frac{2}{3} < y \le 1\ \}$, $T_3^1 = \{(x,y) \in T_3|\ \frac{1}{2} < y \le \frac{2}{3}\ \}$, $T_3^2 = \{(x,y) \in T_3|\ \frac{1}{3} < y \le \frac{1}{2}\ \}$, and $T_3^3 = \{(x,y) \in T_3|\ x \le y \le \frac{1}{3}\ \}$.

In order to define a partition of T_2, let us consider a sequence $a_0, a_1, \ldots$ of reals defined as follows:

$$a_0 \overset{\text{def}}{=} \tfrac{1}{2} \quad \text{and} \quad a_i \overset{\text{def}}{=} \tfrac{1}{3} + \tfrac{1}{6 \times 2^{i-1}+3} \text{ for } i = 1, 2, \ldots. \tag{1}$$

For example, $a_1 = \frac{1}{3} + \frac{1}{9} = \frac{4}{9}$, $a_2 = \frac{1}{3} + \frac{1}{15} = \frac{2}{5}$, and $a_3 = \frac{1}{3} + \frac{1}{27} = \frac{10}{27}$. Note that $\frac{1}{3} < a_i \le \frac{1}{2}$ holds for any $i \ge 0$. Let ϵ be a positive real that will be used as a parameter in the analysis of our algorithms, and ϕ a function from $\mathbf{R}^+$ to $\mathbf{Z}$ defined as follows:

$$\phi(t) \overset{\text{def}}{=} \lceil \log_2 (2/(9t) - 1) \rceil. \tag{2}$$

By using above definitions, the partition of subset T_2 is described as follows.

Definition 2 (Partition of T_2). Given parameter $\epsilon > 0$, subset T_2 is partitioned into three subsets T_2^0, T_2^1, T_2^2, as follows:

$$T_2^0 \overset{\text{def}}{=} \bigcup_{j=0}^{\phi(\epsilon)-1} \{(x,y)|\ a_{j+1} < x \le a_j, 1 - a_j < y \le 1\}$$
$$\cup \{(x,y)|\ \tfrac{1}{3} < x \le a_{\phi(\epsilon)},\ 1 - a_{\phi(\epsilon)} < y \le 1\}$$

$$T_2^1 \overset{\text{def}}{=} \bigcup_{j=0}^{\phi(\epsilon)-1} \{(x,y)|\ a_{j+1} < x \le a_j, \tfrac{1}{2} < y \le 1 - a_j\}$$
$$\cup \{(x,y)|\ \tfrac{1}{3} < x \le a_{\phi(\epsilon)},\ \tfrac{1}{2} < y \le 1 - a_{\phi(\epsilon)}\}$$

$$T_2^2 \overset{\text{def}}{=} \{(x,y) \in T_2|\ \tfrac{1}{3} < y \le \tfrac{1}{2}\}.$$

The overall partitioning of T into 9 subsets is also illustrated in Figure 1 (a). Note that, for example, if the given item is $(\frac{4}{9}, \frac{4}{10})$ then it belongs to subset T_2^2 since $\frac{1}{3} < \frac{4}{9} \le \frac{1}{2}$ and $\frac{1}{3} < \frac{4}{10} \le \frac{1}{2}$. As for the minimum size of items in each subset, we have the following two lemmas (proofs are omitted here).

Lemma 1. *If $i > \log_2\left(\frac{2}{9\epsilon} - 1\right)$ then $\frac{1}{3} \times (1 - a_i) > \frac{2}{9} - \frac{\epsilon}{2}$.*

Lemma 2. *1) The size of items in T_0 is greater than $\frac{4}{9}$; 2) the size of items in T_1 is greater than $\frac{1}{4}$; and 3) the size of items in T_2^0 is greater than $\frac{2}{9} - \frac{\epsilon}{2}$.*

2.3   Strips

In the proposed algorithms, each bin is split into small *strips*, and a given item is packed into a strip of an appropriate size. Before describing the ways of splitting of each bin into strips, let us consider the following infinite set X of reals:

$$X \stackrel{\text{def}}{=} \{ \tfrac{1}{i \times 2^j} \mid i = 3, 4, 5, 7, 9, 11, 13 \text{ and } j = 0, 1, 2, \dots \}.$$

By using X, two (infinite) sets of reals, X_{long} and X_{med}, are defined as follows:

$$X_{\text{long}} \stackrel{\text{def}}{=} X \cup \{ \tfrac{1}{2} \} \tag{3}$$

$$X_{\text{med}} \stackrel{\text{def}}{=} X \cup \{ a_1, a_2, \dots, a_{\phi(\epsilon)} \} \tag{4}$$

where $a_i = \tfrac{1}{3} + \tfrac{1}{6 \times 2^{i-1} + 3}$ for $i \geq 1$. In what follows, given $0 < x \leq 1$, let $\tilde{x}$ denote the smallest element in X_{long} that is greater than or equal to x, and let $\hat{x}$ denote the smallest element in X_{med} that is greater than or equal to x. Note that for any $x < \tfrac{1}{3}$, $x/\tilde{x}\ (= x/\hat{x}) > \tfrac{3}{4}$.

Three types of strips used in the proposed algorithms are defined as follows.

Definition 3 (Strips). A subregion of a bin is said to be a **long strip** if it has a fixed width 1 and a height that is drawn from set X_{long}. A subregion of a bin is said to be a **short strip** if it has a fixed width $\tfrac{1}{2}$ and a height that is drawn from set X_{long}. A subregion of a bin is said to be a **medium strip** if 1) it has a fixed width (resp. height) $\tfrac{2}{3}$ and a height (resp. width) that is drawn from set X, or 2) it has a width (resp. height) $1 - a_i$ and a height (resp. width) a_i, for some $1 \leq i \leq \phi(\epsilon)$.

3   The First Algorithm

3.1   Algorithm Description

In this section we describe our first on-line algorithm. The algorithm is described in an event-driven manner; i.e., it describes the way of packing an input item (x, y) into an active bin of an appropriate size.

Algorithm A_1

If $(x, y) \in T_0 \cup T_1$, then *open* a new bin, put the item into the bin, and *close* the bin; otherwise (i.e., if $(x, y) \in T_2 \cup T_3$) execute the following operations:

Case 1: if $(x, y) \in T_2^0 \cup T_3^0$, then obtain an unused long strip of height $\tilde{x}$ (as an active strip) by calling get_long_strip(x), put the item into the returned strip, and *close* the strip.

Case 2: if $(x, y) \in T_2^1 \cup T_3^1$, then obtain an unused medium strip of height $\hat{x}$ (as an active strip) by calling get_medium_strip(x), put the item into the returned strip, and *close* the strip.

Case 3: if $(x, y) \in T_2^2 \cup T_3^2$, then obtain an unused short strip of height $\tilde{x}$ (as an active strip) by calling get_short_strip(x), put the item into the returned strip, and *close* the strip.

Case 4: if $(x,y) \in T_3^3$, then put the item into an active long strip of height $\tilde{x}$ if such a strip with enough space is available. If there exist no such strips, after *closing all* active long strips of height $\tilde{x}$, obtain an unused long strip of height $\tilde{x}$ by calling **get_long_strip**(x), and put the item into the returned strip. Note that the strip is *not closed* at this time.

Figure 1 (b) illustrates the correspondence of items with strips, in algorithm A_1. The performance of the algorithm, in terms of the occupation ratio of each closed strip, is estimated as follows (proofs of Lemmas 4 and 5 are omitted).

Lemma 3 (Long strips). *Any closed long strip of size S is occupied by items with total size at least $(\frac{4}{9} + \epsilon)S$ if the height of the strip is $\frac{1}{2}$, and with size at most $\frac{S}{2}$, if the height is less than $\frac{1}{2}$.*

Proof. By Lemma 2, a closed long strip of height $\frac{1}{2}$ is filled with items of size at least $\frac{2}{9} - \frac{\epsilon}{2}$; i.e., the occupation ratio of the bin is at least $\frac{4}{9} - \epsilon$. On the other hand, by the same lemma, it is shown that a closed long strip of height $\frac{1}{i}$, for $i \geq 3$, is filled with items of total size more than $\frac{2}{3} \times \frac{1}{i+1}$, that is at least $\frac{1}{2}$. Hence the lemma follows.

Lemma 4 (Short strips). *Any closed short strip of size S is occupied by an item with size at least $\frac{4}{9}S$ if the height of the strip is $\frac{1}{2}$, and with size at most $\frac{S}{2}$, if the height is less than $\frac{1}{2}$.*

Lemma 5 (Medium strip). *1) A medium strip with height a_i for $1 \leq i \leq \phi(\epsilon)$, is filled with an item with size more than $\frac{1}{6}$ $(= \frac{1}{3} \times \frac{1}{2})$; and 2) a medium strip with height $\frac{1}{i}$, for some $i \geq 3$, is filled with an item with size more than $\frac{1}{2(i+1)}$ $(= \frac{1}{i+1} \times \frac{1}{2})$.*

3.2 How to Split a Bin into Strips ?

Next, we consider the ways of splitting (unused) bins into strips. In the following, we give a formal description of three procedures, **get_long_strip**, **get_medium_strip**, and **get_short_strip**, that have been used in the proposed algorithm.

procedure get_long_strip(x)

1. Recall that $\tilde{x}$ denotes the smallest element in X_{long} that is greater than or equal to x. If there is no unused long strip with height $\tilde{x}2^j$ (< 1) for any $j \geq 0$, then execute the following operations:
   (a) Let j' be an integer such that $\tilde{x}2^{j'} < 1 \leq \tilde{x}2^{j'+1}$. Note that $\tilde{x}2^{j'} = \frac{1}{2}, \frac{1}{3}, \frac{1}{5}, \frac{1}{7}, \frac{1}{9}, \frac{1}{11}$, or $\frac{1}{13}$.
   (b) Open an unused bin, and partition it into $\frac{1}{\tilde{x}2^{j'}}$ long strips with height $\tilde{x}2^{j'}$ each.
2. Select one of the "lowest" strips, say Q, with height $\tilde{x}2^j$ (< 1) for some $j \geq 0$. Let $\tilde{x}2^{j''}$ be the height of strip Q.

3. If $j'' = 0$ (i.e., if the height of Q is $\tilde{x}$) then return Q as an active strip; otherwise, partition Q into $j'' + 1$ strips of heights $\tilde{x}2^{j''-1}, \tilde{x}2^{j''-2}, \ldots, \tilde{x}, \tilde{x}$, respectively, and return a strip of height $\tilde{x}$ as an active one.

procedure get_short_strip(x)

1. If there is an unused short strip of height $\tilde{x}$, then return it as an active one.
2. Otherwise, obtain an unused long strip with height $\tilde{x}$ by calling procedure get_long_strip(x), partition the obtained one into two unused short strips with height $\tilde{x}$ each, and return one of them as an active one.

procedure get_medium_strip(x)

1. Recall that $\hat{x}$ denotes the smallest element in X_{med} that is greater than or equal to x. If there is an unused *horizontal* medium strip with height $\hat{x}$ (or, *vertical* medium strip with width $\hat{x}$), then return it as an active one.
2. Otherwise, if $\hat{x} > \frac{1}{3}$, then open an unused bin, partition it into two horizontal medium strips of height $\hat{x}$ and two vertical medium strips of width $\hat{x}$ as in Figure 2 (a), and return one of the four strips as an active one.
3. Otherwise, i.e., if $\hat{x} \leq \frac{1}{3}$, execute the following operations:
   (a) If there is no unused medium strip with height $\hat{x}2^j$ (< 1) for any $j \geq 0$, then open an unused bin, partition it into $\frac{1}{\hat{x}2^{j'}}$ horizontal medium strips of height $\hat{x}2^{j'}$ and $\lfloor \frac{1}{3\hat{x}2^{j'}} \rfloor$ vertical medium strips of width $\hat{x}2^{j'}$ as in Figure 2 (b), where j' is an integer such that $\hat{x}2^{j'} < 1 \leq \hat{x}2^{j'+1}$.
   (b) Select one of the "lowest" medium strips, say Q, with height $\hat{x}2^j$ (< 1) for some $j \geq 0$. Let $\hat{x}2^{j''}$ be the height of strip Q.
   (c) If $j'' = 0$ (i.e., if the height of Q is $\hat{x}$) then return Q as an active strip; otherwise, partition Q into $j'' + 1$ strips of heights $\hat{x}2^{j''-1}, \hat{x}2^{j''-2}, \ldots, \hat{x}, \hat{x}$, respectively, and return a strip of height $\hat{x}$ as an active one.

By Lemma 5 and get_medium_strip, we immediately have the following lemma.

Lemma 6. *Any closed bin that is split into medium strips is filled with items with total size at least $\frac{1}{2}$.*

3.3   Analysis

Theorem 1. $R_{A_1}^{\infty} \leq 2.6112 + \epsilon$ *for any $\epsilon > 0$.*

Proof. Let L be a sequence of n items, consisting of n_0 items in T_0, n_1 items in T_1, where $n_0 + n_1 \leq n$. Note that algorithm A_1 requires b_0 ($= n_0$) bins for items in T_0 and b_1 ($= n_1$) bins for items in T_1. Suppose that it also uses b_2 (≥ 0) bins split into long strips of height $1/2$, and b_3 (≥ 0) bins split into medium strips of height more than $1/3$. Let S be the total size of items in L, and S_4 the total size of items that are not packed into the above $n_0 + n_1 + b_2 + b_3$ bins; i.e., items of height at most $1/3$.

If ϵ is a constant, algorithm A_1 uses at most $n_0 + n_1 + b_2 + b_3 + 2S_4 + O(1)$ bins and OPT requires at least $\max\{n_0 + \max\{n_1, b_2\}, S\}$ bins. Hence $R_{A_1}^{\infty} \leq$

$\{n_0 + n_1 + b_2 + b_3 + 2S_4\}/\{\max\{n_0 + \max\{n_1, b_2\}, S\}\}$. Since $S_4 \leq S - \frac{4}{9}n_0 - (\frac{4}{9} - \epsilon)b_2 - \frac{2}{3}b_3 - \frac{n_1}{4}$ by Lemmas 3, 4, and 6, we have

$$R_{A_1}^\infty \leq \frac{2S + \frac{1}{9}n_0 + \frac{1}{2}n_1 + (\frac{1}{9} + \epsilon')b_2 - \frac{1}{3}b_3}{\max\{n_0 + \max\{n_1, b_2\}, S\}}.$$

When $n_0 + \max\{n_1, b_2\} < S$, $R_{A_1}^\infty \leq 2 + (\frac{1}{S})\{\frac{1}{9}n_0 + (\frac{1}{9} + \epsilon')b_2 + \frac{1}{2}n_1\}$. If $n_1 \leq b_2$, $R_{A_1}^\infty \leq 2 + (1/S)\{\frac{1}{9}n_0 + (\frac{11}{18} + \epsilon')b_2\} \leq 2 + \frac{11}{18} + \epsilon'$, and if $n_1 > b_2$, $R_{A_1}^\infty \leq 2 + (1/S)\{\frac{1}{9}n_0 + (\frac{11}{18} + \epsilon')n_1\} \leq 2 + \frac{11}{18} + \epsilon'$. On the other hand, when $n_0 + \max\{n_1, b_2\} \geq S$, it is at most $R_{A_1}^\infty \leq 2 + \{\frac{1}{9}n_0 + (\frac{1}{9} + \epsilon')b_2 + \frac{1}{2}n_1\}/\{n_0 + \max\{n_1, b_2\}\}$. If $n_1 \leq b_2$, $R_{A_1}^\infty \leq 2 + \{\frac{1}{9}n_0 + (\frac{11}{18} + \epsilon')b_2\}/(n_0 + b_2) \leq 2 + \frac{11}{18} + \epsilon'$, and if $n_1 > b_2$, $R_{A_1}^\infty \leq 2 + \{\frac{1}{9}n_0 + (\frac{11}{18} + \epsilon')n_1\}/(n_0 + n_1) \leq 2 + \frac{11}{18} + \epsilon'$. Hence the theorem follows.

Theorem 2. $R_{A_1}^\infty \geq 2.5624$.

Proof. Consider the following eight items

$A : (\frac{1}{2} + \delta, \frac{1}{2} + \delta)$ $B : (\frac{1}{3} + \delta, \frac{2}{3} - \delta)$ $C : (\frac{1}{3} + \delta, \frac{1}{3} + \delta)$ $D : (\frac{1}{11} + \delta, \frac{1}{2} + \delta)$
$E : (\frac{1}{13} + \delta, \frac{1}{2} + \delta)$ $F : (\frac{1}{11} + \delta, \frac{1}{11} + \delta)$ $G : (\frac{1}{11} + \delta, \frac{1}{13} + \delta)$ $H : (\frac{1}{13} + \delta, \frac{1}{13} + \delta)$

where δ is a sufficiently small constant such that item B is in T_2^0.

Let L be a sequence of items consisting of n copies of items A, C, F, H, and $2n$ copies of items B, D, E, G. Note that since $\frac{1}{11} + \frac{1}{13} = \frac{24}{143} = \frac{1}{6} - \frac{1}{853}$, by selecting δ to be sufficiently small, we can pack one copy of A, C, F, H and two copies of B, D, E, G, as in Figure 3. Hence $OPT(L) = n$.

On the other hand, in algorithm A_1, 1) n copies of item A require n bins; 2) $2n$ copies of item B require n bins, since B is classified into T_2^0; 3) n copies of item C require $\frac{n}{4}$ bins, since C is classified into T_2^1; 4) $2n$ copies of item D require $\frac{2n}{13}$ bins; 5) $2n$ copies of item E require $\frac{2n}{16}$ bins; 6) n copies of item F require $\frac{n}{100}$ bins; 7) $2n$ copies of item G require $\frac{2n}{120}$ bins; and 8) n copies of item H require $\frac{n}{144}$ bins. Hence, in total, it requires at least

$$n + n + \frac{n}{4} + \frac{2n}{13} + \frac{2n}{16} + \frac{n}{100} + \frac{2n}{120} + \frac{n}{144} > 2.5624n$$

bins. Hence the theorem follows.

4   The Second Algorithm

In this section, we propose an extension of the first algorithm. The basic idea behind the extension is to pack an item in T_1 and item(s) in T_3 in the same bin, as much as possible. Recall that under the "on-line" setting, we cannot use any knowledge about the input sequence L in the algorithm.

4.1 Algorithm Description

In our second algorithm, called Algorithm A_2, when an item in T_1 is packed into a new bin, the bin is split into two subregions; a square region of size $\frac{2}{3} \times \frac{2}{3}$ that will be filled with the item in T_1, and the remaining "L-shaped" region that is reserved for possible usage by items in T_3 (note that the reserved L-shaped region will never be used, in the worst case). On the other hand, when an item in T_3 is packed into a new bin, under a certain condition described below, the bin is partitioned into two subregions, and the square region of size $\frac{2}{3} \times \frac{2}{3}$ is reserved for possible usage by items in T_1 (note again that the reserved square region will never be used, in the worst case).

In order to balance the advantage and disadvantage of the above scenario, we partition 20% of bins devoted to items in T_3 for the above use; the remaining 80% of bins are used in the same manner to Algorithm A_1. A formal description of the algorithm is given as follows.

Algorithm A_2

Case 1: If $(x, y) \in T_0 \cup T_2$, then it acts as in Algorithm A_1.

Case 2: If $(x, y) \in T_1$, then put the item into an active bin if there is an active bin that has been split into two subregions, and the small square region has not accommodated any item; if there is no such bin, then open a new bin, split it into two subregions, and pack the item into the small square region.

Case 3: If $(x, y) \in T_3$, then it obtains a strip of height $\tilde{x}$ ($= \hat{x}$) by calling procedures described in Section 3.2 with a minor modification such that an unused L-shaped region is a possible candidate for the splitting into strips, and pack the item into a strip to satisfy the following conditions: 80% of bins dedicated to items in T_3 are filled with closed strips; and 20% of such bins are filled with strips in such a way that there remains an available square region of size $\frac{2}{3} \times \frac{2}{3}$.

4.2 Analysis

Theorem 3. $R_{A_2}^\infty \leq 2.565 + \epsilon$ *for any* $\epsilon > 0$.

Proof. Let L be a sequence of n items. In this proof, we use the same symbols as in the proof of Theorem 1; e.g., n_0 and n_1 denote the numbers of items in T_0 and T_1, respectively.

Recall that algorithm A_1 uses at most $n_0 + n_1 + b_2 + b_3 + 2S_4 + O(1)$ bins and an optimal algorithm OPT requires at least $\max\{n_0 + \max\{n_1, b_2\}, S\}$ bins.

If $n_1 \leq \frac{S_4}{2}$, then since all items in T_1 are packed into bins that is split into small square region and an L-shaped region, the "average" occupation ratio of every closed bin can be increased from at least $\frac{1}{4}$ to at least $\frac{4}{9}$ (note that the average occupation ratio of bins with strips of height at most $\frac{1}{3}$, on the other hand, reduces from $\frac{1}{2}$ to $\frac{2}{5}$); i.e., the worst case ratio is bounded by $\frac{5}{2} = 2.5$. Hence, in the following, we assume that $n_1 > \frac{S_4}{2}$.

If $n_1 > \frac{S_4}{2}$, we can imaginally suppose that the number of bins split into strips of height at most $\frac{1}{3}$ each reduces from $2S_4$ to αS_4 for some constant $\alpha < 2$, by using the splitting method illustrated in Figure 5. If the number of (imaginal) bins used by items in T_3 is αS_4, then by using a similar argument to Theorem 1, we have

$$R_{A_2}^{\infty} \leq 2 + \frac{11}{36}\alpha$$

(note that if $\alpha = 2$, we have the same bound with Theorem 1). The number of (imaginal) bins reduces from $2S_4$ to at most $\frac{24}{13}$ if all items are packed into strips of height $\frac{1}{3}$; at most $\frac{30}{17}$ ($< \frac{24}{13}$) if all items are packed into strips of height $\frac{1}{4}$; and at most $\frac{48}{26}$ ($< \frac{24}{13}$) if all items are packed into strips of height at most $\frac{1}{5}$. Hence we have $\alpha \leq \frac{24}{13}$ for any sequence L; i.e.,

$$R_{A_2}^{\infty} \leq 2 + \frac{11}{36} \times \frac{24}{13} = 2 + \frac{22}{39} \leq 2.565.$$

Hence the theorem follows.

Finally, by using a similar technique to Theorem 2 we have the following theorem (we may simply consider a sequence consisting of items A, B and C).

Theorem 4. $R_{A_2}^{\infty} \geq 2.25$.

5 Concluding Remarks

In this paper, we proposed two on-line algorithms for solving the two-dimensional bin packing problem with rotatable rectangular items.

The proposal of a nontrivial lower bound is an important open problem to be solved (currently best lower bounds on the worst case ratio of an optimal algorithm for "rotatable" setting are the same as the one-dimensional case; i.e., it is at least 1.5401 for unbounded algorithms [7], and at least 1.69103 for bounded algorithms [5]). An extension to higher dimensional cases (e.g., three-dimensional box packing) would also be promising.

Acknowledgments: This research was partially supported by the Ministry of Education, Science and Culture of Japan, and Research for the Future Program from Japan Society for the Promotion of Science (JSPS): Software for Distributed and Parallel Supercomputing (JSPS-RFTF96P00505).

References

1. D. Blitz, A. van Vliet, and G. J. Woeginger. Lower bounds on the asymptotic worst-case ratio of on-line bin packing algorithms. unpublished manuscript, 1996.
2. D. Coppersmith and P. Raghavan. Multidimensional on-line bin packing: Algorithms and worst case analysis. *Oper. Res. Lett.*, 8:17–20, 1989.
3. J. Csirik and A. van Vliet. An on-line algorithm for multidimensional bin packing. *Oper. Res. Lett.*, 13:149–158, 1993.
4. M. R. Garey and D. S. Johnson. *Computers and Intractability: A Guide for the Theory of NP-Completeness*. Freeman, San Francisco, CA, 1979.

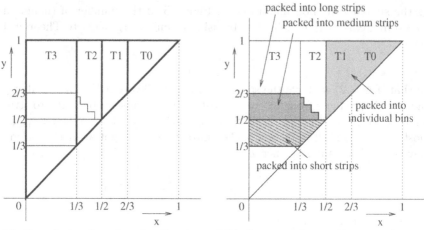

(a) Boundaries for the classification of items by the values of x- and y-coordinates.

(b) Correspondence of items with strips.

Fig. 1. Classification of items.

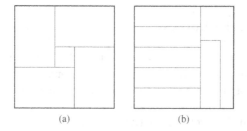

(a)   (b)

Fig. 2. Partition of a bin into medium strips.

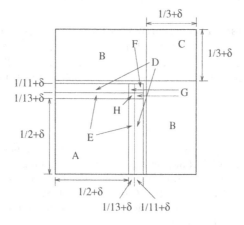

Fig. 3. A bad instance to give a lower bound.

5. C. C. Lee and D. T. Lee. A simple on-line bin packing algorithm. *J. Assoc. Comput. Mach.*, 32:562–572, 1985.
6. M. B. Richey. Improved bounds for harmonic-based bin packing algorithms. *Discrete Appl. Math.*, 34:203–227, 1991.
7. A. van Vliet. An improved lower bound for on-line bin packing algorithms. *Information Processing Letters*, 43:277–284, 1992.

Better Bounds on the Accommodating Ratio for the Seat Reservation Problem

(Extended Abstract)

Eric Bach[1*], Joan Boyar[2**], Tao Jiang[3,4***], Kim S. Larsen[2**], and Guo-Hui Lin[4,5†]

[1] Computer Sciences Department, University of Wisconsin – Madison,
1210 West Dayton Street, Madison, WI 53706-1685.
bach@cs.wisc.edu
[2] Department of Mathematics and Computer Science,
University of Southern Denmark, Odense, Denmark.
{joan,kslarsen}@imada.sdu.dk
[3] Department of Computer Science, University of California,
Riverside, CA 92521.
jiang@cs.ucr.edu
[4] Department of Computing and Software, McMaster University,
Hamilton, Ontario L8S 4L7, Canada.
[5] Department of Computer Science, University of Waterloo,
Waterloo, Ontario N2L 3G1, Canada.
ghlin@math.uwaterloo.ca

Abstract. In a recent paper [J. Boyar and K.S. Larsen, The seat reservation problem, *Algorithmica*, 25(1999), 403–417], the *seat reservation problem* was investigated. It was shown that for the *unit price problem*, where all tickets have the same price, all "fair" algorithms are at least $1/2$-accommodating, while no fair algorithm is more than $(4/5+O(1/k))$-accommodating, where k is the number of stations the train travels. In this paper, we design a more dextrous adversary argument, such that we improve the upper bound on the accommodating ratio to $(7/9+O(1/k))$, even for fair randomized algorithms against oblivious adversaries. For deterministic algorithms, the upper bound is lowered to approximately .7699. It is shown that better upper bounds exist for the special cases with $n = 2$, 3, and 4 seats. A concrete on-line deterministic algorithm FIRST-FIT and an on-line randomized algorithm RANDOM are also examined for the special case $n = 2$, where they are shown to be asymptotically optimal.

* Supported in part by NSF Grant CCR-9510244.
** Part of this work was carried out while the author was visiting the Department of Computer Sciences, University of Wisconsin – Madison. Supported in part by SNF (Denmark), in part by NSF (U.S.) grant CCR-9510244, and in part by the ESPRIT Long Term Research Programme of the EU under project number 20244 (ALCOM-IT).
*** Supported in part by NSERC Research Grant OGP0046613, a CITO grant, and a UCR startup grant.
† Supported in part by NSERC Research Grant OGP0046613 and a CITO grant.

D.-Z. Du et al. (Eds.): COCOON 2000, LNCS 1858, pp. 221–231, 2000.

Keywords: The seat reservation problem, on-line algorithms, accommodating ratio, adversary argument.

1   Introduction

In many train transportation systems, passengers are required to buy seat reservations with their train tickets. The ticketing system must assign a passenger a single seat when that passenger purchases a ticket, without knowing what future requests there will be for seats. Therefore, the seat reservation problem is an on-line problem, and a competitive analysis is appropriate.

Assume that a train with n seats travels from a start station to an end station, stopping at $k \geq 2$ stations, including the first and the last. The start station is station 1 and the end station is station k. The seats are numbered from 1 to n. Reservations can be made for any trip from a station s to a station t as long as $1 \leq s < t \leq k$. Each passenger is given a single seat number when the ticket is purchased, which can be any time before departure. The algorithms (ticket agents) may not refuse a passenger if it is possible to accommodate him when he attempts to make his reservation. That is, if there is any seat which is empty for the entire duration of that passenger's trip, the passenger must be assigned a seat. An on-line algorithm of this kind is *fair*.

The algorithms attempt to maximize income, i.e., the sum of the prices of the tickets sold. Naturally, the performance of an on-line algorithm will depend on the pricing policies for the train tickets. In [6] two pricing policies are considered: one in which all tickets have the same price, the *unit price problem*; and one in which the price of a ticket is proportional to the distance traveled, the *proportional price problem*. This paper focuses on fair algorithms for the unit price problem.

The seat reservation problem is closely related to the problem of optical routing with a number of wavelengths [1,5,9,14], call control [2], interval graph coloring [12] and interval scheduling [13]. The off-line version of the seat reservation problem can be used to solve the following problems [8]: minimizing spill in local register allocation, job scheduling with start and end times, and routing of two point nets in VLSI design. Another application of the on-line version of the problem could be assigning vacation bungalows (mentioned in [15]).

The performance of an on-line algorithm $\mathcal{A}$ is usually analyzed by comparing with an optimal off-line algorithm. For a sequence of requests, I, define the *competitive ratio of algorithm $\mathcal{A}$ applied to I* to be the income of algorithm $\mathcal{A}$ over the optimal income (achieved by the optimal off-line algorithm). The *competitive ratio of algorithm $\mathcal{A}$* is the infimum over all possible sequences of requests. The definition of the *accommodating ratio of algorithm $\mathcal{A}$* [6,7] is similar to that of the competitive ratio, except that the only sequences of requests allowed are sequences for which the optimal off-line algorithm could accommodate all requests therein. This restriction is used to reflect the assumption that the decision as to how many cars the train should have is based on expected ticket demand. Note that since the input sequences are restricted and this is a maximization problem, the accommodating ratio is always at least as large as the competitive ratio. Formally, the accommodating ratio is defined as follows:

Definition 1. *Let* $\mathrm{earn}_{\mathcal{A}}(I)$ *denote how much a fair on-line algorithm* $\mathcal{A}$ *earns with the request sequence* I*, and let* $\mathrm{value}(I)$ *denote how much could be earned if all requests in the sequence* I *were accommodated. A fair on-line algorithm* $\mathcal{A}$ *is* c*-accommodating if, for any sequence of requests which could all have been accommodated by the optimal fair off-line algorithm,* $\mathrm{earn}_{\mathcal{A}}(I) \geq c \cdot \mathrm{value}(I) - b$*, where* b *is a constant which does not depend on the input sequence* I*. The accommodating ratio for* $\mathcal{A}$ *is the supremum over all such* c*.*

Note that in general the constant b is allowed to depend on k. This is because k is a parameter to the problem, and we quantify first over k.

1.1   Previous Results

First, we note that in the case where there are enough seats to accommodate all requests, the optimal off-line algorithm runs in polynomial time [10] since it is a matter of coloring an interval graph with the minimum number of colors. As interval graphs are *perfect* [11], the size of the largest clique is exactly the number of colors needed. Let (k mod 3) denote the residue of k divided by 3, then we have the following known results:

Lemma 1. [6] *Any fair (deterministic or randomized) on-line algorithm for the unit price problem is at least 1/2-accommodating.*

Lemma 2. [6] *No fair (deterministic or randomized) on-line algorithm for the unit price problem (k ≥ 6) is more than* $f(k)$*-accommodating, where*

$$f(k) = (8k - 8(k \bmod 3) - 9)/(10k - 10(k \bmod 3) - 15).$$

Lemma 2 shows that no fair on-line algorithm has an accommodating ratio much better than $4/5$. It is also proven in [6] that the algorithms FIRST-FIT and BEST-FIT are at most $k/(2k-6)$-accommodating, which is asymptotically $1/2$. Let $r_{\mathcal{A}}$ denote the asymptotic accommodating ratio of a fair on-line algorithm $\mathcal{A}$ as k approaches infinity, then the above two lemmas tell that $1/2 \leq r_{\mathcal{A}} \leq 4/5$. The lower bound $1/2$ is also the upper bound for FIRST-FIT and BEST-FIT. This leaves an open problem whether or not a better algorithm exists [6].

1.2   Our Contributions

In the next section, we lower the asymptotic upper bound on the accommodating ratio from $4/5$ to $7/9$, and later show that this upper bound holds, even for fair randomized algorithms against oblivious adversaries. For deterministic algorithms, the upper bound is further lowered to approximately .7699. Furthermore, it is shown that better upper bounds exist for the special cases with $n = 2$, 3, and 4 seats. As a positive result, a concrete on-line deterministic algorithm

FIRST-FIT is examined with regards to the unit price problem for the special case $n = 2$. We show that FIRST-FIT is c_k-accommodating, where c_k approaches $3/5$ as k approaches infinity, and that it is asymptotically optimal. We also examine an on-line randomized algorithm RANDOM for the special case $n = 2$. We show that RANDOM is $3/4$-accommodating and that is also asymptotically optimal, in the randomized sense.

Due to the space constraint, we only provide the proofs for some upper bound results in this extended abstract. Detailed proofs of all other results can be found in the full manuscript [3].

2   A General Upper Bound

The upper bound on the accommodating ratio is lowered from $4/5$ to $7/9$, and later to approximately .7699.

Theorem 1. *No deterministic fair on-line algorithm for the unit price problem* $(k \geq 9)$ *is more than* $f(k)$-*accommodating, where*

$$
f(k) = \begin{cases} (7k - 7(k \bmod 6) - 15)/(9k - 9(k \bmod 6) - 27), \\ \qquad\qquad \text{when } (k \bmod 6) = 0, 1, 2; \\ (14k - 14(k \bmod 6) + 27)/(18k - 18(k \bmod 6) + 27), \\ \qquad\qquad \text{when } (k \bmod 6) = 3, 4, 5. \end{cases}
$$

Proof. The proof of this theorem is an adversary argument, which is a more dextrous design based on the ideas in the proof of Lemma 2 in [6]. Assume that n is divisible by 2. The adversary begins with $n/2$ requests for the intervals $[3s + 1, 3s + 3]$ for $s = 0, 1, \cdots, \lfloor(k-3)/3\rfloor$. Any fair on-line algorithm is able to satisfy this set of $\lfloor k/3 \rfloor \cdot n/2$ requests. Suppose that after these requests are satisfied, there are q_i seats which contain both interval $[3i+1, 3i+3]$ and interval $[3i + 4, 3i + 6]$, $i = 0, 1, \cdots, \lfloor(k-6)/3\rfloor$. Then there are exactly q_i seats which are empty from station $3i + 2$ to station $3i + 5$.

In the following, rather than considering each q_i at a time (as in [6]), we consider q_{2i}, q_{2i+1} together for $i = 0, 1, \cdots, \lfloor(k-9)/6\rfloor$. Let $p_i = q_{2i} + q_{2i+1}(\leq n)$. We distinguish between two cases:

- Case 1: $p_i \leq 5n/9$; and
- Case 2: $p_i > 5n/9$.

In the first case $p_i \leq 5n/9$, the adversary proceeds with $n/2$ requests for the interval $[6i+2, 6i+5]$ and $n/2$ requests for the interval $[6i+5, 6i+8]$. For these n additional requests, the on-line algorithm can accommodate exactly p_i of them. Figure 1(a) shows this configuration. Thus, for those $2n$ requests whose starting station $s \in [6i + 1, 6i + 6]$, the on-line algorithm accommodates $n + p_i$ of them.

In the second case $p_i > 5n/9$, the adversary proceeds with $n/2$ requests for the interval $[6i+3, 6i+7]$, followed by $n/2$ requests for interval $[6i+2, 6i+4]$ and $n/2$ requests for the interval $[6i + 6, 6i + 8]$. For these $3n/2$ additional requests, the on-line algorithm can accommodate exactly $3n/2 - p_i$ of them. Figure 1(b)

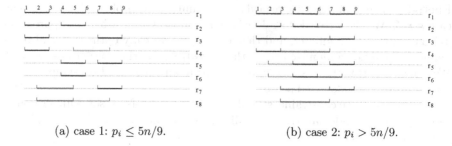

(a) case 1: $p_i \leq 5n/9$.   (b) case 2: $p_i > 5n/9$.

Fig. 1. Example configurations for the two cases.

shows this configuration. Thus, for the $5n/2$ requests whose starting station $s \in [6i+1, 6i+6]$, the on-line algorithm accommodates $5n/2 - p_i$ of them.

In this way, the requests are partitioned into $\lfloor (k-3)/6 \rfloor + 1$ groups; each of the first $\lfloor (k-3)/6 \rfloor$ groups consists of either $2n$ or $5n/2$ requests and the last group consists of either n (if $(k \bmod 6) \in \{0,1,2\}$) or $n/2$ (if $(k \bmod 6) \in \{3,4,5\}$) requests. For each of the first $\lfloor (k-3)/6 \rfloor$ groups, the on-line algorithm can accommodate up to a fraction $7/9$ of the requests therein. This leads to the theorem. More precisely, let S denote the set of indices for which the first case happens, and let $\bar{S}$ denote the set of indices for which the second case happens. When $(k \bmod 6) \in \{0,1,2\}$, the accommodating ratio of this fair on-line algorithm applied to this sequence of requests is

$$\frac{n + \sum_{i \in S}(n + p_i) + \sum_{i \in \bar{S}}(5n/2 - p_i)}{n + \sum_{i \in S} 2n + \sum_{i \in \bar{S}} 5n/2} \leq \frac{n + \sum_{i \in S} 14n/9 + \sum_{i \in \bar{S}} 35n/18}{n + \sum_{i \in S} 2n + \sum_{i \in \bar{S}} 5n/2}$$

$$= \frac{1 + \sum_{i \in S} 14/9 + \sum_{i \in \bar{S}} 35/18}{1 + \sum_{i \in S} 2 + \sum_{i \in \bar{S}} 5/2}$$

$$\leq \frac{1 + 14((k-6-(k \bmod 6))/6)/9}{1 + 2((k-6-(k \bmod 6))/6)}$$

$$= \frac{7k - 7(k \bmod 6) - 15}{9k - 9(k \bmod 6) - 27},$$

where the last inequality holds because in general $a/b = c/d < 1$ and $a < c$ imply that $(e + ax + cy)/(e + bx + dy) \leq (e + a(x+y))/(e + b(x+y))$. When $(k \bmod 6) \in \{3,4,5\}$, the ratio is

$$\frac{n/2 + \sum_{i \in S}(n + p_i) + \sum_{i \in \bar{S}}(5n/2 - p_i)}{n/2 + \sum_{i \in S} 2n + \sum_{i \in \bar{S}} 5n/2} \leq \frac{n/2 + \sum_{i \in S} 14n/9 + \sum_{i \in \bar{S}} 35n/18}{n/2 + \sum_{i \in S} 2n + \sum_{i \in \bar{S}} 5n/2}$$

$$= \frac{1/2 + \sum_{i \in S} 14/9 + \sum_{i \in \bar{S}} 35/18}{1/2 + \sum_{i \in S} 2 + \sum_{i \in \bar{S}} 5/2}$$

$$\leq \frac{1/2 + 14((k-(k \bmod 6))/6)/9}{1/2 + 2((k-(k \bmod 6))/6)}$$

$$= \frac{14k - 14(k \bmod 6) + 27}{18k - 18(k \bmod 6) + 27}.$$

This completes the proof.   □

When k, the number of stations, approaches infinity, the $f(k)$ in the above theorem converges to $7/9 \approx .7778$. Therefore, for any fair on-line algorithm $\mathcal{A}$, we have $1/2 \leq r_{\mathcal{A}} \leq 7/9$. The following theorem shows that $r_{\mathcal{A}}$ is marginally smaller than $7/9$.

Theorem 2. *Any deterministic fair on-line algorithm $\mathcal{A}$ for the unit price problem ($k \geq 9$) has its asymptotic accommodating ratio $\frac{1}{2} \leq r_{\mathcal{A}} \leq \frac{7-\sqrt{22}}{3} < .7699$.*

Proof. This theorem is proven using the following modification to the proof of Theorem 1. After algorithm $\mathcal{A}$ accommodates the set of $\lfloor k/3 \rfloor \cdot n/2$ requests, we distinguish between two cases: $p_i \leq 5(\sqrt{22} - 4)n/6$ and $p_i > 5(\sqrt{22} - 4)n/6$.

The second case can be analyzed similarly to Case 2 in the proof of Theorem 1, yielding that algorithm $\mathcal{A}$ accommodates at most a fraction $(7 - \sqrt{22})/3$ of the subset of $5n/2$ requests whose starting station $s \in [6i + 1, 6i + 6]$.

For the first case, we define p_i' to be the number of seats which contain none of the three intervals $[6i + 1, 6i + 3]$, $[6i + 4, 6i + 6]$ and $[6i + 7, 6i + 9]$. If $p_i' \leq (5 - \sqrt{22})n/2$, then the adversary proceeds with $n/2$ requests for the interval $[6i + 2, 6i + 8]$. Then, algorithm $\mathcal{A}$ can accommodate only $n + p_i'$ requests among the subset of $3n/2$ requests whose starting station $s \in [6i + 1, 6i + 6]$, which is at most a fraction $(7 - \sqrt{22})/3$. If $p_i' > (5 - \sqrt{22})n/2$, then the adversary proceeds first with p_i' requests for the interval $[6i + 2, 6i + 8]$ and then with $n/2 - p_i'$ requests for the interval $[6i + 2, 6i + 5]$ and $n/2 - p_i'$ requests for the interval $[6i + 5, 6i + 8]$. Obviously, algorithm $\mathcal{A}$ can accommodate only $p_i - p_i'$ of them, or equivalently, it can only accommodate $n + p_i - p_i'$ requests among the subset of $2n - p_i'$ requests whose starting station $s \in [6i + 1, 6i + 6]$, which is at most a portion of $(7 - \sqrt{22})/3$ again.

An argument similar to that in the proof of Theorem 1 says that the asymptotic accommodating ratio of algorithm $\mathcal{A}$ is at most $(7 - \sqrt{22})/3 \approx .7699$. $\square$

It is unknown if extending this "grouping" technique to more than two q_i's would lead to a better upper bound.

3   Upper Bounds for Small Values of n

In this section, we show better upper bounds for small values of n. This also demonstrates the power of the adversary argument and the "grouping" technique. However, we note that in train systems, it is unlikely that a train has a small number of seats. So the bounds obtained here are probably irrelevant for this application, but they could be relevant for others such as assigning vacation bungalows.

Trivially, for $n = 1$, any on-line algorithm is 1-accommodating. As a warm up, let us show the following theorem for $n = 2$.

Theorem 3. *If $n = 2$, then no deterministic fair on-line algorithm for the unit price problem ($k \geq 9$) is more than $f(k)$-accommodating, where*

$$f(k) = \begin{cases} (3k - 3(k \bmod 6) - 6)/(5k - 5(k \bmod 6) - 18), \\ \qquad\qquad\qquad \text{when } (k \bmod 6) \in \{0, 1, 2\}; \\ (3k - 3(k \bmod 6) + 6)/(5k - 5(k \bmod 6) + 6), \\ \qquad\qquad\qquad \text{when } (k \bmod 6) \in \{3, 4, 5\}. \end{cases}$$

Proof. The adversary begins with one request for the interval $[3s + 1, 3s + 3]$ for each $s = 0, 1, \cdots, \lfloor (k - 3)/3 \rfloor$. After these requests are satisfied by the on-line algorithm, consider for each $i = 0, 1, \cdots, \lfloor (k - 9)/6 \rfloor$ how the three requests $[6i + 1, 6i + 3]$, $[6i + 4, 6i + 6]$, and $[6i + 7, 6i + 9]$ are satisfied. Suppose that the intervals $[6i + 1, 6i + 3]$, $[6i + 4, 6i + 6]$, and $[6i + 7, 6i + 9]$ are placed on the same seat. Then the adversary proceeds with a request for the interval $[6i + 3, 6i + 7]$ and then requests for each of the intervals $[6i + 2, 6i + 4]$ and $[6i + 6, 6i + 8]$. The on-line algorithm will accommodate the first request, but fail to accommodate the last two. In the second case, suppose only two adjacent intervals (among $[6i + 1, 6i + 3]$, $[6i + 4, 6i + 6]$, and $[6i + 7, 6i + 9]$) are placed on the same seat, say $[6i + 1, 6i + 3]$ and $[6i + 4, 6i + 6]$, then the adversary proceeds with three requests for the intervals $[6i + 2, 6i + 4]$, $[6i + 3, 6i + 5]$, and $[6i + 5, 6i + 8]$. The on-line algorithm will accommodate the first request but fail to accommodate the last two. In the last case, only the intervals $[6i + 1, 6i + 3]$ and $[6i + 7, 6i + 9]$ are placed on the same seat. Then the adversary proceeds with two requests for the intervals $[6i + 2, 6i + 5]$ and $[6i + 5, 6i + 8]$. The on-line algorithm will fail to accommodate both of them.

It then follows easily that the accommodating ratio of the on-line algorithm applied to this sequence of requests is at most $f(k)$ for $k \geq 9$. □

A specific on-line algorithm called FIRST-FIT always processes a new request by placing it on the first seat which is unoccupied for the length of the journey. It has been shown [6] that FIRST-FIT is at most $k/(2k - 6)$-accommodating for general n divisible by 3. However, in the following, we will show that for $n = 2$, FIRST-FIT is c-accommodating, where c is at least $3/5$. Combining this with the previous theorem, FIRST-FIT is $(3/5 + \epsilon)$-accommodating, where $\epsilon > 0$ approaches zero as k approaches infinity. This means that for $n = 2$, FIRST-FIT is an asymptotically optimal on-line algorithm.

Theorem 4. FIRST-FIT *for the unit price problem is c-accommodating, where $c \geq 3/5$.*

Proof. See the full manuscript [3]. □

When $n = 3, 4$, we have the following theorems.

Theorem 5. *If $n = 3$, no deterministic fair on-line algorithm for the unit price problem ($k \geq 11$) is more than $f(k)$-accommodating, where*

$$f(k) = (3k - 3((k - 3) \bmod 8) - 1)/(5k - 5((k - 3) \bmod 8) - 14)$$

Proof. See the full manuscript [3]. □

Theorem 6. *If $n = 4$, no deterministic fair on-line algorithm for the unit price problem $(k \geq 9)$ is more than $f(k)$-accommodating, where*

$$
f(k) = \begin{cases}
(5k - 5(k \bmod 6) - 6)/(7k - 7(k \bmod 6) - 18), \\
\qquad\qquad \text{when } (k \bmod 6) \in \{0, 1, 2\}; \\
(5k - 5(k \bmod 6) + 12)/(7k - 7(k \bmod 6) + 12), \\
\qquad\qquad \text{when } (k \bmod 6) \in \{3, 4, 5\}.
\end{cases}
$$

Proof. See the full manuscript [3].   □

Generally, let $f_n(k)$ denote the upper bound on the accommodating ratio of the deterministic fair on-line algorithms for instances in which there are n seats and k stations. Let f_n denote the asymptotic value of $f_n(k)$. So far, we have $f_2 = 3/5$, $f_3 \leq 3/5$, $f_4 \leq 5/7$, and $f_n < .7699$ for $n \geq 5$. Are $f_n(k)$ and f_n non-increasing in n? So far we don't know the answer.

It is known that the accommodating ratio for FIRST-FIT cannot be better than $k/(2k - 6)$ when $n = 3$ [6]. So it is interesting to examine for $n = 3$, if there is some algorithm whose asymptotic accommodating ratio is better than $1/2$ or not. The problem of finding a better fair on-line algorithm for the unit price problem, for the general case $n \geq 3$, is still open.

4   Randomized Algorithms

In this section, we examine the accommodating ratios for randomized fair on-line algorithms for the unit price problem, by comparing them with an oblivious adversary. Some results concerning randomized fair on-line algorithms for the proportional price problem can be found in [4].

Theorem 7. *No randomized fair on-line algorithm for the unit price problem $(k \geq 9)$ is more than $f(k)$-accommodating, where*

$$
f(k) = \begin{cases}
(7k - 7(k \bmod 6) - 15)/(9k - 9(k \bmod 6) - 27), \\
\qquad\qquad \text{when } (k \bmod 6) = 0, 1, 2; \\
(14k - 14(k \bmod 6) + 27)/(18k - 18(k \bmod 6) + 27), \\
\qquad\qquad \text{when } (k \bmod 6) = 3, 4, 5.
\end{cases}
$$

Proof. The oblivious adversary behaves similarly to the adversary in the proof of Theorem 1. The sequence of requests employed by the oblivious adversary depends on the expected values of $p_i = q_{2i} + q_{2i+1}$, which are defined in the proof of Theorem 1. The oblivious adversary starts with the same sequence as the adversary in the proof of Theorem 1. Then, for each $i = 0, 1, \cdots, \lfloor (k - 9)/6 \rfloor$, it decides on Case 1 or Case 2, depending on the expected value $E[p_i]$ compared with $5n/9$. By generating corresponding requests, the linearity of expectations implies that the expected number of requests accommodated by the randomized algorithm is at most a fraction $f(k)$ of the total number of requests.   □

Although it is straight forward to show that Theorem 1 holds for randomized algorithms, as in the above, one cannot use the same argument and show the same for the other theorems. In fact, one can show that Theorem 3 is false when considering randomized algorithms against oblivious adversaries. The most obvious randomized algorithm to consider for this problem is the one we call RANDOM. When RANDOM receives a new request and there exists at least one seat that interval could be placed on, RANDOM chooses randomly among the seats which are possible, giving all possible seats equal probability.

Theorem 8. *For* $n = 2$, RANDOM *for the unit price problem is at least* $\frac{3}{4}$-*accommodating.*

Proof. See the full manuscript [3]. □

This value of $\frac{3}{4}$ is, in fact, a very tight lower bound on RANDOM's accommodating ratio when there are $n = 2$ seats.

Theorem 9. *For* $n = 2$, RANDOM *for the unit price problem* $(k \geq 3)$ *is at most* $f(k)$-*accommodating, where*

$$f(k) = \frac{3}{4} + \frac{1}{4(k - ((k-1) \bmod 2))}.$$

Proof. We first give the request $[1, 2]$ and then the requests $[2i, 2i + 2]$ for $i \in \{1, \ldots, \lfloor (k-2)/2 \rfloor\}$. If k is odd, we then give the request $[k-1, k]$.

RANDOM will place each of these requests, and, since there is no overlap, each of them is placed on the first seat with probability $\frac{1}{2}$.

Now we continue the sequence with $[2i+1, 2i+3]$ for $i \in \{0, \ldots, \lfloor (k-3)/2 \rfloor\}$.

Each interval in this last part of the sequence overlaps exactly two intervals from earlier and can therefore be accommodated if and only if these two intervals are placed on the same seat. This happens with probability $\frac{1}{2}$.

Thus, all requests from the first part of the sequence, and expected about half of the requests for the last part, are accepted. More precisely we obtain:

$$\frac{\frac{k+1-((k-1) \bmod 2)}{2} + \frac{1}{2} \cdot \frac{k-1-((k-1) \bmod 2)}{2}}{k - ((k-1) \bmod 2)} = f(k).$$

□

The accommodating ratio of $\frac{3}{4}$ for RANDOM with $n = 2$ seats does not extend to more seats. In general, one can show that RANDOM's accommodating ratio is bounded above by approximately $17/24 = 0.70833$. The request sequence used to prove this is very similar to the sequence given to FIRST-FIT and BEST-FIT in the proof of Theorem 3 in [6], but it is given in a different order, to make the accommodating ratio lower.

Theorem 10. *The accommodating ratio for* RANDOM *is at most* $\frac{17k+14}{24k}$, *for the unit price problem, when* $k \equiv 2 \pmod 4$.

Proof. See the full manuscript [3]. □

For other k, not congruent to 2 modulo 4, similar results hold. Using the same sequence of requests (and thus not using the last stations) gives upper bounds of the form $\frac{17k-c_1}{24k-c_2}$ for constants c_1 and c_2 which depend only on the value of k (mod 4).

Acknowledgments

The second author would like to thank Faith Fich for interesting discussions regarding the seat reservation problem with $n = 2$ seats. Guo-Hui Lin would like to thank Professor Guoliang Xue for many helpful suggestions.

References

1. B. Awerbuch, Y. Bartal, A. Fiat, S. Leonardi and A. Rosén, On-line competitive algorithms for call admission in optical networks, in *Proceedings of the 4th Annual European Symposium on Algorithms (ESA'96)*, LNCS 1136, 1996, pp. 431–444.
2. B. Awerbuch, Y. Bartal, A. Fiat and A. Rosén, Competitive non-preemptive call control, in *Proceedings of the 5th Annual ACM-SIAM Symposium on Discrete Algorithms (SODA'94)*, 1994, pp. 312–320.
3. E. Bach, J. Boyar, T. Jiang, K.S. Larsen and G.-H. Lin, Better bounds on the accommodating ratio for the seat reservation problem, PP-2000-8, Department of Mathematics and Computer Science, University of Southern Denmark (2000).
4. E. Bach, J. Boyar and K.S. Larsen, The accommodating ratio for the seat reservation problem, PP-1997-25, Department of Mathematics and Computer Science, University of Southern Denmark, May, 1997.
5. A. Bar-Noy, R. Canetti, S. Kutten, Y. Mansour and B. Schieber, Bandwidth allocation with preemption, in *Proceedings of the 27th Annual ACM Symposium on Theory of Computing (STOC'95)*, 1995, pp. 616–625.
6. J. Boyar and K.S. Larsen, The seat reservation problem, *Algorithmica*, 25(4) (1999), 403–417.
7. J. Boyar, K.S. Larsen and M.N. Nielsen, The accommodating function — a generalization of the competitive ratio, in *Proceedings of the Sixth International Workshop on Algorithms and Data Structures (WADS'99)*, LNCS 1663, 1999, pp. 74–79.
8. M.C. Carlisle and E.L. Lloyd, On the k-coloring of intervals, in *Advances in Computing and Information – ICCI'91*, LNCS 497, 1991, pp. 90–101.
9. J.A. Garay, I.S. Gopal, S. Kutten, Y. Mansour and M. Yung, Efficient on-line call control algorithms, *Journal of Algorithms*, 23 (1997) 180–194.
10. F. Gavril, Algorithms for minimum coloring, maximum clique, minimum covering by cliques, and maximum independent set of a chordal graph, *SIAM Journal on Computing*, 1(2) (1972), 180–187.
11. T.R. Jensen and B. Toft, *Graph coloring problems*, Wiley, New York, 1995.
12. H.A. Kierstead and W.T. Trotter, Jr., An extremal problem in recursive combinatorics *Congressus Numerantium*, 33 (1981) 143–153.
13. R.J. Lipton and A. Tomkins, On-line interval scheduling, in *Proceedings of the 5th Annual ACM-SIAM Symposium on Discrete Algorithms (SODA'94)*, 1994, pp. 302–311.

14. P. Raghavan and E. Upfal, Efficient routing in all-optical networks, in *Proceedings of the 26th Annual ACM Symposium on Theory of Computing (STOC'94)*, 1994, pp. 134–143.

15. L. Van Wassenhove, L. Kroon and M. Salomon, Exact and approximation algorithms for the operational fixed interval scheduling problem, Working paper 92/08/TM, INSEAD, Fontainebleau, France, 1992.

Ordinal On-Line Scheduling on Two Uniform Machines*

Zhiyi Tan and Yong He

Department of Mathematics, Zhejiang University
Hangzhou 310027, P.R. China
heyong@math.zju.edu.cn

Abstract. We investigate the ordinal on-line scheduling problem on two uniform machines. We present a comprehensive lower bound of any ordinal algorithm, which constitutes a piecewise function of machine speed ratio $s \geq 1$. We further propose an algorithm whose competitive ratio matches the lower bound for most of $s \in [1, \infty)$. The total length of the intervals of s where the competitive ratio does not match the lower bound is less than 0.7784 and the biggest gap never exceeds 0.0521.

1 Introduction

In this paper, we consider ordinal on-line scheduling problem on two uniform machines with objective to minimize the *makespan* (i.e., the last job completion time). We are given n independent jobs $J = \{p_1, \cdots, p_n\}$, which has to be assigned to two uniform machines $M = \{M_1, M_2\}$. We identify the jobs with their processing times. Machine M_1 has speed $s_1 = 1$ and machine M_2 has speed $s_2 = s \geq 1$. If p_i is assigned to machine M_j, then p_i/s_j time units are required to process this job. Both machines and jobs are available at time zero, and no preemption is allowed. We further assume that jobs arrive one by one and we know nothing about the value of the processing times but the order of the jobs by processing times. Hence without loss of generality, we suppose $p_1 \geq p_2 \geq \cdots \geq p_n$. We are asked to decide the assignment of all the jobs to some machine at time zero by utilizing only ordinal data rather than the actual magnitudes. For convenience, we denote the problem as $Q2|ordinal\ on-line|C_{max}$.

Competitive analysis is a type of worst-case analysis where the performance of an on-line algorithm is compared to that of the optimal off-line algorithm [11]. For an on-line algorithm A, let C^A denote the makespan of the solution produced by the algorithm A and C^{OPT} denote the minimal makespan in an off-line version. Then the competitive ratio of the algorithm A is defined as the smallest number c such that $C^A \leq cC^{OPT}$ for all instances. An algorithm with

---

* This research is supported by National Natural Science Foundation of China (grant number: 19701028) and 973 National Fundamental Research Project of China (Information Technology and High-Performance Software).

D.-Z. Du et al. (Eds.): COCOON 2000, LNCS 1858, pp. 232–241, 2000.

a competitive ratio c is called c-*competitive algorithm*. An on-line deterministic (randomized) algorithm A is called the best possible (or optimal) algorithm if there is no deterministic (randomized) on-line algorithm for the discussed problem with a competitive ratio better than that of A.

Problems with ordinal data exist in many fields of combinatorial optimization such as matroid, bin-packing, and scheduling [7, 8, 9, 1]. Algorithm which utilizes only ordinal data rather than actual magnitudes are called *ordinal algorithm*. For ordinal scheduling, Liu et al. [8] gave thorough study on $Pm||C_{max}$. Because ordinal algorithm is, in some extent, stricter than the non-clairvoyant one in that we must assign all jobs at time zero, classical on-line algorithms depending on machine loads, such as List Scheduling (LS), are of no use. For $m = 2, 3$, [8] has presented respective optimal ordinal algorithms. In the same paper, an algorithm with competitive ratio $101/70$ was given while the lower bound is $23/16$ for $m = 4$, and an algorithm with competitive ratio $1 + (m-1)/(m + \lceil m/2 \rceil) \leq 5/3$ was developed while the lower bound is $3/2$ for general $m > 4$.

On the other hand, our discussed problem also belongs to a kind of *semi* on-line scheduling, a variant of on-line where we do have some partial knowledge on job set which makes the problem easier to solve than standard on-line scheduling problems. In our problem we know the order of jobs by their processing times. Although there are many results on semi on-line scheduling problem on identical machines [2, 5, 6, 10], to the authors knowledge, little is known about uniform machine scheduling. Due to the above motivation, this paper considers the scheduling problem $Q2|ordinal\ on-line|C_{max}$. We will give the lower bound of any ordinal algorithm, which is, denoted by c_{low}, a piecewise function dependent on speed ratio $s \in [1, \infty)$. We further present one algorithm *Ordinal* whose competitive ratio matches the lower bound for the majority of the value of machine speed ratio s. The total length of the intervals of s where the competitive ratio does not match the lower bound is less than 0.7784, and the biggest gap between them never exceeds 0.0521.

Another related problem is the classical clairvoyant on-line scheduling problem $Q2||C_{max}$, which is deeply studied. Cho and Sahni [3] has shown that competitive ratio of LS algorithm is $(1 + \sqrt{5})/2$. Epstein et al. [4] have proved that the parametric competitive ratio of LS is $\min\left\{\frac{2s+1}{s+1}, 1 + \frac{1}{s}\right\}$, and LS is thus the best possible on-line algorithm for any s. They further proved that randomization does not help for $s \geq 2$, presented a simple memoryless randomized algorithm with a competitive ratio of $(4-s)(1+s)/4 \leq 1.5625$, and devised barely random algorithms with competitive ratio at most 1.53 for any $s < 2$.

The paper is organized as follows. Section 2 gives the lower bound dependent on s. Section 3 presents the algorithm *Ordinal* and proves its competitive ratio.

2   Parametric Lower Bound

This section presents the parametric lower bound of any deterministic ordinal algorithm applied to $Q2|ordinal\ on-line|C_{max}$. We generalize the results in the following two theorems.

Theorem 1. *For any* $s \in \left[1, \frac{5+\sqrt{265}}{20}\right) \cup \left[\frac{1+\sqrt{7}}{3}, \infty\right)$, *the competitive ratio of any deterministic ordinal algorithm is at least* $c_{low}(s)$, *where*

$$
c_{low}(s) = \begin{cases}
\frac{2s+2}{3} & s \in [1, \frac{5+\sqrt{265}}{20}) \approx [1, 1.064), \\
\frac{2s+1}{2s} & s \in [\frac{1+\sqrt{7}}{3}, \frac{1+\sqrt{5}}{2}) \approx [1.215, 1.618), \\
\frac{s+1}{2} & s \in [\frac{1+\sqrt{5}}{2}, \frac{3+\sqrt{57}}{6}) \approx [1.618, 1.758), \\
\frac{3s+2}{3s} & s \in [\frac{3+\sqrt{57}}{6}, 2) \approx [1.758, 2), \\
\frac{s+2}{3} & s \in [2, \frac{1+\sqrt{10}}{2}) \approx [2, 2.081), \\
\frac{4s+3}{4s} & s \in [\frac{1+\sqrt{10}}{2}, \frac{1+\sqrt{13}}{2}) \approx [2.081, 2.303), \\
\frac{s+3}{4} & s \in [\frac{1+\sqrt{13}}{2}, \frac{5+\sqrt{345}}{10}) \approx [2.303, 2.357), \\
\frac{5s+4}{5s} & s \in [\frac{5+\sqrt{345}}{10}, \frac{1+\sqrt{17}}{2}) \approx [2.357, 2.562), \\
\frac{s+4}{5} & s \in [\frac{1+\sqrt{17}}{2}, \frac{3+\sqrt{159}}{6}) \approx [2.562, 2.602), \\
\frac{6s+5}{6s} & s \in [\frac{3+\sqrt{159}}{6}, \frac{3+2\sqrt{6}}{3}) \approx [2.602, 2.633), \\
\frac{s}{2} & s \in [\frac{3+2\sqrt{6}}{3}, 1+\sqrt{3}) \approx [2.633, 2.732), \\
\frac{s+1}{s} & s \in [1+\sqrt{3}, \infty) \approx [2.732, \infty).
\end{cases}
$$

Theorem 2. *For any* $s \in [\frac{5+\sqrt{265}}{20}, \frac{1+\sqrt{7}}{3})$, *the competitive ratio of any deterministic ordinal algorithm is at least* $c_{low}(s)$, *where*

$$
c_{low}(s) = \begin{cases}
\frac{5s+2}{5s} & s \in [\frac{5+\sqrt{265}}{20}, s_2(2)), \\
\frac{(3k-4)s+(2k-2)}{4k-5} & s \in [s_2(k-1), s_1(k)), \quad k = 3, 4, \ldots, \\
\frac{(2k+1)s+k}{(2k+1)s} & s \in [s_1(k), s_2(k)), \quad k = 3, 4, \ldots,
\end{cases}
$$

and for each $k \geq 3$,

$$
s_1(k) = \frac{(2k+1)(2k-3) + \sqrt{(2k+1)^2(2k-3)^2 + 4k(2k+1)(3k-4)(4k-5)}}{2(2k+1)(3k-4)}
$$

is the positive root of the equation

$$
\frac{(2k+1)s+k}{(2k+1)s} = \frac{(3k-4)s+2k-2}{4k-5}
$$

and

$$
s_2(k) = \frac{(2k-1)(2k+1) + \sqrt{(2k+1)^2(2k-1)^2 + 4k(2k+1)(3k-1)(4k-1)}}{2(2k+1)(3k-1)}
$$

is the positive root of the equation

$$
\frac{(2k+1)s+k}{(2k+1)s} = \frac{(3k-1)s+2k}{4k-1}.
$$

From the definition of $s_1(k)$ and $s_2(k)$, we can see that $s_1(k-1) < s_2(k-1) < s_1(k) < s_2(k)$ and $\lim_{k \to \infty} s_2(k) = \frac{1+\sqrt{7}}{3}$, so the lower bound is well-defined for any $s \geq 1$ through Theorem 1 and 2.

The proofs will be completed by adversary method. We analyze the performance of the best possible algorithm applied to several instances. For easier reading and understanding, we show Theorem 1 by proving the following Lemmas 1-7 separately.

Lemma 1. *For any $s \in [1, \frac{5+\sqrt{265}}{20})$ and any ordinal on-line algorithm A, there exists some instance such that $C^A/C^{OPT} \geq c_{low}(s) = (2s+2)/3$.*

Proof. If an algorithm A assigns p_1 to M_1 and p_2, p_3, p_4 to M_2, consider the instance with job set $\{p_1 = p_2 = 1, p_3 = p_4 = 1/s\}$. It follows that $C^A \geq (1 + 2/s)/s$, $C^{OPT} = 2/s$, $C^A/C^{OPT} \geq (s+2)/(2s) > (2s+2)/3$. If A assigns p_1 and at least one of p_2, p_3, p_4 to M_1, consider the instance with job set $\{p_1 = 1, p_2 = p_3 = p_4 = 1/(3s)\}$. It implies that $C^A \geq 1 + 1/(3s)$, $C^{OPT} = 1/s$, $C^A/C^{OPT} \geq (3s+1)/3 > (2s+2)/3$. So we know that the competitive ratio will be greater than $(2s+2)/3$ as long as A assigns p_1 to M_1.

Now we analyze the case that A assigns p_1 to M_2. Firstly, we deduce that both p_2 and p_3 cannot be assigned to M_2. Because otherwise consider the instance $\{p_1 = 1, p_2 = p_3 = 1/(2s)\}$, it implies that $C^A \geq (1 + 1/(2s))/s$, $C^{OPT} = 1/s$, $C^A/C^{OPT} \geq (2s+1)/(2s) > (2s+2)/3$. Denote the instance by (I1) for future use. Secondly, p_4 must be assigned to M_2. Otherwise we have the instance $\{p_1 = p_2 = p_3 = 1, p_4 = (2-s)/s\}$, which follows that $C^A \geq 1 + 2/s$, $C^{OPT} = 2/s$, $C^A/C^{OPT} = (s+2)/2 > (2s+2)/3$. Denote the instance by (I2). Thirdly, if A assigns at least one of p_5 and p_6 to M_2, consider $\{p_1 = 1, p_2 = \cdots = p_6 = 1/(5s)\}$. It is clear that $C^A \geq (1 + 2/(5s))/s$, $C^{OPT} = 1/s$, $C^A/C^{OPT} \geq (5s+2)/(5s) > (2s+2)/3$. Lastly, we are left to consider the case where A assigns p_1, p_4 to M_2 and p_2, p_3, p_5, p_6 to M_1. At this moment, we choose $\{p_1 = p_2 = p_3 = 1, p_4 = p_5 = p_6 = 1/s\}$. Therefore, $C^A \geq 2 + 2/s$, $C^{OPT} = 3/s$, $C^A/C^{OPT} \geq (2s+2)/3$.

In summary, we have proved for any s in the given interval, any algorithm A is such that $C^A/C^{OPT} \geq (2s+2)/3$. $\square$

Lemma 2. *For any $s \in [\frac{1+\sqrt{7}}{3}, \frac{1+\sqrt{5}}{2})$ and any ordinal on-line algorithm A, there exists some instance such that $C^A/C^{OPT} \geq c_{low}(s) = (2s+1)/(2s)$.*

Proof. Similarly we can get the result by considering the following series of instances: $\{p_1 = 1, p_2 = \cdots = p_5 = 1/(4s)\}$, $\{p_1 = \cdots = p_4 = 1, p_5 = 2s-2\}$, $\{p_1 = 1, p_2 = \cdots = p_6 = 1/(5s)\}$, $\{p_1 = \cdots = p_5 = 1, p_6 = 2s-3\}$, (I1), (I2), $\{p_1 = p_2 = p_3 = 1, p_4 = p_5 = (2-s)/(2s)\}$, and $\{p_1 = 1, p_2 = \cdots = p_5 = 1/(4s)\}$. $\square$

Before going to other lemmas, consider two instances. If algorithm A assigns p_1 to M_1, let $p_1 = 1$ and no other job arrives, then we know that $C^A/C^{OPT} \geq s$. If algorithm A assigns p_1 and p_2 to M_2, let $p_1 = 1, p_2 = 1/s$ and no other job comes, then $C^A/C^{OPT} \geq (s+1)/s$. Because s and $(s+1)/s$ are no less than $c_{low}(s)$ defined in Theorem 1 as long as $s \geq (1+\sqrt{5})/2$, in the remainder of the proof of Theorem 1, we assume that M_1 processes p_2 and M_2 processes p_1 by any algorithm A.

Lemma 3. *For any* $s \in [\frac{1+\sqrt{5}}{2}, 2)$ *and any ordinal on-line algorithm* A, *there exists some instance such that*

$$\frac{C^A}{C^{OPT}} \geq c_{low}(s) = \begin{cases} \frac{s+1}{2} & s \in [\frac{1+\sqrt{5}}{2}, \frac{3+\sqrt{57}}{6}), \\ \frac{3s+2}{3s} & s \in [\frac{3+\sqrt{57}}{6}, 2). \end{cases}$$

Proof. If an algorithm A assigns p_3 to M_1, consider the instance $\{p_1 = p_2 = 1, p_3 = s - 1\}$. Then $C^A/C^{OPT} \geq s$. Therefore A should assign p_3 to M_2. If p_4 is assigned to M_1, consider the instance $\{p_1 = p_2 = 1, p_3 = p_4 = 1/s\}$. We know $C^A/C^{OPT} \geq (s+1)/2$. If p_4 is assigned to M_2, Consider $\{p_1 = 1, p_2 = p_3 = p_4 = 1/(3s)\}$. We have $C^A/C^{OPT} \geq (3s+2)/(3s)$. Thus we know that $C^A/C^{OPT} \geq \min\{(s+1)/2, (3s+2)/(3s)\}$. The result can be got by comparing the two functions. □

Lemma 4. *For any* $s \in [\frac{2+2\sqrt{6}}{3}, \infty)$ *and any ordinal on-line algorithm* A, *there exists some instance such that*

$$\frac{C^A}{C^{OPT}} \geq c_{low}(s) = \begin{cases} \frac{s}{2} & s \in [\frac{2+2\sqrt{6}}{3}, 1 + \sqrt{3}), \\ \frac{s+1}{s} & s \in [1 + \sqrt{3}, \infty). \end{cases}$$

Proof. If p_2 is assigned to M_1 by an algorithm A, the instance $\{p_1 = p_2 = 1\}$ deduces $C^A/C^{OPT} \geq s/2$. As illustrated in the content before Lemma 3, $C^A/C^{OPT} \geq (s+1)/s$, the lemma can be proved by comparing $s/2$ with $(s+1)/s$. □

By considering the following two instances: $\{p_1 = 1, p_2 = p_3 = 1/(s-1)\}$ and $\{p_1 = p_2 = 1, p_3 = p_4 = 1/s\}$, we conclude that M_1 should process p_2 and M_2 should process p_1, p_3, p_4 in any algorithm A in order to get the desired lower bound for $2 \leq s < (2 + 2\sqrt{6})/3$. Hence we assume it holds in the rest of the proof.

Lemma 5. *For any* $s \in [2, \frac{1+\sqrt{13}}{2})$ *and any ordinal on-line algorithm* A, *there exists some instance such that*

$$\frac{C^A}{C^{OPT}} \geq c_{low}(s) = \begin{cases} \frac{s+2}{3} & s \in [2, \frac{1+\sqrt{10}}{2}), \\ \frac{4s+3}{4s} & s \in [\frac{1+\sqrt{10}}{2}, \frac{1+\sqrt{13}}{2}). \end{cases}$$

Proof. Similar to the proof of Lemma 4, we can get the results by the instances $\{p_1 = p_2 = 1, p_3 = p_4 = p_5 = (s-1)/3\}$ and $\{p_1 = 1, p_2 = \cdots = p_5 = 1/(4s)\}$. □

Lemma 6. *For any* $s \in [\frac{1+\sqrt{13}}{2}, \frac{1+\sqrt{17}}{2})$ *and any ordinal on-line algorithm A, there exists some instance such that*

$$\frac{C^A}{C^{OPT}} \geq c_{low}(s) = \begin{cases} \frac{s+3}{4} & s \in [\frac{1+\sqrt{13}}{2}, \frac{5+\sqrt{345}}{10}), \\ \frac{5s+4}{5s} & s \in [\frac{5+\sqrt{345}}{10}, \frac{1+\sqrt{17}}{2}). \end{cases}$$

Proof. The instances $\{p_1 = p_2 = 1, p_3 = p_4 = p_5 = (s-1)/3\}$, $\{p_1 = p_2 = 1, p_3 = \cdots = p_6 = (s-1)/4\}$, and $\{p_1 = 1, p_2 = \cdots = p_6 = 1/(5s)\}$ can reach the goal. $\qquad\square$

Lemma 7. *For any* $s \in [\frac{1+\sqrt{17}}{2}, \frac{2+2\sqrt{6}}{3})$ *and any ordinal on-line algorithm A, there exists some instance such that*

$$\frac{C^A}{C^{OPT}} \geq c_{low}(s) = \begin{cases} \frac{s+4}{5} & s \in [\frac{1+\sqrt{17}}{2}, \frac{3+\sqrt{159}}{6}), \\ \frac{6s+5}{6s} & s \in [\frac{3+\sqrt{159}}{6}, \frac{2+2\sqrt{6}}{3}). \end{cases}$$

Proof. By considering the following instances, we can have the results: $\{p_1 = p_2 = 1, p_3 = p_4 = p_5 = (s-1)/3\}, \{p_1 = p_2 = 1, p_3 = \cdots = p_6 = (s-1)/4\}, \{p_1 = p_2 = 1, p_3 = \cdots = p_7 = (s-1)/5\}$ and $\{p_1 = 1, p_2 = \cdots = p_7 = 1/(6s)\}$. $\qquad\square$

Through Lemma 1 to 7, we have finished the proof of Theorem 1. To prove Theorem 2, we rewrite it as follows:

Theorem 3. *For any* $l(l \geq 3)$ *and* $s \in \left[\frac{5+\sqrt{265}}{20}, \frac{1+\sqrt{7}}{3}\right)$, *the competitive ratio of any deterministic ordinal algorithm is at least* $c_{low}(s, l)$, *where*

$$c_{low}(s, l) = \begin{cases} \frac{5s+2}{5s} & s \in [\frac{5+\sqrt{265}}{20}, s_2(2)), \\ \frac{(3k-4)s+(2k-2)}{4k-5} & s \in [s_2(k-1), s_1(k)), \quad k = 3, 4, \cdots, l \\ \frac{(2k+1)s+k}{(2k+1)s} & s \in [s_1(k), s_2(k)), \quad k = 3, 4, \cdots, l, \\ \frac{3s+2}{4} & s \in [s_2(l), \frac{1+\sqrt{7}}{3}). \end{cases}$$

Because $\{\frac{(3k-4)s+2k-2}{4k-5}\}$ is an increasing sequence of k and

$$lim_{k\to\infty}\frac{(3k-4)s + 2k - 2}{4k - 5} = \frac{3s + 2}{4},$$

one can easily verify the equivalence of Theorem 2 and Theorem 3 as $l \to \infty$.

Proof. We prove the results by induction. For $l = 3$, by using the similar methods in the proof of Lemma 1 and Lemma 2, it is not difficult to prove

$$\frac{C^A}{C^{OPT}} \geq c_{low}(s) = \begin{cases} \frac{5s+2}{5s} & s \in [\frac{5+\sqrt{265}}{20}, \frac{5+\sqrt{145}}{15}), \\ \frac{3s+2}{4} & s \in [\frac{5+\sqrt{145}}{15}, \frac{1+\sqrt{7}}{3}). \end{cases}$$

But it is a little bit different from the desired bound described in the theorem, to reach the goal, we need more careful analysis. By the same arguments as in the proof of Lemma 1 and 2, we only need consider the following assignment: A assigns p_2, p_3, p_5 to M_1, and p_1, p_4, p_6 to M_2. Because for other assignments, the ratio is no less than $c_{low}(s)$ already.

Instead of using the instances which we did in the proof of Lemma 2 to achieve the lower bound of $(3s+2)/4$, we now distinguish three cases according to the assignment of the next two jobs p_7, p_8.

Case 1 Algorithm A assigns p_7 to M_2. By considering the instance $\{p_1 = 1, p_2 = \cdots = p_7 = 1/(6s)\}$, we have $C^A \geq (1 + 1/(2s))/s$, $C^{OPT} = 1/s$, $C^A/C^{OPT} \geq (2s+1)/(2s)$.

Case 2 Algorithm A assigns p_7, p_8 to M_1. By considering the instance $\{p_1 = p_2 = p_3 = 1, p_4 = \cdots = p_8 = (2-s)/(4s-1)\}$, we have $C^A \geq (5s+4)/(4s-1)$, $C^{OPT} = 7/(4s-1)$, $C^A/C^{OPT} \geq (5s+4)/7$.

Case 3 Algorithm A assigns p_7 to M_1 and p_8 to M_2. The instance $\{p_1 = 1, p_2 = \cdots = p_8 = 1/(7s)\}$ deduces that $C^A \geq 1/s(1 + 3/(7s))$, $C^{OPT} = 1/s$ and $C^A/C^{OPT} \geq (7s+3)/(7s)$.

By straight arithmetic calculation, we have

$$\frac{C^A}{C^{OPT}} \geq c_{low}(s,3) = \begin{cases} \frac{5s+2}{5s} & s \in [\frac{5+\sqrt{265}}{20}, \frac{3+\sqrt{65}}{10}) = [\frac{5+\sqrt{265}}{20}, s_2(2)), \\ \frac{5s+4}{7} & s \in [\frac{3+\sqrt{65}}{10}, \frac{3+\sqrt{69}}{10}) = [s_2(2), s_1(3)), \\ \frac{7s+3}{7s} & s \in [\frac{3+\sqrt{69}}{10}, \frac{7+\sqrt{301}}{21}) = [s_1(3), s_2(3)), \\ \frac{3s+2}{4} & s \in [\frac{7+\sqrt{301}}{21}, \frac{1+\sqrt{7}}{3}) = [s_2(3), \frac{1+\sqrt{7}}{3}). \end{cases}$$

Hence the theorem holds for $l = 3$. By induction, we assume that the theorem is true for all $l < k$, and further we can assume M_1 processes $\{p_2, p_3, p_5, p_7, \cdots, p_{2k-1}\}$ and M_2 processes $\{p_1, p_4, p_6, \cdots, p_{2k}\}$ in any algorithm A with the desired bound. For $l = k$ we distinguish three cases according to the assignment of the next two jobs:

Case 1 Algorithm A assigns p_{2k+1} to M_2. By consider the instance $\{p_1 = 1, p_2 = \cdots = p_{2k+1} = 1/(2ks)\}$, we have that $C^A \geq 1/s(1 + 1/(2s))$, $C^{OPT} = 1/s$, and $C^A/C^{OPT} \geq (2s+1)/(2s)$.

Case 2 Algorithm A assigns p_{2k+1}, p_{2k+2} to M_1. By considering $\{p_1 = p_2 = p_3 = 1, p_4 = \cdots = p_{2k+2} = (2-s)/((2k-2)s - 1) \stackrel{\triangle}{=} x\}$, we have that $C^A \geq 2 + kx$, $C^{OPT} = 1 + (2k-2)x$, and $C^A/C^{OPT} \geq \frac{(3k-4)s+(2k-2)}{4k-5}$.

Case 3 Algorithm A assigns p_{2k+1} to M_1 and p_{2k+2} to M_2. By considering $\{p_1 = 1, p_2 = \cdots = p_{2k+2} = 1/((2k+1)s)\}$, we have that $C^A \geq (1 + k/((2k+1)s))/s$, $C^{OPT} = 1/s$, and $C^A/C^{OPT} \geq ((2k+1)s + k)/((2k+1)s)$.

We thus have shown that $C^A/C^{OPT} \geq \min\{\frac{(3k-4)s+(2k-2)}{4k-5}, \frac{(2k+1)s+k}{(2k+1)s}\}$, and the proof is completed. $\square$

3   The Algorithm *Ordinal*

In this section, we present an ordinal algorithm for $Q2|ordinal\ on-line|C_{max}$. The algorithm consists of an infinite series of procedures. For any s, it definitely chooses exactly one procedure to assign the jobs. First we give the definition of procedures.

Procedure(0): Assign all jobs to M_2.

Procedure(1): Assign jobs in the subset $\{p_{3i+2}, p_{3i+3}|i \geq 0\}$ to M_1; Assign jobs in the subset $J = \{p_{3i+1}|i \geq 0\}$ to M_2.

Procedure(l), $l \geq 2$: Assign jobs in the subset $\{p_{li+2}|i \geq 0\}$ to M_1; Assign jobs in the subset $\{p_1\} \cup \{p_{li+3}, p_{li+4}, \ldots, p_{li+l+1}|i \geq 0\}$ to M_2.

Algorithm *Ordinal*
1. If $s \geq 1 + \sqrt{3}$, assign jobs by Procedure(0).
2. If $s \in [s(l-1), s(l))$, assign jobs by Procedure(l), where

$$s(l) = \begin{cases} 1 & l = 0, \\ \frac{1+\sqrt{13}}{4} & l = 1, \\ \frac{l^2-1+\sqrt{(l^2-1)^2+2l^3(l+1)}}{l(l+1)} & l \geq 2. \end{cases}$$

In the above definition, $s(l)$ is the positive root of equation $\frac{(l+1)s+l}{(l+1)s} = \frac{ls+2}{2l}$ for any $l \geq 2$. Because it is an increasing sequence of l and $\lim_{l \to \infty} s(l) = 1 + \sqrt{3}$, the algorithm is well-defined for all $s \geq 1$.

Theorem 4. *The parametric competitive ratio of the algorithm Ordinal is*

$$c^{ordinal}(s) = \begin{cases} \frac{2s+2}{3} & s \in [s(0), s(1)), \\ \frac{ls+l-1}{ls} & s \in [s(l-1), s'(l)), \ l = 2,3,\ldots, \\ \frac{ls+2}{2l} & s \in [s'(l), s(l)), \ l = 2,3,\ldots, \\ \frac{s+1}{s} & s \geq 1 + \sqrt{3}, \end{cases}$$

where $s'(l) = \frac{l-1+\sqrt{(l-1)^2+2l(l-1)}}{l}$ $(l \geq 2)$ is the positive root of $\frac{ls+l-1}{ls} = \frac{ls+2}{2l}$.

Proof. We first show that $C^{ordinal}/C^{OPT} \leq c^{ordinal}(s)$. Let α_1 and α_2 be the completion times of M_1 and M_2 after processing all jobs by our algorithm respectively, and b be the total processing time of all jobs. Obviously, $C^{OPT} \geq b/(s+1)$ and $C^{OPT} \geq p_1/s$, $C^{ordinal} = \max\{\alpha_1, \alpha_2\}$. We prove the ratio by distinguishing the speed ratio s.

Case 1 $s > 1 + \sqrt{3}$. In this case, *Ordinal* chooses Procedure(0). It is clear that $\alpha_1 = 0$ and $\alpha_2 = \frac{b}{s} = \frac{s+1}{s}\frac{b}{s+1} \leq \frac{s+1}{s}C^{OPT}$. This is the desired ratio.

Case 2 $s \in [s(0), s(1))$. In this case, *Ordinal* chooses Procedure(1). We prove the case of $n = 3k$, the other cases of $n = 3k+1$ and $3k+2$ can be proved similarly. By the rule of the procedure, we have

$$\alpha_1 = \sum_{i=0}^{k-1}(p_{3i+2} + p_{3i+3}) \leq \frac{2}{3}\sum_{i=0}^{k-1}(p_{3i+1} + p_{3i+2} + p_{3i+3})$$

$$= \frac{2b}{3} = \frac{2(s+1)}{3}\frac{b}{s+1} \leq \frac{2(s+1)}{3}C^{OPT},$$

$$\alpha_2 = \frac{1}{s}\sum_{i=0}^{k-1}p_{3i+1} \leq \frac{p_1}{s} + \frac{1}{3s}\sum_{i=2}^{n}p_i = \frac{2p_1}{3s} + \frac{b}{3s} \leq \frac{3s+1}{3s}C^{OPT}.$$

Note that for $s \in [s(0), s(1))$, we have $\frac{3s+1}{3s} \leq \frac{2(s+1)}{3}$, we thus get the desired ratio for this case.

Case 3 $s \in [s(1), s(2))$. At this moment, we need analyze Procedure(2). We prove the case of $n = 2k$, the another case of $n = 2k+1$ can be proved similarly. In fact, we have

$$\alpha_1 = \sum_{i=0}^{k-1} p_{2i+2} \leq \frac{b}{2} \leq \frac{s+1}{2} C^{OPT},$$

$$\alpha_2 = \frac{1}{s}\left(p_1 + \sum_{i=0}^{k-2} p_{2i-3}\right) \leq \frac{p_1}{s} + \frac{1}{2s}\sum_{i=2}^{n} p_i = \frac{p_1}{2s} + \frac{b}{2s} \leq \frac{2s+1}{2s} C^{OPT}.$$

By comparing $(s+1)/2$ and $(2s+1)/(2s)$, we know that

$$\frac{C^{ordinal}}{C^{OPT}} \leq \begin{cases} \frac{2s+1}{2s} & s \in [s(1), s'(2)), \\ \frac{s+1}{2} & s \in [s'(2), s(2)), \end{cases}$$

where $s'(2)$ is defined in the theorem.

Case 4 $s \in [s(l-1), s(l))$, $(l \geq 3)$. Consider Procedure(l) as follows. We only prove the case of $n = kl + 1$ as above. Obviously we have $\alpha_1 = \sum_{i=0}^{k-1} p_{il+2} \leq p_2 + \frac{1}{l}\sum_{i=3}^{n} p_i$. To get the ratio between α_1 and C^{OPT}, we distinguish two subcases according to the assignment of the first two jobs in the optimum.

Subcase 1 p_1 and p_2 are processed in the same machine in the optimal schedule. Then $C^{OPT} \geq (p_1 + p_2)/s$. Therefore we get

$$\alpha_1 \leq \frac{l-2}{2l}(p_1 + p_2) + \frac{b}{l} \leq \frac{ls+2}{2l} C^{OPT}.$$

Subcase 2 p_1 and p_2 are not processed in the same machine in the optimal schedule. Then $C^{OPT} \geq p_2$ and

$$\alpha_1 \leq \frac{l-2}{l} p_2 + \frac{b}{l} \leq \frac{l+s-1}{l} C^{OPT}.$$

Now we remain to get the ratio between α_2 and C^{OPT}. By straight computation, we have

$$\alpha_2 = \frac{1}{s}\left(p_1 + \sum_{i=1}^{k-1}\sum_{j=3}^{l+1} p_{il+j}\right) \leq \frac{p_1}{s} + \frac{l-1}{ls}\sum_{i=2}^{n} p_i$$

$$\leq \frac{p_1}{ls} + \frac{(l-1)b}{ls} \leq \left(\frac{1}{l} + \frac{(l-1)(s+1)}{ls}\right) C^{OPT}$$

$$= \frac{ls+l-1}{ls} C^{OPT}.$$

Note for $l \geq 3$, and $s > 2$, we have $(ls+l-1)/(ls) > (l+s-1)/l$, and $(ls+2)/(ls) > (l+s-1)/l$, which imply that $C^{ordinal}/C^{OPT} \leq \max\{\frac{ls+l-1}{ls}, \frac{ls+2}{2l}\}$. The desired ratio can thus be obtained.

To show the algorithm *Ordinal* is $c^{ordinal}$-competitive algorithm, we consider the following instances. If $s \in [s(0), s(1))$, let $p_1 = p_2 = p_3 = 1, p_4 = p_5 = p_6 = 1/s$. If $s \in [s(l-1), s'(l))$, let $p_1 = 1, p_2 = \cdots = p_l = 1/(ls)$. If $s \in [s'(l), s(l))$, let $p_1 = p_2 = 1, p_3 = \cdots = p_l = 2/(ls)$. If $s \geq 1 + \sqrt{3}$, let $p_1 = 1, p_2 = 1/s$.   □

Comparing the lower bound with the competitive ratio of *Ordinal*, we conclude that our algorithm is optimal in the following intervals of s

$$\left[1, \frac{5 + \sqrt{265}}{20} \right] \cup \left[\frac{1 + \sqrt{7}}{3}, 2 \right] \cup \left[\frac{1 + \sqrt{10}}{2}, \frac{3 + \sqrt{33}}{4} \right] \cup \left[1 + \sqrt{3}, \infty \right),$$

the total length of the intervals where the ratio does not match the lower bound is less than 0.7784 and the biggest gap is approximately 0.0520412 which occurs at $s = \frac{35 + \sqrt{8617}}{112} \approx 1.14132$.

References

[1] Agnetis, A.: No-wait flow shop scheduling with large lot size. Rap 16.89, Dipartimento di Informatica e Sistemistica, Universita Degli Studi di Roma "La Sapienza", Rome, Italy, 1989.

[2] Azar, Y., Regev, O.: Online bin stretching. *Proc. of RANDOM'98*, 71-82(1998).

[3] Cho, Y., Sahni, S.: Bounds for list scheduling on uniform processors. *SIAM J. Computing* **9**, 91-103(1980).

[4] Epstein, L., Noga, J., Seiden, S. S., Sgall, J., Woeginger, G.J.: Randomized online scheduling for two related machines. *Proc. of the 10th ACM-SIAM Symp. on Discrete Algorithms*, 317-326(1999).

[5] He, Y., Zhang, G.: Semi on-line scheduling on two identical machines. *Computing*, **62**, 179-197(1999).

[6] Kellerer, H., Kotov, V., Speranza, M., Tuza, Z.: Semi online algorithms for partition problem. *Oper. Res. Letters*, **21**, 235-242(1997).

[7] Lawler, E. L.: Combinatorial Optimization: Networks and Matroids. Holt, Rinehart and Winston, Toronto, 1976.

[8] Liu, W. P., Sidney, J. B., Van Vliet, A.: Ordinal algorithms for parallel machine scheduling. *Oper. Res. Letters*, **18**, 223-232(1996).

[9] Liu, W. P., Sidney, J. B.: Bin packing using semi-ordinal data. *Oper. Res. Letters*, **19**, 101-104(1996).

[10] Seiden, S., Sgall, J., Woeginger, G.: Semi-online scheduling with decreasing job sizes. Technical Report, TU Graz, Austria, 1999.

[11] Sleator, D., Tarjan, R.E.: Amortized efficiency of list update and paging rules. *Communications of ACM* **28**, 202-208(1985).

Agents, Distributed Algorithms, and Stabilization

Sukumar Ghosh*

The University of Iowa, IA 52242, USA
ghosh@cs.uiowa.edu

Abstract. This note illustrates the use of agents in the stabilization of distributed systems. The goal is to build stabilizing systems on the Internet, where the component processes are not under the control of a single administration. Two examples are presented to illustrate the idea: the first is that of mutual exclusion on a unidirectional ring, and the second deals with the construction of a DFS spanning tree.

1 Introduction

An agent [7] is a program that can migrate from one node to another, perform various types of operations at these nodes, and can take autonomous routing decisions. In contrast with messages that are passive, an agent is an active entity that can be compared with a messenger. In the past few years, agents have proven to be an attractive tool in the design of various types of distributed services.

This note examines the use of agents in the design of stabilizing distributed systems [4]. Conventional stabilizing systems expect processes to run predefined programs that have been carefully designed to guarantee recovery from all possible bad configurations. However, in an environment like the Internet, where processes are not under the control of a single administration, expecting every process to run predefined programs for the sake of stabilization is unrealistic. This paper explores an alternative mechanism for stabilizing a distributed system, in which processes are capable of accommodating visiting agents.

Consider a network of processes, where each process communicates with its immediate neighbors using messages. Each process, in addition to sending out its own state, can read incoming messages, and update its own state (using a set of rules). The state of the network consists of the states of each of the processes, and can be classified into the categories *legal* and *illegal*. Each update action modifies the state of the network. A stabilizing system guarantees that regardless of the starting state, the network reaches a legal state in a bounded number of steps, and remains in the legal state thereafter. Convergence and closure are the two cornerstones of stabilizing systems [1].

---

* This research was supported in part by the National Science Foundation under grant CCR-9901391, and by the Alexander Von Humboldt Foundation, Germany while the author was visiting the University of Dortmund.

D.-Z. Du et al. (Eds.): COCOON 2000, LNCS 1858, pp. 242–251, 2000.

We assume that in addition to the ongoing activities in the above network of processes, one or more processes can send out agents that can migrate from one process to another. While the individual processes maintain the closure of legal configurations, agents help the system converge to a legal configuration as soon as it detects that the configuration is illegal. The detection involves taking a total or partial snapshot of the system state. Corrective actions by the agent involve the modification of shared variables of one or more processes that the agent is scheduled to visit.

In seeking a solution to the problem of stabilization with agents, we make the following assumption:

[**Non-interference**] The normal operation of the system neither depends on, nor is influenced by the presence of agents.

Only at convenient intervals, agents are sent out to initiate a "clean-up phase," but otherwise the operation of the system continues as usual. The individual processes are oblivious to the presence of the agent(s). We do not take into consideration any minor slowdown in the operation of a node due to the sharing of the resources by the visiting agent.

Our solution can be viewed as a stabilizing extension of a distributed system as proposed Katz & Perry [6]. However, agents make the implementation straightforward, and sometimes elegant. We consider two well-known problems to illustrate our idea: the first deals with stabilizing mutual exclusion on a unidirectional ring first illustrated by Dijkstra [4], and the second is a stabilizing algorithm for constructing a DFS spanning tree for a given graph.

2   Model and Notations

Consider a strongly connected network of processes. The processes communicate with one another through messages. Any process can send out a message containing its current state to its immediate neighbors, and receive similar messages from its immediate neighbors. Each channel is of unit capacity. Furthermore, channels are FIFO, i.e. through any channel, messages are received in the order they are sent.

The program for each process consists of a set of rules. Each rule is a guarded action of the form $g \rightarrow A$, where g is a boolean function of the state of that process and those of its neighbors received via messages, and A is an action that is executed when g is true. An action by a process consists of operations that update its own state, followed by the sending of messages that contain the updated state. The execution of each rule is atomic, and it defines a step of a computation. Furthermore, when multiple rules are applicable, any one of them can be chosen for execution. A *computation* is a sequence of atomic steps. It can be finite or infinite.

The other component of the algorithm is the program that the agent executes every time it visits a node. The variables of the agent will be called *briefcase variables*. A guard is a boolean function of the briefcase variables, and the variables of the process that the agent is currently visiting. An action consists of one or more of the following three operations: (i) modification of its briefcase variables, (ii) modification of the variables of the process that it is currently visiting, and (iii) identification of the *next process* to visit. This next process can either be predetermined, or it can be computed on-the-fly, and will be represented by the variable *next*. We will use guarded commands to express the agent program.

A privileged process, in addition to everything that an ordinary process does, can test for the arrival of an agent as a part of evaluating its guard, and can send out an agent (with appropriate initialization) as a part of its action. These agents use the same channel that the messages use. However, agents always receive a higher priority over messages, as represented by the following assumption:

[**Atomicity**] At any node, the visit of an agent is an atomic event. Once an agent arrives at a node, the node refrains from sending or receiving any message until the agent leaves that node.

Thus from the node's point of view, an agent's visit consists of an indivisible set of operations. Exceptions may be made for privileged processes only. This assumption is important to make any total or partial snapshot of the system state taken by the agent a meaningful or consistent one. For further details about this issue, we refer the readers to [2]

3   Mutual Exclusion

We consider here the problem of stabilizing mutual exclusion on a unidirectional ring, first illustrated by Dijkstra in [4]. The system consists of a unidirectional ring of N processes numbered 0 through $N - 1$. Process i can only read the state of its left neighbor $(i - 1)$ *mod* N in addition to its own state. The state of process i is represented by the boolean $b(i)$. Process 0 is a privileged process, and behaves differently from other processes. A process is said to have a token, when its guard is true. We need to design a protocol, so that regardless of the starting configuration, (1) eventually a state is reached in which exactly one process has a token, (2) the number of tokens always remains one thereafter, and (3) in an infinite computation, every process receives a token infinitely often.

Consider the program of the individual processes first. Process i has a boolean variable $b(i)$ that is sent out to its neighbor $(i + 1)$ using a message.

(**Program for process** $i : i \neq 0$) $b(i) \neq b(i - 1) \rightarrow b(i) := \neg b(i)$
(**Program for process** 0) $b(0) = b(N - 1) \rightarrow b(0) := \neg b(0)$

It can be easily shown that the number of tokens is greater than 0, and this number never increases. Therefore, once the system is in a legal state (i.e., it has a single token) it continues in that state.

The states of the processes may be corrupted at unpredictable moments. To steer the system into a legal configuration, process 0 sends out an agent from time to time. This is done after process 0 executes its current action, if any, and appropriately initializes the briefcase variables. Furthermore, the following condition must hold:

Condition 1. Until the agent returns home, process 0 suspends subsequent actions[1].

The briefcase consists of two booleans, B and *token*, that are initialized by process 0 when the agent is sent out. As the agent traverses the ring, it determines if the number of tokens is greater than one. In that case, the agent modifies the state of a process to possibly reduce the number of tokens. The program for the agent consists of the three rules ME1, ME2, and ME3 and is shown below.

Program for the agent while visiting node i

Initially, $token = false$, $B = b(0)$ (initialization done by process 0)

if

    (ME1) $B \neq b(i) \land \neg token \rightarrow B := b(i); token := true; next := (i+1) \bmod N$
    (ME2) $B \neq b(i) \land token \rightarrow b(i) := B; next := (i+1) \bmod N$
    (ME3) $B = b(i) \rightarrow next := (i+1) \bmod N$

fi

Note that the briefcase variables can also be badly initialized, or corrupted. However, the agent is guaranteed to return to process 0, which initializes the briefcase variables. In the remainder of this section, we assume that the briefcase variables have correct values.

Theorem 1. When the agent returns to process 0, the system returns to a legal configuration.

Proof. When the agent returns to process 0 , the following two cases are possible.

Case 1 : $token = true$.

This implies that the agent has detected a token at some process j, Thereafter, the state $b(k)$ of any other process k $(k > j)$ that had a token when the agent visited it, has been complemented, making that token disappear. Thus, in the state *observed* by the agent, the values of $b(i)$ are the same for all $i \geq j$, and are the complement of the values of $b(i)$ for all $i < j$. This means that the system is in a legal configuration.

Case 2 : $token = false$.

In this case, the agent has not been able to detect any token. Since process 0 did not execute a move in the mean time, in the *observed state*, the values of $b(i)$ for every process i are identical. This means that process has a token, and the system is in a legal configuration. □

---

[1] To monitor the arrival of the agent, we rely on a built-in timeout mechanism that is an integral part of any message-passing system. We elaborate on this further in the concluding section.

It takes only one traversal of the ring by the agent to stabilize the system, where a traversal begins and ends at process 0. However, if a transient failure hits a process immediately after the agent's visit, then the resulting illegal configuration may persist until the next traversal of the agent.

It is important here to note that unlike the original solution in [4], the space complexity of the processes as well as the agent is constant. This makes the solution scalable to rings of arbitrary size. In terms of space complexity, it is comparable to [5] that uses only three bits per process, and no agents. However, the communication complexity, defined as the number of bits that are sent (and received) to move the token from one process to the next, is substantially lower (in fact, it is 1-bit) in our solution, as opposed to $O(N)$ bits in [5].

Why can't we get rid of all these, and define a process having a token only when the agent visits that process? This violates the condition of non-interference. Since the agents are periodically invoked, the system will come to a standstill during the interim period, when the agent is absent!

4   DFS Tree Generation

4.1   The Algorithm

Consider a connected network of processes. The network is represented by a graph $G = (V, E)$, where V denotes a set of nodes representing processes, and E denotes the set of undirected edges joining adjacent processes, through which interprocess communication takes place. A special process 0 is designated as the *root*. Every process i ($i \neq 0$) has a *parent* process $P(i)$ that is chosen from its neighbors. By definition, the root does not have a parent. The spanning tree consists of all the processes and the edges connecting them with their parents. By definition, a spanning tree is acyclic. However, transient failures may corrupt the value of $P(i)$ from one or more process i, leading to the formation of cycles in the subgraph, or altering the tree to another spanning tree that is not DFS, making the resulting configuration illegal. The individual processes will remain oblivious to it. So, we will entrust the root node with the task of sending out agents to detect such irregularities. If required, the agent will appropriately change the parents of certain processes to restore the spanning tree.

For the convenience of presentation, we will use the following abbreviations:

$child(i) \equiv \{j : P(j) = i\}$
$N(i) \equiv \{j : (i, j)\}$
$friend(i) \equiv \{j : j \in N(i) \wedge j \neq P(i) \wedge P(j) \neq i)$

A node $k \in friend(i)$ will also be called a non-tree neighbor of node i.

One of the key issues here is that of graph traversal. Assume that each process has a boolean flag f. Initially, $f = false$. This flag is set whenever the process is visited by the agent. Then, the following two rules will help an agent traverse the graph down the spanning tree edges, and eventually return to the root. Note that the set of briefcase variables is empty.

(DFS1) $\exists j \in child(i) : \neg f(j) \rightarrow next := j$
(Note: After reaching j, the agent will set $f(j)$ to *true*)

(DFS2) $\forall j \in N(i) : f(j) \rightarrow next := P(i)$

This will also maintain closure of the legal configuration. The boolean flag $f(j)$ has to be reset before a subsequent traversal begins.

Unfortunately, this traversal may be affected when the DFS spanning tree is corrupted. For example, if the parent links form a cycle, then no tree edge will connect this cycle to the rest of the graph. Accordingly, using DFS1, the nodes in the cycle will be unreachable for the agent, and the traversal will remain incomplete. As another possibility, if the agent reaches one of these nodes contained in a cycle before the cycle is formed, then the agent is trapped, and cannot return to the root using DFS2.

To address the first problem, we add the following rule to "force open" a path to the unreachable nodes.

(DFS3) $\forall j \in child(i) : f(j) \wedge \exists k \in friend(i) : \neg f(k) \rightarrow next := k$
(Note: After reaching k, the agent will set $f(k)$ to true, and $P(k)$ to i, i.e., i will adopt k as a child).

If the unreachable nodes form a cycle, then this rule will break it. This rule will also help restore the legal configuration, when the spanning tree is acyclic, but not a DFS tree.

To address the second problem, we ask the agent to keep track of the number of nodes visited while returning to the root via the parent link. For this purpose, the briefcase of the agent will include a non-negative integer counter C. Whenever the agent moves from a parent to a child using DFS1 or DFS3, C is reset to 0, and and when the agent moves from a node to its parent using DFS2, C is incremented by 1. (Note: This will modify rules DFS1 and DFS2). When C exceeds a predetermined value bigger than the size N of the network, a new parent has to be chosen, and the counter has to be reset.

(DFS4) $(C > N) \wedge \exists k \in friend(i) \rightarrow P(i) := k; C := 0, next := k$

This will break the cycle that trapped the agent.

The only remaining concern is about the flag f being corrupted, and appropriately initializing them before a new traversal begins. For this purpose, we include a boolean flag R to the briefcase to designate the round number of the traversal. Before a new traversal begins, the root complements this flag. When $f(i) \neq R$, the node i is considered unvisited, and to indicate that node i has been visited, $f(i)$ is set to R. This leads us to the overall agent program for stabilizing a DFS spanning tree.

The program adds a extra variable *pre* to the briefcase of the agent. It stores the identity of the previous node that it is coming from. A slightly folded version of the program is shown below.

Program for the agent while visiting node i

if $f(i) \neq R \rightarrow f(i) := R$ **fi**;
if $pre \in friend(i) \rightarrow P(i) := pre$ **fi**;
if

  (DFS1) $\exists j \in child(i) : f(j) \neq R \rightarrow next := j; C := 0$
  (DFS2) $\forall j \in N(i) : f(j) = R \wedge (C < N \vee friend(i) = \emptyset) \rightarrow$
      $next := P(i); C := C + 1$
  (DFS3) $\forall j \in child(i) : f(j) = R \wedge \exists k \in N(i) : f(k) \neq R \rightarrow$
      $next := k; C := 0$
  (DFS4) $(C > N) \wedge \exists k \in friend(i) \rightarrow next := k; P(i) := k; C := 0$

fi;
$pre := i$

4.2 Correctness Proof

Define j to be a *descendant* of node i, if $j \in child(i)$, or j is a descendant of $child(i)$. In a legal configuration, the following three conditions must hold:

1. $\forall i \in V$ ($i \neq root$), the edges joining i with $P(i)$ form a tree.
2. $j \in friend(i) \Rightarrow j$ is a descendant of node i.
3. $\forall i, j \in V, f(i) = f(j)$.

When the first condition does not hold, the tree edges form more than one disconnected subgraph or segment. The segment containing the root will be called the *root segment*.

When the agent returns to the root, $\forall i \in V : f(i) = R$. During the subsequent traversal, the root complements R, and $\forall i \in V : f(i) \neq R$ holds. If for some node j this is not true, then we say that node j has an incorrect flag. Note that if only the first two conditions hold but not the third one, then any application using the DFS tree will remain unaffected.

Lemma 1. In a DFS tree, when the agent returns to a node after visiting all its children, every non-tree neighbor is also visited.

Proof. Follows from the definition of DFS traversal.   □

Lemma 1 shows that $DFS1$ and $DFS2$ maintain the closure of the legal configuration.

Lemma 2. Starting from an arbitrary configuration, the agent always returns to the root.

Proof. If the agent is in the root segment, and C does not exceed N during this trip when the agent moves towards the root, then $DFS2$ guarantees that the statement holds.

If C exceeds N, then either (1) C is corrupted, or (2) the agent is trapped into a loop. In Case 1, the number of disconnected segments possibly increases (using DFS3), and the agent moves to a segment away from the root segment.

However, this eventually triggers $DFS4$. The execution of $DFS4$ breaks the loop, connects it with another segment, and routes the agent to that segment. At the same time, C is re-initialized. In a bounded number of steps, the number of segments is reduced to one, which becomes the root segment. Using $DFS2$, the agent returns to the root. ◻

Lemma 3. If no node i has an incorrect flag $f(i)$, then the DFS tree is restored in at most one traversal of the graph.

Proof. We consider three cases here.

Case 1. Assume that the root segment consisting of all the nodes in V and their parents form a spanning tree, but it is not a DFS tree. This allows for unvisited non-tree neighbors even after all the children have been visited. This triggers $DFS3$, the repeated application of which transforms the given spanning tree into a DFS tree.

Case 2. Assume that the agent is in the root segment, and there is no tree edge connecting it to the rest of the graph. In this case, eventually $DFS3$ will be executed, and that will establish a new tree edge connecting the root segment with the rest of the graph. This will augment the size of the root segment, and the number of segments will be reduced by one. Repeated application of this argument leads to the fact that the root segment will include all the nodes before the agent completes its traversal.

Case 3 Assume that the agent is in a segment away from the root segment, and in that segment, the edges connecting nodes with their parents form a cycle. In this case, within a bounded number of steps, $DFS4$ will be executed, causing the agent to move away from the current segment by following the newly updated parent link, and re-initializing C. This will also reduce the number of segments by one. In a finite number of steps, the agents will return to the root segment, and eventually the DFS tree will be restored. ◻

Lemma 4. If there are nodes with incorrect flags, then after each traversal by the agent, the number of nodes with incorrect flags is decremented by at least one.

Proof. Assume that there are some nodes with incorrect flags when the agent begins its traversal. This means that there are some nodes that have not been visited by the agent, but their flags incorrectly show that they have been visited. We will demonstrate that the flag of at least one such node is set correctly, before the current traversal ends and the next traversal begins.

Using $DFS2$, the agent moves from a node to its parent, when every neighbor has been visited. This may include a neighbor k that has not yet been visited by the agent, but its flag has been incorrectly set as visited (i.e., $f(k) = R$). However, during the next traversal, R is complemented, and so $f(k) \neq R$ holds - this corresponds to k being unvisited, and the flag is now correct. ◻

An unpleasant consequence of Lemma 4 is that the agent may need upto $N - 1$ traversals to set all the flags correctly. When it encounters as little as one incorrect flag, it can potentially abandon the task of stabilizing the rest of the graph during that traversal, and postpone it to the next round.

Theorem 2. In at most N traversals by the agent, the DFS tree returns to a legal configuration.

Proof. From Lemma 4, it may take at most $(N-1)$ traversals to set all the flags correctly. From Lemma 3, it follows that within the next traversal, the DFS tree is reconstructed. $\square$

4.3   Complexity Issues

The agent model is stronger than the classical message passing or shared memory models. Note that we are measuring the stabilization time in terms of the number of traversals made by the agent, where a traversal begins when the agent leaves the root, and ends when it returns to the root. It is possible for the agent to be corrupted in the middle of a traversal, or the state variables of the processes to be corrupted while the agent is in transit. In such cases, the stabilization time will be measured after the failure has occurred, and the agent has returned to the root, which is guaranteed by Lemma 2.

It is feasible to cut down the stabilization time to one traversal with a high probability by using signatures. The agent, while visiting a process i for the first time during a traversal, will append its signature to $f(i)$. A failure at that node will make that signature unrecognizable with a high probability. If the agent cannot recognize its own signature at any node, then it will consider that node to be unvisited.

We make another observation while comparing the space complexity of this solution with those of the more traditional stabilizing spanning tree algorithms like [3]. In [3], each process has to keep track of the size of the network, that requires $log_2 N$ bits per process. In our solution, the space complexity of each process is constant - only the agent needs to know the value of N, and will need $log_2 N$ bits in its briefcase. It appears that the N copies of $log_2 N$ bits has been traded by just one copy of $log_2 N$, which is a favorable trade, and contributes to the scalability. The additional space complexity for the briefcase variable *pre* should not exceed $log_2 \delta$ bits (where δ is the maximum degree of a node), since the choice of a parent is limited to only the immediate neighbors of a node.

5   Conclusion

Agents can be a powerful tool in the design of distributed algorithms on systems with a heterogeneous administrative structure. In contrast with *self-stabilization* where processes execute the stabilizing version of the program, here processes execute the original program, but are ready to accommodate corrective actions by one or more visiting agents.

Since the task of inconsistency detection and consistency restoration is delegated to a separate entity, agent-based stabilization is conceptually simpler that self-stabilization. However, in some cases, it can cause new and tricky problems, like the trapping of an agent in the DFS spanning tree algorithm. On the other hand, we believe that algorithms for controlled recovery, like fault-containment algorithms, can be painlessly built into an existing distributed system by using agents.

To send and receive agents, the privileged process relies on a timeout mechanism. In case an agent does not return within the timeout period, it is assumed to be lost, and a new agent is generated. A side effect is the possibility of multiple agents patroling the network simultaneously. For some protocols (like the mutual exclusion protocol), this does not pose a problem, except that it wastes network resources. In others, the agent management system has to take the responsibility of eventually reducing the number of agents to one. A presentation of such a protocol is outside the scope of this paper.

Since the agents are invoked periodically, a possible criticism is that the illegal configuration may persist for a "long period." However, in the classical asynchronous self-stabilizing systems also, we do not put an upper bound on the real time that a process will take to detect or correct an illegal configuration. The frequency of invocation of the agent will be determined by the failure probability, the maximum duration for which a faulty behavior can be tolerated, and the failure history of the system.

One of our future goals is to demonstrate that besides being viable, agents make it easier to design stabilizing algorithms for distributed system when compared with the classical approach. In non-reactive systems, the agent basically walks through a distributed data structure, and modifies it as and when necessary. In reactive systems, the critical issues are taking a distributed snapshot of a *meaningful part* of the system, and making appropriate alterations to the states of some processes on-the-fly (i.e., a total or partial distributed reset) such that legitimacy is restored.

References

1. Arora, A., Gouda, M.G.: Closure and Convergence: A Foundation of Fault-tolerant Computing. IEEE Transactions on Software Engineering, Volume 19 (1993) 1015-1027.
2. Chandy K.M., Lamport L.: Distributed Snapshots: Determining Global States of a Distributed System. ACM Trans. Computer Systems, Volume 3, No.1 (1985) 63-75.
3. N. S. Chen, H. P. Yu, S. T. Huang.: A Self-Stabilizing Algorithm for Constructing a Spanning Tree. Information Processing Letters, Volume 39 (1991) 147-151.
4. Dijkstra, E.W.: Self-Stabilizing Systems In Spite of Distributed Control. Communications of the ACM, Volume 17 (1974) 643-644.
5. Gouda M.G., Haddix F.: The Stabilizing Token Ring in Three Bits. Journal of Parallel and Distributed Computing, Volume 35 (1996) 43-48.
6. Katz, S., Perry K.: Self-Stabilizing Extensions for Message-passing Systems. Proc. 9th Annual Symposium on Principles of Distributed Computing (1990) 91-101.
7. Gray Robert, Kotz D., Nog S., Rus D., Cybenko G.: Mobile Agents for Mobile Computing. Proceedings of the Second Aizu International Symposium on Parallel Algorithms/Architectures Synthesis, Fukushima, Japan (1997).

A Fast Sorting Algorithm and Its Generalization on Broadcast Communications[*]

Shyue-Horng Shiau and Chang-Biau Yang[**]

Department of Computer Science and Engineering,
National Sun Yat-Sen University,
Kaohsiung, Taiwan 804, R.O.C.,
{shiaush, cbyang}@cse.nsysu.edu.tw,
http://par.cse.nsysu.edu.tw/~cbyang

Abstract. In this paper, we shall propose a fast algorithm to solve the sorting problem under the *broadcast communication model*(BCM). The key point of our sorting algorithm is to use successful broadcasts to build broadcasting layers logically and then to distribute the data elements into those logic layers properly. Thus, the number of *broadcast conflicts* is reduced. Suppose that there are n input data elements and n processors under BCM are available. We show that the average time complexity of our sorting algorithm is $\Theta(n)$. In addition, we expand this result to the generalized sorting, that is, finding the first k largest elements with a sorted sequence among n elements. The analysis of the generalization builds a connection between the two special cases which are maximum finding and sorting. We prove that the average time complexity for finding the first k largest numbers is $\Theta\left(k + \log\left(n - k\right)\right)$.

1 Introduction

One of the simplest parallel computation models is the *broadcast communication model* (BCM)[1,2,5,6,14,13,4,9,7,11,12,15]. This model consists of some processors sharing one common channel for communications. Each processor in this model can communicate with others only through this shared channel. Whenever a processor broadcast messages, any other processor can hear the broadcast message via the shared channel. If more than one processor wants to broadcast messages simultaneously, a *broadcast conflict* occurs. When a conflict occurs, a conflict resolution scheme should be invoked to resolve the conflict. This resolution scheme will enable one of the broadcasting processors to broadcast successfully. The Ethernet, one of the famous local area networks, is an implementation of such a model.

The time required for an algorithm to solve a problem under BCM includes three parts: (1) resolution time: spent to resolve conflicts, (2) transmission time:

---

[*] This research work was partially supported by the National Science Council of the Republic of China under NSC89-2213-E110-005.

[**] To whom all correspondence should be sent.

D.-Z. Du et al. (Eds.): COCOON 2000, LNCS 1858, pp. 252–261, 2000.

spent to transmit data, (3) computation time: spent to solve the problem. For the sorting problem, it does not seem that transmission time and computation time can be reduced. Therefore, to minimize resolution time is the key point to improve the time complexity.

The probability concept, proposed by Martel [7], is to estimate a proper broadcasting probability, which can reduce conflict resolution time. Another idea to reduce conflict resolution time is the layer concept proposed by Yang [13]. To apply the layer concept for finding the maximum among a set of n numbers [11], we can improve the time complexity from $\Theta(\log^2 n)$ to $\Theta(\log n)$ [12]. As some research has also been done on the BCM when processors have a known order, our paper is the case that processors have a unkown order.

Some researchers [8,2,4,5,15,10] have solved the sorting algorithm under BCM. The selection sort can be simulated as a straightforward method for sorting under BCM. The method is to have all processors broadcast their elements and to find the maximum element repeatedly. In other words, the maximum elements which are found sequentially form the sorted sequence. Levitan and Foster [5,6] proposed a maximum finding algorithm, which is nondeterministic and requires $O(\log n)$ successful broadcasts in average. Martel [7], and Shiau and Yang [11] also proposed maximum finding algorithms, which require $O(\log n)$ and $\Theta(\log n)$ time slots (including resolution time slots) in average respectively. Thus, the time required for the straightforward sorting method based on the repeated maximum finding is $O(n \log n)$.

Martel et al. [8] showed the sorting problem can be solved in $O(n)$. They also proved that the expected time for finding the first k largest numbers is $O(k + \log n)$. They used a modified version of selection sort, i.e., sort by repeatedly finding maximum based on the results from previous iterations. It can be viewed as a modified Quicksort. Until now, Martel's algorithm is the fastest one for solving the sorting problem under BCM. However, the constants associated with $O(n)$ and $O(k + \log n)$ bounds proved by Martel et al. [8] are not very tight.

This paper can be divided into two main parts. In the first part, we shall propose a sorting algorithm under BCM. In the algorithm, we make use of the maximum finding method [11] to build layers and reduce conflicts. After the process of maximum finding terminates, we can use these successful broadcasts to build broadcasting layers logically and then to distribute the data elements into those logic layers properly. Thus, conflict resolution time can be reduced. And we shall give the time complexity analysis of our algorithm. The average number of time slots (including conflict slots, empty slots, and slots for successful broadcast) required for sorting n elements is T_n, where $\frac{7}{2}n - \frac{1}{2} \leq T_n \leq \frac{23}{6}n - \frac{7}{6}$. Here, it is assumed that each time slot requires constant time. As we can see that the bound of our time complexity is very tight.

In the second part, we shall generalize the result of the first part. We prove that the average number of time slots required for finding the first k largest numbers with a sorted sequence is T_k^n, where $\frac{7}{2}k + 4\ln(n - (k - 2)) - \left(\frac{1}{2} + 4\ln 2\right) \leq T_k^n \leq \frac{23}{6}k + 5\ln(n - (k - 1)) - \frac{7}{6}$. The time complexity is also very tight. By the analysis of the generalization, we figure out the connection between maximum finding and sorting, which are two special cases of our generalization.

On the high level view, our sorting algorithm and generalization algorithm have a similar approach as that proposed by Martel *et al.* [8]. On the low level view, our algorithms are based on the layer concept, thus they can be analyzed easily and a tighter bound can be obtained. And Lemma 1 of this paper can be regarded as a generalization for the work which has the same high level approach. If there is another maximum finding algorithm under BCM whose bound can be analyzed , not only it can be applied into our sorting and generalized sorting approaches, but also the approaches based on it can be analyzed immediately by our lemmas in this paper.

This paper is organized as follows. In Section 2, we review the previous results for our maximum finding[11]. Based on the maximum finding, we present the sorting algorithm and its analysis in Section 3 and 4. In Section 5, we present the generalization of the sorting. In Section 6, we provide the analysis of the generalization. Finally, Section 7 concludes the paper.

2   The Previous Results for Maximum Finding

We shall first briefly review the maximum finding algorithm under BCM, which is proposed and analyzed by Shiau and Yang [11]. In the maximum finding algorithm, each processor holds one data element initially. The key point of the maximum finding algorithm is to use broadcast conflicts to build broadcasting layers and then to distribute the data elements into those layers. For shortening the description in our algorithm, we use the term "data element" to represent "the processor storing the data element" and to represent the data element itself at different times if there is no ambiguity. When a broadcast conflict occurs, a new upper layer is built up. Each data element which joins the conflict flips a coin with equal probabilities. That is, the probability of getting a head is $\frac{1}{2}$. All which get heads continue to broadcast in the next time slot, and bring themselves up to the new upper layer. This new layer becomes the active layer. The others which get tails abandon broadcasting, and still stay on the current layer. At any time, only the processors which are alive and on the active layer may broadcast.

In this paper, we shall use the notation $\langle x_0, x_1, x_2, \cdots, x_t \rangle$ to denote a linked list, where x_t is the head of the list. Suppose $L_1 = \langle x_0, x_1, x_2, \cdots, x_t \rangle$ and $L_2 = \langle y_0, y_1, y_2, \cdots, y_{t'} \rangle$, then $\langle L_1, L_2 \rangle$ represents the concatenation of L_1 and L_2, which is $\langle x_0, x_1, x_2, \cdots, x_t, y_0, y_1, y_2, \cdots, y_{t'} \rangle$.

The maximum finding algorithm can be represented as a recursive function: $Maxfind(C, L_m)$, where C is a set of data elements (processors). Initially, each processor holds one alive data element and sets an initial linked list $L_m = \langle x_0 \rangle$, where $x_0 = -\infty$.

Algorithm Maximum-Finding: $Maxfind(C, L_m)$
Step 1: Each alive data element in C broadcasts its value. Note that a data element is alive if its value is greater than the head of L_m. Each element smaller than or equal to the head drops out (becomes dead).

Step 2: There are three possible cases:

    Case 2a If a broadcast conflict occurs, then do the following.

        Step 2a.1: Each alive data element in C flips a coin. All which get heads form C' and bring themselves to the upper layer, and the others form C'' and stay on the current layer.

        Step 2a.2: Perform $L_m = Maxfind(C', L_m)$.

        Step 2a.3: Perform $L_m = Maxfind(C'', L_m)$.

        Step 2a.4: Return L_m.

    Case 2b: If exactly one data element successfully broadcasts its value y, then return $\langle L_m, y \rangle$.

    Case 2c: If no data element broadcasts, then return L_m. (C is empty or all data elements of C are dead.)

In the above algorithm, we can get a sequence of successful broadcasts, which is represented by the linked list $L_m = \langle x_0, x_1, x_2, \cdots, x_{t-1}, x_t \rangle$, where $x_{i-1} < x_i$, $1 \leq i \leq t$ and $x_0 = -\infty$. Note that L_m is held by each processor and the head x_t is the maximum. A new upper layer is built up when the recursive function $Maxfind$ is called and the current layer is revisited after we return from $Maxfind$. In Case 2c, there are two possible situations. One is that no element goes up to the upper layer. The other is that no alive element remains on the current layer after the current layer is revisited because all elements go up to the upper layer, or some elements broadcast successfully on upper layers and they kill all elements on the current layer.

Shiau and Yang [11] proved that the total number of time slots, including conflict slots, empty slots and slots for successful broadcasts, is M_n in average, where $4 \ln n - 4 < M_n < 5 \ln n + \frac{5}{2}$. Grabner *et al.* [3] gave a precise asymptotic formula that $M_n \sim (\frac{\pi^2}{3 \ln 2}) \ln n \sim (4.74627644\ldots) \ln n$.

Here we use the lemma

$$\tfrac{4}{n} \leq M_n - M_{n-1} \leq \tfrac{5}{n},$$

which was proved by Yang *et al.* [11], rather than the precise asymptotic formula $M_n \sim (\frac{\pi^2}{3 \ln 2}) \ln n \sim (4.74627644\ldots) \ln n$. This is done because it is hard to apply the precise asymptotic formula and its fluctuation into the proof directly.

3   The Sorting Algorithm

In our sorting algorithm, each processor holds one data element initially. The key point of our algorithm is to use successful broadcasts which occur within the progress of maximum finding to build broadcasting layers and then to distribute the data elements into those layers properly.

When a series of successful broadcasts occur within the progress of maximum finding, some correspondent layers are built up. All processors which participated in the maximum finding are divided into those layers. Each layer performs the sorting algorithm until the layer is empty or contains exactly one data element.

Our algorithm can be represented as a recursive function: $Sorting(M)$, where M is a set of data elements (processors). In the algorithm, each processor holds one data element and maintains a linked list L_s. Our sorting algorithm is as follows:

Algorithm Sorting: $Sorting(M)$
Step 1: Each data element in M broadcasts its value.
Step 2: There are three possible cases:
    Case 2a: If exactly one data element successfully broadcasts its value y, then return $\langle y \rangle$.
    Case 2b: If no data element broadcasts, then return $NULL$.
    Case 2c: If a broadcast conflict occurs, then perform
       $L_m = Maxfind(M, \langle x_0 \rangle)$, where $L_m = \langle x_0, x_1, x_2, \cdots, x_{t-1}, x_t \rangle$ is a series of successful broadcasts and $x_{i-1} < x_i$, $1 \le i \le t$ and $x_0 = -\infty$.
       Step 2c.1: Set $L_s = NULL$.
       Step 2c.2: For $i = t$ down to 1
             Set $L_s = \langle x_i, L_s \rangle$.
             do $L_s = \langle Sorting(M_i), L_s \rangle$,
                where $M_i = \{x | x_{i-1} < x < x_i\}$.
       Step 2c.3: Return L_s.

In the above algorithm, x_1 is the element of the first successful broadcast, and x_t is the element of the last successful broadcast and also the largest element in M.

For example, suppose that M has 2 data elements $\{1, 2\}$. We can get a series of successful broadcasts L_m after $Maxfind(M, \langle x_0 \rangle)$ is performed. The first successful broadcast may be one of the two elements randomly, each case having probability $\frac{1}{2}$. If the first successful broadcast is 2, then $L_m = \langle x_0, 2 \rangle$, and the final sorted linked list is $\langle x_0, Sorting(M_1), 2 \rangle$, where $M_1 = \{x | x_0 < x < x_1\} = \{x | -\infty < x < 2\} = \{1\}$. If the first successful broadcast is 1, then $L_m = \langle x_0, 1, 2 \rangle$, and the final sorted linked list is $\langle x_0, Sorting(M_1), 1, Sorting(M_2), 2 \rangle$, where $M_1 = \{x | x_0 < x < x_1\} = \{x | -\infty < x < 1\} = NULL$, $M_2 = \{x | x_1 < x < x_2\} = \{x | 1 < x < 2\} = NULL$.

Martel *et al.* [8] showed the sorting problem can be solved in $O(n)$. They also proved that the expected time for finding first k largest numbers is $O(k + \log n)$. However, the constants associated with $O(n)$ and $O(k + \log n)$ bounds proved by Martel *et al.* [8] are not very tight.

On the high level view, our sorting algorithm has a similar approach as that proposed by Martel *et al.* [8]. On the low level view, our algorithm is based on the layer concept, thus it can be analyzed easily and a tighter bound can be obtained. And the analysis can be applied to other sorting algorithms which have the same high level approach under BCM.

4   Analysis of the Sorting Algorithm

In this section, we shall prove that the average time complexity of our sorting algorithm is $\Theta(n)$, where n is the number of input data elements. Suppose that there are n data elements held by at least n processors in which each processor holds at most one data element. Let T_n denote the average number of time slots, including conflict slots, empty slots and slots for successful broadcasts, required when the algorithm is executed. When there is zero or one input data element for the algorithm, one empty slot or one slot for successful broadcast is needed. Thus $T_0 = 1$ and $T_1 = 1$.

If $n = 2$, we have the following recursive formula:

$$T_2 = M_2 + R_2, \text{ where } R_2 = \left\{ \begin{array}{c} \frac{1}{2}[\quad T_1] \\ + \frac{1}{2}[T_0 + T_0] \end{array} \right\}. \tag{1}$$

We shall explain the above two equations. In the first equation, the first term M_2 is the average number of time slots required for the maximum finding algorithm while $n = 2$. The second term R_2 is the average number of time slots required for finishing the sorting after the maximum finding ends.

In the equation for R_2, the first successful broadcast may be either one of the 2 elements randomly, each case having probability $\frac{1}{2}$. The subterm of first term, T_1, arises when the first successful broadcast is the largest element. Thus, the layer between $x_0 = -\infty$ and the largest element contains one element, which is the second largest element, and needs T_1 time slots.

In second term, the subterm, $[T_0 + T_0]$, arises when the first successful broadcast is the second largest element and the second successful broadcast is the largest element. Thus, the layers between $x_0 = -\infty$, the second and the largest element are both empty and needs T_0 time slots in both layers.

Shiau and Yang [11] have proved that $M_2 = 5$, then we have

$$T_2 = 5 + \{\frac{1}{2}[1] + \frac{1}{2}[1 + 1]\} = \frac{13}{2}.$$

If $n = 3$, we also have the following recursive formula:

$$T_3 = M_3 + R_3, \text{ where } R_3 = \left\{ \begin{array}{c} \frac{1}{3}[\quad T_2] \\ + \frac{1}{3}[T_0 + T_1] \\ + \frac{1}{3}[R_2 + T_0] \end{array} \right\}. \tag{2}$$

In the equation for R_3, the first successful broadcast may be any one of the 3 elements randomly, each case having probability $\frac{1}{3}$. The subterm of first term, T_2, arises when the first successful broadcast is the largest element. Thus, the layer between $x_0 = -\infty$ and the largest element contains two elements, which is the second and third largest element, and needs T_2 time slots for sorting.

In the second term, the subterm, $[T_0 + T_1]$, arises when the first successful broadcast is the second largest element and the second successful broadcast is the largest element. Thus, the layer between the second and the largest element is empty and needs T_0 time slots. And the layer between $x_0 = -\infty$ and the second largest element contains one element and needs T_1 time slots.

In third term, the subterm, $[R_2 + T_0]$, arises when the first successful broadcast is the third largest element. Thus, after the first successful broadcast, there are two possible successful broadcast sequences, each having probability $\frac{1}{2}$. The first case is that the second successful broadcast is the largest data element. Thus, the layer between the first two successful broadcasts contains only one data element which is the second largest data element. The second case is that the second successful broadcast is the second largest data element, and the largest data element will be broadcast in the third successful broadcast. Thus, the layers between the first, second and third successful broadcasts, are both empty. Comparing with R_2 in Eq.(1), we can find that the above two cases are equivalent to those in R_2. Thus, R_2 represents the number of time slots required for the layers bounded by the sequence of successful broadcasts L_m in finding the maximum. Note that the time slots required for L_m is counted in M_3. Finally, the layer between $x_0 = -\infty$ and the third largest element is empty and needs T_0 time slots.

Generalizing Eq.(2), we have

$$T_n = M_n + R_n, \text{ where } R_n = \begin{cases} \frac{1}{n}[& T_{n-1}] \\ +\frac{1}{n}[T_0 & +T_{n-2}] \\ +\frac{1}{n}[R_2 & +T_{n-3}] \\ +\cdots & \\ +\frac{1}{n}[R_{k-1} & +T_{n-k}] \\ +\cdots & \\ +\frac{1}{n}[\ R_{n-1} & +T_0] \end{cases} . \tag{3}$$

In the above equations, M_n represents the average number of time slots required for finding the maximum among n data elements, and R_n represents the average number of time slots required for finishing sorting after the maximum is found.

In the equation for R_n, the first successful broadcast may be any one of the n elements randomly, each case having probability $\frac{1}{n}$. The subterm T_{n-1} of the second term arises when the first successful broadcast is the largest element. Thus, the layer between $x_0 = -\infty$ and the largest element contains $n-1$ elements and needs T_{n-1} time slots for sorting.

In the second term, the subterm, $T_0 + T_{n-2}$, arises when the first successful broadcast is the second largest element and the second successful broadcast is the largest element. Thus, the layer between the second and the largest element is empty and needs T_0 time slots. And the layer between $x_0 = -\infty$ and the second largest element contains $n - 2$ elements and needs T_{n-2} time slots for sorting.

In the kth term, the subterm, $R_{k-1} + T_{n-k}$, arises when the first successful broadcast is the kth largest element. Thus, R_{k-1} is the number of time slots required for finishing the sorting on the $k - 1$ elements after the maximum is found. And the layer between $x_0 = -\infty$ and the first successful broadcast element contains $n - k$ data elements and needs T_{n-k} time slots for sorting $n - k$ data elements.

Substituting $R_k = T_k - M_k$, $2 \leq k \leq n - 1$ into Eq.(3), we have

$$T_n = M_n + \left\{ \begin{array}{ll} \frac{1}{n}[& T_{n-1}] \\ + \frac{1}{n}[T_0 & +T_{n-2}] \\ + \frac{1}{n}[(T_2 - M_2) & +T_{n-3}] \\ + \cdots \\ + \frac{1}{n}[(T_{k-1} - M_{k-1}) & +T_{n-k}] \\ + \cdots \\ + \frac{1}{n}[(T_{n-1} - M_{n-1}) & +T_0] \end{array} \right\}. \tag{4}$$

Lemma 1.

$$\frac{1}{n+1}T_n - \frac{1}{3}T_2 = \sum_{3 \leq k \leq n} \frac{1}{k+1}(M_k - M_{k-1}).$$

Lemma 2.

$$4\left(\frac{1}{3} - \frac{1}{n+1}\right) \leq \sum_{3 \leq k \leq n} \frac{1}{k+1}(M_k - M_{k-1}) \leq 5\left(\frac{1}{3} - \frac{1}{n+1}\right).$$

Theorem 1.

$$\frac{7}{2}n - \frac{1}{2} \leq T_n \leq \frac{23}{6}n - \frac{7}{6}.$$

5   Analysis of the Generalization

T_n in Eq.(4) can be regarded as T_n^n. Therefore generalizing Eq.(4), we obtain

$$T_k^n = M_n + \frac{1}{n} \left\{ \begin{array}{ll} [& T_{k-1}^{n-1}] \\ + [T_0^0 & +T_{k-2}^{n-2}] \\ + [(T_2^2 - M_2) & +T_{k-3}^{n-3}] \\ + \cdots \\ + [(T_{k-3}^{k-3} - M_{k-3}) & +T_2^{n-(k-2)}] \\ + [(T_{k-2}^{k-2} - M_{k-2}) & +T_1^{n-(k-1)}] \\ + [(T_{k-1}^{k-1} - M_{k-1}) & +0] \\ + [(T_k^k - M_k) & +0] \\ + \cdots \\ + [(T_k^{n-1} - M_{n-1}) & +0] \end{array} \right\}.$$

In the above equations, the first term, M_n, is the average number of time slots required for the maximum finding algorithm on n data elements.

The second term, $\frac{1}{n}$, is one of the cases that each data element in M_n successfully broadcasts its value. Then, the first successful broadcast may be one of the n elements randomly, each case having probability $\frac{1}{n}$.

The third term $\left[T_{k-1}^{n-1}\right]$ is that the first successful broadcast is the largest data element in the progress of the maximum finding algorithm. Since we have the largest data element, we need to find the $k-1$ largest data elements among the $n-1$ data elements, which is presented by T_{k-1}^{n-1}.

The fourth term $\left[T_0^0 + T_{k-2}^{n-2}\right]$ is that the first successful broadcast is the second largest data element. Obviously the second successful broadcast is the largest data element. T_0^0 represents a $NULL$ layer because it is empty between the largest and second largest data elements. And the subterm T_{k-2}^{n-2} represents that we need to find the $k-2$ largest data elements among $n-2$ data elements after the largest and second largest data elements broadcasting.

The $(k+3)$th term $\left[(T_{k-1}^{k-1} - M_{k-1}) + 0\right]$ is that the first successful broadcast is the kth largest data element. Thus, $(T_{k-1}^{k-1} - M_{k-1})$ is the number of time slots required for finishing the sorting on the $k-1$ elements after the maximum is found. And since the first k, $k \geq 1$, largest elements was found in the layers upper than the first successful broadcast element, thus it dose not need any time slots for sorting data elements in the lower layer.

We omit the explanation of the rest of the equation, since it is quite similar to the above cases.

Lemma 3. *Let $L_n = M_n - M_{n-1}$, then*

$$T_k^n - T_k^{n-1} = L_n + \frac{1}{n}L_{n-1} + \frac{1}{n-1}L_{n-2} + \cdots + \frac{1}{n-(k-2)}L_{n-(k-1)}.$$

Lemma 4.

$$T_k^k + 4\left[\ln\left(n - (k-2)\right) - \ln 2\right] \leq T_k^n \leq T_k^k + 5\ln\left(n - (k-1)\right).$$

Theorem 2.

$$\frac{7}{2}k + 4\ln\left(n - (k-2)\right) - \left(\frac{1}{2} + 4\ln 2\right) \leq T_k^n \leq \frac{23}{6}k + 5\ln\left(n - (k-1)\right) - \frac{7}{6}.$$

Martel *et al.* [8] showed that the expected time for finding the first k largest numbers is $O(k + \log n)$. However, the constants associated with $O(k + \log n)$ bounds proved by Martel *et al.* [8] are not very tight. And they did not point out the relationship between the two special cases maximum finding and sorting.

By Theorem 2, we prove that the average time complexity of finding the first k numbers is $\Theta(k + \log(n - k))$. And the connection between the two special cases, maximum finding and sorting, is built by the analysis of our generalization.

6   Conclusion

The layer concept [13] can help us to reduce conflict resolution when an algorithm is not conflict-free under BCM. In this paper, we apply the layer concept to solve the sorting problem and get a tighter bound. The total average number of time slots, including conflict slots, empty slots and slots for successful broadcasts, is $\Theta(n)$. And the generalization of our sorting algorithm can be analyzed that the average time complexity for selecting k largest numbers is $\Theta(k + \log(n - k))$.

On the high level view, our sorting algorithm has a similar approach as that proposed by Martel et al. [8]. On the low level view, our algorithm is based on the layer concept, thus it can be analyzed easily and a tighter bound can be obtained.

We are interested in finding other maximum finding algorithms which can be applied into our sorting approach and can get better performance. That is one of our future works. And the other future work is to find the native bound of the sorting approach under BCM.

References

1. Capetanakis, J.I.: Tree algorithms for packet broadcast channels. IEEE Transactions on Information Theory, **25(5)** (May 1997) 505–515
2. Dechter, R., Kleinrock, L.: Broadcast communications and distributed algorithms. IEEE Transactions on Computers, **35(3)** (Mar. 1986) 210–219
3. Grabner, P.J., Prodinger, H.: An asymptotic study of a recursion occurring in the analysis of an algorithm on broadcast communication. Information Processing Letters, **65** (1998) 89–93
4. Huang, J.H., Kleinrock, L.: Distributed selectsort sorting algorithm on broadcast communication. Parallel Computing, **16** (1990) 183–190
5. Levitan, S.: Algorithms for broadcast protocol multiprocessor. Proc. of 3rd International Conference on Distributed Computing Systems, (1982) 666–671
6. Levitan, S.P., Foster C.C.: Finding an extremum in a network. Proc. of 1982 International Symposium on Computer Architechure, (1982) 321–325
7. Martel, C.U.: Maximum finding on a multi access broadcast network. Information Processing Letters, **52** (1994) 7–13
8. Martel, C.U., Moh, M.: Optimal prioritized conflict resolution on a multiple access channel. IEEE Transactions on Computers, **40(10)** (Oct. 1991) 1102–1108
9. Martel, C.U., Moh W.M., Moh T.S.: Dynamic prioritized conflict resolution on multiple access broadcast networks. IEEE Transactions on Computers, **45(9)** (1996) 1074–1079
10. Ramarao, K.V.S.: Distributed sorting on local area network. IEEE Transactions on Computers, **C-37(2)** (Feb. 1988) 239–243
11. Shiau, S.H., Yang, C.B.: A fast maximum finding algorithm on broadcast communication. Information Processing Letters, **60** (1996) 81–96
12. Shiau, S.H., Yang, C.B.: The layer concept and conflicts on broadcast communication. Journal of Chang Jung Christian University, **2(1)** (June 1998) 37–46
13. Yang, C.B.: Reducing conflict resolution time for solving graph problems in broadcast communications. Information Processing Letters, **40** (1991) 295–302
14. Yang, C.B.,Lee, R.C.T., Chen, W.T.: Parallel graph algorithms based upon broadcast communications. IEEE Transactions on Computers, **39(12)** (Dec. 1990) 1468–1472
15. Yang, C.B., Lee, R.C.T., Chen, W.T.: Conflict-free sorting algorithm broadcast under single-channel and multi-channel broadcast communication models. Proc. of International Conference on Computing and Information, (1991) 350–359

Efficient List Ranking Algorithms on Reconfigurable Mesh*

Sung-Ryul Kim and Kunsoo Park

School of Computer Science and Engineering,
Seoul National University,
Seoul 151-742, Korea
{kimsr, kpark}@theory.snu.ac.kr

Abstract. List ranking is one of the fundamental techniques in parallel algorithm design. Hayashi, Nakano, and Olariu proposed a deterministic list ranking algorithm that runs in $O(\log^* n)$ time and a randomized one that runs in $O(1)$ expected time, both on a reconfigurable mesh of size $n \times n$. In this paper we show that the same deterministic and randomized time complexities can be achieved using only $O(n^{1+\epsilon})$ processors, where ϵ is an arbitrary positive constant < 1. To reduce the number of processors, we adopt a reconfigurable mesh of high dimensions and develop a new technique called *path embedding*.

1 Introduction

The *reconfigurable mesh* [11] or *RMESH* combines the advantages of the mesh with the power and flexibility of a dynamically reconfigurable bus structure. Recently, efficient algorithms for the RMESH have been developed for graph problems [2,3,15,12], image problems [8,12], geometric problems [6,11,12], and some other fundamental problems in computer science [2,4,7,?].

The problem of *list ranking* is to determine the *rank* of every node in a given linked list of n nodes. The *rank* of a node is defined to be the number of nodes following it in the list. In weighted list ranking every node of a linked list has a weight. The problem of weighted list ranking is to compute, for every node in the list, the sum of weights of the nodes following it in the list. Since the unweighted version is a special case of the weighted version of the list ranking problem, we will solve the weighted version.

The list ranking problem has been studied on RMESH. On an RMESH of size $n \times n$, an $O(\log n)$ time algorithm for the list ranking problem using a technique called *pointer jumping* [16] is implicit in [11,14]. Using a complicated algorithm, Olariu, Schwing, and Jhang [14] showed that the problem can be solved in $O\left(\frac{\log n}{\log m}\right)$ time on an RMESH of size $mn \times n$ where $m \leq n$. Hayashi, Nakano, and Olariu proposed a deterministic list ranking algorithm [5] running in $O(\log^* n)$ time and a randomized one running in $O(1)$ time, both on an RMESH of size $n \times n$.

---

* This work was supported by the Brain Korea 21 Project.

D.-Z. Du et al. (Eds.): COCOON 2000, LNCS 1858, pp. 262–271, 2000.

In this paper we generalize the result in [5] to d-dimensional RMESH for $d \geq 3$. As a result, we show that the same deterministic and randomized time complexities as in [5] can be achieved using only $O(n^{1+\epsilon})$ processors, where $0 < \epsilon < 1$. In [5], a technique called *bus embedding* is used. Bus embedding is easy on a 2-dimensional RMESH but it does not seem to be easy on a d-dimensional RMESH for $d \geq 3$. We develop a new technique, which we call *path embedding*, that can be used to implement the bus embedding on an RMESH of high dimensions.

2   RMESH

An RMESH of size $n \times n$ consists of an $n \times n$ array of processors arranged as a 2-D grid and connected by a reconfigurable bus system. See Fig. 1 for an example. Processors of the $n \times n$ array are denoted by $P_{i,j}$, $1 \leq i, j \leq n$. Each processor has four ports, denoted by N, S, E, and W. Ports E and W are built along dimension x and ports N and S are built along dimension y. Through these ports processors are connected to the reconfigurable bus.

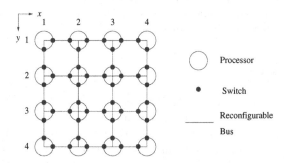

Fig. 1. The RMESH architecture

Different configurations of the reconfigurable bus system can be established by setting different local connections among ports within each processor. For example, if each processor connects E and W (N and S) ports together, row buses (column buses) are established. In a bus configuration, maximal connected components of the configuration are called *subbuses* of the configuration.

In one time step a processor may perform a single operation on words of size $O(\log n)$, set any of its switches, or write data to or read data from the subbus. We assume that in each subbus at most one processor is allowed to write to the subbus at one time. As in [9,10,11], we assume that data placed on a subbus reaches all processors connected to the subbus in unit time.

We assume that each processor knows its row and column indices. In addition, we also want processors to know their ranks in some total ordering on the mesh. Of the widely used orderings we describe the *row-major* ordering and the *snake-like* ordering. In the row-major ordering, we number processors left to right in

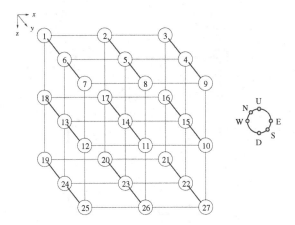

Fig. 2. A $3 \times 3 \times 3$ RMESH indexed in the snake-like ordering

each row, beginning with the top row and ending with the bottom row. In the snake-like ordering, we simply reverse every other row, so that processors with consecutive indices are adjacent. We can assume that a processor initially knows its positions in the row-major ordering and the snake-like ordering because they can be computed from its indices in constant time.

Reconfigurable meshes of high dimensions can be constructed in a similar way. For example, in a 3-D RMESH, processors are denoted by $P_{i,j,k}$ and each processor has six ports, denoted by U, D, N, S, E, and W, where the new ports U and D are built along dimension z. Also, the row-major and the snake-like orderings can be generalized to high dimensions. An example $3 \times 3 \times 3$ RMESH indexed in the snake-like ordering is shown in Fig. 2.

In our algorithms we will make use of the algorithm in [1] that sorts n elements in constant time on an RMESH of size $n^{1+\epsilon}$ where ϵ is an arbitrary positive constant less than or equal to 1. Also in our algorithms we will solve the following version of the packing problem: Given an n-element set $X = \{x_1, x_2, \ldots, x_n\}$ and a collection of predicates $P_1, P_2, \ldots, P_m$, where each x_i satisfies exactly one of the predicates and the predicate that each x_i satisfies is already known, permute the elements of X in such way that for every $1 \le j \le m$ the elements of X satisfying P_j occur consecutively. It is clear that this version of the packing problem is easier than sorting and can be solved using the sorting algorithm.

3   Previous Algorithms

In this section we describe the deterministic and randomized algorithms by Hayashi, Nakano, and Olariu [5]. Later, we will use these algorithms as subroutines, so we describe only the parts of the algorithms relevant to this paper.

In [5], the input list L is given by the pairs $(i, l(i))$ at $P_{1,i}$, $1 \le i \le n$, where $l(i)$ is the node following node i in L. The pair (t, t) is given for the last node t in L. In the weighted list ranking problem the weight w_i associated with each

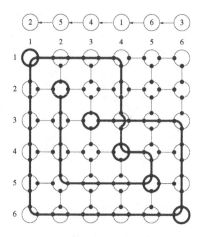

Fig. 3. An example bus embedding

node i is also given to $P_{1,i}$. For a given linked list $L = \{(i, l(i)) : 1 \leq i \leq n\}$, the corresponding bus configuration can be obtained by assigning node i to the diagonal processor $P_{i,i}$. See Fig. 3. The fact that $l(i)$ follows i in the list is captured by a subbus connecting $P_{i,i}$ and $P_{l(i),l(i)}$. The subbus is formed by connecting $P_{i,i}$ and $P_{i,l(i)}$ by a row bus and then connecting $P_{i,l(i)}$ and $P_{l(i),l(i)}$ by a column bus. The bus configuration formed in this way is called the *bus embedding* of L. A bus embedding for L can be easily formed with two parallel broadcast steps.

Consider a subset S of L. The *S-shortcut* of L is the linked list obtained from L by retaining the nodes in S only. Specifically, an ordered pair (p, q) belongs to the S-shortcut if both p and q belong to S and no nodes of L between p and q belong to S. If (p, q) belongs to the S-shortcut, the q-*sublist* is the sublist of L starting at $l(p)$ and ending at q. A subset S of L is called a *good ruling set* if it satisfies the following conditions:

1. S includes the last node of L,
2. S contains at most $n^{2/3}$ nodes, and
3. No p-sublist for $p \in S$ contains more than $n^{1/2}$ nodes.

It is shown in [5] that if a good ruling set of L is available, then the list ranking problem can be solved in constant time. Next, we describe how a good ruling set can be obtained.

3.1   The Deterministic Good Ruling Set Algorithm

In [5], an algorithm for ranking an n-node list on an RMESH of size $2n \times 2n$ is given and the constant factor in the number of processors can be removed by reducing the input size using pointer jumping. For convenience, we will describe the algorithm on an RMESH of size $2n \times 2n$. The algorithm consists of $O(\log^* n)$

stages and each stage obtains a smaller shortcut using the shortcut of the previous stage. If we allocate $k \times k$ processors for each node in the list, we can have k parallel buses between two nodes (which are called a *bundle* of k buses). The idea is that as the shortcuts get smaller, larger bundles of buses are available. And when a larger bundle is available, a much smaller shortcut can be obtained in the next stage. Initially, a bundle consisting of 2 buses are available.

We need a few definitions here. A *k-sample* of a linked list L is a set of nodes obtained by retaining every k-th node in L, that is, nodes whose unweighted ranks are $0, k, 2k, \ldots, \lfloor \frac{n-1}{k} \rfloor k$. Let L/k denote the shortcut consisting of a k-sample.

linked list

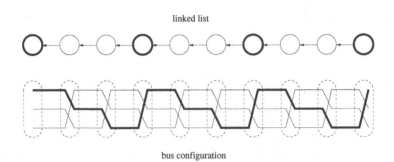

bus configuration

Fig. 4. Illustrating the concept of bundle of buses with cyclic shift

We describe how L/k can be obtained from L using a bundle of k buses. Assume that a bundle of k buses are available. See Fig. 4 for a conceptual description. The k buses are shifted cyclically as the buses pass each node, as illustrated in Fig. 4. If a signal is sent from the topmost bus in the last node of L, the signal passes through the topmost bus of every k-th subsequent node. The desired k-sample consists of such nodes.

Assume that a bundle of k buses is available for each node in a sublist. We use the following *parallel prefix remainder* technique. Let $p_1(=2), p_2(=3), p_3(= 5), \ldots, p_q$ be the first q prime numbers such that $p_1 + p_2 + \cdots + p_q \leq k < p_1+p_2+\cdots+p_q+p_{q+1}$. Since k buses are available, we can form q separate bundles of $p_1, p_2, \ldots,$ and p_q buses. Hence, for a list L, q linked lists $L/p_1, L/p_2, \ldots, L/p_q$ can be computed in constant time. Having done that, we retain in the sample only the nodes common to all the q linked lists $L/p_1, L/p_2, \ldots, L/p_q$. In other words, we select node p only if its rank $r(p)$ in L satisfies:

$$r(p) \bmod p_1 = r(p) \bmod p_2 = \cdots = r(p) \bmod p_q = 0.$$

For such node p, $r(p) \bmod p_1 p_2 \cdots p_q = 0$. Hence, using the bundle of k buses, we can obtain $L/(p_1 p_2 \cdots p_q)$ from L in constant time, where $p_1+p_2+\cdots+p_q \leq k < p_1 + p_2 + \cdots + p_q + p_{q+1}$.

Now, [5] estimates that $p_1 p_2 \cdots p_q = 2^{\Theta(\sqrt{k \log k})}$. Let $K(t)$ be an integer such that $L/K(t)$ is computed in the t-th iteration. Then $K(t+2) = 2^{\Theta(\sqrt{K(t+1) \log(k+1)})}$

$> 2^{\Theta(K(t))}$ and $K(1) = 2$ holds. Since $K(t) \geq \sqrt{n}$ is sufficient, the number of iterations is bounded by $O(\log^* n)$.

3.2 The Randomized Good Ruling Set Algorithm

The randomized good ruling set algorithm in [5] is the following.

> **Step 1.** Each $P_{i,i}$ randomly selects node i as a sample in S with probability $n^{-5/12}$.
>
> **Step 2.** Check whether S is a good ruling set. If S is a good ruling set, the algorithm terminates; otherwise repeat Step 1.

It is easy to see that Step 2 of the algorithm can be implemented to work in constant time using the bus embedding and packing. It is proven in [5] that the expected number of selections before the algorithm terminates is $O(1)$.

4   Reducing the Number of Processors

In this section we show how we can reduce the number of processors to $O(n^{1+\epsilon})$ by adopting an RMESH of high dimensions. First, fix a constant d, which is a natural number larger than 1. We will implement the deterministic and randomized algorithms on a $(d+1)$-dimensional RMESH of size $cn^{1/d} \times cn^{1/d} \times \cdots \times cn^{1/d}$ for some constant c, which consists of $c^{d+1}n^{1+1/d}$ processors.

We first enhance each processor in RMESH so that there are C ports in each direction where C is a constant. We will call a processor enhanced in such a way a *superprocessor*. For example, in 2-D, a processor has C ports in each direction named $E(1), E(2), \ldots, E(C), W(1), \ldots, S(C)$. It should be noted that this is not a drastic diversion from the definition of RMESH since a constant number of regular processors can simulate a superprocessor if C is a constant. We will call an RMESH consisting of superprocessors a *super-RMESH*. We denote by $SP_{i,j,\ldots,k}$ the superprocessor at location $(i, j, \ldots, k)$ in a super-RMESH. Also, we denote by $SP(i)$ the superprocessor whose index is i in the snake-like ordering in a super-RMESH.

We assume that the input is given in the first n superprocessors, indexed in the snake-like ordering, of a super-RMESH of size $n^{1/d} \times n^{1/d} \times \cdots n^{1/d}$, which can be simulated by an RMESH of size $cn^{1/d} \times cn^{1/d} \times \cdots \times cn^{1/d}$ for some constant c.

Given a list L, a subset S of L is a d-good ruling set of L if

1. S includes the last node of L,
2. S contains at most $n^{3/2d}$ nodes, and
3. no p-sublist for $p \in S$ contains more than $2^{(d-1)/d}$ nodes.

Each of our two algorithms consists of two parts. In the first part, we obtain the d-good ruling set of L using the bus embedding. In the second part, we rank L recursively, using the d-good ruling set obtained in the first part. We will describe the second part first.

To use recursion in the second part, we extend the list ranking problem slightly so that a number of lists $L_1, L_2, \ldots, L_m$ are to be individually ranked, where the total number of nodes do not exceed n. On RMESH, the nodes are given one node per processor in processors $P(1)$ through $P(n)$ so that the nodes in each list appear consecutively in the snake-like ordering with the boundaries between the lists marked.

We assume, for simplicity, that exactly n nodes are given in the extended list ranking problem. Let s_j and t_j be the first and the last node in each L_j, respectively. To reduce the problem to normal list ranking, we connect t_j with s_{j+1} for each $1 \leq j < m$. Then we rank the resulting list by a normal list ranking algorithm. To obtain ranks in each list, we subtract the rank of t_j from the rank of each node in L_j. The details are omitted.

4.1   The Second Part

Since the second parts of the deterministic and the randomized algorithms are analogous, we describe the second part of the deterministic algorithm only.

Assume we have a shortcut S of L. We first show that if we can rank all p-sublists for $p \in S$ in time T and rank the shortcut S in time T', then we can rank L in time $T + T' + D$, where D is a constant independent of L. First, rank the p-sublists for $p \in S$ in time T. Let s_p be the first node in a p-sublist. Let $r_p(s_p)$ be the weighted rank of s_p in the p-sublist and let W_p be $r_p(s_p) + w$, where w is the weight of s_p. Rank S using W_p as the weight of p in S. Let R_p be the computed rank of p. Now, use packing to move the nodes in a sublist to adjacent superprocessors and broadcast R_p among the superprocessors holding the nodes in the p-sublist. Now, the rank of node i in the p-sublist is $r_p(i) + R_p$, where $r_p(i)$ is the rank of node i in the p-sublist.

Assume that $d = 2$. We have n nodes on a super-RMESH of size $n^{1/2} \times n^{1/2} \times n^{1/2}$. Assume that we have obtained a 2-good ruling set S in the first part. Now, S contains at most $n^{3/4}$ nodes and each p-sublist for $p \in S$ contains at most $n^{1/2}$ nodes. All the p-sublists can be ranked simultaneously in $O(\log^* n)$ time by distributing the sublists to $n^{1/2}$ submeshes of size $n^{1/2} \times n^{1/2}$ using the deterministic algorithm in [5].

All we have to do now is to rank S. We apply the d-good ruling set algorithm to S and obtain a 2-good ruling set S' of S. Now, the p'-sublists for $p' \in S'$ can be ranked as before using the deterministic algorithm in [5]. S' contains at most $n^{9/16}$ nodes. To rank S', we apply the d-good ruling set algorithm once again to S' and obtain a 2-good ruling set S'' of S'. Now, S'' contains at most $n^{27/64} < n^{1/2}$ nodes and S'' and the sublists of S' and S'' can be ranked using the deterministic algorithm in [5]. The time for the second part is $O(\log^* n)$.

Assume that $d > 2$. Assume that we have obtained the d-good ruling set S in the first part. Each p-sublist for $p \in S$ contains at most $n^{(d-1)/d}$ nodes and hence all the sublists can be recursively ranked simultaneously in $O(\log^* n)$ time by distributing the sublists to a d-dimensional super-RMESH of size $n^{1/d} \times n^{1/d} \times \cdots \times n^{1/d}$. Also, S has at most $n^{3/2d} < n^{(d-1)/d}$ nodes and can be ranked recursively. The time for the second part is $O(\log^* n)$.

4.2   Bus Embedding

The main difficulty in generalizing the list ranking algorithms is that the bus embedding itself is not easy in high dimensions. To construct a bus embedding, we develop a new technique called *path embedding*, which can be used along with any *data movement* algorithm. By a data movement algorithm we mean an algorithm where a data item is treated as one object and nothing is assumed about its internal representation. The number of processors increases slightly with path embedding but the technique introduces only constant factor in the number of processors if the data movement algorithm works in constant time.

Algorithm Path-Embed

Stage 1. Let C be the number of steps of the data movement algorithm. Run the data movement algorithm while using the j-th ports in each direction of a superprocessor in the j-th step of the data movement algorithm. Each superprocessor retains the bus configuration and remembers every bus operation it performs.

Stage 2. If a superprocessor read an item in the j-th step and wrote the item in the k-th step, $j < k$, the superprocessor connects its $X(j)$ and $Y(k)$ ports together, where $X(j)$ is the direction it read the node in the j-th step and $Y(k)$ is the direction it wrote the node in the k-th step. Repeat for all possible values of j and k.

Now we can implement the bus embedding in constant time as follows. We assume that there are C ports in each direction of a superprocessor, where C is the number of steps of the sorting algorithm.

Algorithm Embed

Step 1. Identify the first node in the list by sorting the nodes $(i, l(i))$ by $l(i)$ and checking for the unique integer between 1 and n that is not present among the $l(i)$'s. Let $(s, l(s))$ be the first node.

Step 2. Temporarily replace the last node (t, t) by (t, s), that is $l(t) = s$. Run algorithm **Path-Embed** with the sorting algorithm that sorts the nodes $(i, l(i))$ by $l(i)$ (into the snake-like ordering).

Step 3. For each i that is not the first or the last node in the list, $SP(i)$ connects its $X(1)$ port and $Y(C)$ port, where $X(1)$ is the direction where it wrote its node, $(i, l(i))$, in the first step of the sort and $Y(C)$ is the direction it read i in the final step of the sort.

4.3   The Deterministic d-Good Ruling Set Algorithm

In this subsection, we assume that n nodes are given on a $(d + 1)$-dimensional super-RMESH of size $(2n)^{1/d} \times (2n)^{1/d} \times \cdots \times (2n)^{1/d}$. We first show that we

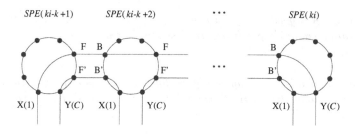

Fig. 5. Implementing the cyclic shift modulo k with superprocessors

can obtain a d-good ruling set in $O(\log^* n)$ time on a $(d+1)$-dimensional super-RMESH of size $(2n)^{1/d} \times (2n)^{1/d} \times \cdots \times (2n)^{1/d}$. First we use sorting to allocate two superprocessors $SP(2i-1)$ and $SP(2i)$ for each node i.

Assume that we have k superprocessors $SP(ki-k+1)$ through $SP(ki)$ allocated for each node (initially $k=2$). $SP(ki)$ will be called the topmost superprocessor for node i. Now, a bundle of k buses and a cyclic shift modulo k can be constructed by modifying Algorithm Embed. To implement the cyclic shift modulo k we add four more ports F, F', B, and B' to each superprocessor such that for $SP(i)$, F and F' connects $SP(i)$ with $SP(i+1)$ and B and B' connects $SP(i)$ with $SP(i-1)$. See Fig. 5 for an illustration of the cyclic shift.

Now we can implement the parallel prefix remainder technique described in section 3 using the cyclic shift. We can obtain a d-good ruling set in $O(\log^* n)$ time on a $(d+1)$-dimensional super-RMESH of size $(2n)^{1/d} \times (2n)^{1/d} \times \cdots \times (2n)^{1/d}$.

Theorem 1. *A d-good ruling set can be computed in $O(\log^* n)$ time on a $(d+1)$-dimensional super-RMESH of size $(2n)^{1/d} \times (2n)^{1/d} \times \cdots \times (2n)^{1/d}$, where d is a constant.*

Theorem 2. *The list ranking can be performed in $O(\log^* n)$ time on a $(d+1)$-dimensional RMESH of size $cn^{1/d} \times cn^{1/d} \times \cdots \times cn^{1/d}$, where c and d are constants.*

4.4   The Randomized d-Good Ruling Set Algorithm

Our randomized d-good ruling set algorithm on a $(d+1)$-dimensional super-RMESH is the following. Note that Step 2 can be performed in constant time using the bus embedding and packing.

Step 1. Each $SP(i)$ randomly selects node i as a sample in S with probability $n^{(5-4d)/4d}$.

Step 2. Check whether S is a d-good ruling set. If S is a d-good ruling set, the algorithm terminates; otherwise repeat Step 1.

The proofs of the following theorems are omitted.

Theorem 3. *A d-good ruling set can be computed in constant expected time on a $(d + 1)$-dimensional super-RMESH of size $n^{1/d} \times n^{1/d} \times \cdots \times n^{1/d}$, where d is a constant.*

Theorem 4. *The list ranking can be performed in constant expected time on a $(d + 1)$-dimensional RMESH of size $cn^{1/d} \times cn^{1/d} \times \cdots \times cn^{1/d}$, where c and d are constants.*

References

1. Chen, Y. C. and Chen, W. T.: Constant Time Sorting on Reconfigurable Meshes. IEEE Transactions on Computers. Vol. 43, No. 6 (1994) 749–751
2. Chen, G. H. and Wang, B. F.: Sorting and Computing Convex Hull on Processor Arrays with Reconfigurable Bus System. Information Science. 72 (1993) 191–206
3. Chen, G. H., Wang, B. F., and Lu, C. J.: On Parallel Computation of the Algebraic Path Problem. IEEE Transactions on Parallel and Distributed Systems. Vol. 3, No. 2, Mar. (1992) 251–256
4. Hao, E., MacKenzie, P. D., and Stout, Q. F.: Selection on the Reconfigurable Mesh. Proc. of fourth Symposium on the Frontiers of Massively Parallel Computation. (1992) 38–45
5. Hayashi, T., Nakano, K., and Olariu, S.: Efficient List Ranking on the Reconfigurable Mesh, with Applications. Proc. of 7th International Simposium on Algorithms and Computation, ISAAC'96. (1996) 326-335
6. Jang, J. and Prasanna, V. K.: Efficient Parallel Algorithms for Geometric Problems on Reconfigurable Mesh. Proc. of International Conference on Parallel Processing. 3 (1992) 127–130
7. Jang, J. and Prasanna, V. K.: An Optimal Sorting Algorithm on Reconfigurable Mesh. International Parallel Processing Symposium. (1992) 130–137
8. Jenq, J. F. and Sahni, S.: Reconfigurable Mesh Algorithms for the Hough Transform. Proc. of International Conference on Parallel Processing. 3 (1991) 34–41
9. Li, H. and Maresca, M.: Polymorphic-torus Network. IEEE Transactions on Computers. 38 (1989) 1345–1351
10. Maresca, M.: Polymorphic Processor Arrays. IEEE Transactions on Parallel and Distributed Systems. 4 (1993) 490–506
11. Miller, R., Prasanna Kumar, V. K., Reisis, D. I., and Stout, Q. F.: Parallel Computations on Reconfigurable Mesh. Technical Report # 229. USC, Mar. (1922)
12. Miller, R. and Stout, Q. F.: Efficient Parallel Convex Hull Algorithms. IEEE Transactions on Computer. Vol 37, No. 12, Dec. (1988) 1605–1618
13. Niven, I., Zuckerman, H. S., and Montgomery, H. L.: An Introduction to the Theory of Numbers, Fifth Edition. (1991) John Wiley & Sons
14. Olariu, S., Schwing, J. L., and Jhang, J.: Fundamental Algorithms on Reconfigurable Meshes. Proc. of 29th Annual Allerton Conference on Communication, Control, and Computing. (1991) 811–820
15. Wang, B. F. and Chen, G. H.: Constant Time Algorithms for the Transitive Closure and Some Related Graph Problems on Processor Arrays with Reconfigurable Bus System. IEEE Transactions on Parallel and Distributed Systems. Vol. 1, No. 4, Oct. (1990) 500–507
16. Wylie, J. C.: The Complexity of Parallel Computation. Doctoral Thesis, Cornell University. (1979)

Tripods Do Not Pack Densely

Alexandre Tiskin

Department of Computer Science, University of Warwick,
Coventry CV4 7AL, United Kingdom
tiskin@dcs.warwick.ac.uk

Abstract. In 1994, S. K. Stein and S. Szabó posed a problem concerning simple three-dimensional shapes, known as semicrosses, or tripods. By definition, a tripod is formed by a corner and the three adjacent edges of an integer cube. How densely can one fill the space with non-overlapping tripods of a given size? In particular, is it possible to fill a constant fraction of the space as the tripod size tends to infinity? In this paper, we settle the second question in the negative: the fraction of the space that can be filled with tripods of a growing size must be infinitely small.

1 Introduction

In [SS94, Ste95], S. K. Stein and S. Szabó posed a problem concerning simple three-dimensional polyominoes, called "semicrosses" in [SS94], and "tripods" in [Ste95].

A *tripod* of size n is formed by a corner and the three adjacent edges of an integer $n \times n \times n$ cube (see Figure 1). How densely can one fill the space with non-overlapping tripods of size n? In particular, is it possible to keep a constant fraction of the space filled as $n \to \infty$? Despite their simple formulation, these two questions appear to be yet unsolved.

In this paper, we settle the second question in the negative: the density of tripod packing has to approach zero as tripod size tends to infinity. It is easy to prove (see [SS94]) that this result implies similar results in dimensions higher than three.

Instead of dealing with the problem of packing tripods in space directly, we address an equivalent problem, also introduced in [SS94, Ste95]. In this alternative setting, tripods of size n are to be packed without overlap, so that their corner cubes coincide with one of the unit cells of an $n \times n \times n$ cube. We may also assume that all tripods have the same orientation. If we denote by $f(n)$ the maximum number of non-overlapping tripods in such a packing, then the maximum fraction of space that can be filled with arbitrary non-overlapping tripods is proportional to $f(n)/n^2$ (see [SS94] for a proof). The only known values of function f are $f(1)$ up to $f(5)$: 1, 2, 5, 8, 11. It is easy to see that for all n, $n \le f(n) \le n^2$. Stein and Szabó's questions are concerned with the upper bound of this inequality. They can now be restated as follows: what is the asymptotic behaviour of $f(n)/n^2$ as $n \to \infty$? In particular, is $f(n) = o(n^2)$, or is $f(n)$ bounded away from 0?

D.-Z. Du et al. (Eds.): COCOON 2000, LNCS 1858, pp. 272–280, 2000.
© Springer-Verlag Berlin Heidelberg 2000

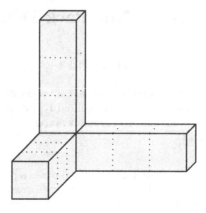

Fig. 1. A tripod of size $n = 4$

In this paper we show that, in fact, $f(n) = o(n^2)$, so the fraction of the space that can be filled with tripods of a growing size is infinitely small. Our proof methods are taken from the domain of extremal graph theory. Our main tools are two powerful, widely applicable graph-theoretic results: Szemerédi's Regularity Lemma, and the Blow-up Lemma.

2 Preliminaries

Throughout this paper we use the standard language of graph theory, slightly adapted for our convenience. A graph G is defined by its set of nodes $V(G)$, and its set of edges $E(G)$. All considered graphs are simple and undirected. *The size* of a graph G is the number of its nodes $|V(G)|$; the *edge count* is the number of its edges $|E(G)|$. A graph H is *a subgraph* of G, denoted $H \subseteq G$, iff $V(H) \subseteq V(G)$, $E(H) \subseteq E(G)$. Subgraph $H \subseteq G$ is *a spanning subgraph* of G, denoted $H \sqsubseteq G$, iff $V(H) = V(G)$.

A complete graph on r nodes is denoted K_r. The term *k-partite graph* is synonymous with "*k-coloured*". We write *bipartite* for "2-partite", and *tripartite* for "3-partite". Complete bipartite and tripartite graphs are denoted K_{rs} and K_{rst}, where r, s, t are sizes of the colour classes. Graph $K_2 = K_{11}$(short for $K_{1,1}$) is a single edge; we call its complement $\bar{K}_2$ (an empty graph on two nodes) *a nonedge*. Graph $K_3 = K_{111}$ is called *a triangle*; graph K_{121} is called *a diamond* (see Figure 2). We call a k-partite graph *equi-k-partite*, if all its colour classes are of equal size.

The density of a graph is the ratio of its edge count to the edge count of a complete graph of the same size: if $G \sqsubseteq K_n$, then

$$\text{dens}(G) = |E(G)|/|E(K_n)| = |E(G)|/\binom{n}{2}.$$

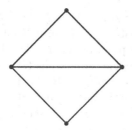

Fig. 2. The diamond graph K_{121}

The *bipartite density* of a bipartite graph $G \sqsubseteq K_{mn}$ is

$$\mathrm{dens}_2(G) = |E(G)|/|E(K_{mn})| = |E(G)|/(mn).$$

Similarly, the *tripartite density* of a tripartite graph $G \sqsubseteq K_{mnp}$ is

$$\mathrm{dens}_3(G) = |E(G)|/|E(K_{mnp})| = |E(G)|/(mn + np + pm).$$

Let H be an arbitrary graph. Graph G is called H-*covered*, if every edge of G belongs to a subgraph isomorphic to H. Graph G is called H-*free*, if G does not contain any subgraph isomorphic to H. In particular, we will be interested in triangle-covered diamond-free graphs.

Let us now establish an upper bound on the density of an equitripartite diamond-free graph.

Lemma 1. *The tripartite density of an equitripartite diamond-free graph G is at most* $3/4$.

Proof. Denote $|V(G)| = 3n$. By Dirac's generalisation of Turán's theorem (see e.g. [Gou88, p. 300]), we have $|E(G)| \le (3n)^2/4$. Since $|E(K_{nnn})| = 3n^2$, the theorem follows trivially. $\square$

The upper bound of $3/4$ given by Lemma 1 is not the best possible. However, that bound will be sufficient to obtain the results of this paper. In fact, any constant upper bound strictly less than 1 would be enough.

3   The Regularity Lemma and the Blow-Up Lemma

In most definitions and theorem statements below, we follow [KS96, Kom99, Die00].

For a graph G, and node sets $X, Y \subseteq V(G)$, $X \cap Y = \emptyset$, we denote by $G(X, Y) \subseteq G$ the bipartite subgraph obtained by removing from G all nodes except those in $X \cup Y$, and all edges adjacent to removed nodes. Let F be a bipartite graph with colour classes A, B. Given some $\epsilon > 0$, graph F is called ϵ-*regular*, if for any $X \subseteq A$ of size $|X| \ge \epsilon \cdot |A|$, and any $Y \subseteq B$ of size $|Y| \ge \epsilon \cdot |B|$, we have

$$|\mathrm{dens}_2(F(X, Y)) - \mathrm{dens}_2(F)| \le \epsilon.$$

Let G denote an arbitrary graph. We say that G admits *an ϵ-partitioning of order m*, if $V(G)$ can be partitioned into m disjoint subsets of equal size, called *supernodes*, such that for all pairs of supernodes A, B, the bipartite subgraph $G(A, B)$ is ϵ-regular. The ϵ-regular subgraphs $G(A, B)$ will be called *superpairs*.

For different choices of supernodes A, B, the density of the superpair $G(A, B)$ may differ. We will distinguish between superpairs of "low" and "high" density, determined by a carefully chosen threshold. For a fixed d, $0 \leq d \leq 1$, we call a superpair $G(A, B)$ *a superedge*, if $\mathrm{dens}_2(G(A, B)) \geq d$, and *a super-nonedge*, if $\mathrm{dens}_2(G(A, B)) < d$. Now, given a graph G, and its ϵ-partitioning of order m, we can build a high-level representation of G by a graph of size m, which we will call *a d-map of G*. The d-map M contains a node for every supernode of G. Two nodes of M are connected by an edge, if and only if the corresponding supernodes of G are connected by a superedge. Thus, edges and nonedges in M represent, respectively, superedges and super-nonedges of G. For a node pair (edge or nonedge) e in G, we denote by $\mu(e)$ the corresponding pair in the d-map M. We call $\mu : E(G) \to E(M)$ *the mapping function*. The union of all superedges $\mu^{-1}(E(M)) \subseteq E(G)$ will be called *the superedge subgraph of G*. Similarly, the union of all super-nonedges in G will be called *the super-nonedge subgraph of G*.

We rely on the following fact, which is a restricted version of the Blow-up Lemma (see [Kom99]).

Theorem 1 (Blow-up Lemma).
Let $d > \epsilon > 0$. Let G be a graph with an ϵ-partitioning, and let M be its d-map. Let H be a subgraph of M with maximum degree $\Delta > 0$. If $\epsilon \leq (d - \epsilon)^{\Delta}/(2 + \Delta)$, then G contains a subgraph isomorphic to H.

Proof. See [Kom99].   □

Since we are interested in diamond-free graphs, we take H to be a diamond. We simplify the condition on d and ϵ, and apply the Blow-up Lemma in the following form: if $\epsilon \leq (d - \epsilon)^3/5$, and G is diamond-free, then its d-map M is also diamond-free.

Our main tool is Szemerédi's Regularity Lemma. Informally, it states that any graph can be transformed into a graph with an ϵ-partitioning by removing a small number of nodes and edges. Its precise statement, slightly adapted from [KS96], is as follows.

Theorem 2 (Regularity Lemma).
Let G be an arbitrary graph. For every $\epsilon > 0$ there is an $m = m(\epsilon)$ such that for some $G_0 \subseteq G$ with $|E(G) \setminus E(G_0)| \leq \epsilon \cdot |V(G)|^2$, graph G_0 admits an ϵ-partitioning of order at most m.

Proof. See e.g. [KS96, Die00].   □

The given form of the Regularity Lemma is slightly weaker than the standard one. In particular, we allow to remove a "small" number of nodes and edges from the graph G, whereas the standard version only allows to remove a "small"

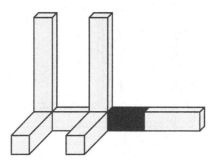

Fig. 3. An axial collision

number of nodes (with adjacent edges), and then a "small" number of superpairs. In our context, the difference between two versions is insignificant.

Note that if $|E(G)| = o(|V(G)|^2)$, the statement of the Regularity Lemma becomes trivial. In other words, the Regularity Lemma is only useful for dense graphs.

4   Packing Tripods

Consider a packing of tripods of size n in an $n \times n \times n$ cube, of the type described in the Introduction (no overlaps, similar orientation, corner cubes coinciding with $n \times n \times n$ cube cells). A tripod in such a packing is uniquely defined by the coordinates of its corner cube (i, j, k), $0 \le i, j, k < n$. Moreover, if two of the three coordinates (i, j, k) are fixed, then the packing may contain at most one tripod with such coordinates — otherwise, the two tripods with an equal pair of coordinates would form an *axial collision*, depicted in Figure 3.

We represent a tripod packing by an equitripartite triangle-covered graph $G \sqsubseteq K_{nnn}$ as follows. Three color classes $U = \{u_i\}$, $V = \{v_j\}$, $W = \{w_k\}$, $0 \le i, j, k < n$, correspond to the three dimensions of the cube. A tripod (i, j, k) is represented by a triangle $\{(u_i, v_j), (v_j, w_k), (w_k, u_i)\}$. To prevent axial collisions, triangles representing different tripods must be edge-disjoint. Hence, if m is the number of tripods in the packing, then the representing graph G contains $3m$ edges.

We now prove that the graph G is diamond-free. In general, G might contain a triangle with three edges coming from three different tripods; such a triangle would give rise to three diamonds. To prove that such a situation is impossible, we must consider, apart from axial collisions, also *simple collisions*, depicted in Figure 4.

Lemma 2. *A tripod packing graph is diamond-free.*

Proof. It is sufficient to show that the tripod packing graph does not contain any triangles apart from those representing tripods. Suppose the contrary: there is a triangle $\{(u_i, v_j), (v_j, w_k), (w_k, u_i)\}$, which does not represent any tripod. Then

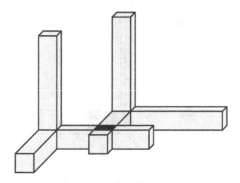

Fig. 4. A simple collision

its three edges must come from triangles representing three different tripods; denote these tripods (i, j, k'), (i, j', k), (i', j, k), where $i \neq i'$, $j \neq j'$, $k \neq k'$. Consider the differences $i' - i$, $j' - j$, $k' - k$, all of which are non-zero. At least two of these three differences must have the same sign; without loss of generality assume that $i' - i$, $j' - j$ are of the same sign. Thus, we have either $i' < i$, $j' < j$, or $i' > i$, $j' > j$. In both cases, the tripods (i, j', k), (i', j, k) collide. Hence, our assumption must be false, and the triangle $\{(u_i, v_j), (v_j, w_k), (w_k, u_i)\}$ not representing any tripod cannot exist. Therefore, no triangles in G can share an edge — in other words, G is diamond-free.   □

Thus, tripod packing graphs are equitripartite, triangle-covered and diamond-free. Note that these graph properties are invariant under any permutation of graph nodes within colour classes, whereas the property of a tripod packing being overlap-free is not invariant under permutation of indices within each dimension. Hence, the converse of Lemma 2 does not hold. However, even the loose characterisation of tripod packing graphs by Lemma 2 is sufficient to obtain our results.

The following theorem is a special case of an observation attributed to Szemerédi by Erdős (see [Erd87], [CG98, p. 48]). Since Szemerédi's proof is apparently unpublished, we give an independent proof of our special case.

Theorem 3. *Consider an equitripartite, triangle-covered, diamond-free graph of size n. The maximum density of such a graph tends to 0 as $n \to \infty$.*

Proof. Suppose the contrary: for some constant $d > 0$, and for an arbitrarily large n (i.e. for some $n \geq n_0$ for any n_0), there is a tripartite, triangle-covered, diamond-free graph G of size n, such that $\mathrm{dens}_3(G) \geq d > 0$. The main idea of the proof is to apply the Regularity Lemma and the Blow-up Lemma to the graph G. This will allow us to "distil" from G a new graph, also triangle-covered and diamond-free, with tripartite density higher than $\mathrm{dens}_3(G)$ by a constant factor λ. Repeating this process, we can raise the density to $\lambda^2 d$, $\lambda^3 d$, etc., until the density becomes higher than 1, which is an obvious contradiction.

Let us now fill in the details of the "distilling" process. We start with a constant γ, $0 < \gamma < 1$; its precise numerical value will be determined later. Select a constant $\epsilon > 0$, such that $\epsilon \leq (\gamma d - \epsilon)^3/5$, as required by the Blow-up Lemma. By the Regularity Lemma, graph G admits an ϵ-regular partitioning, the order of which is constant and independent of the size of G. Denote by M the γd-map of this partitioning, and let $\mu : G \rightarrow M$ be the mapping function.

Consider the superedge subgraph $\mu^{-1}(E(M))$. Let $G_\triangle \sqsubseteq \mu^{-1}(E(M)) \sqsubseteq G$ be a spanning subgraph of G, consisting of all triangles *completely contained* in $\mu^{-1}(M)$; in other words, each triangle in $G_\triangle$ is completely contained in some supertriangle of G. We claim that $G_\triangle$ contains a significant fraction of all triangles (and, hence, of all edges) in G. Indeed, the bipartite density of a super-nonedge is by definition at most γd, hence the super-nonedge subgraph has at most $3\gamma d \cdot n^2$ edges. Every triangle not completely contained in $\mu^{-1}(E(M))$ must have at least one edge in the super-nonedge subgraph; since triangles in G are edge-disjoint, the total number of such triangles cannot exceed $3\gamma d \cdot n^2$. By initial assumption, the total number of triangles in G is at least $d \cdot n^2$, therefore the number of triangles in $G_\triangle$ must be at least $(1 - 3\gamma)d \cdot n^2$. By selecting a sufficiently small γ, we can make the number of triangles in $G_\triangle$ arbitrarily close to $d \cdot n^2$. For the rest the proof, let us fix the constant γ within the range $0 < \gamma < 1/12$, e.g. $\gamma = 1/24$. As a corresponding ϵ we can take e.g. $\epsilon = (\gamma d/2)^3/5 = d^3/(5 \cdot 48^3)$.

It only remains to observe that, since graph G is diamond-free, its γd-map M is diamond-free by the Blow-up Lemma. By Lemma 1, $\text{dens}(M) \leq 3/4$. This means that among all superpairs of G, the fraction of superedges is at most $3/4$. All edges of $G_\triangle$ are contained in superedges of G, therefore the average density of a superedge in $G_\triangle$ is at least $4/3 \cdot \text{dens}(G_\triangle)$. In particular, there must be some superpair in $G_\triangle$ with at least such density. Since $G_\triangle$ consists of edge-disjoint supertriangles, this superpair is contained in a unique supertriple $F \sqsubseteq G_\triangle$, with

$$\text{dens}_3(F) \geq 4/3 \cdot \text{dens}_3(G_\triangle) \geq 4/3 \cdot (1 - 3\gamma)d =$$
$$4/3 \cdot (1 - 3 \cdot 1/24)d = 7/6 \cdot d.$$

In our previous notation, we have $\lambda = 7/6 > 1$.

We define the supertriple F to be our new "distilled" equitripartite triangle-covered diamond-free graph. Graph size has only been reduced by a constant factor, equal to the size of the ϵ-partitioning. By taking the original graph G large enough, the "distilled" graph F can be made arbitrarily large. Its density $\text{dens}_3(F) \geq \lambda d = 7/6 \cdot d > d$. By repeating the whole process, we can increase the graph density to $(7/6)^2 \cdot d$, $(7/6)^3 \cdot d$, ... , and eventually to values higher than 1, which contradicts the definition of density (in fact, values higher than $3/4$ will already contradict Lemma 1). Hence, the initial assumption of existence of arbitrarily large equitripartite triangle-covered diamond-free graphs with constant positive density must be false. Negating this assumption, we obtain our theorem.   □

The solution of Stein and Szabó's problem is now an easy corollary of Lemma 2 and Theorem 3.

Corollary 1. *Consider a tripod packing of size n. The maximum density of such a packing tends to 0 as $n \to \infty$.*

5   Conclusion

We have proved that the density of a tripod packing must be infinitely small as its size tends to infinity. Since the Regularity Lemma only works for dense graphs, the question of determining the precise asymptotic growth of a maximum tripod packing size remains open.

Nevertheless, we can obtain an upper bound on this growth. Let d be the maximum density of tripod packing of size n. In our proof of Theorem 3, it is established that the maximum density of tripod packing of size $m(d^3/(5 \cdot 48^3)) \cdot n$ is at most $6/7 \cdot d$, where $m(\cdot)$ is the function defined by the Regularity Lemma. In [ADL+94, DLR95] it is shown that for tripartite graphs, $m(t) \leq 4^{t^{-5}}$. These two bounds together yield a desired upper bound on d as a function of n, which turns out to be a rather slow-growing function. By applying the "descending" technique from [SS94], we can also obtain an upper bound on the size of a maximal r-pod, which can tile (without any gaps) an r-dimensional space. The resulting bound is a fast-growing function of r. In [SS94] it is conjectured that this bound can be reduced to $r - 2$ for any $r \geq 4$. The conjecture remains open.

6   Acknowledgement

The author thanks Sherman Stein, Chris Morgan and Mike Paterson for fruitful discussion.

References

[ADL+94]   N. Alon, R. A. Duke, H. Lefmann, V. Rödl, and R. Yuster. The algorithmic aspects of the regularity lemma. *Journal of Algorithms*, 16(1):80–109, January 1994.

[CG98]   F. Chung and R. Graham. *Erdős on Graphs: His Legacy of Unsolved Problems*. A K Peters, 1998.

[Die00]   R. Diestel. *Graph Theory*. Number 173 in Graduate Texts in Mathematics. Springer, second edition, 2000.

[DLR95]   R. A. Duke, H. Lefmann, and V. Rödl. A fast approximation algorithm for computing the frequencies of subgraphs in a given graph. *SIAM Journal on Computing*, 24(3):598–620, 1995.

[Erd87]   P. Erdős. Some problems on finite and infinite graphs. In *Logic and Combinatorics. Proceedings of the AMS-IMS-SIAM Joint Summer Research Conference, 1985*, volume 65 of *Contemporary Mathematics*, pages 223–228, 1987.

[Gal98]   D. Gale. *Tracking the Automatic Ant, and Other Mathematical Explorations*. Springer, 1998.

[Gou88]   R. Gould. *Graph Theory*. The Benjamin/Cummings Publishing Company, 1988.

[Kom99]   J. Komlós. The Blow-up Lemma. *Combinatorics, Probability and Computing*, 8:161–176, 1999.

[KS96]   J. Komlós and M. Simonovits. Szemerédi's Regularity Lemma and its applications in graph theory. In D. Miklós, V. T. Sós, and T. Szőnyi, editors, *Combinatorics: Paul Erdős is Eighty (Part II)*, volume 2 of *Bolyai Society Mathematical Studies*. 1996.

[SS94]   S. K. Stein and S. Szabó. *Algebra and Tiling: Homomorphisms in the Service of Geometry*. Number 25 in The Carus Mathematical Monographs. The Mathematical Association of America, 1994.

[Ste95]   S. K. Stein. Packing tripods. *The Mathematical Intelligencer*, 17(2):37–39, 1995. Also appears in [Gal98].

An Efficient k Nearest Neighbor Searching Algorithm for a Query Line

Subhas C. Nandy*

Indian Statistical Institute, Calcutta - 700 035, India

Abstract. In this paper, we present an algorithm for finding k nearest neighbors of a given query line among a set of points distributed arbitrarily on a two dimensional plane. Our algorithm requires $O(n^2)$ time and space to preprocess the given set of points, and it answers the query for a given line in $O(k + \log n)$ time, where k may also be an input at the query time. Almost a similar technique is applicable for finding the k farthest neighbors of a query line, keeping the time and space complexities invariant. We also discuss some constrained version of the problems where the preprocessing time and space complexities can be reduced keeping the query times unchanged.

Key words : nearest- and farthest-neighbor, query line, arrangement, duality.

1 Introduction

Given a set $P = \{p_1, p_2, \ldots, p_n\}$ of n points arbitrarily distributed on a plane, we study the problem of finding the k nearest neighbors of a query line ℓ (in the sense of perpendicular distance). The problem of finding the nearest neighbor of a query line was initially addressed in [3]. An algorithm of preprocessing time and space $O(n^2)$ was proposed in that paper which can answer the query in $O(\log n)$ time. Later, the same problem was solved using geometric duality in [8], with the same time and space complexities. The space complexity of the problem has recently been improved to $O(n)$ [11]. The preprocessing time of that algorithm is $O(n \log n)$, but the query time is $O(n^{.695})$. In the same paper, it is shown that if the query line passes through a specified point, that information can be used to construct a data structure in $O(n \log n)$ time and space, so that the nearest neighbor query can be answered in $O(\log^2 n)$ time. Apart from a variation of the proximity problems in computational geometry, this problem is observed to be important in the area of pattern classification and data clustering [11], Astrophysical database maintenance and query [2], linear facility location, to name a few.

---

* The work was done when the author was visiting School of Information Science, Japan Advanced Institute of Science and Technology, Ishikawa 923-1292, Japan

D.-Z. Du et al. (Eds.): COCOON 2000, LNCS 1858, pp. 281–290, 2000.

In this paper, we address a natural generalization of the above problem where the objective is to report the k nearest neighbors of a query line in the same environment. We shall use geometric duality for solving this problem. The preprocessing time and space required for creating the necessary data structure is $O(n^2)$, which maintains the levels of the arrangement of lines corresponding to the duals of the points in P [1,4,6]. The query time complexity is $O(k + \log n)$, where k is an input at the query time. Almost a similar technique works for finding the k farthest neighbors of a query line keeping the time and space complexities invariant. The farthest neighbor query is also useful in Statistics, where the objective is to remove the outliers from a data set.

We also show that in the following two constrained cases, the preprocessing time can be reduced keeping the query times invariant.

(i) If k is a fixed constant, the k farthest neighbors of a query line can be reported in $O(k + \log n)$ time on a preprocessed data structure of size $O(kn)$, which can be constructed in $O(kn + n\log n)$ time.

(ii) When the query line is known to pass through a fixed point q and k is known prior to the preprocessing, we use a randomized technique to construct a data structure of size $O(kn)$ in $O(kn + \min(n\log^2 n, kn\log n))$ time, which reports k nearest neighbors of such a query line in $O(k + \log n)$ time. Thus, for the nearest neighbor problem (i.e., $k = 1$), our algorithm is superior in terms of both space required and query time, compared to the algorithm proposed in [11] for this constrained case, keeping the preprocessing time invariant.

2   Geometric Preliminaries

First we mention that if the query line is vertical, we maintain an array with the points in P, sorted with respect to their x coordinates. This requires $O(n)$ space and can be constructed in $O(n\log n)$ time. Now, given a vertical query line we position it on the x-axis by performing a binary search. To find the k-nearest neighbors, a pair of scans (towards left and right) are required in addition. So, the query time complexity is $O(k + \log n)$. It is easy to understand that the k-farthest neighbors can be obtained in $O(k)$ time by scanning the array from its left and right ends.

In this paper, we shall consider the non-vertical query line. We use geometric duality for solving these problems. It maps (i) a point $p = (a, b)$ of the primal plane to the line $p^* : y = ax - b$ in the dual plane, and (ii) a non-vertical line $\ell : y = mx - c$ of the primal plane to the point $\ell^* = (m, c)$ in the dual plane. Needless to say, a point p is below (resp., on, above) a line ℓ in the primal plane if and only if p^* is above (resp., on, below) ℓ^* in the dual plane.

Let H be the set of dual lines corresponding to the points in P. Let $\mathcal{A}(H)$ denote the arrangement of the set of lines H. The number of vertices, edges and faces in $\mathcal{A}(H)$ are all $O(n^2)$ [4]. Given a query line ℓ, the problem of finding its nearest or farthest point can be solved as follows:

Nearest neighbor algorithm : [8] Consider the point ℓ^* which is the dual of the line ℓ. Next use the point location algorithm of [5] to locate the cell of $\mathcal{A}(H)$ containing ℓ^* in $O(\log n)$ time. As the cells of the arrangement $\mathcal{A}(H)$ are split into trapezoids, the line p_i^* and p_j^* lying vertically above and below ℓ^*, can be found in constant time. The distance of a point p and the line ℓ in the primal plane can be obtained from the dual plane as follows :

Let $x(\ell^*)$ be the x-coordinate of the point ℓ^* in the dual plane which corresponds to the slope of the line ℓ in the primal plane. We draw a vertical line from the point ℓ^* which meets the line p^* at a point $\alpha(\ell^*, p^*)$ in the dual plane. Now, the perpendicular distance of the point p and the line ℓ in the primal plane is equal to $\frac{d(\ell^*, \alpha(\ell^*, p^*))}{x(\ell^*)}$, where $d(.,.)$ denotes the distance between two points.

Thus, the point nearest to ℓ in the primal plane will be any one of p_i or p_j depending on whether $d(\ell^*, \alpha(\ell^*, p_i^*)) <$ or $> d(\ell^*, \alpha(\ell^*, p_j^*))$, and it can be computed in constant time. In order to report the nearest neighbor from any arbitrary query line, one needs to maintain the arrangement $\mathcal{A}(H)$, which can be created in $O(n^2)$ time and space [5].

Farthest neighbor algorithm : Here we construct the lower and upper envelope of the lines in H corresponding to the duals of the points in P, and store them in two different arrays, say L_L and L_U. This requires $O(n \log n)$ time [7]. Now, given the line ℓ, or equivalently the dual point ℓ^*, we draw a vertical line at $x(\ell^*)$ which hits the line segments $\lambda_1 \in L_L$ and $\lambda_2 \in L_U$, and they can be located in $O(\log n)$ time. If λ_1 and λ_2 are respectively portions of p_i^* and p_j^*, then the farthest neighbor of ℓ is either of p_i and p_j which can be identified easily. Thus for the farthest neighbor problem, the preprocessing time and space required are $O(n \log n)$ and $O(n)$ respectively, and the query can be answered in $O(\log n)$ time.

In this paper, we consider the problem of locating the k nearest neighbors of a query line. As a preprocessing, we construct the following data structure which stores the arrangement of the dual lines in H. It is defined using the concept of *levels* as stated below. The same data structure can answer the k farthest neighbors query also with the same efficiency.

Definition 1. [4] A point π in the dual plane is at level θ $(0 \leq \theta \leq n)$ if there are exactly θ lines in H that lie strictly below π. The θ-level of $\mathcal{A}(H)$ is the closure of a set of points on the lines of H whose levels are exactly θ in $\mathcal{A}(H)$, and is denoted as $L_\theta(H)$. See Figure 1 for an illustration.

Clearly, $L_\theta(H)$ is a poly-chain from $x = -\infty$ to $x = \infty$, and is monotone increasing with respect to the x-axis. In Figure 1, a demonstration of levels in the arrangement $\mathcal{A}(H)$ is shown. Here the thick chain represents $L_1(H)$. Among the vertices of $L_1(H)$, those marked with empty (black) circles are appearing in level 0 (2) also. Each vertex of the arrangement $\mathcal{A}(H)$ appears in two consecutive levels, and each edge of $\mathcal{A}(H)$ appears in exactly one level. We shall store $L_\theta(H)$, $0 \leq \theta \leq n$ in a data structure as described in the next section.

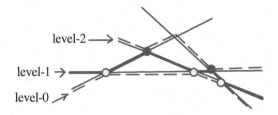

Fig. 1. Demonstration of levels in an arrangement of lines

3   Algorithm

Below we describe the data structure, called *level*-structure, and briefly explain our algorithm.

3.1   *Level*-Structure

It is a balanced binary search tree T, called the *primary structure*, whose nodes correspond to the *level-id* of the arrangement $\mathcal{A}(H)$. To avoid confusion, we shall use the term *layer* to represent (different depth) levels of the tree T. The root of T corresponds to the *layer-0* and the *layer-id* increases as we proceed towards the leaf level.

Initially, for each node of T (representing a level, say θ, of $\mathcal{A}(H)$) we create a *secondary structure* $T(\theta)$ which is an array containing edges of $L_\theta(H)$ in left to right order. Next, to facilitate our search process, we augment the secondary structures for all the non-leaf nodes as follows :

Consider an edge $e \in \mathcal{A}(H)$ at a certain level, say θ. Let θ' be the level which appears as the left child of level θ in T, and $V_{\theta'}^e$ ($\in \mathcal{A}(H)$) be the set of vertices whose projections on the x-axis are completely covered by the projection of e on x-axis. If $|V_{\theta'}^e| > 1$ then we leave the leftmost vertex of $V_{\theta'}^e$ and vertically project the next vertex on e, and do the same for every alternate vertex of $V_{\theta'}^e$ as shown in Figure 2. The same action is taken for all the edges of level θ and the edges in its left successor level θ'. All the projections create new vertices at level θ, and as a result, the number of edges in level θ is also increased. This augmentation yields the following lemma :

Lemma 1. *After the augmentation step, the projection of an edge of the parent level can overlap on the projections of at most two edges at its left child level.*

Proof : The following three situations may arise.

$|V_{\theta'}^e| = 0$. Here edge e is completely covered by an edge at level θ' (Figure 2a).
$|V_{\theta'}^e| = 1$. Here edge e overlaps with exactly two edges at level θ' (Figure 2b).
$|V_{\theta'}^e| > 1$. In this case the truth of the lemma follows as we have projected every alternate vertex of $V_{\theta'}^e$ on e (Figure 2c).   □

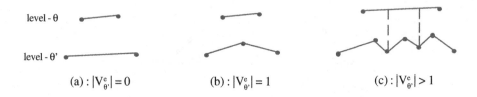

level - θ

level - θ'

(a) : $|V_{\theta'}^e| = 0$ (b) : $|V_{\theta'}^e| = 1$ (c) : $|V_{\theta'}^e| > 1$

Fig. 2. Augmentation of *secondary* structure

A similar action is taken for all the edges of level θ with respect to the vertices of level θ'' appearing as the right child of θ in T. Now, we clearly mention the following points :

(1) A result similar to Lemma 1 holds for the projection of each edge at level θ on the edges at its right child level θ''.

(2) Lemma 1 and the result stated in item (1) remains valid for all edges from layer $(\lceil (\log n) \rceil - 2)$ up to the root layer with the enhanced set of vertices.

(3) The proposed augmentation step causes an increase in the number of edges at the level θ to at most $|E_\theta| + (|E_{\theta'}| + |E_{\theta''}|)/2$, where E_θ is the set of edges at level θ in $\mathcal{A}(H)$.

From the original secondary structure at each level of the *level*-structure, we create a new secondary structure as follows :

The secondary structures appearing in the leaf level of T will remain as it is.

We propagate the vertices appearing in a leaf of T to its parent as stated above, and construct a linear array with the *edges* formed by the original vertices on that level and the vertices contributed by its both left child and right child. This array is also ordered with respect to the x-coordinates of the end points of the enhanced set of edges, and this new array will now serve the role of the secondary structure.

For a node in the non-leaf layer of T, the original vertices in that level as well as the vertices inherited from its successors are propagated to its parent layer in a similar procedure. For each non-leaf node, the original secondary structure is replaced by a new secondary structure with the edges formed by the enhanced set of vertices.

This method of propagation of vertices is continued in a bottom-up manner till the root layer of T is reached.

The original secondary structures in the non-leaf layers of the *level*-structures are no more required.

The augmented structure obtained in this manner is called $\mathcal{A}^*(H)$.

Lemma 2. *The number of vertices generated in the secondary structure of all the levels of $\mathcal{A}^*(H)$ is $O(n)$ in the worst case.*

Proof : Consider only the set of vertices of $\mathcal{A}^*(H)$ which appear in those levels which belong to the leaf layer of T. Only half of these vertices (in total) are propagated to the nodes in different levels at the predecessor layer of the leaves of T. Among the set of vertices, which has been propagated from the leaves of T to the current layer, at most half of them may further be propagated to their predecessor layer in T. This process may continue until the root (layer) is reached. But if k be the total number of vertices at all the leaves, their contribution to all the layers of the tree is at most $k + \frac{k}{2} + \frac{k}{4} + \ldots$, which may be at most $2k$ in the worst case. The same is true for the original vertices of $\mathcal{A}(H)$ appearing in the different levels at the non-leaf layers of T also. Thus it is clear that, this type of propagation does not cause the number of vertices of $\mathcal{A}^*(H)$ more than twice the original number of vertices of $\mathcal{A}(H)$. □

Next, for each edge of a level, say θ, appearing in a non-leaf layer of T in the enhanced structure $\mathcal{A}^*(H)$, we keep four pointers. A pair of pointers L_1 and L_2 point to two edges in the level θ' which appear as left child of θ in T and the other pair R_1 and R_2 point to a pair of edges in the level θ'' appearing in the right child of θ. By Lemma 1 and subsequent discussions, an edge appearing in a level, which corresponds to a non-leaf node in T, can not overlap on more than two edges of its left (resp. right) child level (in T) with respect to the vertical projections. If the x-range of an edge e (appearing at level θ) is completely covered by that of an edge e' of its left (resp. right) child level θ' (θ'') of $\mathcal{A}^*(H)$, both the pointers L_1 and L_2 (resp. R_1 and R_2) attached to e point to e'. Otherwise, L_1 and L_2 (resp. R_1 and R_2) of the edge e point to two adjacent edges e_1 and e_2 in the level θ' (θ''), such that the common vertex of e_1 and e_2 lies in the x-interval of the edge e.

3.2   Sketch of the Algorithms

Given a query line ℓ, we compute its dual point ℓ^*. In order to find the cell of $\mathcal{A}^*(H)$ containing ℓ^*, we proceed as follows :

We choose the level θ_{root} of $\mathcal{A}^*(H)$ attached to the root of T, and do a binary search on the secondary structure $T(\theta_{root})$ attached to it to locate the edge e which spans horizontally on ℓ^*. Now, compare ℓ^* and e to decide whether to proceed towards the left or the right child of θ_{root} in the next layer. Without loss of generality, let us assume that search proceeds towards the left child and θ_{left} be the level attached to it. Note that, L_1 and L_2 pointers uniquely determine the edge $e' \in T(\theta_{left})$ by doing at most 2 comparisons. If search proceeds towards the right of the root, we may need to use the pointers R_1 and R_2 attached to e. Similarly proceeding, the cell of $\mathcal{A}^*(H)$ containing ℓ^* may be easily reached in $O(\log n)$ steps.

Reporting of k nearest neighbors

After locating ℓ^* in the appropriate cell of $\mathcal{A}^*(H)$, let $e_a \in T(\theta)$ and $e_b \in T(\theta-1)$ be the two edges of $\mathcal{A}^*(H)$ which are vertically above and below ℓ^* respectively.

We compute the vertical distances of e_a and e_b from ℓ^*, and report the nearest one. Without loss of generality, let e_a be the nearest neighbor of ℓ^*. Next, we need to reach an edge at level $\theta + 1$ (the inorder successor of level θ in T) whose horizontal interval spans ℓ^*. If e_b is closer to ℓ^* then we need to locate the edge in level $\theta - 2$ (the inorder predecessor of level $\theta - 1$ in T) which horizontally spans ℓ^*.

In order to reach an appropriate edge in the inorder predecessor or successor of the current level during reporting, we maintain two stacks, namely DN and UP. They contain the edges (along with the level-id) of $\mathcal{A}^*(H)$ in different layers from the root to the relevant level ($\theta - 1$ and θ) whose x-interval span ℓ^*. The initial configuration of the stacks is as follows :

- Both the stacks will have same set of elements from the root layer up to certain layer corresponding to the level, say θ^* ($= \theta$ or $\theta - 1$) depending on which one appear in a non-leaf layer. Then one of these stacks will contain few more elements from level θ^* to either of θ and $\theta - 1$ which appear in the leaf layer.

These will help while moving from one level to the next level (above or below) in $\mathcal{A}^*(H)$, as described in the following lemma.

Lemma 3. *After locating a pair of edges in two consecutive levels, say $\theta - 1$ and θ, of $\mathcal{A}^*(H)$, whose x-interval contains ℓ^*, the edges in the k levels vertically below (above) $\theta - 1$ (θ) can be reported in $O(k + \log n)$ time.*

Proof : Without loss of generality, we shall consider the visiting of the edges in k consecutive levels above the level θ whose x-interval contains ℓ^*.

If the level $\theta + 1$, which is the inorder successor of level θ in T, is

- in its right subtree, then we traverse all the levels that appear along the path from θ to $\theta + 1$ in T. In each move to the next level, we can reach an appropriate edge whose x-interval is spanning ℓ^*, using either of (L_1 and L_2) and (R_1 and R_2) in constant time. We need to store all these edges in the stack UP, for backtracking, if necessary.
- in some predecessor layer, then we may need to backtrack along the path through which we reached from $\theta + 1$ to θ during forward traversal. If δ be the difference in the layer-id of the levels θ and $\theta + 1$ in T, then the number of elements popped from the UP stack is δ.

Note that, if it needs to proceed to the level $\theta + 2$, we may again move δ layers towards leaf. Needless to say, the edges on that path will be pushed in the UP stack. But such a forward movement may again be required after visiting all the levels in the right subtree rooted at level $\theta + 1$. Thus apart from reporting, this extra traversal in T may be required at most twice, and the length of the path may at most be $O(\log n)$. Hence the proof of the lemma.

In order to visit the edges in k consecutive levels below $\theta - 1$, whose x-interval contains ℓ^*, we need to use the DN stack.   □

Thus Lemmas 2 and 3 lead to the following result stating the time and space complexities of our algorithm.

Theorem 1. *Given a set of points on a plane, they can be preprocessed in $O(n^2)$ time and space such that the problem of locating k nearest neighbors for a given query line can be solved in $O(k + \log n)$ time.*   □

Reporting of k farthest neighbors

For the farthest neighbor problem, one needs to traverse from the root to the top-most and the bottom-most layers along the tree T. This also requires maintaining the two stacks UP and DN, whose role is same as that of the earlier problem. Here also, for the root layer only, a binary search is required in the secondary structure to locate an edge e which spans horizontally ℓ^*. From the next layer onwards, the pointers maintained with each edge will locate the appropriate edge in the corresponding level. Finally, either (i) an edge from the top-most layer or (ii) an edge from the bottom-most layer will be reported as the farthest edge from ℓ^*. In order to get the next farthest neighbor, one may consult its inorder successor or predecessor depending on the case (i) or (ii). This traversal can be made in amortized constant time using our proposed *level*-structure $\mathcal{A}^*(H)$. The process continues until the k farthest neighbors are reported. Thus we have the following theorem.

Theorem 2. *Given a set of points on a plane, they can be preprocessed in $O(n^2)$ time and space such that the problem of locating k farthest neighbors for a given query line can be solved in $O(k + \log n)$ time.*

4 Constrained Query

In this section, we shall discuss the complexity of the k nearest neighbors and the k farthest neighbors problems, if k is known in advance (i.e., k is not an input at the query time).

4.1 Farthest k Neighbors Problem

In this case, we need to maintain only (i) k levels from bottom starting from level-1 up to the level-k, denoted by $L_{\leq k}(H)$, and (ii) k levels at the top starting from level-$(n-k+1)$ up to level-n, denoted by $L_{\geq (n-k+1)}(H)$, of the arrangement $\mathcal{A}(H)$. Here level-1 and level-n correspond to the lower and upper envelope of the lines in H respectively.

Lemma 4. *The number of edges in the $L_{\leq k}(H)$ of an arrangement $\mathcal{A}(H)$ of n lines in the plane is $O(nk)$, and it can be computed in $O(nk + n\log n)$ time.*

Proof : The counting part of the lemma follows from Corrolary 5.17 of [12]. The time complexity of the construction follows from [6].   □

Next, we project the edges of level-k on the edges of level-$(k-1)$ as described in Subsection 3.1. As the horizontal interval of an edge in level $(k-1)$ may be spanned by at most two edges of level-k (see Lemma 1), with each edge of level $(k-1)$, we attach two pointers Q_1 and Q_2 to point to a pair of appropriate edges of level-k as described in Subsection 3.1. The enhanced set of edges in level-$(k-1)$ is then projected on the edges of level-$(k-2)$. This process is repeated till we reach level-1.

The same technique is adopted for constructing $L_{\geq(n-k+1)}(H)$. The projection of edges from one level on the edges of its next higher level is continued from level-$(n-k+1)$ upwards till we reach the upper envelope (i.e., the level-n) of the arrangement $\mathcal{A}(H)$.

As stated in Lemma 2, the number of edges in the enhanced arrangement will remain $O(nk)$. Next, during the k farthest neighbors query for a given line ℓ, the edges of level-1 and level-n of the augmented $\mathcal{A}(H)$ spanning the x-coordinate of ℓ^* can be found using binary search. But, one can reach an edge from level-i to the edge at level-$(i+1)$ (level-$(i-1)$) in $L_{\leq k}(H)$ $(L_{\geq(n-k+1)}(H))$ spanning the x-coordinate of ℓ^* using the two pointers Q_1, Q_2 attached with the edge at level-i, in constant time. Thus by Lemma 4, and the above discussions, we have the following theorem :

Theorem 3. *If k is known in advance, a data structure with a given set P of n points can be made in $O(nk + n\log n)$ time and $O(nk)$ space, such that for any query line, its k farthest neighbors can be answered in $O(k + \log n)$ time.*   □

4.2   Nearest k Neighbors Problem when Query Line Passes through a Specified Point

If the query line ℓ passes through a specified point q, then the dual point ℓ^* of the line ℓ will always lie on the line $h = q^*$, dual of the point q.

Now consider the set $H = \{h_1, h_2, \ldots, h_n\}$ of lines in the dual plane, where $h_i = p_i^*$. Let $H' = \{h'_1, h'_2, \ldots, h'_n\}$ denote the set of half lines, where h'_i is the portion of h_i above h. Similarly, $H'' = \{h''_1, h''_2, \ldots, h''_n\}$ is defined below the line h. We denote the $(\leq k)-level$ above (resp. below) h by $L_{\leq k}(H', h^+)$ (resp. $L_{\leq k}(H'', h^-)$).

In [10], the *zone theorem* for line arrangement [9] is used to show that for the set of half lines H' (H'') the complexity of $L_{\leq k}(H', h^+)$ $(L_{\leq k}(H'', h^-)$ is $O(nk)$. A randomized algorithm is proposed in [10] which computes $L_{\leq k}(H', h^+)$ $(L_{\leq k}(H'', h^-)$ in $O(kn + \min(n\log^2 n, kn\log n))$ time.

Note that, here also, we need to enhance $L_{\leq k}(H', h^+)$ $(L_{\leq k}(H'', h^-))$ by projecting the accumulated set of edges of its one level to the next lower level, and maintaining two pointers for each edge of the enhanced arrangement as stated in the earlier subsection.

In order to answer the k nearest neighbor query for a line ℓ where $\ell^* \in h$, we first report the nearest neighbor of ℓ by locating the edges e_1 and e_2, spanning the x-coordinate of ℓ^*, by using binary search on the level-1 of $L_{\leq k}(H', h^+)$ and $L_{\leq k}(H'', h^-)$ respectively. To report the next nearest neighbor, we move to the next level of either $L_{\leq k}(H', h^+)$ or $L_{\leq k}(H'', h^-)$ using the pointers attached with e_1 or e_2. This process is repeated to report k nearest neighbors.

Theorem 4. *If the query line is known to pass through a specified point q, and k is known in advance, then for a given set of n points we can construct a data structure of size $O(nk)$ in $O(nk + \min(n\log^2 n, kn\log n))$ time, which can report the k nearest neighbors of an arbitrary query line in $O(k + \log n)$ time.*

Acknowledgment : The author wishes to acknowledge Prof. T. Asano and T. Harayama for helpful discussions.

References

1. P. K. Agarwal, M. de Berg, J. Matousek and O. Schwarzkopf, *Constructing levels in arrangements and higher order Voronoi diagram*, SIAM J. on Computing, vol. 27, pp. 654-667, 1998.
2. B. Chazelle, *Computational Geometry : Challenges from the Applications*, CG Impact Task Force Report, 1994.
3. R. Cole and C. K. Yap, *Geometric retrieval problems*, Proc. 24th. IEEE Symp. on Foundation of Comp. Sc., pp. 112-121, 1983.
4. H. Edelsbrunner, *Algorithms in Combinatorial Geometry*, Springer, Berlin, 1987.
5. H. Edelsbrunner, L. J. Guibas and J. Stolfi, *Optimal point location in monotone subdivision*, SIAM J. on Computing, vol. 15, pp. 317-340, 1986.
6. H. Everett, J. -M. Robert and M. van Kreveld, *An optimal algorithm for the $(\leq k)$-levels, with applications to separation and transversal problems*, Int. J. Comput. Geom. with Appl., vol. 6, pp. 247-261, 1996.
7. J. Hershberger, *Finding the upper envelope of n line segments in $O(n\log n)$ time*, Information Processing Letters, vol. 33, pp. 169-174, 1989.
8. D. T. Lee and Y. T. Ching, *The power of geometric duality revisited*, Information Processing Letters, vol. 21, pp. 117-122, 1985.
9. K. Mulmuley, *Computational Geometry : An Introduction Through Randomized Algorithms*, Prentice Hall, Englewood Cliffs, NJ.
10. C. -S. Shin, S. Y. Shin and K. -Y. Chwa, *The widest k-dense corridor problems*, Information Processing Letters, vol. 68, pp. 25-31, 1998.
11. P. Mitra and B. B. Chaudhuri, *Efficiently computing the closest point to a query line*, Pattern Recognition Letters, vol. 19, pp. 1027-1035, 1998.
12. M. Sharir and P. K. Agarwal, *Davenport-Schinzel Sequence and Their Geometric Applications*, Cambridge University Press, 1995.

Tetrahedralization of Two Nested Convex Polyhedra

Cao An Wang and Boting Yang

Department of Computer Science,
Memorial University of Newfoundland,
St.John's, NF, Canada A1B 3X5
wang@cs.mun.ca

Abstract. In this paper, we present an algorithm to tetrahedralize the region between two nested convex polyhedra without introducing Steiner points. The resulting tetrahedralization consists of linear number of tetrahedra. Thus, we answer the open problem raised by Chazelle and Shouraboura [6]: "whether or not one can tetrahedralize the region between two nested convex polyhedra into linear number of tetrahedra, avoiding Steiner points?". Our algorithm runs in $O((n+m)^3 \log(n+m))$ time and produces $2(m + n - 3)$ tetrahedra, where n and m are the numbers of the vertices in the two given polyhedra, respectively.

1 Introduction

The tetrahedralization in 3D is a fundamental problem in computational geometry [1], [3], [7], [9], [10]. Unlike its 2D counterpart, where a simple polygon with n vertices is always triangulatable and each triangulation contains exactly $n - 2$ triangles, a simple polyhedron with n vertices may not be tetrahedralizable [11] and the number of tetrahedra in its tetrahedralizations may vary from $O(n)$ to $O(n^2)$ [4] even if it is tetrahedralizable. Naturally, one will investigate the following two problems: to identify some classes of polyhedra which are tetrahedralizable and to find tetrahedralization methods for these polyhedra which will produce a linear number of tetrahedra or even optimal number of tetrahedra.

So far, very few non-trivial classes of simple polyhedra are known to be tetrahedralizable (without introducing Steiner points). One is so-called two 'side-by-side' convex polyhedra and the other is so-called two 'nested' convex polyhedra [8], [2], [6].

For the second class, Bern showed that the number of tetrahedra in the area between two nested convex polyhedra is bounded by $O(n \log n)$, where n is the number of vertices in the polyhedra. Chazelle and Shouraboura [6] presented an algorithm which produces $O(n)$ tetrahedra by introducing Steiner points. They raised an open question that "whether or not one can tetrahedralize the region between two nested convex polyhedra into linear number of tetrahedra, avoiding Steiner points?".

D.-Z. Du et al. (Eds.): COCOON 2000, LNCS 1858, pp. 291–298, 2000.

In this paper, we present an algorithm to tetrahedralize the region between two nested convex polyhedra into linear number of tetrahedra, without introducing any Steiner points. The basic idea comes from the following analysis. The 'dome removal' method invented by Chazelle [5] and Bern [2] has been pushed into its limit in the 'two nested convex polyhedra' case. In more detail, a 'dome' with tip v of P is the difference space between P and the convex hull of $(P - \{v\}) \cup Q$. After a dome is removed, some vertices S'_Q will lie on $CH((P - \{v\}) \cup Q)$, thus the subsequent dome removal process shall avoid to choose a tip vertex belonging to S'_Q. For simplicity, Bern uses $P - Q$ to denote the remainder of P and Q as well as the remainder after the subsequent removals of these domes. The method is to remove one dome at a time with the tip vertices in the portion of P of $P - Q$ until $P - Q$ contains only the vertices of Q. Note that a dome can be decomposed into k tetrahedra if it contains k triangle facets in its bottom. Thus, the number of tetrahedra produced by the method depends on the number of vertices in the domes. While this method takes only n remove steps for $n = |P| + |Q|$, it may result in $O(n \log n)$ tetrahedra even by removing these domes whose tips have the smallest degree in the surface of $P - Q$. The reason that the above bound is difficult to improve is as follows. In the worst case, the vertices with smaller degrees in the surface of $P - Q$ may lie on the Q of $P - Q$ and which forces the method to choose the vertices of P of $P - Q$ with larger degrees. Bern showed that if the number of $P - Q$ vertices is i, then the degree of the minimum-degree vertex can be $\frac{6n}{i}$. Overall the bound will be $n(2 + 6H_n)$, where H_n is the n-th Harmonic number (less than $\ln n$). Our observation is that Bern's method totally depends on the degree of the tip vertices of the domes. We shall avoid this. Note that the well-known one-vertex emission tetrahedralization of a convex polyhedron always yields a linear number of tetrahedra, more precisely $2n - degree(v) - 4$, where n is the number of vertices of the polyhedron and v is the emission vertex. In the nested polyhedra case, the one-vertex emission tetrahedralization must cope with the visibility of the vertex due the presence of obstacle polyhedron. We shall combine the dome removal method with the one-vertex emission method, i.e., each time we obtain a 'maximum dome' which contains all visible vertices of P from the tip v and some vertices of Q visible from v. The tetrahedra formed by this method will consist of two types: one is determined by v and a triangle facet of $P - Q$ visible to v and the other is determined by v and a *bridge* triangle (which is a inner triangle facet spans on P and Q of $P - Q$). We shall prove that the total number of bridge triangle faces is bounded by the size of P and Q. Thus, while the maximum dome removal method remains $O(n)$ steps the number of tetrahedra produced by this method is linearly proportional to the sizes of P and Q.

2   Preliminary

Let P and Q be two convex polyhedra. Q is *nested* in P if Q is entirely contained inside P, where they may share some common triangle facets, edges, or vertices.

Let f_P and f_Q be the number of facets in P and Q, respectively, where $f_P = n$ and $| f_Q | = m$. Let G_P and G_Q denote the corresponding surface graphs.

A vertex $v \epsilon P$ is *visible* to a vertex $w \epsilon P \cup Q$ if and only if the interior of the line segment $\overline{vw}$ does not intersect Q (as point set). Let P^v denote the vertex set of P visible to v in the presence of Q, and let P_b^v denote the subset of P^v such that each element in P_b^v has at least one incident facet in G_P not being entirely visible to v. Let Q^v denote the vertex set of Q visible from v, and let Q'^v be the subset of Q^v which lies on the $CH(P_b^v \cup Q^v)$ ($=CH((P - P^v) \cup P_b^v \cup Q^v)$). Let Q_b^v denote the subset of Q'^v such that each element in Q_b^v has at least one adjacent vertex in G_Q which does not belong to $CH(P_b^v \cup Q^v)$.

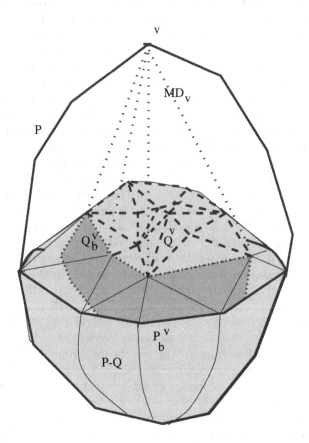

Fig. 1. For the definitions.

Definition: A *maximum dome* with tip vertex v (denoted by MD_v) is the closed area between convex hulls $CH(P^v)$ and $CH(P_b^v \cup Q^v)$ such that $CH(P_b^v \cup Q^v)$ contain at least one vertex of Q^v as inner vertex or at least one facet of Q^v. Let $P - Q$ represent the remainder of P and Q as well as the remainder after the subsequent removal of a maximum dome.

Definition: A tetrahedralization T_v of P is *vertex v emission* if T_v is obtained by connecting v to every vertex in $P - \{v\}$ by an edge. In the nested polyhedra P and Q, the tetrahedra obtained by connecting v to every vertex in $P^v \cup Q'^v$ form a maximum dome.

The lemma below follows from the definition of MD and one-vertex emission tetrahedralization.

Lemma 1. *Let MD_v be a maximum dome in nested P and Q. Then, the number of tetrahedra produced by v-emission method in MD_v is*

$$f_P^v + f_Q^v - degree(v) + |\; P_b^v \cup Q_b^v \;|,$$

where f_P^v is the number of facets in P of $P - Q$, each of them is entirely visible to v, and f_Q^v is the number of facets in Q of $P - Q$, each of them is entirely visible to v and belongs to $CH(P_b^v \cup Q^v)$.

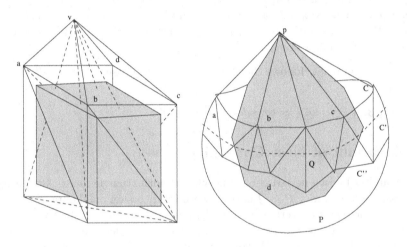

Fig. 2. For the proof of the lemma.

Lemma 2. *Let $P - Q$ be the remainder after the removal of all these domes which have the tip-vertex of degree less than six. There exists at least one maximum dome in $P - Q$ with tip-vertex belonging to Q, where the bottom surface of the dome either consists of some facets of Q in $P - Q$ or the interior of the bottom surface contains some vertices of Q in $P - Q$.*

Proof. (Sketched) It is easy to see that there always exists a vertex of P with degree of five or less when P and Q do not share any vertices and faces. (This can be proved using Euler's formula for P with more than 30 vertices.) By the assumption of the lemma, there must exist at least one shared vertex in $P - Q$ because otherwise we can continue to remove these domes with the tip-vertices of degree less than six in P until either $P - Q$ becomes the desired case or P has at most 30 vertices and P and Q still do not share any vertex. The later case will easily result in a linear-number of tetrahedra in $P - Q$. For the former case, we shall only discuss the case that P and Q share exactly one vertex, say p. For the other cases, the following analysis can be applied with a more detail analysis. Due to the blockage of Q, the visible area and the non-visible area in P from vertex p must form a closed chain, say C'. C' crosses a sequence of edges in P so that the vertices of these edges visible from p form a chain C and the vertices of these edges not visible from p form a chain C''. (Refer to the right-hand side of Figure 2.) All vertices of C are connected to p by edges in $P - Q$ due to our removal process. In such a $P - Q$, there must either exist three vertices in C, say a, b, and c, such that b is visible to both a and c, and $\overline{ac}$ crosses Q or no such three vertices so that $\overline{ac}$ crosses Q. In the latter case, the convex hull of P in $P - Q$ would not cross Q. Then, by choosing a vertex in C, say b, as tip-vertex, the corresponding maximum dome will be the desired dome. In the former case, let C''' denote the chain consisting of all these vertices in P of $P - Q$ such that every vertex in C''' is visible to b and one of its neighboring vertices is not visible to b. Then, the convex hull of the vertices in C''' must cross Q since at least one edge of the convex hull, say $\overline{ac}$ crosses Q. Then, b as the tip-vertex determines the desired maximum dome.

The lemma below follows the fact that P_b^v and Q_b^v are subsets of P^v and Q^v, respectively.

Lemma 3. $| P_b^v \cup Q_b^v |$ *is bounded by* $| P^v \cup Q^v |$.

3   Invariant of Nest Convexity

Let P and Q be two nested convex polyhedra with triangle faces. After a MD_v process, the remainder $P - Q$ is still two nested convex polyhedra. They share the faces (or edges or vertices) formed by Q'^v. The outer surface of $P - Q$ is the convex hull of Q and $(P - P^v) \cup P_b^v$ which includes the bridge faces, and the inner surface is still Q. However, the visible part of the inner surface must exclude the faces of Q on the convex hull $CH(Q \cup ((P - P^v) \cup P_b^v))$.

The following figures show how a maximum dome with tip vertex v is removed from convex polyhedron P. It is easy to see that after the removal, the remainder $P - Q$ is still a nested convex polyhedra case.

The property of nest convexity invariant with respect to the maximum dome-removal operation ensures the correctness of our decomposition. Lemma 1, Lemma 2, and Lemma 3 ensure the number of tetrahedra produced by our maximum dome-removal decomposition is linear.

4   The Algorithm

MaxD-REMOVE(P, Q)

   Input: Two nested convex polyhedra P and Q (a straight-line planar graph in DCEL and vertex-edge incident list).

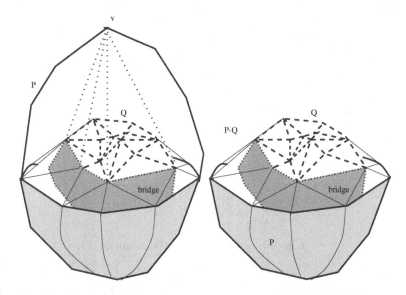

Fig. 3. The maximum dome with tip v.

Output: $T = (t_1, t_2, ..., t_k)$ (a sequence of tetrahedra), $\cup_{i \in k} t_i = P - Q$.

1. While $P - Q \neq \emptyset$ Do
   (a) $v \leftarrow select(P - Q)$; (* the bottom of MD_v contains at least one vertex of Q or a facet of Q. This can be done due to Lemma 2 *)
   (b) Find **MaxDome**$(v, P - Q)$
   (c) Update $P - Q$;
2. EndDo.

   MaxDome$(v, P - Q)$

1. Find P^v and Q^v in $P - Q$;
2. Find the maximum-dome MD_v (i.e., find $CH(P_b^v \cup Q)$: the bottom and $CH(P^v)$: the top);
3. $T \leftarrow v$-emission tetrahedralization of the maximum-dome;

End-MaxD-REMOVE.

Lemma 4. *The number of bridge-faces created in the process of maximum dome removal is bounded by* $O(m + n)$.

Proof. The number of bridge-faces created in an MD removal is bounded by the number of edges in the chains formed by P_b^v and Q_b^v. Thus, we only need show that each edge in (chains of) P_b^v and Q_b^v created by an MD_v cannot appear in (chains of) P_b^u and Q_b^u created by a later MD_u. That is, such an edge is removed in only one MD. We shall consider the following two cases as well as

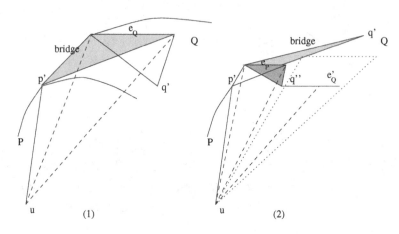

Fig. 4. For the proof of the lemma.

some subcases. (1) If and edge e_Q of Q_b^v is visible to u of P, then by the nest-convexity invariant of $P - Q$ triangle facet $\triangle e_Q q'$ for q' of Q and triangle facet $\triangle e_Q p'$ for p' of P in $P - Q$ are visible to u. Then, tetrahedra $up'e_Q$ and $uq'e_Q$ will belong to the maximum dome MD_u. Then, e_Q is not on the new Q_b^u and e_Q will not appear in the later MD-removal. (2) If e_P of P_b^v is visible to u, then by the nest-convexity invariant of $P - Q$ vertex p' of portion P is visible to u. Furthermore (i) if u is also visible to q', then both tetrahedra $up'e_P$ and $uq'e_P$ are belong to the MD_u and e_P will not belong to the new P_b^u; (ii) if u is not visible to q', then e_P must form a new bridge face with other vertex of portion Q in $P - Q$, say q''. But, both triangle facets $\triangle e_P p'$ and $\triangle e_P q''$ are visible to u and they will be removed. The chain formed by the new P_b^u will not contain e_P. In any case, the edges in P_b^v and Q_b^v cannot appear in P_b^u and Q_b^u. This implies that the number of bridge-facets created and removed in the process corresponds the number of surface facets in $P - Q$.

Theorem 1. *Algorithm $MaxD - REMOVE$ produces $2(m + n - 3)$ tetrahedra for the region between two nested convex polyhedra with m and n vertices and the algorithm runs $O((m + n)^3 \log(m + n))$ time.*

5   Remarks

We have shown a method, called 'maximum-dome removal', to tetrahedralize the region between two nested convex polyhedra into a linear number of tetrahedra. The method is simple and we are implementing the algorithm. The time complexity of our algorithm can be further improved into $O((m + n)^2)$ by more detail analysis, which is too long to included in this conference version.

Acknowledgments

This work is partially supported by NSERC grant OPG0041629 and ERA grant in Memorial University of NFLD.

References

1. D.Avis and H.ElGindy, Triangulating point sets in space, *Discrete and Computational Geometry* , 2(1987)99–111.
2. M.Bern, Compatible tetrahedralizations, In *Proc. 9th ACM Symp. Computational Geometry* 281–288, 1993.
3. M.Bern and D.Eppstein, Mesh generation and optimal triangulation, In *Computing in Euclidean Geometry* , (D. Du and F. Hwang eds.), World Scientific Publishing Co. 1992.
4. B.Chazelle, Convex partitions of polyhedra: a lower bound and worst-case optimal algorithm, *SIAM J. Computing* , 13(1984)488–507.
5. B.Chazelle and L.Palios, Triangulating a nonconvex polytope, *Discrete and computational geometry* , 5(1990)505–526.
6. B.Chazelle and N.Shouraboura, Bounds on the size of tetrahedralizations, In *Proc. 10th ACM Symp. Computational Geometry* , 231–239, 1994.
7. H.Edelsbrunner, F.Preparata, and D.West, Tetrahedralizing point sets in three dimensions, *Journal of Symbolic Computation* , 10(1990)335–347.
8. J.Goodman and J.Pach, Cell Decomposition of Polytopes by Bending, *Israel J. of Math*, 64(1988)129–138.
9. J.Goodman and J.O'Rourke, *Handbook of Discrete and Computational Geometry*, CRC Press, New York, 1997.
10. J.Rupert and R.Seidel, On the difficulty of tetrahedralizing 3-dimensional nonconvex polyhedra, *Discrete and Computational Geometry*, 7(1992)227–253.
11. E.Schonhardt, Uber die Zerlegung von Dreieckspolyedern in Tetraeder, *Math. Annalen* 98(1928)309–312.

Efficient Algorithms for Two-Center Problems for a Convex Polygon

(Extended Abstract)

Sung Kwon Kim[1] and Chan-Su Shin[2]

[1] Dept. of Computer Engineering, Chung-Ang University, Seoul 156–756, Korea.
skkim@cau.ac.kr
[2] Dept. of Computer Science, HKUST, Clear Water Bay, Hong Kong.
cssin@cs.ust.hk

Abstract. Let P be a convex polygon with n vertices. We want to find two congruent disks whose union covers P and whose radius is minimized. We also consider its discrete version with centers restricted to be at vertices of P. Standard and discrete two-center problems are respectively solved in $O(n \log^3 n \, \log \log n)$ and $O(n \log^2 n)$ time. Furthermore, we can solve both of the standard and discrete two-center problems for a set of points in convex positions in $O(n \log^2 n)$ time.

1 Introduction

Let A be a set of n points in the plane. The *standard two-center problem* for A is to cover A by a union of two congruent closed disks whose radius is as small as possible. The standard two-center problem has been studied extensively. Sharir [14] firstly presented a near-linear algorithm running in $O(n \log^9 n)$ time, and Eppstein [8] subsequently proposed a randomized algorithm with expected $O(n \log^2 n)$ running time. Chan [3] recently gave a randomized algorithm that runs in $O(n \log^2 n)$ time with high probability, as well as a deterministic algorithm that runs in $O(n \log^2 n (\log \log n)^2)$ time. As a variant, we consider the *discrete two-center problem* for A that finds two congruent closed disks whose union covers A and whose centers are at points of A. Recently, this problem was solved in $O(n^{4/3} \log^5 n)$ time by Agarwal et al. [1].

In this paper, we consider some variants of the two-center problems. Let P be a convex polygon with n vertices in the plane. We wan to find two congruent closed disks whose union covers P (its boundary and interior) and whose radius is minimized. We also consider its *discrete* version with centers restricted to be at some vertices of P. See Figure 1. Compared with the two-center problems for points, differences are that (1) points to be covered by two disks are the vertices of P in convex positions (not in arbitrary positions) and (2) two disks should cover the edges of P as well as its vertices. By points in *convex positions*, we mean that the points form the vertices of a convex polygon. (1) suggests our problems are most likely easier than the point-set two-center problems, but (2) tells us that they could be more difficult.

D.-Z. Du et al. (Eds.): COCOON 2000, LNCS 1858, pp. 299–309, 2000.

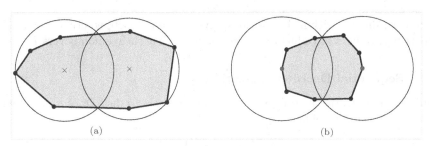

Fig. 1. Two center problems for a convex polygon. (a) Standard problem. (b) Discrete problem.

We assume in this paper that the vertices of P are in *general circular position*, meaning that no four or more vertices are co-circular. Our results are summarized in Table 1. (Some improve the authors' previous work of [15].)

	Standard 2-center		Discrete 2-center	
Convex polygon	$O(n \log^3 n \log \log n)$	Thm. 3	$O(n \log^2 n)$	Thm. 5
Points in convex positions	$O(n \log^2 n)$	Thm. 7	$O(n \log^2 n)$	Thm. 6
Points in arbitrary positions	$O(n \log^2 n \ (\log \log n)^2)$	[3]	$O(n^{4/3} \log^5 n)$	[1]

Table 1. The summary of the results.

Most of our algorithms presented in this paper are based on parametric search technique proposed by Megiddo [13]. To apply the technique, we need to design efficient sequential and parallel algorithms, A_s and A_p, for the corresponding decision problem: for a minimum parameter r^*, decide whether a given parameter $r \geq r^*$ or $r < r^*$. If A_s takes $O(T_s)$ time, and A_p takes $O(T_p)$ time using $O(P)$ processors, parametric search results in a sequential algorithm for the optimization problem with running time $O(PT_p + T_sT_p \log P)$. Cole [7] described an improvement technique that achieves $O(PT_p + T_s(T_p + \log P))$ time, assuming that the parallel algorithm satisfies a "bounded fan-in/out" requirement.

2   Algorithm for Standard Two-Center Problem for a Convex Polygon

Let P be a convex polygon with n vertices. The vertices are numbered $0, 1, \ldots, n-1$[1] counterclockwise and an edge is denoted by specifying its two end points, e.g., $(i, i+1)$. Let $\langle a, b \rangle$ be the counterclockwise sequence of vertices $a, a+1, \ldots, b$, if $a \leq b$, and $a, a+1, \ldots, n-1, 0, \ldots, b$, otherwise.

Let r^* be the minimum value such that two disks of radius r^* cover P. Since parametric search technique will be employed, we will give a sequential algorithm

---

[1] All vertex indices are taken modulo n.

and a parallel algorithm that decides, given a value $r > 0$, whether $r \geq r^*$, i.e., whether two disks of radius r can be drawn so that their union covers P.

2.1   Sequential Decision Algorithm

Our decision algorithm starts with drawing for each i a disk D_i of radius r with center at vertex i. Define

$$\mathcal{I}(\langle a, b \rangle) = \bigcap_{i=a}^{b} D_i.$$

Define m_i to be the index such that $\mathcal{I}(\langle i, m_i \rangle) \neq \emptyset$ and $\mathcal{I}(\langle i, m_i + 1 \rangle) = \emptyset$. The index m_i is the "counterclockwise farthest" vertex from i so that a disk of radius r can include the vertices in $\langle i, m_i \rangle$. Note that the m_i's are monotone increasing, i.e., there is no pair of $\langle i, m_i \rangle$ and $\langle j, m_j \rangle$ such that $\langle i, m_i \rangle \subset \langle j, m_j \rangle$.

Let $\mathcal{J}_i = \mathcal{I}(\langle i, m_i \rangle)$. Then any point in $\mathcal{J}_i$ can be the center of a disk of radius r that covers $\langle i, m_i \rangle$. We check each of the following cases.

(i) If there are two vertices i and j such that $i, m_i \in \langle j, m_j \rangle$ and $j, m_j \in \langle i, m_i \rangle$, then any two disks of radius r, one with center in $\mathcal{J}_i$ and the other with center in $\mathcal{J}_j$, can cover P, and the decision algorithm returns "yes". To check for a fixed i whether there is an index j such that $i, m_i \in \langle j, m_j \rangle$ and $j, m_j \in \langle i, m_i \rangle$, it is sufficient to consider the case of $j = m_i$ only, due to the monotonicity of m-values.

(ii) If there are i and j such that $m_j + 1 = i$ and $j \in \langle i, m_i \rangle$, then $\langle i, m_j \rangle$ is guaranteed to be covered by two disks of radius r with centers in $\mathcal{J}_i$ and in $\mathcal{J}_j$ and the edge (m_j, i) of P is "missing". So, we need check whether it can be covered by two disks of radius r with centers in $\mathcal{J}_i$ and in $\mathcal{J}_j$. To do this, find a disk C of radius r with center on the edge (m_j, i) and touching $\mathcal{J}_i$ at a point p and checks whether C intersects $\mathcal{J}_j$. One disk of radius r with center p and the other with center at any point in $C \cap \mathcal{J}_j$ together cover the edge (m_j, i) as well as the other boundary of P. If such point p can be found and $C \cap \mathcal{J}_j \neq \emptyset$, then the decision algorithm returns "yes". Again, for a fixed i, it does not need to consider all j's such that $m_j = i - 1$ and $j \in \langle i, m_i \rangle$. It is easy to see that considering the case of $j = m_i$ is sufficient.

(iii) Finally, if there are i and j such that $m_i + 1 = j$ and $m_j + 1 = i$, then two edges (m_i, j) and (m_j, i) are "missing". To decide whether these two edges can simultaneously be covered by two disks of radius r with centers in $\mathcal{J}_i$ and in $\mathcal{J}_j$, find as above a disk C_1 of radius r with center on (m_j, i) and touching $\mathcal{J}_j$, and a disk C_2 of radius r with center on (m_j, i) and touching $\mathcal{J}_i$. Similarly, find two disk C_1', C_2' of radius r with centers on (m_i, j) touching $\mathcal{J}_j$ and $\mathcal{J}_i$, respectively. Now, check whether both $C_1 \cap C_1' \cap \mathcal{J}_j$ and $C_2 \cap C_2' \cap \mathcal{J}_i$ are nonempty. If both are nonempty, two disks of radius r, one with center in $C_1 \cap C_1' \cap \mathcal{J}_j$ and the other with center in $C_2 \cap C_2' \cap \mathcal{J}_i$, can cover both edges as well as the other boundary of P, and thus the decision algorithm returns "yes".

The decision returns "no" if none of the three cases above returns "yes".

Implementation. Compute m_0 and $\mathcal{J}_0$, and then $\mathcal{J}_{m_0}$ and $\mathcal{J}_{m_0+1}$. For $i > 0$, $\mathcal{J}_i$, $\mathcal{J}_{m_i}$, and $\mathcal{J}_{m_i+1}$ can be obtained from $\mathcal{J}_{i-1}$, $\mathcal{J}_{m_{i-1}}$ and $\mathcal{J}_{m_{i-1}+1}$, by adapting the algorithm in [9] for maintaining convex hulls of a simple chain while points are dynamically inserted into and deleted from the ends of the chain. An intermixed sequence of $O(\ell)$ insertions and deletions on chains with a total of $O(\ell)$ vertices can be processed in time $O(\ell \log \ell)$ [9].

Whenever $\mathcal{J}_i$, $\mathcal{J}_{m_i}$, and $\mathcal{J}_{m_i+1}$ are obtained, we check whether any of three cases (i), (ii) and (iii) returns "yes". Each of these checks can be done in $O(\log n)$ time, provided that $\mathcal{J}_i$, $\mathcal{J}_{m_i}$, and $\mathcal{J}_{m_i+1}$ are given. For this we need a constant number of the following operations.

(O1) Find a touching point between the intersection of $O(n)$ congruent disks and another congruent disk with center on a specific edge of P.

(O2) Decide whether a congruent disk with center on a specific edge of P intersects an intersection of $O(n)$ congruent disks. If yes, find the intersection points on the boundary.

Both (O1) and (O2) can be done in $O(\log n)$ time by simply performing binary searches on the boundary of the intersection of $O(n)$ congruent disks. We assume here that the boundaries of the intersections are stored in arrays or balanced trees for binary searches to be applicable.

As a whole, we have the following theorem.

Theorem 1. *Given a convex polygon P with n vertices and a value $r > 0$, we can decide in $O(n \log n)$ time whether two disks of radius r can cover P.*

2.2   Parallel Decision Algorithm

Our parallel algorithm builds a segment tree on the sequence, $D_0, \ldots, D_{n-1}$, of radius r disks centered at vertices. Each internal node[2] of the tree corresponds to a *canonical subsequence* of the vertices that are assigned to the leaves of the subtree rooted at the node. For each canonical subsequence S, compute the intersection of the disks, $\mathcal{I}(S) = \cap_{i \in S} D_i$. The boundary of $\mathcal{I}(S)$ is divided into two (upper and lower) chains by its leftmost and rightmost points, and these two chains are stored into two arrays of arcs[3] sorted according to their x-coordinates.

Constructing the segment tree, computing $\mathcal{I}(S)$ for each canonical subsequence S, and storing their chains into arrays can be carried out in $O(\log n)$ time using $O(n \log n)$ processors, as explained in [3].

Computing m_i for each i can be done in divide-and-conquer manner. Compute $m_{\lfloor n/2 \rfloor}$ by noting $m_{\lfloor n/2 \rfloor} = \ell$ such that $\mathcal{I}(\langle \lfloor n/2 \rfloor, \ell \rangle) \neq \emptyset$ and $\mathcal{I}(\langle \lfloor n/2 \rfloor, \ell + 1 \rangle) = \emptyset$. Then due to the monotonicity of the m-values, $m_i \in \langle 0, m_{\lfloor n/2 \rfloor} \rangle$ for $i \in \langle 0, \lfloor n/2 \rfloor - 1 \rangle$, and $m_i \in \langle m_{\lfloor n/2 \rfloor}, n - 1 \rangle$ for $i \in \langle \lfloor n/2 \rfloor + 1, n - 1 \rangle$. Thus, they can be recursively obtained. Our parallel algorithm for this step performs $O(\log n)$ recursions each consisting of $O(n)$ decisions on the emptiness of $\mathcal{I}(\langle a, b \rangle)$ for different a and b. Since, given a and b, whether $\mathcal{I}(\langle a, b \rangle) = \emptyset$ can be decided

---

[2] To avoid confusion, *nodes* instead of vertices will be used in trees.

[3] Instead of edges, we use *arcs* to denote sides of an intersection of disks.

in $O(\log n \log \log n)$ time using $O(\log n)$ processors [3], the whole of this step takes $O(\log^2 n \log \log n)$ time using $O(n \log n)$ processors.

Checking the first case (i) in parallel is easy. For each i, decide whether $m_{m_i} \in \langle i, m_i \rangle$. The algorithm returns "yes" if any one of these n decisions is true.

For case (ii), collect all i's such that $m_{m_i} = i - 1$, and for each such i decide whether edge $(i - 1, i)$ can be covered by two disks of radius r with centers in $\mathcal{J}_i$ and in $\mathcal{J}_{m_i}$.

For case (iii), collect all i's such that $m_{m_i+1} = i-1$, and for each such i decide whether both edges $(m_i, m_i + 1)$ and $(i - 1, i)$ can be simultaneously covered by two disks of radius r with centers in $\mathcal{J}_i$ and in $\mathcal{J}_{m_i+1}$.

For any of cases (ii) and (iii), as in the sequential algorithm, our parallel algorithm also needs to compute $\mathcal{J}_i$ for each i. Computing all $\mathcal{J}_i$'s and storing them explicitly requires $\Omega(n^2)$ work and space, which is far from the bounds of our parallel algorithm. Instead, whenever a $\mathcal{J}_i$ is needed, we compute it from the canonical subsequences of $\langle i, m_i \rangle$ with the help of the segment tree.

Assume that $\langle i, m_i \rangle$ is a union of $k = O(\log n)$ canonical subsequences $S_1, \ldots, S_k$, and also assume that $S_1, \ldots, S_k$ are indexed counterclockwise so that $i \in S_1, m_i \in S_k$, and S_ℓ and $S_{\ell+1}$ for $1 \le \ell \le k - 1$ appear consecutively on the boundary of P. Then, $\mathcal{J}_i = \mathcal{I}(S_1) \cap \ldots \cap \mathcal{I}(S_k)$. If $i > m_i$, then $\langle i, m_i \rangle$ is divided into $\langle i, n - 1 \rangle$ and $\langle 0, m_i \rangle$. $\langle i, m_i \rangle$ is a union of the canonical subsequences for $\langle i, n - 1 \rangle$ and the canonical subsequences for $\langle 0, m_i \rangle$.

Let $\beta(\mathcal{J}_i)$ be the upper chain of $\mathcal{J}_i$ and U_j be the upper chain of $\mathcal{I}(S_j)$ for $1 \le j \le k$. We show how $\beta(\mathcal{J}_i)$ is constructed from the U_j. The computation of its lower chain is completely symmetric.

Build a minimum-height binary tree whose leaves correspond to $U_1, \ldots, U_k$ in a left-to-right order. Let v be a node of the tree and its canonical subsequence be $\langle a, b \rangle$. Let $\mathcal{I}_v = \mathcal{I}(S_a) \cap \ldots \cap \mathcal{I}(S_b)$ and $\beta(\mathcal{I}_v)$ be its upper chain. Then, $\beta(\mathcal{I}_v)$ consists of some subchains of $U_a, U_{a+1}, \ldots, U_b$[4] and thus can be represented by a set of tuples, where each tuple says which subchain of which U_j contributes to $\beta(\mathcal{I}_v)$.

In a bottom-up fashion, compute $\beta(\mathcal{I}_v)$ for each node v of the minimum-height binary tree. For each leaf v corresponding to U_j,
$\beta(\mathcal{I}_v) = \{(j, 1, x^-, x^+, 1, |U_j|)\}$, where x^- (resp., x^+) is the x-coordinate of the leftmost (resp., rightmost) point of U_j. For each internal node v, $\beta(\mathcal{I}_v) = \{(j_1, p_1, x_1^-, x_1^+, r_1^-, r_1^+), \ldots, (j_t, p_t, x_t^-, x_t^+, r_t^-, r_t^+)\}$, where t is the number of subchains of U_j's that consist of $\beta(\mathcal{I}_v)$. Each tuple represents a subchain – which U_j it originally comes from, the rank of its leftmost arc in its original U_j, its x-range, and the ranks of its leftmost and rightmost arcs in $\beta(\mathcal{I}_v)$. For example, the first tuple $\beta(\mathcal{I}_v)$ means the following three things:

1. the subchain of U_{j_1} in the x-range $[x_1^-, x_1^+]$ appears in $\beta(\mathcal{I}_v)$;
2. its leftmost (resp., rightmost) arc is bounded by a vertical line $x = x_1^-$ (resp., $x = x_1^+$) and its leftmost arc is a part of the p_1-th arc in U_{j_1}; and

---

[4] A subchain of U_j is a part of U_j between any two points on it.

3. its leftmost (resp., rightmost) arc becomes the r_1^--th (resp., r_1^+-th) arc in $\beta(\mathcal{I}_v)$.

Lemma 1. *Assume that v has two children w and z. Given $\beta(\mathcal{I}_w)$ and $\beta(\mathcal{I}_z)$, $\beta(\mathcal{I}_v)$ can be obtained in $O(\log n \cdot (|\beta(\mathcal{I}_w)| + |\beta(\mathcal{I}_z)| + \log n))$ sequential time.*

Proof. Let $f = \beta(\mathcal{I}_w)$ and $g = \beta(\mathcal{I}_z)$. Let S^w and S^z be the canonical subsequences of w and z, respectively. In other words, $\mathcal{I}_w = \cap_{i \in S^w} D_i$ and $\mathcal{I}_z = \cap_{i \in S^z} D_i$. Let $|S^w|$ and $|S^z|$ be the number of vertices in S^w and S^z. Let $[x^-, x^+]$ be the common x-range of f and g. The intersection points, if any, of f and g must be in the vertical strip bounded by two lines $x = x^-$ and $x = x^+$. So, we may assume that f and g are functions over $[x^-, x^+]$. Note that f and g consist of several arcs of radius r. Lemma 1 will be proved through a series of lemmas whose proofs are given in the full paper [11].

Lemma 2. *f and g intersect at most twice.*

Since both f and g are upper chains of the intersection of congruent disks, they are upward convex with one highest point. Define $h(x) = f(x) - g(x)$ over $[x^-, x^+]$.

Lemma 3. *If both f and g are either increasing or decreasing over $[x^-, x^+]$, $f(x^-) > g(x^-)$, and $|S^w| \geq 3$ and $|S^z| \geq 3$, then h over $[x^-, x^+]$ is a downward convex function with one lowest point.*

Using Lemma 3, we can prove the following lemma.

Lemma 4. *If each of f and g is either increasing or decreasing over $[x^-, x^+]$, then their intersections, if any, can be found in $O(\log n \cdot (|\beta(\mathcal{I}_w)| + |\beta(\mathcal{I}_z)| + \log n))$ sequential time.*

To compute the intersection points between f and g, find x^f and x^g, the x-coordinates of the highest points of f and g, respectively. Such points, x^f and x^g, can be computed in time $O(|\beta(\mathcal{I}_w)| \cdot \log n)$ and $O(|\beta(\mathcal{I}_z)| \cdot \log n)$, respectively.

Wlog, assume that $x^f < x^g$. Then, $[x^-, x^+]$ is divided into three subranges $[x^-, x^f]$, $[x^f, x^g]$, and $[x^g, x^+]$. In each subrange, f is either increasing or decreasing and the same is true for g. Then Lemma 4 applies to compute intersections of f and g over each subrange. For details, see [11].

After locating the intersection points between f and g, it is easy to determine which tuples from $\beta(\mathcal{I}_w)$ and $\beta(\mathcal{I}_z)$ are eligible to be in $\beta(\mathcal{I}_v)$. Thus, the tuples of $\beta(\mathcal{I}_v)$ can be computed in additional $O(|\beta(\mathcal{I}_w)| + |\beta(\mathcal{I}_z)|)$ time.

This completes the proof of Lemma 1.

Given all $\beta(\mathcal{I}_w)$ for the nodes w of some level of the minimum-height tree, if k processors are available, then all $\beta(\mathcal{I}_v)$ for the nodes v of the next level (toward the root) can be obtained in $O(\log n \cdot (|\beta(\mathcal{I}_w)| + \log n)) = O((k + \log n) \log n)$ time by Lemma 1. Note that $|\beta(\mathcal{I}_w)| = O(k)$ for any node w. $|\beta(\mathcal{I}_w)|$ is the number of tuples, not the number of arcs in it. The minimum-height tree with k leaves has $O(\log k)$ levels.

Lemma 5. *For any i, $\beta(\mathcal{J}_i)$, in the form of a set of tuples, can be constructed in $O((k + \log n) \log k \log n)$ time using $O(k)$ processors.*

Back to checking case (ii) or case (iii), assign $k = O(\log n)$ processors to each i satisfying the condition of (ii) or (iii), and construct $\mathcal{J}_i$, in the form of two sets of tuples, one for its upper chain and the other for its lower chain, in $O(\log^2 n \log \log n)$ time by Lemma 5. Since there are at most n such i's, we need a total of $O(n \log n)$ processors.

The remaining part of (ii) or (iii) can be done using a constant number of operations (O1) and (O2) on $\mathcal{J}_i$. Each of (O1) and (O2), using the two sets of tuples, can now be executed in $O(\log^2 n)$ sequential time. As a result, we have the theorem.

Theorem 2. *Given a convex polygon P with n vertices and a value $r > 0$, we can decide in $O(\log^2 n \log \log n)$ time using $O(n \log n)$ processors whether two disks of radius r cover P.*

Combining Theorems 1 and 2, we have proved the theorem.

Theorem 3. *Given a convex polygon P with n vertices, we can compute the minimum radius r^* in $O(n \log^3 n \log \log n)$ time such that two disks of radius r^* cover P.*

Proof. By parametric search technique, we immediately have an $O(n \log^4 n \log \log n)$ time algorithm. Our parallel algorithm can easily be made to satisfy the Cole's requirement [7]. The time complexity is now reduced to $O(n \log^3 n \log \log n)$.

3   Algorithm for Discrete Two-Center Problem for a Convex Polygon

Let P be a convex polygon with n vertices. We want to find the minimum value r^* and two vertices i, j such that two disks of radius r^* with centers i and j cover P. Since again parametric search technique will be employed, we will give a sequential algorithm and a parallel algorithm that decides, given a value $r > 0$, whether $r \geq r^*$.

3.1   Sequential Decision Algorithm

Let ∂P denote the boundary of P. For two points a, b on ∂P, $\langle\!\langle a, b \rangle\!\rangle$ denotes the subchain of ∂P that walks counterclockwise starting at a and ending at b.

Given $r > 0$, draw a disk D_i of radius r with center at vertex i for each i. Then, $D_i \cap \partial P$ may be disconnected and thus may consist of several subchains of P. One of them contains i, which will be denoted by $\partial_i = \langle\!\langle a_i, b_i \rangle\!\rangle$.

Lemma 6. *$r \geq r^*$ iff there exist two vertices i and j such that $\partial P = \partial_i \cup \partial_j$.*

Computing the end points a_i, b_i of ∂_i for each i can be done by simulating sequentially our parallel algorithm in Section 3.2. The simulation takes $O(n \log n)$ time for preprocessing and for a query i answers a_i and b_i in $O(\log n)$ time. Alternatively, the algorithm in [6] can be used. It, given $r > 0$, preprocesses P in $O(n \log n)$ time, and, for a query point c, one can find in $O(\log n)$ the first (clockwise and counterclockwise) hit point on ∂P when a circle of radius r with center at c is drawn.

After having ∂_i for all i, by Lemma 6 we need to decide whether there exist two vertices i and j such that $\partial P = \partial_i \cup \partial_j$. This is an application of *minimum circle cover* algorithm in [2,10,12] that runs in $O(n \log n)$ time. In the minimum circle cover problem, we are given a set arcs of a circle and want to find the minimum number of arcs that covers the circles (i.e., whose union is the circle).

Theorem 4. *Given a convex polygon P with n vertices and a value $r > 0$, we can decide in $O(n \log n)$ time whether two disks of radius r with centers at vertices of P can cover P.*

3.2   Parallel Decision Algorithm

Our parallel algorithm first computes ∂_i for each i by locating a_i and b_i. Build a segment tree on the sequence of radius r disks centered at vertices, $D_0, \ldots, D_{n-1}$, $D_0, \ldots, D_{n-2}$. For each node v with canonical subsequence S, compute the intersection of the disks, $\mathcal{I}_v = \cap_{i \in S} D_i$. The boundary of $\mathcal{I}_v$ is divided into two (upper and lower) chains by its leftmost and rightmost points, and the upper and lower chains are stored into two arrays, $\beta^+(\mathcal{I}_v)$ and $\beta^-(\mathcal{I}_v)$, respectively, of arcs sorted according to their x-coordinates. Given a point q, whether $q \in \mathcal{I}_v$ can be decided by doing binary searches on $\beta^+(I_v)$ and on $\beta^-(I_v)$.

To compute $\langle\!\langle a_i, b_i \rangle\!\rangle$, it is sufficient to find the minimal $\langle\!\langle c_i, d_i \rangle\!\rangle \supseteq \langle\!\langle a_i, b_i \rangle\!\rangle$, where c_i and d_i are vertices of P.

For a given vertex i, d_i can be searched using the segment tree as follows: Choose the left one from the two D_i's at the leaves of the tree. We go upward along the leaf-to-root path starting at the chosen D_i, and locate the first node on the path with a right child z' satisfying $i \notin \mathcal{I}_{z'}$. Starting at z', we now search downward. At a node with left child w and right child z, search under the following rule: Go to z if $i \in \mathcal{I}_w$, and go to w, otherwise. It is easy to verify that when search is over at a leaf v, we have $d_i = v$.

Searching c_i is completely symmetric.

Implementation. Constructing the segment tree and computing $\beta^+(\mathcal{I}_v)$ and $\beta^-(\mathcal{I}_v)$ for each v can be carried out in $O(\log n)$ time using $O(n \log n)$ processors [3] as in Section 2.2.

For a fixed i, c_i and d_i can be found in $O(\log^2 n)$ sequential time as we visit $O(\log n)$ nodes and at each node v we spend $O(\log n)$ time doing binary search to decide whether $i \in \mathcal{I}_v$. This type of search (i.e., searching the same value in several sorted lists along a path of a binary tree) can be improved by an application of fractional cascading [4,5], saving a logarithmic time to achieve $O(\log n)$ search time.

From c_i and d_i, we compute $\partial_i = \langle\langle a_i, b_i \rangle\rangle$ by computing the intersection points between a circle of radius r with center i and two edges $(c_i, c_i + 1), (d_i - 1, d_i)$. With ∂_i for all i, an application of parallel minimum circle cover algorithms in [2,10] runs in $O(\log n)$ time using $O(n)$ processors.

Theorem 5. *Given a convex polygon P with n vertices, we can compute the minimum radius r^* in $O(n \log^2 n)$ time such that two disks of radius r^* with centers at vertices of P cover P.*

Proof. By Theorem 4, our sequential decision algorithm takes $O(n \log n)$ time.

After constructing the segment tree and computing $\beta^+(\mathcal{I}_v)$ and $\beta^-(\mathcal{I}_v)$ for each v in our parallel algorithm, sort the vertices of P and the vertices of $\beta^+(\mathcal{I}_v) \cup \beta^-(\mathcal{I}_v)$ for all nodes v by their x-coordinates. This makes the application of fractional cascading and computation of c_i and d_i no longer depend on r, by using ranks rather than x-coordinates. So, these steps can be done sequentially.

Since the Cole's improvement [7] again can be applied to the parallel steps (i.e., constructing the segment tree, sorting, and applying parallel circle cover algorithm, that take $O(\log n)$ time using $O(n \log n)$ processors), the time bound can be achieved.

Our idea for Theorem 5 can be easily extended to the discrete two-center problem for points when the points are the vertices of a convex polygon.

Theorem 6. *Given a set A of n points that are vertices of a convex polygon, we can compute the minimum radius r^* in $O(n \log^2 n)$ time such that two disks of radius r^* with centers at points of A cover A.*

4   Algorithm for Standard Two-Center Problem for a Set of Points in Convex Position

Suppose that a set A of n points in convex positions are given. We assume that as input a convex polygon with vertices at the points of A is given, so the points are sorted counterclockwise. We want to find two disks of radius r^* whose union covers A. As usual, the points are numbered $0, \dots, n-1$ counterclockwise.

For any $i, j \in A = \langle 0, n-1 \rangle$, let r_{ij}^1 be the radius of the smallest disk containing $\langle i, j-1 \rangle$ and let r_{ij}^2 be the radius of the smallest disk containing $\langle j, i-1 \rangle$. Then, $r^* = \min_{i,j \in A} \max\{r_{ij}^1, r_{ij}^2\}$.

Let $r_0^* = \min_{j \in A} \max\{r_{0,j}^1, r_{0,j}^2\}$. Let k be the index such that $r_0^* = \max\{r_{0,k}^1, r_{0,k}^2\}$. Then, A is separated into $\langle 0, k-1 \rangle$ and $\langle k, n-1 \rangle$.

Lemma 7. *For any i, j such that $i, j \in \langle 0, k-1 \rangle$ or $i, j \in \langle k, n-1 \rangle$, $\max\{r_{ij}^1, r_{ij}^2\} > r_0^*$.*

By Lemma 7 we need to consider pairs of i, j such that $i \in \langle 0, k-1 \rangle$ and $j \in \langle k, n-1 \rangle$. We now apply divide-and-conquer with $A_1 = \langle 0, k-1 \rangle$ and $A_2 = \langle k, n-1 \rangle$. (i) Pick the medium-index vertex $\lfloor k/2 \rfloor$ from A_1 and (ii) find the vertex

k' such that $\max\{r^1_{\lfloor k/2 \rfloor, k'}, r^2_{\lfloor k/2 \rfloor, k'}\} = \min_{j \in \langle k, n-1 \rangle} \max\{r^1_{\lfloor k/2 \rfloor, j}, r^2_{\lfloor k/2 \rfloor, j}\}$. A_1 is separated into $A_{11} = \langle 0, \lfloor k/2 \rfloor - 1 \rangle$ and $A_{12} = \langle \lfloor k/2 \rfloor, k-1 \rangle$, and A_2 is separated into $A_{21} = \langle k, k'-1 \rangle$ and $A_{22} = \langle k', n-1 \rangle$. By Lemma 7, we consider pairs of points between A_{11} and A_{21}, and between A_{12} and A_{22}. (iii) Repeat recursively (i)–(ii) with $A_1 = A_{11}$ and $A_2 = A_{21}$, and with $A_1 = A_{12}$ and $A_2 = A_{22}$. Recursion stops when A_1 consists of a single point ℓ, and at that time $\min_{j \in A_2} \max\{r^1_{\ell,j}, r^2_{\ell,j}\}$ can be found as in (ii).

Since a smallest disk containing a convex chain can be found in linear time, $\max\{r^1_{ij}, r^2_{ij}\}$ for any i, j can be computed in linear time. Since $r^1_{0,1}, \ldots, r^1_{0,n-1}$ is increasing and $r^2_{0,1}, \ldots, r^2_{0,n-1}$ is decreasing, $\max\{r^1_{0,1}, r^2_{0,1}\}, \ldots, \max\{r^1_{0,n-1}, r^2_{0,n-1}\}$ is decreasing and then increasing. So, r^* and k can be found in $O(n \log n)$ time by binary search, in which each decision takes $O(n)$ time. Similarly, (ii) takes $O(n \log n)$ time. Since our algorithm invokes at most $O(\log n)$ recursions, and after recursion stops a single point $\ell \in A_1$ requires additional $O(|A_2| \log |A_2|)$ time, we can conclude the following theorem.

Theorem 7. *Given a set A of n points that are vertices of a convex polygon, we can compute the minimum radius r^* in $O(n \log^2 n)$ time such that two disks of radius r^* cover A.*

References

1. P. K. Agarwal and M. Sharir and E. Welzl, The discrete 2-center problem, *Proc. 13th Ann. ACM Symp. Comput. Geom.*, pp. 147–155, 1997.
2. L. Boxer and R. Miller, A parallel circle-cover minimization algorithm, *Inform. Process. Lett.*, 32, pp. 57–60, 1989.
3. T. M. Chan, More planar two-center algorithms, *Computational Geometry: Theory and Applications*, 13, pp. 189-198, 1999
4. B. Chazelle, and L. J. Guibas, Fractional cascading: I. A data structuring technique, *Algorithmica*, 1, pp. 133–162, 1986.
5. B. Chazelle, and L. J. Guibas, Fractional cascading: II. Applications, *Algorithmica*, 1, pp. 163–191, 1986.
6. S.-W. Cheng, O. Cheong, H. Everett, and R. van Oostrum, Hierarchical vertical decompositions, ray shooting, and circular arc queries in simple polygons, *Proc. 15th Ann. ACM Symp. Comput. Geom.*, pp. 227–236, 1999.
7. R. Cole, Slowing down sorting networks to obtain faster sorting algorithms, *J. ACM*, 34, pp. 200–208, 1987.
8. D. Eppstein, Faster construction of planar two-centers, *Proc. 8th ACM-SIAM Symp. Discrete Algorithms*, pp. 131–138, 1997.
9. J. Friedman, J. Hershberger, and J. Snoeyink, Compliant motion in a simple polygon, *Proc. 5th Ann. ACM Symp. Comput. Geom.*, pp. 175–186, 1989.
10. S. K. Kim, Parallel algorithms for geometric intersection graphs, Ph.D. Thesis, Dept. of Comput. Sci. Eng., U. of Washington, 1990.
11. S. K. Kim, and C.-S. Shin, Efficient algorithms for two-center problems for a convex polygon, *HKUST-TCSC-1999-16*, Dept. of Comp. Sci., HKUST, 1999.
12. C. C. Lee and D. T. Lee, On a circle-cover minimization problem, *Inform. Process. Lett.*, 18, pp. 109–115, 1984.

13. N. Megiddo, Applying parallel computation algorithms in the design of serial algorithms, *J. ACM*, 30, pp. 852–865, 1983.

14. M. Sharir, A Near-Linear Algorithm for the Planar 2-Center Problem, *Proc. 12th Ann. ACM Symp. Comput. Geom.*, pp. 106–112, 1996.

15. C.-S. Shin, J.-H. Kim, S. K. Kim, and K.-Y. Chwa, Two-center problems for a convex polygon, *Proc. 6th Euro. Symp. on Algo.*, Vol. 1461, pp. 199–210, 1998.

On Computation of Arbitrage for Markets with Friction

Xiaotie Deng[1,*], Zhongfei Li[2,**], and Shouyang Wang[3,***]

[1] Department of Computer Science
City University of Hong Kong
Hong Kong, P. R. China
Deng@cs.cityu.edu.hk

[2] Institute of Systems Science
Chinese Academy of Sciences, Beijing, P.R. China
zfli@iss02.iss.ac.cn

[3] Academy of Mathematical and Systems Science
Chinese Academy of Sciences, Beijing, P.R. China
swang@iss04.iss.ac.cn

Abstract. We are interested in computation of locating arbitrage in financial markets with frictions. We consider a model with a finite number of financial assets and a finite number of possible states of nature. We derive a negative result on computational complexity of arbitrage in the case when securities are traded in integer numbers of shares and with a maximum amount of shares that can be bought for a fixed price (as in reality). When these conditions are relaxed, we show that polynomial time algorithms can be obtained by applying linear programming techniques. We also establish the equivalence for no-arbitrage condition & optimal consumption portfolio.

1   Introduction

The function of arbitrage in an economy is most vividly described as an *invisible hand* that led economic states move to equilibrium (Adam Smith in 1776 [30]). Arbitrage is commonly defined as a profit-making opportunity at no risk. In other words, it is an opportunity for an investor to construct a portfolio modification at non-positive costs that is guaranteed to generate a non-negative return later. Economic states with arbitrage are believed not to exist (at least not for any significant duration of time) in a normal situation. Accordingly, arbitrage-free assumption has been a fundamental principle in the studies of mathematical economics and finance [3,4,19,24,32]. Natually, characterizing arbitrage (or no-arbitrage) states for a set of economic parameter would be important here. The exact form would depend on the specific economic models we would consider. But

* Research partially supported by a CERG grant of Hong Kong UGC and an SRG grant of City University of Hong Kong
** Research partially supported by Natural Science Foundation of China
*** Research partially supported by Natural Science Foundation of China

D.-Z. Du et al. (Eds.): COCOON 2000, LNCS 1858, pp. 310–319, 2000.
© Springer-Verlag Berlin Heidelberg 2000

luckily, in most cases, it has been quite straight forward to do so, especially under usual assumptions of frictionless and complete market conditions [24,13,23,31].

Going into a different direction, the purpose of our study is to understand the computational complexity involving in characterization of arbitrage states (see [11,17,25] for the importance of computation in finance or economics). To deal with this issue, we follow a model of the market in [31] with a finite number of financial assets, $1, 2, \cdots, n$, with a finite number of possible states of nature, $s_1, s_2, \ldots, s_m$. Assets are traded at date 0 and returns are realized at date 1. Let $R = (r_{ij})$ be an $n \times m$ matrix where r_{ij} is the payoff to financial asset i if state s_j occurs. Let $P = (p_1, p_2, \ldots, p_m)^T$ be the price vector, where p_i be the price of asset i $(i = 1, 2, \ldots, m)$ and T denotes the transposition. A portfolio is a vector $x = (x_1, x_2, \ldots, x_m)^T$, where x_i is the number of shares invested in asset i, $i = 1, 2, \ldots, m$. Under perfect market conditions, an arbitrage opportunity is a solution x such that $x^T P \leq 0$, $x^T R \geq 0$, and $-x^T P + x^T R e > 0$, where e is the vector of all ones. In a frictionless market, the problem is solvable by linear programming (see, e.g., [8]).

Since linear programming is known to be solvable in polynomial time, copmutationally this justifies the general belief that any arbitrage possibility will be shortlived since it will be identified very quickly, and be taken advantage of, and subsequently bring economic states to a no-arbitrage one. In real life, however, there are various forms of friction conditions imposed on trading rules. Some assumptions in the previous analysis may not always hold. For example, ask-bid prices of the same stock are often different (ask-bid spread). At stock exchange listings, one offers or bids for a stock at a certain price at a certain volume; and one can only buy or sell a stock at an integer amount. We want to know whether these rules, institutionalized by human, create difficulties for arbitrage, and thus create difficulties for general equilibrium to hold. If this is the case, understanding the rationals behind the motives of making arbitrage difficult may provide us with new insights in understanding the role of the financial market in the economy. Indeed, we show that under the above requirements (integrality of shares traded and the capacity constraints in the number of shares asked/bid), the problem to find out an arbitrage portfolio (under a given price vector) is computationally hard. We present this result in Section 2. One may notice that removing any one of the two constraints will result in a polynomial time solution via linear programming techniques. Obviously, the ask/bid amount and the corresponding price by investors cannot be changed. However, the integrality condition may be removed in principle, or at least be weakened by redividing shares so that the gain of arbitrage will be insignificant. We should consider the case without integrality conditions in the subsequent sections.

In Section 3, we consider transaction costs, another important factor in deviation from frictionless market. Friction complicates any theory one is to develop. Partially because of this, friction is generally ignored in asset pricing models although it is often brought up in informal discussion and debates. Recently, a few authors have started to incorporate transaction costs into financial market models [2,7,12,21]. In [21], Pham and Touzi established the equivalence between the no local arbitrage condition and the existence of an equivalent probability measure satisfying a generation of the martingale property as well as the equivalence of the no local free lunch and the no local arbitrage conditions. However, Their work is a complicated mathematical condition that may not always lead to an efficient algorithm to solve it. In Section 3, we establish a necessary and sufficient condition for arbitrage-free asset prices, in presence of transaction costs, for a financial market with finite assets and finite states of nature. In comparison, our result is of the classic type: the existence of positive implicit state-claim prices that satisfy a linear inequality system, which can be easily solved by linear program techniques.

In Section 4, we study a optimal consumption-portfolio selection problem in financial markets with friction under uncertainty. The absence of arbitrage is employed to characterize the existence or a necessary and sufficient condition of optimal consumption-portfolio policies when the financial assets and the state of nature are finite and when the friction is subject to proportional transaction costs. Historically optimal consumption and portfolio selection problems have been investigated by a number of authors [14,18,6,15,16,29,8,28,27,26]. Typically, continuous time problem is studied using the approach of stochastic dynamic programming, and under the assumption of frictionless market. The non-differentiability of transaction cost functions makes analysis of our problem complicated. We are able to establish a necessary and sufficient condition for a positive consumption bundle to be consumer optimal for some investor is that a transformed market exhibits no arbitrage.

Section 5 contains a few concluding remarks and suggestions for possible future extensions.

2   NP-Completeness of Arbitrage under Integral Conditions

Theorem 1 *In general, it is strongly NP-hard to find an arbitrage portfolio.*

The reduction is from 3SAT, a well known NP-hard problem [11]. Consider an instance of 3SAT: the variables are $X = \{x_1, x_2, \cdots, x_n\}$, and the clauses are $C_j = \{l_{j1}, l_{j2}, l_{j3}\} \subseteq X$, $j = 1, 2, \ldots, m$. We create two assets a_i (ask) and b_i (bid) for each variable, $i = 1, 2, \cdots, n$, and another asset a_0. The events are related to the variables and the clauses.

1. At event e_0, the payoff of a_0 is $n + 1$ and the payoffs to all other assets are -1 (so that a_0 must be bought);
2. at event f_i $(i = 1, 2, \cdots, n)$, the payoff of a_0 is -1, the payoff of a_i is 1 the payoff of b_i is -1 and the payoffs to all other assets are zero (either a_i is bought or b_i is sold);
3. at event g_i $(i = 1, 2, \cdots, n)$, the payoff of a_0 is 1 the payoff of a_i is -1, the payoff of b_i is 1, and the payoffs to all other assets are zero (not both a_i is bought and b_i is sold);

4. at event h_j $(j = 1, 2, \cdots, m)$, the payoff of a_0 is -1 if $x_i \in C_j$, the payoff to a_i is 1, if $\bar{x}_i \in C_j$, the payoff to b_i is -1 (at least one asset corresponding to C_j is traded).

With more careful analysis we can show that, for event e_0, we may choose the payoff to be one for a_0, and zero for all other assets.

3   A Necessary and Sufficient Condition for No-Arbitrage Pricing of Assets with Transaction Costs

In addition to the notations we introduced above, we use λ_i and μ_i respectively for the transaction cost rates for purchasing and selling one share of asset i $(i = 1, 2, \ldots, n)$, where $0 \leq \lambda_i, \mu_i < 1$.

The asset market is now completely described by the quadruple $\{P, R, \lambda, \mu\}$, where P is the price vector, R the payoff matrix, $\lambda = (\lambda_1, \lambda_2, \ldots, \lambda_n)^T$ the vector of transaction costs for purchasing, and $\mu = (\mu_1, \mu_2, \ldots, \mu_n)^T$ the vector of transaction costs for selling.

Without loss of generality, we ignore any redundant assets, omit any states with zero probability, and assume that R has rank $n \leq m$. Markets are said to be complete if $n = m$.

A portfolio modification is a vector $x = (x_1, x_2, \ldots, x_n)^T \in R^n$, where x_i is the number of shares of security i modified by a investor. If $x_i > 0$, additional asset i is bought for the amount of x_i; and if $x_i < 0$, additional asset i is sold for the amount of $-x_i$.

We introduce the functions defined on R by

$$\tau_i(z) = \begin{cases} (1 + \lambda_i)z, & \text{if } z > 0 \\ (1 - \mu_i)z, & \text{otherwise.} \end{cases}, \quad i = 1, 2, \ldots, n.$$

and the function τ mapping R^n into R by

$$\tau(x) = \sum_{i=1}^{n} p_i \tau_i(x_i), \quad \forall x \in R^n \tag{1}$$

The total cost including the proportional transaction cost of the portfolio modification $x = (x_1, x_2, \ldots, x_n)^T$ is then $\tau(x)$ and the vector of state-dependent payoffs on the modification is $R^T x$. If $\tau(x) = 0$, then the modification is costless, while if $\tau(x) < 0$, then the modification generates a positive date-0 cash flow.

For simplicity, we also define, for any $y = (y_1, y_2, \ldots, y_k)^T, z = (z_1, z_2, \ldots, z_k)^T \in R^k$ (where $k \geq 2$),

$$y \geq z \text{ if and only if } y_i \geq z_i (i = 1, 2, \ldots, k),$$
$$y > z \text{ if and only if } y \geq z \text{ and } y \neq z,$$
$$y \gg z \text{ if and only if } y_i > z_i (i = 1, 2, \ldots, k),$$

and denote

$$R_+^k = \{y \in R^k : y > 0\}, \ R_{++}^k = \{y \in R^k : y \gg 0\}.$$

Definition 1 *The market $\{P, R, \lambda, \mu\}$ has an arbitrage opportunity if there exists a portfolio modification $x = (x_1, x_2, \ldots, x_n)^T$ such that the following three conditions hold*

$$R^T x \geq 0, \tag{2}$$
$$\tau(x) \leq 0, \tag{3}$$
$$R^T x \neq 0 \text{ or } \tau(x) \neq 0. \tag{4}$$

In other words, the market exhibits no-arbitrage if there exists no portfolio modification $x = (x_1, x_2, \ldots, x_n)^T$ such that

$$\left(-\tau(x), x^T R\right)^T > 0. \tag{5}$$

To characterize the no-arbitrage condition, we need some lemmas.

A function $\psi : R^k \to R$ is sublinear if, for any $x, y \in R^k$ and $\lambda \in R_+$,

$$\psi(x + y) \leq \psi(x) + \psi(y) \quad \text{and} \quad \psi(\lambda x) = \lambda \psi(x).$$

It can be easily verify that

Lemma 1 *The cost functions $\tau, \tau_i, i = 1, 2, \ldots, n$ are sublinear and hence convex.*

For any topological vector space Y, denote by Y^* the topological dual space of Y. A set $C \subset Y$ is said to be a cone if $\lambda c \in C$ for any $c \in C$ and $\lambda \geq 0$ and a convex cone if in addition $C + C \subset C$. A cone C is said to be pointed if $C \cap (-C) = \{0\}$. For a set $C \subset Y$, its topological closure is denoted by cl C, its generated cone is the set

$$\text{cone } C := \{\lambda c : \ \lambda \geq 0, \ c \in C\},$$

its dual cone is defined as

$$C^+ := \{f \in Y^* : \ f(c) \geq 0, \ \forall c \in C\},$$

and the quasi-interior of C^+ is the set

$$C^{+i} := \{f \in Y^* : \ f(c) \gg 0, \ \forall c \in C \setminus \{0\}\}.$$

Recall that a base of a cone C is a convex subset B of C such that

$$0 \notin \text{cl } B \text{ and } C = \text{cone } B.$$

We will make reference to the following cone separation theorem [5].

Lemma 2 *Let Q and C be two closed convex cones in a locally convex vector space and let C be pointed and have a compact base. If $Q \cap (-C) = \{0\}$, then there exists $h \in C^{+i}$ such that $h \in Q^{+}$.*

Define the subset of R^{m+1}

$$K = \left\{ y = (y_0, y_1, \ldots, y_m)^T : y_0 \leq -\tau(x), (y_1, \ldots, y_m) = x^T R, x \in R^n \right\}.$$

Then, we have a property of K and a elementary equivalent definition of the no arbitrage as follows.

Lemma 3 *K is a closed convex cone.*

Proof. By the sublinearity of functional τ, it is clear that K is a convex cone. By the continuity of τ and the full row rank of R, it can be verified that K is closed.

Lemma 4 *The market $\{P, R, \lambda, \mu\}$ exhibits no-arbitrage if and only if*

$$K \cap R_{+}^{m+1} = \{0\}.$$

Proof.

$$K \cap R_{+}^{m+1} \neq \{0\} \tag{6}$$
$$\Leftrightarrow \quad \exists 0 \neq y \in K \cap R_{+}^{m+1} \tag{7}$$
$$\Leftrightarrow \exists y \in R^{m+1}, x \in R^n \text{ s.t. } 0 < y \leq \left(-\tau(x), \ x^T R\right)^T \tag{8}$$
$$\Leftrightarrow \qquad \text{there exists arbitrage.} \tag{9}$$

Now we state and prove our main result.

Theorem 2 *The market $\{P, R, \lambda, \mu\}$ exhibits no-arbitrage if and only if there exists $q = (q_1, q_2, \ldots, q_m)^T \in R_{++}^m$ such that*

$$- \mu_i p_i \leq \sum_{j=1}^{m} r_{ij} q_j - p_i \leq \lambda_i p_i, \quad i = 1, 2, \ldots, n. \tag{10}$$

Proof. Sufficiency: Assume that there exists $q = (q_1, q_2, \ldots, q_m)^T \in R_{++}^m$ such that (10) holds. If the market $\{P, R, \lambda, \mu\}$ had a arbitrage opportunity, then there would exist a portfolio x such that expressions (2)–(4) hold. Hence,

$$- \tau(x) + x^T R q > 0. \tag{11}$$

On the other hand, multiplying the first inequality of (10) by $x_i > 0$ and the second by $x_i \leq 0$ yields

$$x_i \sum_{j=1}^{m} r_{ij} q_j - x_i p_i \leq \begin{cases} \lambda_i x_i p_i, & \text{if } x_i > 0, \\ -\mu_i x_i p_i, & \text{otherwise.} \end{cases}$$

That is,

$$x_i \sum_{j=1}^{m} r_{ij} q_j \le \begin{cases} (1+\lambda_i) x_i p_i, & \text{if } x_i > 0 \\ (1-\mu_i) x_i p_i, & \text{otherwise} \end{cases} = p_i \tau_i(x_i).$$

Hence,

$$\tau(x) = \sum_{i=1}^{n} p_i \tau_i(x_i) \ge \sum_{i=1}^{n} x_i \sum_{j=1}^{m} r_{ij} q_j = \sum_{j=1}^{m} q_j \sum_{i=1}^{n} x_i r_{ij} = (x^T R) q,$$

i.e,

$$-\tau(x) + x^T R q \le 0$$

which contradicts expression (11). Therefore, the market $\{P, R, \lambda, \mu\}$ exhibits no-arbitrage.

Necessity: Assume that the market $\{P, R, \lambda, \mu\}$ exhibits no-arbitrage. Then, by Lemma 4,

$$(-K) \cap (-R_+^{m+1}) = \{0\},$$

where K is a closed convex cone by Lemma 3. Since R_+^{m+1} is a closed convex pointed cone and has a compact base, by Lemma 2, there exists $\bar{q} = (q_0, q) = (q_0, q_1, \ldots, q_m)^T \in (R_+^{m+1})^{+i} = R_{++}^{m+1}$ such that $\bar{q}^T y \le 0$ for all $y \in K$. Hence,

$$-q_0 \tau(x) + x^T R q \le 0, \ \forall x \in R^n$$

because $(-\tau(x), x^T R)^T \in K$ for all $x \in R^n$. Without loss of generality, we assume that $q_0 = 1$. Then,

$$-\sum_{i=1}^{n} p_i \tau_i(x_i) + \sum_{j=1}^{m} q_j \sum_{i=1}^{n} r_{ij} x_i \le 0, \ \forall x \in R^n. \tag{12}$$

Taking $x_i = \pm 1, x_j = 0, j \ne i$ in (12) yields the desired result (10).

Definition 2 *A vector $q \in R^m$ is said to be a state-price vector if it satisfies the expression (10).*

With this terminology we can re-state Theorem 2 as: there is no-arbitrage if and only if there is a strictly positive state-price vector.

In addition, from the proof of theorem 2 we have

Corollary 1 *If $q \in R^m$ is a state-price vector, then, for any portfolio modification x,*

$$x^T R q \le \tau(x).$$

That is, a discount value of the expected payoff of any modification under a specially chosen "risk-neutral" probability world is not more than its total cost.

Definition 3 *A portfolio modification x is said to be riskless if it has the same return in each state, i.e.*

$$x^T R = r(1, 1, \ldots, 1) \quad \text{for some} \quad r \in R.$$

Corollary 2 *If there exists a riskless asset or portfolio, then the sum of components of any state-price vector is not more than the discount on riskless borrowing.*

Proof. Let $\bar{x}$ be a riskless portfolio with $\bar{x}^T R = (1, 1, \ldots, 1)$. Then the cost $\tau(\bar{x})$ of the portfolio is the discount on riskless borrowing. Let q be a state-price vector. Then, by Corollary 1, one has

$$\tau(\bar{x}) \geq (\bar{x}^T R)q = (1, 1, \ldots, 1)q = \sum_{i=1}^{n} q_i.$$

4 Optimal Consumption Portfolio

Consider now the following optimal consumption-portfolio selection problem

$$(CP) \quad \begin{cases} \max_x & U(c_1, c_2, \ldots, c_m) \\ \text{subject to:} & \sum_{i=1}^{n} p_i \tau_i(x_i) \leq W \\ & c_j = \sum_{i=1}^{n} r_{ij} x_i + e_j, j = 1, 2, \ldots, m \\ & c_j \geq 0, j = 1, 2, \ldots, m \end{cases}$$

where $W > 0$ is the investor's initial wealth, $e = (e_1, e_2, \ldots, e_m)^T \in R_+^m$ is the vector of his or her endowment in every state at date 1, $c = (c_1, c_2, \ldots, c_m)^T$ is a vector of his or her consumption in every state at date 1, and $U : R_+^m \to R$ is his or her utility function which is assumed throughout the paper to be concave and strictly increasing.

The investor's budget-feasible set (i.e., the set of feasible solutions to (CP)) is

$$B = \{(c, x) : \ c \in R_+^m, \ x \in R^n, \ c = e + R^T x, \ \tau(x) \leq W\}.$$

We establish that (the proof is omitted here and will be given in the full paper):

Theorem 3 *If problem (CP) has an optimal solution, then the market $\{R, P, \lambda, \mu\}$ exhibits no-arbitrage. Conversely, if U is continuous and the market $\{R, P, \lambda, \mu\}$ has no-arbitrage, then there exists an optimal solution to problem (CP).*

We comment that the restrictions we imposed in the beginning of the section can be removed to by making a transformation procedure to change the states and thus the payoff matrix R. We can obtain a similar necessary and sufficient condition as stated below and details will be given in the full paper.

Theorem 4 *An allocation* $c^* = (c_1^*, c_2^*, \ldots, c_m^*)^T > 0$ *will be an optimal consumption for some investor with differentiable, strictly concave, state-independent, vNM utility function in the market* $\{R, P, \lambda, \mu\}$ *if and only if the market* $\{\bar{R}, P, \lambda, \mu\}$ *exhibits no-arbitrage.*

5 Conclusion

In this paper, we first establish a negative computational complexity result for arbitrage under realistic financial market situations. Then, we move on by dropping the integrality constraints (justifiable up to a certain extent with the bounded rationality principle) to derive a necessary and sufficient condition for the no-arbitrage market with transaction costs. The condition can be computed in polynomial time using linear programming.

Another dimension in the analysis of economic theory is the utility preferences of the participants. This determines, among all the no-arbitrage conditions, one that is a general equilibrium state under which no participant has the incentive to shift to a different allocation of its investment, under the frictionless assumption. Assuming risk aversion of participants, Green and Srivastava [13] showed that the equilibrium state of the market is equivalent to no-arbitrage. We studied an optimal consumption and portfolio selection problem in a finite-state financial market with transaction costs. We characterized the existence or a necessary and sufficient condition of an optimal consumption-portfolio policy by no-arbitrage of the original market as well as a transformed market.

References

1. Allingham, M., 1991. Arbitrage, St. Martin's Press, New York.
2. Ardalan, K., The no-arbitrage condition and financial markets with transaction costs and heterogeneous information: the bid-ask spread, *Global Finance Journal*, 1999, 10(1): 83.
3. Arrow, K. J., Debreu, G., Existence of an equilibrium for a competitive economy, *Econometrica*, 1954, 22: 265.
4. Black, F., Sholes, M., The pricing of options and corporate liabilities, *Journal of Political Economy*, 1973, 81: 637.
5. Borwein, J.M., 1977. Proper Efficient Points for Maximizations with Respect to Cones. *SIAM Journal on Control and Optimization* 15, 1977, pp.57–63.
6. J. C. Cox and C. F. Huang, *Optimal consumption and portfolio policies when asset prices follow a diffusion process*, Journal of Economic Theory **49** (1989), 33–83.
7. Dermody, J. C., Prisman, E. Z., No arbitrage and valuation in market with realistic transaction costs, *Journal of Financial and Quantitative Analysis* 1993, 28(1): 65.
8. J. B. Detemple and F. Zapatero, *Optimal consumption-portfolio policies with habit formation*, Mathematical Finance **2** (1992), no. 4, 251–274.
9. Duffie, D., *Dynamic asset pricing theory*, 2nd ed., Princeton University Press, New Jersey, 1996.
10. P. Dybvig and S. Ross, Arbitrage, in The New Palgrave: A Dictionary of Economics, 1, J. Eatwell, M. Milgate and P. Newman, eds., London: McMillan, 1987,, pp. 100–106.

11. Garey, M. R., Johnson, D. S., *Computers and Intractability: a Guide to the Theory of NP-completeness*, New York : W.H. Freeman, 1979.
12. Garman, M. B., Ohlson, J. A., Valuation of risky assets in arbitrage-free economies with transactions costs, *Journal of Financial Economics*, 1981, 9: 271.
13. R. C. Green, and S. Srivastava, "Risk Aversion and Arbitrage", *The Journal of Finance*, Vol. 40, pp. 257–268, 1985.
14. N. H. Hakansson, *Optimal investment and consumption strategies under risk for a class of utility functions*, Econometrica **38** (1970), no. 5, 587–607.
15. H. He and N. D. Pearson, *Consumption and portfolio policies with incomplete markets and short-sale constrains: the infinite dimensional case*, Journal of Economic Theory **54** (1991), 259–304.
16. H. He and N. D. Pearson, *Consumption and portfolio policies with incomplete markets and short-sale constrains: the finite dimensional case*, Mathematical Finance **1** (1991), no. 3, 1–10.
17. Hirsch, M. D. , Papadimitriou, C. H., Vavasis, S. A., Exponential lower bounds for finding brouwer fixed points, *Journal of Complexity*, 1979, 5: 379.
18. I. Karatzas, J. P. Lehoczky, and S. E. Shreve, *Optimal portfolio and consumption decisions for a "small investor" on a finite horizon*, SIAM Journal on Control and Optimization **25** (1987), no. 6, 1557–1586.
19. Merton, R. C., 1973. Theory of Rational Option Pricing. Bell Journal of Economics and Management Science 4, 141–183.
20. R. C. Merton, *Optimum consumption and portfolio rules in a continuous-time model*, Journal of Economic Theory **3** (1971), 373–413.
21. H. Pham and N. Touzi, The fundamental theorem of asset pricing with cone constraints, Journal of Mathematical Economics, 31, 265–279, 1999.
22. R. T. Rockafellar, *Convex Analysis*, Princeton, New Jersey, Princeton University Press, 1970.
23. Ross, S. A., The arbitrage theory of capital asset pricing, *Journal of Economic Theory*, 1976, 13: 341.
24. S. A. Ross, "Return, Risk and Arbitrage", *Risk and Return in Finance*, Friend and Bicksler (eds.), Cambridge, Mass., Ballinger, pp. 189–218, 1977.
25. Scarf, H., *The Computation of Economic Equilibria*, New Haven: Yale University Press, 1973.
26. S. P. Sethi, *Optimal consumption and investment with bankruptcy*, Kluwer, Boston, 1997.
27. H. Shirakawa, *Optimal consumption and portfolio selection with incomplete markets and upper and lower bound constaints*, Mathematical Finance **4** (1994), no. 1, 1–24.
28. H. Shirakawa and H. Kassai, *Optimal consumption and arbitrage in incomplete, finite state security markets*, Journal of Operations Research **45** (1993), 349–372.
29. S. E. Shreve, H. M. Soner, and G.-L. Xu, *Optimal investment and consumption with two bounds and transaction costs*, Mathematical Finance **1** (1991), no. 3, 53–84.
30. Smith, A., *An Inquiry into the Nature and Causes of the Wealth of Nations*, Modern Library, 1937.
31. Spremann, K., The simple analytics of arbitrage, *Capital Market Equilibria* (ed. Bamberg, G., Spremann, K.), New York: Springer-Verlag, 1986, pp.189–207.
32. Wilhelm, J., *Arbitrage Theory*, Berlin: Springer-Verlag, 1985.

On Some Optimization Problems in Obnoxious Facility Location*

Zhongping Qin[1], Yinfeng Xu[2], and Binhai Zhu[3]

[1] Dept. of Mathematics, Huazhong University of Science and Technology, China
[2] School of Management, Xi'an Jiaotong University, Xi'an, China
[3] Dept. of Computer Science, City University of Hong Kong, Kowloon, Hong Kong.
bhz@cs.cityu.edu.hk

Abstract. In this paper we study the following general MaxMin-optimization problem concerning undesirable (obnoxious) facility location: Given a set of n sites S inside a convex region P, construct m garbage deposit sites V_m such that the minimum distance between these sites V_m and the union of S and V_m, $V_m \cup S$, is maximized. We present a general method using Voronoi diagrams to approximately solve two such problems when the sites S's are points and weighted convex polygons (correspondingly, V_m's are points and weighted points and the distances are L_2 and weighted respectively). In the latter case we generalize the Voronoi diagrams for disjoint weighted convex polygons in the plane. Our algorithms run in polynomial time and approximate the optimal solutions of the above two problems by a factor of 2.

1 Introduction

In the area of environmental operations management we usually face the problem of locating sites to deposit (nuclear and/or conventional) wastes. Because of the radiation and pollution effect we do not want to locate such a site too close to a city. Moreover, we cannot even locate two sites too close to each other (even if they are far from cities where human beings reside) — this is especially the case when the sites are used to deposit nuclear wastes as the collective radiation becomes stronger.

Another related application is something to do with satellite placement. In the sky we already have quite a lot of satellites and certainly the number is increasing year by year. When we launch a new satellite, we certainly need to place it at such a location which is far from the existing ones as long as there is no other constraints — putting them too close would affect the communication quality.

This problem arises in many other applications in practice. However, how to find efficient solutions for these kinds of problems is not an easy task and this

---

* The authors would like to acknowledge the support of research grants from NSF of China, Research Grants Council of Hong Kong SAR, China (CERG Project No. CityU1103/99E) and City University of Hong Kong

D.-Z. Du et al. (Eds.): COCOON 2000, LNCS 1858, pp. 320–329, 2000.

presents a new challenge for computational geometers. In this paper we study several versions of this problem. We call this general problem the Undesirable Facility Location Problem (UFL, for short) and formulate two versions of the problem as follows.

I. Undesirable Point Placement Problem (UPP). Given a convex polygonal domain P and a set of point sites S in P, find the location of m undesirable or obnoxious facilities in P, which consist of a point set V_m, so as to

$$\text{Maximize } \{\hat{d}(V_m, V_m \cup S)\}$$

where the maximum is over all the possible positions of the points in V_m and $\hat{d}(V_m, V_m \cup S)$ defined as

$$\hat{d}(V_m, V_m \cup S) = \min_{\{x \neq q\} \wedge \{x \in V_m\} \wedge \{q \in V_m \cup S\}} \{d(x, q)\}$$

with $d(x, q)$ being the Euclidean distance between x and q.

Note that the Undesirable Point Placement problem (UPP) is evidently a MaxMin-optimization problem. We can see that the solution maximizes the distances between the undesirable facilities and the distances between the ordered pairs of points from the undesirable facilities and the sites. Therefore, the damage or pollution to the sites is low.

In reality, we know that the importance, areas, or the capacity of enduring damages of various cities are different. We take this into consideration and make the problem more general and practical, we introduce the weighted function $w(q)$ for point q, extend the point sites to disjoint weighted convex polygons and permit the metric to be either L_2 or L_1. Our objective is to distinguish the difference of importance of various sites, capture the real geographic conditions and to satisfy various needs in practice. We therefore formulate the following more general problem.

II. Undesirable Weighted Point Placement Problem (UWPP). Given a convex polygonal domain P and a set of disjoint convex polygonal sites S in P where each point q in a polygon $s \in S$ has weight $w(q) = w(s)$ ($w(s) \geq 1$ is associated with the polygon s), find the location of m undesirable facilities in P, which consist of a point set V_m and have the same weight $1 \leq w_0 \leq w(s), s \in S$, so as to

$$\text{Maximize } \{\hat{d}_w(V_m, V_m \cup S)\}$$

where the maximum is over all the possible positions of the points in V_m and $\hat{d}_w(V_m, V_m \cup S)$ is defined as

$$\hat{d}_w(V_m, V_m \cup S) = \min_{\{x \neq q\} \wedge \{x \in V_m\} \wedge \{q \in V_m \cup S\}} \{d_w(x, q)\}$$

with $d_w(x, q) = \frac{1}{w(q)} d(x, q)$, $d(x, q)$ being the Euclidean distance in between x and q.

We remark that the Undesirable Weighted Point Placement Problem (UWPP) is a much more complex MaxMin-optimization problem. If set V_m is chosen such that the objective function of UWPP reaches its optimal value (which is called the m-optimal value of UWPP, denoted as d_m^*) then we call the thus chosen set V_m a m-extreme set of UWPP, denoted as V_m^*.

It is obvious that in a solution to UWPP the undesirable facilities are farther away from those sites with larger weights. Hence the sites with larger weights can get more protect against the undesirable facilities. Although we formulate the problem as the Undesirable Weighted Point Placement Problem in the plane, it can be formulated as problem in other fields like packing satellites in the sky or packing transmitters on the earth.

It should be noted that finding the optimal solutions, V_m^*, for either UPP or UWPP is not an easy task. The exhaustive search method is not possible as the spaces of the solutions to the problems are continuous regions in R^{2m} and the minimizing step of the objective function concerns some continuous regions in R^2. So far we have not been able to obtain an algorithm to find an optimal solution for either of the two problems even if it is of exponential time complexity. The difficulty of the problem can also be seen from the following. It is not yet known how to pack as many as possible unit circles in a given special polygon P, like a square. In this setting, S is empty and we want to determine the maximum set V_m such that the distance $\hat{d}(V_m, V_m)$ is at least two [13].

Although the problem is hard to solve, in practice, usually an approximate solution is acceptable as long as the solution is efficient. In this paper we present a general incremental Voronoi diagram algorithm which approximate the optimal solution of the two problems by a factor of 2.

2   The Incremental Voronoi Diagram Algorithm for UPP

In this section we present an incremental Voronoi diagram algorithm for approximating the Undesirable Point Placement Problem (UPP). The basic ideas of the algorithm are as follows: (1). We successively choose the points of V_m from a discrete set. At i-th step we choose the point which is the farthest away from the points in V_{i-1} chosen previously and all sites in S. (2). The point in $V_i - V_{i-1}$ which is the farthest from $V_{i-1} \cup S$ is always among the Voronoi vertices of $Vor(V_{i-1} \cup S)$, vertices of polygon P and the intersection points of the Voronoi edges of $Vor(V_{i-1} \cup S)$ and P.

Basically, the algorithm uses a strategy of computing points in V_m from a finite discrete field rather than from a continuous field. The algorithm and analysis is as follows.

Algorithm UPP
   Input: A convex polygon P, a set of points S in P.
   Output: The set of points V_m in P.
   Procedure:

1. Initialize $V := \emptyset$.
2. Compute the Voronoi diagram of $(V \cup S)$, $Vor(V \cup S)$.

3. Find the set B consisting of the Voronoi vertices of $(V \cup S)$, the vertices of P and the intersection points between Voronoi edges and the edges of P. Among the points in B, choose the point v which maximizes $d(v, q_v), v \in B$, where $q_v \in V \cup S$ and Vor_region(q_v) contains point v.
4. Update $V := V \cup \{v\}$ and return to Step 2 when $|V| \leq m$.

We have the following theorem regarding the above algorithm.

Theorem 1. *The output V_m generated by Algorithm UPP presents a 2-approximation to the optimal solution for the Undesirable Point Placement Problem (UPP).*

The proof of Theorem 1 is based on the following lemma, which is similar to the methods used in [6,8].

Lemma 2. *Given any convex polygon P and any set S of point sites in P, for any set V_t of t points in P there exists a point $p \in P$ such that $d(p, V_t \cup S) \geq \frac{d^*_{t+1}}{2}$, where d^*_{t+1} is the $(t + 1)$-optimal value of the UPP, where $d(p, V_t \cup S)$ denotes the Euclidean distance between point p and set $V_t \cup S$.*

Proof. Suppose that the lemma is not true. Then there exists a point set V_t in P such that for any $p \in P$,

$$\max d(p, V_t \cup S) < \frac{d^*_{t+1}}{2}.$$

Let r be the value of the left-hand side and we have $r < \frac{d^*_{t+1}}{2}$. For each point $q \in V_t \cup S$, draw a circle centered at q with radius r. By the optimality of r, every pair of points in P and $V_t \cup S$ must be at most r distance away. Therefore, it is obvious that the union of these circles must cover the whole area of P. This implies that one of these circles must cover two points of $V^*_{t+1} \cup S$. Consequently, $d^*_{t+1} = \hat{d}(V^*_{t+1}, V^*_{t+1} \cup S) \leq 2r$. This contradicts the fact that $r < \frac{d^*_{t+1}}{2}$. □

Proof for Theorem 1. For any $k \leq m$ let point $p \in P$ maximizes $d(x, V_k \cup S), x \in P$. We assert that p must be in the set of points consisting of the Voronoi vertices of Vor($V_k \cup S$), the vertices of P and the intersection points between Voronoi edges of Vor($V_k \cup S$) and the edges of P. In other words p must be a boundary point of some Voronoi region of Vor($V_k \cup S$) $\cap P$. If it is not the case then p would fall in the interior of Vor_region(q) $\cap P$ for some $q \in V_k \cup S$. Hence there is one point p' on the boundary of Vor_region(q) $\cap P$ such that $d(p', q) > d(p, q)$ which contradicts the definition of p. Hence for point v_k, the k-th ($k \leq m$) chosen point for set V_m by step 3 of Algorithm UPP, $d(v_k, V_{k-1} \cup S) = \max_{\{x \in P\}} \{d(x, V_{k-1} \cup S)\}$ (recall that $d(v_k, V_{k-1} \cup S)$ denotes the distance between point v_i and set $V_{i-1} \cup S$).

Also, we have the relation

$$d(v_k, V_{k-1} \cup S) \leq d(v_{k-1}, V_{k-2} \cup S), \quad k = 2, 3, \cdots, m.$$

The reason is that all $Vor\_region(q), q \neq v_k$, which belong to $Vor(V_k \cup S) \cap P$ will not grow relative to $Vor\_region(q) \in Vor(V_{k-1} \cup S) \cap P$ after point v_k is inserted. In addition, the point on the boundary of $Vor\_region(v_k) \in Vor(V_k \cup S) \cap P$ which is the farthest to v_k is always on the boundary of $Vor\_region(q) \in Vor(V_k \cup S) \cap P, q \in V_{k-1} \cup S$, which is adjacent to $Vor\_region(v_k) \in Vor(V_k \cup S) \cap P$. So from this relation by induction we can obtain the result that $d(v_k, V_{k-1} \cup S) = d(V_k, V_k \cup S)$. Hence by the above assertion and Lemma 2 $\hat{d}(V_m, V_m \cup S) = d(v_m, V_{m-1} \cup S) = \max_{\{x \in P\}}\{d(x, V_{m-1} \cup S)\} \geq \frac{d_m^*}{2}$. This shows that V_m, the output of Algorithm UPP, is a 2-approximation of V_m^* (the m-extreme set of UPP). □

Time complexity. The algorithm runs in $O(mN \log N)$ time, where $N = |S| + |P| + m$. At each of the m iterations, Step 2 takes $O(N \log N)$ time, Step 3 involves finding the nearest neighbor of each vertex v hence takes $O(N \log N)$ time — we need to perform $O(N)$ point locations each taking logarithmic time. It is possible to improve Step 2 by using the dynamic Voronoi diagram (Delaunay triangulation) algorithm of [3,4] so that over all of the m iterations [1], Step 2 takes $O(N \log N + m \log N)$ time instead of $O(mN \log N)$ time. But this will not change the overall time complexity of the algorithm as Step 3 takes $O(N \log N)$ time at each of the m iterations.

3   A Generalization to the Weighted Plane

In this section, we consider yet another generalization of the problem to the weighted (L_2 or L_1) plane. We make some necessary adjustment to Algorithm UPP so as to approximately solve the Undesirable Weighted Point Placement Problem (UWPP). The basic idea is the same as Algorithm UPP. But the set B in step 3 should be change to the set B' consisting of the Voronoi vertices and dividing points of $Vor(V_t \cup S)$, vertices of P, the intersection points between Voronoi segments of $Vor(V \cup S)$ and the edges of P.

3.1   The Weighted Voronoi Diagram of Convex Polygons in the Plane

We generalize the concept of Voronoi diagram of points in the plane in three aspects: (1) change point sites to disjoint convex polygons, (2) each site has a positive weight which is at least one, (3) under both metric L_2 and L_1 (we will focus on L_2). To the best knowledge of the authors, in metric L_2 the Voronoi diagram of weighted points in the plane is well studied [1], but the Voronoi diagram of weighted convex polygon objects is seldomly studied. In metric L_1

---

[1] In practice, we advocate the use of [7] to update the Voronoi diagram when a new point is added and the point location algorithm of [5,12].

there are only results on the Voronoi diagram of points in the plane [10,11] before this work.

Let S be a set of disjoint convex polygons in the plane which are associated with weights $w(s) \geq 1, s \in S$. The *weighted Voronoi diagram of S* (for short WVD(S)) is a subdivision of the plane consisting of Voronoi regions, Voronoi faces, Voronoi edges, Voronoi segments, Voronoi vertices and Voronoi dividing points. *Vor_region(s)* is defined as

$$Vor\_region(s) = \{x| \frac{1}{w(s)} d(x,s) \leq \frac{1}{w(t)} d(x,t), t \in S\}.$$

A *Voronoi face* of the WVD(S) is a connected component of a thus defined Voronoi region. A *Voronoi edge* is the intersection of two Voronoi faces. A *Voronoi segment* is the maximal portion of a Voronoi edge which can be described by a curve equation. A *Voronoi dividing point* is an endpoint of a Voronoi segment and a *Voronoi vertex* is the intersection of three Voronoi edges. We now focus on L_2 and present the necessary details. In Figure 1, we show an example of three convex polygons with weights 1, 2 and 3 (for convenience, we use 1, 2 and 3 to represent them as well). Note that the Voronoi edge between 2 and 3 which is inside P is composed a set of Voronoi segments.

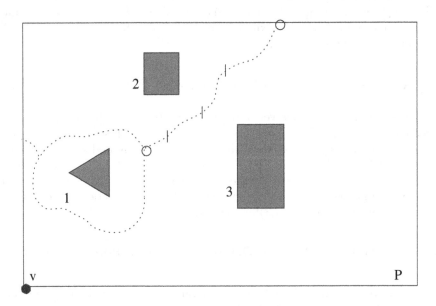

Fig. 1. A weighted Voronoi diagram of three convex polygons.

Lemma 3. *Suppose $S = \{s_1, s_2\}$ consist of two weighted disjoint convex polygons in the plane and $1 \leq w(s_1) < w(s_2)$, then Vor_region(s_1) is a limitary connected region. The boundary of Vor_region(s_1) (which is a Voronoi edge) consists of some Voronoi segments which are parts of conic curves, respectively, i.e.*

1. *A Voronoi segment is a part of a circle determined by equation*

$$\frac{1}{w(p)}d(x,p) = \frac{1}{w(q)}d(x,q)$$

where x is a point of the Voronoi segment and p, q are a pair of vertices of two convex polygons s_1 and s_2.

2. *A Voronoi segment is a part of a hyperbola, parabola or ellipse determined by equation*

$$\frac{1}{w(p)}d(x,p) = \frac{1}{w(e)}d(x,e)$$

where x is a point of the Voronoi segment and p (e) is vertex (edge) convex polygon s_1 (s_2). The segment is a part of a hyperbola, parabola or ellipse when the weight of the vertex p is bigger than, equal to or less than the weight of the edge e.

3. *A Voronoi segment is a straight line segment determined by equation*

$$\frac{1}{w(e_1)}d(x,e_1) = \frac{1}{w(e_2)}d(x,e_2)$$

where x is a point of the Voronoi segment and e_1, e_2 are a pair of edges of two convex polygons s_1 and s_2.

There are some difference between the WVD(S) and the classic Voronoi diagram of points (for short VD(P)) or the weighted Voronoi diagram of points (for short WVD(P)): an Voronoi edge in VD(P) or WVD(P) has just one Voronoi segment which is a part of a straight line (in VD(P)) or a part of circle (in WVD(P)), but a Voronoi edge in WVD(S) contains some Voronoi segments which are of different conic curves. Therefore, in WVD(S) Voronoi dividing points are different from Voronoi vertices. Similar to [2], we make the following definition.

Definition 1. *Given two weighted disjoint convex polygons s_1, s_2 in the plane and $1 \leq w(s_1) < w(s_2)$, $Vor\_region(s_1)$ is defined as the dominance of s_1 over s_2, for short $dom(s_1, s_2)$, and the closure of the complement of $Vor\_region(s_1)$ is called the dominance of s_2 over s_1, for short $dom(s_2, s_1)$.*

The following lemma follows from the definition of Voronoi diagrams and is the basis for constructing WVD(S).

Lemma 4. *Let S be a finite set of weighted convex polygons in the plane and $s \in S$.*

$$Vor\_region(s) = \bigcap_{t \in S-\{s\}} dom(s,t).$$

In general, $Vor\_region(s)$ in WVD(S) may have $O(|S|)$ Voronoi faces, i.e., the region does not need to be connected and its connected parts do not need to be simply connected. Overall, the WVD(S) can be computed in $O(|S|^3)$ time and $O(|S|^2)$ space, using standard techniques in computational geometry [14].

3.2   Algorithm

We now present the approximation algorithm for UWPP.

Algorithm UWPP

Input: A convex polygon P, a set S of disjoint weighted convex polygons in P and each polygon $s \in S$ has a weight $w(s) \geq 1$.
Output: The set of points V_m, all with weight $1 \leq w_0 \leq w(s), s \in S$.
Procedure:

1. Initialize $V := \emptyset$.
2. Compute the weighted Voronoi diagram of $(V \cup S)$, WVD($V \cup S$).
3. Find the set B' consisting of the Voronoi vertices and dividing points of $WVD(V \cup S)$, the vertices of P and the intersection points between Voronoi segments of $WVD(V \cup S)$ and the edges of P. Among the points in B', choose the point v which maximizes $\frac{1}{w(q_v)}d(v, q_v)$, where $v \in B'$, $q_v \in V \cup S$ and $v \in Vor\_region(q_v)$.
4. Update $V := V \cup \{v\}$ and return to Step 2 when $|V| \leq m$.

In Figure 1, if we only add one obnoxious facility then it would be placed at the low-left corner of P. We have the following theorem.

Theorem 5. *The output V_m of Algorithm UWPP is a 2-approximation to the Undesirable Weighted Point Placement Problem.*

First we need to generalize Lemma 2, which is important to the proof of Theorem 1, to the following Lemma 6.

Lemma 6. *Let P and S be as in the definition of UWPP. For any set V_t of t points in P with weight $1 \leq w_0 \leq w(s), s \in S$ there exists a point $x \in P$ such that*

$$\max \frac{1}{w(q_x)}d(x, q_x) \geq \frac{d^*_{t+1}}{2}, x \in P,$$

where $q_x \in V_t \cup S, x \in Vor\_region(q_x)$ and d^*_{t+1} is the $(t+1)$-optimal value of UWPP.

Proof. Suppose the lemma is not true, then there exists a point set V_t in P (note that points in V_t have the same weight $1 \leq w_0 \leq w(s), s \in S$) such that

$$\max \frac{1}{w(q_x)}d(x, q_x) < \frac{d^*_{t+1}}{2}, x \in P$$

where $q_x \in V_t \cup S$ and x is contained in $Vor\_region(q_x)$. Let r be the value of the left-hand side, so we have $r < \frac{d^*_{t+1}}{2}$. For $q \in V_t \cup S$ ($V_t \cup S$ is either a point in V_t or a convex polygon in S), we expand q to $\bar{q}$ such that $\bar{q}$ contains all the points within $w(q)r$ distance (in L_2) to q. By the optimality of r, $\bigcup_{q \in V_t \cup S} \{\bar{q}\}$

contains all the points in P. So there are two points of $V_{t+1}^* \cup S$ (V_{t+1}^* is the $(t+1)$-extreme set of UWPP) contained in some $\bar{q}$ ($q \in V_t$). This implies that the $(t+1)$-optimal value of the UWPP $d_{t+1}^* \leq 2r$, a contradiction with $r < \frac{d_{t+1}^*}{2}$. $\square$

Proof for Theorem 5. Let v_k be the k-th chosen point for set V_m by step 3 of Algorithm UWPP. Similar to the proof of Theorem 1 we also have the result that v_k maximizes $\frac{1}{w(q_x)}L(x, q_x)$, $x \in P$ where $q_x \in V_{k-1} \cup S$ and $x \in Vor\_region(q_x)$. We assert that $\frac{1}{w(q_{v_k})}d(v_k, q_{v_k}) = \hat{d}_w(V_k, V_k \cup S)$. In fact, we have

$$\hat{d}_w(v_{k-1}, q_{v_{k-1}}) \geq \hat{d}_w(v_k, q_{v_k}), k = 1, 2, \cdots, m.$$

The reason is that as $w_0 \leq w(s), s \in S$, all $Vor\_region(q), q \neq v_k$, which belong to $Vor(V_k \cup S)$ will not grow relative to $Vor\_region(q) \in Vor(V_{k-1} \cup S)$ after point v_k is inserted. Also, the point on the boundary of $Vor\_region(v_k) \in Vor(V_k \cup S)$ which is the farthest to v_k is always on the boundary of $Vor\_region(q) \in Vor(V_k \cup S), q \in V_{k-1} \cup S$, which is adjacent to $Vor\_region(v_k) \in Vor(V_k \cup S)$. Based on this fact we can easily prove the above assertion by induction. Therefore, following Lemma 6

$$\hat{d}_w(V_m, V_m \cup S) = \frac{1}{w(q_{v_m})}d(v_m, q_{v_m})$$
$$= \max_{\{x \in P\}}\{\frac{1}{w(q_x)}d(x, q_x)\},$$
$$where\ q_x \in V_{m-1} \cup S\ and\ x \in Vor\_region(q_x)$$
$$\geq \frac{d_m^*}{2}.$$

$\square$

Time complexity. The algorithm runs in $O(mN^3)$ time, where $N = |S| + |P| + m$. At each of the m iterations, Steps 2 and 3 take $O(N^3)$ time and $O(N^2)$ space.

We comment that the WVD(S) can be generalized to L_1 metric. Moreover, the facilities can have different weights (but we then should first place the facility with the largest weight, among those unplaced ones). The details are omitted.

Finally, we comment that it might be possible for us to trade the running time of this algorithm with the approximation factor. We can use the geometric Voronoi diagram of S in L_2, which is planar and can be constructed in $O(N \log N)$ time and $O(N)$ space [9,15,16].

4 Concluding Remarks

In this paper we present a method to solve a MaxMin-optimization problem in obnoxious facility location. The method is general and can be generalized to many interesting cases when the sites are not necessary points. It is an open question whether we can obtain an optimal solution for the problem (even it is of exponential time complexity). Another question is whether the $O(N^3)$ time for computing WVD(S) in Section 3 can be further improved. Finally, all the facilities we consider are points (or weighted points); however, in practice sometimes the facilities might be of some size as well. It is not know whether our method can be generalized to these situations. For example, what if the facilities are unit circles?

References

[1] F. Aurenhammer, Voronoi diagrams: a survey of a fundamental geometric data structures, *ACM Comput. Surveys*, 23(3), pp. 343-405, 1991.

[2] F. Aurenhammer and H. Edelsbrunner, An optimal algorithm for constructing the weighted Voronoi diagram in the plane, *Pattern Recognition*, 17, pp. 251-257, 1984.

[3] J. Boissonnat, O. Devillers, R. Schott, M. Teillaud and M. Yvinec, Applications of random sampling to on-line algorithms in computational geometry, *Disc. Comp. Geom.* 8, pp. 51-71, 1992.

[4] O. Devillers, S. Meiser and M. Teillaud, Fully dynamic Delaunay triangulation in logarithmic expected time per operation, *Comp. Geom. Theory and Appl.* 2, pp. 55-80, 1992.

[5] L. Devroye, E. Mücke and B. Zhu, A note on point location in Delaunay triangulations of random points, *Algorithmica*, special issue on Average Case Analysis of Algorithms, *22(4)*, pp. 477-482, Dec, 1998.

[6] T. Feder and D. Greene, Optimal algorithms for approximate clustering, *Proc. 20th STOC*, pp. 434-444, 1988.

[7] L. Guibas, D. Knuth and M. Sharir, Randomized incremental construction of Delaunay and Voronoi diagrams, *Algorithmica*, 7, pp. 381-413, 1992.

[8] T. Gonzalez, Clustering to minimize the maximum intercluster distance, *Theo. Comput. Sci.*, 38, pp. 293-306, 1985.

[9] R. Klein, K. Mehlhorn and S. Meiser, Randomized incremental construction of abstract Voronoi diagrams, *Comp. Geom. Theory and Appl.* 3, pp. 157-184, 1993.

[10] D.T. Lee and C.K. Wong, Voronoi diagrams in the L_1 (L_∞) metric with two-dimensional storage applications, *SIAM. J Comput.*, 9, pp. 200-211, 1980.

[11] D.T. Lee, Two-dimensional Voronoi diagrams in the L_p metric, *J. ACM*, 27, pp. 604-618, 1980.

[12] E. Mücke, I. Saias and B. Zhu, Fast Randomized Point Location Without Preprocessing in Two and Three-dimensional Delaunay Triangulations, *Comp. Geom. Theory and Appl*, special issue for SoCG'96, *12(1/2)*, pp. 63-83, Feb, 1999.

[13] K. Nurmela and P. Ostergard, Packing up to 50 equal circles in a square, *Disc. Comp. Geom.* 18, pp. 111-120, 1997.

[14] F. Preparata and M. Shamos, Computational Geometry, Springer-Verlag, 1985.

[15] P. Widmayer, Y.F. Wu, and C.K. Wong, On some distance problems in fixed orientations. *SIAM J. Comput.*, 16, pp. 728–746, 1987.

[16] C.K. Yap, An O($n \log n$) algorithm for the Voronoi diagram of a set of curve segment, *Disc. Comp. Geom.* 2, pp. 365-393, 1987.

Generating Necklaces and Strings with Forbidden Substrings

Frank Ruskey and Joe Sawada

University of Victoria, Victoria, B.C. V8W 3P6, Canada,
{fruskey,jsawada}@csr.csc.uvic.ca

Abstract. Given a length m string f over a k-ary alphabet and a positive integer n, we develop efficient algorithms to generate
(a) all k-ary strings of length n that have no substring equal to f,
(b) all k-ary circular strings of length n that have no substring equal to f, and
(c) all k-ary necklaces of length n that have no substring equal to f, where f is an aperiodic necklace.
Each of the algorithms runs in amortized time $O(1)$ per string generated, independent of k, m, and n.

1 Introduction

The problem of generating discrete structures with forbidden sub-structures is an area that has been studied for many objects including graphs (e.g., with forbidden minors), permutations (e.g., which avoid the subsequence 312 of relative values), and trees (e.g., of bounded degree). In this paper we are concerned with generating strings that avoid some particular substring. For example, the set of binary strings that avoid the pattern 11 are known as *Fibonacci strings*, since they are counted by the Fibonacci numbers. The set of circular binary strings that avoid 11 are counted by the Lucas numbers. Within combinatorics, the counting of strings which avoiding particular substrings can be handled with the "transfer matrix method" as explained, for example, in Stanley [5]. The ordinary generating function of the number of such strings is always rational, even in the case of circular strings. In spite of the importance of these objects within combinatorics, we know of no papers that explicitly address the problem of efficiently generating all strings or necklaces avoiding a given substring.

The problem of generating strings with forbidden substrings is naturally related to the classic pattern matching problem, which takes as input a pattern P of length m and a text T of length n, and finds all occurrences of the pattern in the text. Several algorithms perform this task in linear time, $O(n+m)$, including the Boyer-Moore algorithm, the Knuth-Morris-Pratt (KMP) algorithm and an automata-based algorithm (which requires non-linear initialization) [2].

The Boyer-Moore algorithm is not suitable for our purposes since it does not operate in real-time. On the other hand, the automata-based algorithm operates in real-time and the KMP algorithm can be adapted to do so [2].

D.-Z. Du et al. (Eds.): COCOON 2000, LNCS 1858, pp. 330–339, 2000.

Our algorithms are recursive, generating the string from left-to-right and applying the pattern matching as each character is generated. It is straight forward to generate unrestricted strings in such a recursive manner and adding the pattern matching is easy as well. However, when the pattern is taken circularly, the algorithm and its analysis become considerably more complicated.

The algorithm for generating the necklaces uses the recursive scheme introduced in [4]. This scheme has been used to generate other restricted classes of necklaces, such as unlabelled necklaces [4], fixed-density necklaces [3], and bracelets (necklaces that can be turned over), and is ideally suited for the present problem.

Within the context of generating combinatorial objects, usually the primary goal is to generate each object so that the amount of computation is $O(1)$ per object in an amortized sense. Such algorithms are said to be CAT (for Constant Amortized Time). Clearly, no algorithm can be asymptotically faster. Note that we do not take into account the time to print or process each object; rather we are counting the total amount of data structure change that occurs.

The main result of this paper is the development of CAT algorithms to generate:

- all k-ary strings of length n that have no substring equal to f,
- all k-ary circular strings of length n that have no substring equal to f, and
- all k-ary necklaces of length n that have no substring equal to f, given that f is Lyndon word.

Each algorithm has an embedded automata-based pattern matching algorithm. In principle we could use the same approach to generate unlabelled necklaces, fixed density necklaces, bracelets, or other types of necklaces, all avoiding a forbidden necklace pattern. We expect that such algorithms will also be CAT, but the analysis will be more difficult.

In the following section we provide background and definitions for these objects along with a brief description of the automata-based pattern matching algorithm. In Section 3, we outline the details of each algorithm. We analyze the algorithms, proving that they run in constant amortized time, in Section 4.

2   Background

We denote the set of all k-ary strings of length n with no substring equal to f by $\mathbf{I}_k(n, f)$. The cardinality of this set is $I_k(n, f)$. For the remainder of this paper we will assume that the forbidden string f has length m. Clearly if $m > n$, then $I_k(n, f) = k^n$, and if $m = n$ then $I_k(n, f) = k^n - 1$. If $m < n$, then an exact formula will depend on the forbidden substring f, but can be computed using the transfer matrix method. In Section 4 we derive several bounds on the value of $I_k(n, f)$.

We denote the set of all k-ary circular strings of length n with no substring equal to f by $\mathbf{C}_k(n, f)$. The cardinality of this set is $C_k(n, f)$. In this case, we allow the forbidden string to make multiple passes around the circular string.

Thus, if a string α is in $\mathbf{I}_k(n, f)$ and $m > n$, then it is still possible for the string α to contain the forbidden string f. For example, if $\alpha = 0110$ and $f = 11001100$, then α is *not* in the set $\mathbf{C}_k(4, f)$. We prove that $C_k(n, f)$ is proportional to $I_k(n, f)$ in Section 4.1.

Under rotational equivalence, the set of strings of length n breaks down into equivalence classes of sizes that divide n. We define a *necklace* to be the lexicographically smallest string in such an equivalence class of strings under rotation. An aperiodic necklace is called a *Lyndon word*. A word α is called a *pre-necklace* if it is the prefix of some necklace. Background information, including enumeration formulas, for these objects can be found in [4].

The set of all k-ary necklaces of length n with no substring equal to f is denoted $\mathbf{N}_k(n, f)$ and has cardinality $N_k(n, f)$. The set of all k-ary Lyndon words of length n with no substring equal to f is denoted $\mathbf{L}_k(n, f)$ and has cardinality $L_k(n, f)$. Of course for N_k and L_k we consider the string to be circular when avoiding f. The set of all k-ary pre-necklaces of length n with no substring equal to f is denoted $\mathbf{P}_k(n, f)$ and has cardinality $P_k(n, f)$. A standard application of Burnside's Lemma will yield the following formula for $N_k(n, f)$:

$$N_k(n, f) = \frac{1}{n} \sum_{d \mid n} \phi(d) C_k(n/d, f). \tag{1}$$

2.1   The Automata-Based String Matching Algorithm

One of the best tools for pattern recognition problems is the finite automaton. If $f = f_1 f_2 \cdots f_m$ is the pattern we are trying to find in a string α, then a deterministic finite automaton can be created to process the string α one character at a time, in constant time per character. In other words, we can process the string α in *real-time*. The preprocessing steps required to set up such an automaton can be done in time $O(km)$, where k denotes the size of the alphabet (see [1], pg. 334).

The automaton has $m+1$ states, which we take to be the integers $0, 1, \ldots, m$. The state represents the length of the current match. Suppose we have processed t characters in the string $\alpha = a_1 a_2 \cdots a_n$ and the current state is s. The transition function $\delta(s, j)$ is defined so that if $j = a_{t+1}$ matches f_{s+1}, then $\delta(s, j) = s + 1$. Otherwise, $\delta(s, j)$ is the largest state q such that $f_1 \cdots f_q = a_{t-q+2} \cdots a_{t+1}$. If the automaton reaches state m, the only accepting state, then the string f has been found in α. The transition function is efficiently created using an auxiliary function *fail*. The failure function is defined for $1 \leq i \leq m$ such that $fail(i)$ is the length of the longest proper suffix of $f_1 \cdots f_i$ equal to a prefix of f. If there is no such suffix, then $fail(i) = 0$. This fail function is the same as the fail function in the classic KMP algorithm.

3   Algorithms

In this section we describe an efficient algorithm to generate necklaces with forbidden necklace substrings. We start by looking at the simpler problem of generating k-ary strings with forbidden substrings and then consider circular strings, focusing on how to handle the wraparound.

If n is the length of the strings being generated and m is the length of the forbidden substring f, then the following algorithms apply for $2 < m \leq n$. We prove that each algorithm runs in constant amortized time in the following section. In the cases where $m = 1$ or 2, trivial algorithms can be developed. For circular strings and necklaces, if $m > n$, then the forbidden substring can be truncated to a length n string, as long as it repeats in a circular manner after the nth character.

3.1   Generating k-ary Strings

A naïve algorithm to generate all strings in $\mathbf{I}_k(n, f)$ will generate all k-ary strings of length n, and then upon generation of each string, perform a linear time test to determine whether or not it contains the forbidden substring. A simple and efficient approach for generating strings is to construct a length n string by taking a string of length $n-1$ and appending each of the k characters in the alphabet to the end of the string. This strategy suggests a simple recursive scheme, requiring one parameter for the length of the current string. Since this recursive algorithm runs in constant amortized time, the naïve algorithm will take linear time per string generated.

A more advanced algorithm will embed a real-time automata-based string matching algorithm into the string generation algorithm. Since an automata-based string matching algorithm takes constant time to process each character, we can generate each new character in constant time. We store the string being generated in $\alpha = a_1 a_2 \cdots a_n$ and at each step, we maintain two parameters: t and s. The parameter t represents the next position in the string to be filled, and the parameter s represents the state in the finite automata produced for the string f. Recall that each state s is an integer value that represents the length of the current match. Thus if we begin a recursive call with parameters t and s, then $f_1 \cdots f_s = a_{t-s} \cdots a_{t-1}$. We continue generating the current string as long as $s \neq m$. When $t > n$, we print out the string using the function PrintIt(). Pseudocode for this algorithm is shown in Figure 1. The transition function $\delta(s, j)$ is used to update the state s as described in Section 2.1. The initial call is GenStr(1,0).

Following this approach, each node in the computation tree will correspond to a unique string in $\mathbf{I}_k(j, f)$ where j ranges from 1 to n. Since the amount of computation at each node is constant, the total computation is proportional to

$$CompTree_k(n, f) = \sum_{j=1}^{n} I_k(j, f).$$

We can show that this sum is proportional to $I_k(n, f)$, which proves the following theorem. (Recall that the preprocessing required to set up the automata for f takes time $O(km)$. This amount is negligible compared to the size of the computation tree.)

```
procedure GenStr ( t, s : integer );
local j, q : integer;
begin
    if t > n then PrintIt()
    else begin
        for j ∈ {0, 1, ..., k − 1} do begin
            a_t := j;
            q := δ(s, j);
            if q ≠ m then GenStr(t + 1, q);
end; end; end;
```

Fig. 1. An algorithm for generating k-ary strings with no substring equal to f.

Theorem 1. *The algorithm* **GenStr**(t, s) *for generating k-ary strings of length n with no substring equal to f is CAT.*

3.2 Generating k-ary Circular Strings

We now focus on the more complicated problem of generating circular strings with forbidden substring f. To solve this problem we use the previous algorithm, but now we must also check that the wraparound of the string does not yield the forbidden substring. More precisely, if $\alpha = a_1 a_2 \cdots a_n$ is in $\mathbf{I}_k(n, f)$ then the additional substrings we must implicitly test against the forbidden string f are $a_{n-m+1+i} \cdots a_n a_1 \cdots a_i$ for $i = 1, 2, \ldots, m-1$. To perform these additional tests, we could continue the pattern matching algorithm by appending the first $m - 1$ characters to the end of the string. This will result in $m - 1$ additional checks for each generated string, yielding an algorithm that runs in time $O(m I_k(n, f))$. This approach can be tweaked to yield a CAT algorithm for circular strings, but leads to difficulties in the analysis when applied in the necklace context.

If we wish to use the algorithm GenStr(t, s), we need another way to test the substrings starting in the last $m - 1$ positions of α. We accomplish this feat by maintaining a new boolean data structure $match(i, t)$ and dividing the work into two separate steps. In the first step we compare the substring $a_1 \cdots a_i$ against the last i characters in f. If they match, then the boolean value $match(i, i)$ is set to TRUE; otherwise it is set to FALSE. In the second step, we check to see if $a_{n-m+1+i} \cdots a_n$ matches the first $m - i$ characters in f. If they match and $match(i, i)$ is TRUE, then we reject the string. If there is no match for all $1 \leq i \leq m - 1$, then α is in $\mathbf{C}_k(n, f)$.

To execute the first step, we start by initializing $match(i, 0)$ to TRUE for all i from 1 to $m - 1$. We define $match(i, t)$ for $1 \leq i \leq m - 1$ and $i \leq t \leq m - 1$ to be TRUE if $match(i, j - 1)$ is TRUE and $a_t = f_{m-j+t}$. Otherwise $match(i, t)$ is FALSE. This definition implies that $match(i, i)$ will be TRUE if $a_1 \cdots a_i$ matches the last i characters of f. Pseudocode for a routine that sets these values for each t is shown in Figure 2. The procedure SetMatch(t) is called after the t-th character in the string α has been assigned for $t < m$. Thus, if $t < m$, we must

```
procedure SetMatch ( t : integer );
local i : integer;
begin
    for i ∈ {t, t + 1, ..., m − 1} do begin
        if match(i, t − 1) and f_{m−i+t} = a_t then match(i, t) := TRUE;
        else match(i, t) := FALSE;
end; end;
```

Fig. 2. Procedure used to set the values of $match(i, t)$.

```
function CheckSuffix ( s : integer ) returns boolean;
begin
    while s > 0 do begin
        if match(m − s, m − s) then return(FALSE);
        else s := fail(s);
    end;
    return(TRUE);
end;
```

Fig. 3. Function used to test the wraparound of circular strings.

perform additional work proportional to $m − t$ for all strings in $\mathbf{I}_k(t, f)$. We will prove later that this extra work will not affect the asymptotic running time of the algorithm.

To execute the second step, we observe that after the nth character has been generated ($t = n + 1$), the string $a_{n−s+1} \cdots a_n$ is the longest suffix of α to match a prefix of f. Using the array $fail$, as described in Section 2.1, we can find all other suffixes that match a prefix of f in constant time per suffix. Then, for each suffix with length j found equal to a prefix of f, we check $match(m − j, m − j)$. If $match(m − j, m − j)$ is TRUE, then α is not in $\mathbf{C}_k(n, f)$. Note that $I_k(n − j, f)$ is an upper bound on the number of strings where $a_{n−j+1} \cdots a_n$ matches a prefix of f. Thus, for each $1 \leq j \leq m − 1$ the extra work done is proportional to $I_k(n − j, f)$. Pseudocode for the tests required by this second step is shown in Figure 3. The function CheckSuffix(s) takes as input the parameter s which represents the length of the longest suffix of α to match a prefix of f. It returns TRUE if α is in $\mathbf{C}_k(n, f)$ and FALSE otherwise.

Following this approach, we can generate all circular strings with forbidden substrings by adding the routines SetMatch(t) and CheckSuffix(s) to GenStr(t, s). Pseudocode for the resulting algorithm is shown in Figure 4. The initial call is GenCirc(1,0).

Observe that the size of the computation tree will be the same as before; however, in this case, the computation at each node is not always constant. The extra work performed at these nodes is bounded by

```
procedure GenCirc ( t, s : integer );
local j, q : integer;
begin
    if t > n then begin
        if CheckSuffix(s) then PrintIt();
    end else begin
        for j ∈ {0, 1, . . . , k − 1} do begin
            a_t := j;
            q := δ(s, j);
            if t < m then SetMatch(t);
            if q ≠ m then GenCirc(t + 1, q);
end; end; end;
```

Fig. 4. An algorithm for generating k-ary circular strings with no substring equal to f.

$$ExtraWork_k(n, f) \leq \sum_{j=1}^{m-1}(m - j)I_k(j, f) + \sum_{j=1}^{m-1} I_k(n - j, f)$$

The first sum represents the work done by SetMatch and the second the work done by CheckSuffix. In Section 4.1 we show that this extra work is proportional to $I_k(n, f)$. In addition, we also prove that $C_k(n, f)$ is proportional to $I_k(n, f)$. These results prove the following theorem.

Theorem 2. *The algorithm GenCirc(t, s) for generating k-ary circular strings of length n with no substring equal to f is CAT.*

3.3   Generating k-ary Necklaces

Using the ideas from the previous two algorithms, we now outline an algorithm to generate necklaces with forbidden necklace substrings. First, we embed the real-time automata based string matching algorithm into the necklace generation algorithm described in [4]. Then, since we must also test the wraparound for necklaces, we add the same tests as outlined in the circular string algorithm. Applying these two simple steps will yield an algorithm for necklace generation with no substring equal to the forbidden necklace f. Pseudocode for such an algorithm is shown in Figure 5. The additional parameter p in GenNeck(t, p, s), represents the length of the longest Lyndon prefix of the string being generated. Lyndon words can be generated by replacing the test "$n \bmod p = 0$" with the test "$n = p$." The initial call is GenNeck(1,1,0).

To analyze the running time of this algorithm, we again must show that the number of necklaces generated, $N_k(n, f)$, is proportional to the amount of computation done. In this case the size of the computation tree is

$$NeckCompTree_k(n, f) = \sum_{j=1}^{n} P_k(j, f)$$

```
procedure GenNeck ( t, p, s : integer );
local j, q : integer;
begin
    if t > n then begin
        if n mod p = 0 and CheckSuffix(s) then PrintIt();
    end else begin
        a_t := a_{t-p};
        q := δ(s, a_t);
        if t < m then SetMatch(t);
        if q ≠ m then GenNeck(t + 1, p, q);
        for j ∈ {a_{t-p} + 1, ..., k − 1} do begin
            a_t := j;
            q := δ(s, j);
            if t < m then SetMatch(t);
            if q ≠ m then GenNeck(t + 1, t, q);
    end; end; end;
```

Fig. 5. An algorithm for generating k-ary necklaces with no substring equal to f.

However, as with the circular string case, not all nodes perform a constant amount of work. The extra work performed by these nodes is bounded by

$$NeckExtraWork_k(n, f) \leq \sum_{j=1}^{m-1} (m - j) P_k(j, f) + \sum_{j=1}^{m-1} P_k(n - j, f)$$

Note that this expression is the same as the extra work in the circular string case, except we have replaced $I_k(n, f)$ with $P_k(n, f)$.

In Section 4.2 we show that $NeckCompTree_k(n, f)$ and $NeckExtra\text{-}Work_k(n, f)$ are both proportional to $\frac{1}{n} I_k(n, f)$. In addition, we also prove that $N_k(n, f)$ is proportional to $\frac{1}{n} I_k(n, f)$. These results prove the following theorem.

Theorem 3. *The algorithm* GenNeck(t, p, s) *for generating k-ary necklaces of length n with no substring equal to f is CAT, so long as f is a Lyndon word.*

We remark that the algorithm works correctly even if f is not a Lyndon word and appears to be CAT for most strings f.

4   Analysis of the Algorithms

In this section we will state the results necessary to prove that the work done by each of the forbidden substring algorithms is proportional to the number of strings generated. The constants in the bounds derived in this section can be reduced with a more complicated analysis. The algorithms are very efficient in practice. Space limitations prevent us from giving proofs or analyzing the first algorithm, which has the simplest analysis. We do however, need one lemma from that analysis.

Lemma 1. *If $|f| > 2$, then $\sum_{j=1}^{n} I_k(j, f) \leq 3 I_k(n, f)$.*

4.1   Circular Strings

In the circular string algorithm, the size of the computation tree is the same as the previous algorithm, where it was shown to be proportional to $I_k(n, f)$. In this case, however, there is some extra work required to test the wrap-around of the string. Recall that this extra work is proportional to $ExtraWork_k(n, f)$ which is bounded as follows.

$$ExtraWork_k(n, f) \le \sum_{j=1}^{m-1} (m - j) I_k(j, f) + \sum_{j=1}^{m-1} I_k(n - j, f)$$

$$\le \sum_{j=1}^{n} \sum_{t=1}^{j} I_k(t, f) + \sum_{j=1}^{n} I_k(j, f).$$

We now use Lemma 1 to simplify the above bound.

$$ExtraWork_k(n, f) \le 3 \sum_{j=1}^{n} I_k(j, f) + \sum_{j=1}^{n} I_k(j, f) \le 12 I_k(n, f).$$

We have now shown that the total work done by the circular string algorithm is proportional to $I_k(n, f)$. Since the total number of strings generated is $C_k(n, f)$, we must show that $C_k(n, f)$ is proportional to $I_k(n, f)$.

Theorem 4. *If $|f| > 2$, then $3C_k(n, f) \ge I_k(n, f)$.*

These lemmas and Theorem 4 prove Theorem 2.

4.2   Necklaces

To prove Theorem 3, we must show that the computation tree along with the extra work done by the necklace generation algorithm is proportional to $N_k(n, f)$. To get a good bound on $NeckCompTree_k(n, f)$ we need three additional lemmas; the bound of the first lemma does not necessarily hold if f is not a Lyndon word.

Lemma 2. *If f is Lyndon word where $|f| > 2$, then $P_k(n, f) \le \sum_{j=1}^{n} L_k(j, f)$.*

Lemma 3. *If $|f| > 2$, then $L_k(n, f) \le \frac{1}{n} C_k(n, f)$.*

Lemma 4. *If $|f| > 2$, then $\sum_{j=1}^{n} \frac{1}{j} I_k(j, f) \le \frac{8}{n} I_k(n, f)$.*

Applying the previous three lemmas, we show that the computation tree is proportional to $\frac{1}{n} I_k(n, f)$.

$$NeckCompTree_k(n, f) = \sum_{j=1}^{n} P_k(j, f) \le \sum_{j=1}^{n} \sum_{t=1}^{j} L_k(t, f)$$

$$\le \sum_{j=1}^{n} \sum_{t=1}^{j} \frac{1}{t} C_k(t, f) \le \sum_{j=1}^{n} \sum_{t=1}^{j} \frac{1}{t} I_k(t, f)$$

$$\le \sum_{j=1}^{n} \frac{12}{j} I_k(j, f) \le \frac{144}{n} I_k(n, f).$$

This inequality is now used to show that the extra work done by the necklace algorithm is also proportional to $\frac{1}{n}I_k(n, f)$. Recall that the extra work is given by:

$$NeckExtraWork_k(n, f) \leq \sum_{j=1}^{m-1} (m - j)P_k(j, f) + \sum_{j=1}^{m-1} P_k(n - j, f)$$

$$\leq \sum_{j=1}^{n} \sum_{t=1}^{j} P_k(t, f) + \sum_{j=1}^{n} P_k(j, f).$$

Further simplification of this bound follows from the bound on $NeckComp$-$Tree_k(n, f)$ along with Lemma 4.

$$NeckExtraWork_k(n, f) \leq 144 \sum_{j=1}^{n} \frac{1}{j} I_k(j, f) + \frac{144}{n} I_k(n, f)$$

$$\leq \frac{12^3 + 12^2}{n} I_k(n, f).$$

We have now shown that the total computation performed by the necklace generation algorithm is proportional to $\frac{1}{n}I_k(n, f)$. From equation (1), $N_k(n, f) \geq \frac{1}{n}C_k(n, f)$, and since $C_k(n, f) \geq \frac{1}{3}I_k(n, f)$, Theorem 3 is proved.

References

1. A. Aho, J. Hopcroft, and J. Ullman, *The Design and Analysis of Computer Algorithms*, Addisom-Wesley , 1974.
2. D. Gusfield, *Algorithms on Strings, Trees, and Sequences*, Cambridge University Press, 1997.
3. F. Ruskey, J. Sawada, An efficient algorithm for generating necklaces of fixed density, SIAM Journal on Computing, 29 (1999) 671-684.
4. F. Ruskey, J. Sawada, A fast algorithm to generate unlabeled necklaces, 11th Annual ACM-SIGACT Symposium on Discrete Algorithms (SODA), 2000, 256-262.
5. R.P. Stanley, *Enumerative Combinatorics, Volume I*, Wadsworth & Brooks/Cole, 1986.

Optimal Labelling of Point Features in the Slider Model[*]

(Extended Abstract)

Gunnar W. Klau[1] and Petra Mutzel[2]

[1] MPI für Informatik, Saarbrücken, Germany.
Currently TU Wien, Austria.
guwek@mpi-sb.mpg.de
[2] TU Wien, Austria.
mutzel@apm.tuwien.ac.at

Abstract. We investigate the label number maximisation problem (LNM): Given a set of labels Λ, each of which belongs to a point feature in the plane, the task is to find a largest subset Λ_P of Λ so that each $\lambda \in \Lambda_P$ labels the corresponding point feature and no two labels from Λ_P overlap.

Our approach is based on two so-called constraint graphs, which code horizontal and vertical positioning relations. The key idea is to link the two graphs by a set of additional constraints, thus characterising all feasible solutions of LNM. This enables us to formulate a zero-one integer linear program whose solution leads to an optimal labelling.

We can express LNM in both the discrete and the slider labelling model. The slider model allows a continuous movement of a label around its point feature, leading to a significantly higher number of labels that can be placed. To our knowledge, we present the first algorithm that computes provably optimal solutions in the slider model. First experimental results on instances created by a widely used benchmark generator indicate that the new approach is applicable in practice.

1 Introduction

Recently, map labelling has attracted a lot of researchers in computer science due to its numerous applications, *e.g.*, in cartography, geographic information systems and graphical interfaces. A major problem in map labelling is the point-feature label placement, in which the task is to place labels adjacent to point features so that no two labels overlap. In general, the labels are assumed to be axis-parallel rectangles.

Many papers have been published on the label number maximisation problem (for an overview, see the bibliography [9]). We state the problem as follows:

---

[*] This work is partially supported by the Bundesministerium für Bildung, Wissenschaft, Forschung und Technologie (No. 03–MU7MP1–4).

D.-Z. Du et al. (Eds.): COCOON 2000, LNCS 1858, pp. 340–350, 2000.

Problem 1 (Label Number Maximisation, LNM). Given a set $P = \{p_1, \ldots, p_k\}$ of k points in the plane, a set $\Lambda = \{\lambda_1, \ldots, \lambda_l\}$ of l labels, two functions $w, h :$ $\Lambda \rightarrow \mathbb{Q}$ and a function $a : \Lambda \rightarrow P$, find a subset $\Lambda_P \subseteq \Lambda$ of largest cardinality and an assignment $r : \Lambda_P \rightarrow R$, where R is the set of axis-parallel rectangles in the plane, so that the following conditions hold:

(L1) Rectangle $r(\lambda)$ has width $w(\lambda)$ and height $h(\lambda)$ for every $\lambda \in \Lambda_P$.
(L2) Point $a(\lambda)$ lies on the boundary of $r(\lambda)$ for all $\lambda \in \Lambda_P$.
(L3) The open intersection $r(\lambda) \cap r(\mu)$ is empty for all distinct $\lambda, \mu \in \Lambda_P$.

Properties (L1) and (L2) make sure that each label is attached correctly to its point feature and drawn with the given size. Property (L3) forbids overlaps between the labels. We allow, however, that two labels touch each other.

So far, most previous work on map labelling has concentrated on the discrete model, which allows only a finite number of positions per label. The most popular discrete model is the four-position model (see Fig. 1); the two- and one-position models have been introduced rather for theoretical purposes. Christensen, Marks and Shieber [2] have presented a comprehensive treatment of LNM in the four-position model including complexity analysis, heuristic methods and a computational study. They have introduced a procedure for randomly creating labelling instances which has become a widely used benchmark generator in the map labelling literature.

The only practically efficient algorithm for computing provably optimal solutions in the discrete model has been suggested by Verweij and Aardal [8]. They treat the problem as an independent set problem and solve it using a branch-and-cut algorithm. The algorithm is able to optimally label up to 800 point features (using the benchmark generator from [2]) within moderate computation time (about 20 minutes).

More natural than the discrete model is the slider model, which allows a continuous movement of a label around its point feature. Although Hirsch considered this model already in 1982, it was not further investigated until very recently. In [7], van Kreveld, Strijk and Wolff introduce several variations of the slider model (see Fig. 1). They prove NP-hardness of LNM in the four-slider model and suggest a polynomial time algorithm which is able to find a solution that is at least half as good as an optimal solution. Moreover, their computational results show that the slider model is significantly better than the discrete model. The four-slider model allows to place up to 15% more labels in real-world instances and up to 92% more labels in pseudo-random instances.

We will present an algorithm for the label number maximisation problem that works in any of the above mentioned labelling models. We allow several labels per point feature and labels of different sizes. Figure 2 shows a provably optimal labelling for 700 point features computed with a first implementation of our new approach; 699 labels could be placed.

In Sect. 2 we transform LNM into the combinatorial optimisation problem CGF and show that both problems are equivalent. Section 3 contains an integer linear programming formulation for CGF. We show that we can use a feasible solution of the integer linear program to construct a feasible labelling. In particular, this

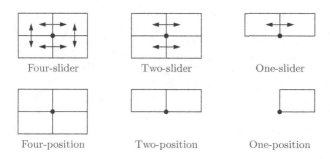

Fig. 1. Axis-parallel rectangular labelling models. A label can be placed in any
of the positions indicated by the rectangles and slid in the directions of the arcs

enables us to find an optimal labelling. We describe a first implementation in
Sect. 4 and discuss our observations of test-runs on instances created by the
benchmark generator presented in [2].

Note that our formulation of LNM allows labels to overlap unlabelled point
features (which is desirable in some applications). We describe in Sect. 5 how we
can exclude these overlaps and investigate in how far our algorithm can satisfy
additional criteria like preferable positions and labels of different importance.
Due to space limitations, we will omit some proofs and which can be found in
the full version of the paper.

2   LNM as a Combinatorial Optimisation Problem

In this section we reformulate the label number maximisation problem as a
problem of combinatorial nature. Our approach uses a pair of *constraint graphs*.
These graphs originated in the area of VLSI design and we have studied them
in previous work for two problems from graph drawing: The two-dimensional
compaction problem [5] and a combined compaction and labelling problem [4].

An important common feature of these problems—and also of LNM—is the
decomposition into a horizontal and a vertical problem component; this observation motivates us to treat both directions separately. Nodes in the *horizontal
constraint graph* $D_x = (V_x, A_x)$ correspond to x-coordinates of objects in the
problem, weighted directed edges to horizontal distance relations between the
objects corresponding to their endpoints. Similarly, the directed graph D_y codes
the vertical relationships.

Definition 1. *A coordinate assignment for a pair of constraint graphs (D_x, D_y)
with $D_x = (V_x, A_x)$, $D_y = (V_y, A_y)$ and arc weights $\omega \in \mathbb{Q}^{|A_x \cup A_y|}$ is a function
$c : V_x \cup V_y \to \mathbb{Q}$. We say c respects an arc set $A \subseteq A_x \cup A_y$ if $c(v_j) - c(v_i) \geq \omega_{ij}$
for all $(v_i, v_j) \in A$.*

We will show later that we can use a pair of constraint graphs together with a
coordinate assignment which respects this pair to construct a feasible labelling—

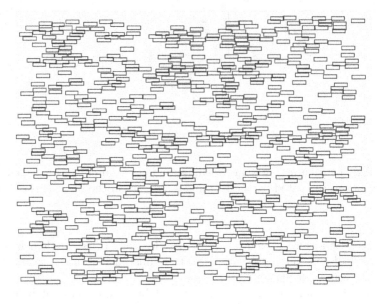

Fig. 2. Optimal labelling of 700 point features in the four-slider model. Instance randomly generated with a standardised procedure described in [2]

as long as the graphs satisfy certain conditions. The following theorem expresses an important connection between constraint graphs and coordinate assignments.

Theorem 1. *Let $D = (V, A)$ be a constraint graph with arc weights $\omega \in \mathbb{Q}^{|A|}$. There exists a coordinate assignment c that respects A if and only if A does not contain a directed cycle of positive weight.*

Proof. For the forward direction assume otherwise. Without loss of generality, let $C = ((v_1, v_2), (v_2, v_3), \ldots, (v_{|C|}, v_1))$ be a directed cycle of positive weight in A. Since c respects A, it must respect in particular the arcs in C. It follows $c(v_2) - c(v_1) \geq \omega_{(1,2)}$, $c(v_3) - c(v_2) \geq \omega_{(2,3)}$, $\ldots$, $c(v_1) - c(v_{|C|}) \geq \omega_{(|C|,1)}$. Summing up the left sides yields zero, summing up the right sides yields $\sum_{a \in C} \omega_a$. We get $0 \geq \sum_{a \in C} \omega_a > 0$, a contradiction.

For the backward direction of the proof, let $\mathcal{A} c \geq \omega$ be the set of inequalities describing the requirements for a coordinate assignment. By Farkas' Lemma, there is a feasible solution if and only if there does not exist a vector $y \geq 0$ with $y^T \mathcal{A} = 0$ and $y^T \omega > 0$. Assume otherwise, *i.e.*, there is such a vector y. Then y corresponds to a flow in A with positive weight. Since all supplies in the corresponding network are zero, this flow must be circular and thus corresponds to a directed cycle of positive weight.   □

In the following, we describe the construction of the pair (D_x, D_y) for a given instance of the label number maximisation problem.

Modelling point features. The positions of the k point features are specified in the input set P. For each $p_i \in P$ with coordinates $x(p_i)$ and $y(p_i)$ we introduce a node x_i in V_x and a node y_i in V_y; one for its x-, one for its y-coordinate. We fix the positions of the point features by inserting four directed paths $P_x = (x_1, \ldots, x_k)$, $P_{-x} = (x_k, \ldots, x_1)$, $P_y = (y_1, \ldots, y_k)$ and $P_{-y} = (y_k, \ldots, y_1)$ with weights $\omega_{x_i x_{i+1}} = x(p_{i+1}) - x(p_i)$, $\omega_{x_{i+1} x_i} = x(p_i) - x(p_{i+1})$, $\omega_{y_i y_{i+1}} = y(p_{i+1}) - y(p_i)$ and $\omega_{y_{i+1} y_i} = y(p_i) - y(p_{i+1})$ for $i \in \{1, \ldots, k-1\}$.

We call the directed edges on these paths *fixed distance arcs* and refer to them as A_F. Figure 3 shows a set of point features and its representation in the constraint graphs.

Lemma 1 (Proof omitted). *A coordinate assignment c that respects A_F results in a correct placement of point features (up to translation).*

Modelling labels. Each label $\lambda \in \Lambda$ has to be represented by a rectangle $r(\lambda)$ of width $w(\lambda)$ and height $h(\lambda)$. Additionally, we have to ensure that λ will be placed correctly with respect to its point feature $a(\lambda)$, *i.e.*, $a(\lambda)$ must lie on the boundary of $r(\lambda)$.

Straightforwardly, we model a label λ by two nodes in V_x and two nodes in V_y, representing the coordinates of $r(\lambda)$. We call these nodes the left, right, bottom and top *limit* of λ and refer to them as l_λ, r_λ, b_λ and t_λ, respectively. We introduce four *label size arcs* $A_L(\lambda) = \{(l_\lambda, r_\lambda), (r_\lambda, l_\lambda), (b_\lambda, t_\lambda), (t_\lambda, b_\lambda)\}$ in order to model the size of $r(\lambda)$. The weights of these arcs are $\omega_{l_\lambda r_\lambda} = w(\lambda)$, $\omega_{r_\lambda l_\lambda} = -w(\lambda)$, $\omega_{b_\lambda t_\lambda} = h(\lambda)$ and $\omega_{t_\lambda b_\lambda} = -h(\lambda)$, see Fig. 4(a).

A label λ must be placed close to its point feature $a(\lambda)$. Let x and y be the nodes representing point $a(\lambda)$ in the constraint graphs. We add four *proximity arcs* $A_P(\lambda) = \{(x, r_\lambda), (l_\lambda, x), (y, t_\lambda), (b_\lambda, y)\}$, as illustrated in Fig. 4(b). These arcs have zero weight and exclude that the point feature $a(\lambda)$ lies outside the rectangle $r(\lambda)$.

The point feature may still lie inside $r(\lambda)$. We disallow this by adding at least one of the four *boundary arcs* $\{(r_\lambda, x), (x, l_\lambda), (t_\lambda, y), (y, b_\lambda)\}$, each of weight zero. Note that these arcs are inverse to the proximity arcs for label λ. If, *e.g.*, (r_λ, x) is present in D_x, it forces—together with its inverse proximity arc (x, r_λ)—the

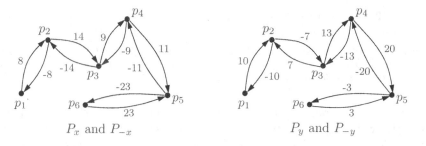

$$P_x \text{ and } P_{-x} \qquad\qquad P_y \text{ and } P_{-y}$$

Fig. 3. Modelling the placement of point features with fixed distance arcs

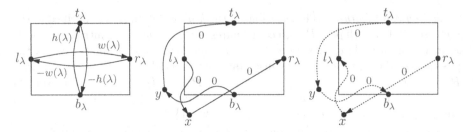

Fig. 4. Modelling labels. (a) Label size arcs. (b) Proximity arcs. (c) Boundary arcs

coordinate of the right side of $r(\lambda)$ to be equal to the coordinate of x; the label has to be placed at its leftmost position. See also Fig. 4(c).

At this point we can influence the labelling model. We define a set $A_B(\lambda)$ containing the boundary arcs for the different models:

Labelling model		Boundary arcs $A_B(\lambda)$
Slider model	Four-slider	$\{(r_\lambda, x), (x, l_\lambda), (t_\lambda, y), (y, b_\lambda)\}$
	Two-slider	$\{(t_\lambda, y), (y, b_\lambda)\}$
	One-slider	$\{(y, b_\lambda)\}$
Discrete model	Four-position	$\{(r_\lambda, x), (x, l_\lambda), (t_\lambda, y), (y, b_\lambda)\}$
	Two-position	$\{(y, b_\lambda), (r_\lambda, x), (x, l_\lambda)\}$
	One-position	$\{(y, b_\lambda), (x, l_\lambda)\}$

For a slider model, at least one arc $a \in A_B(\lambda)$ must be contained in the constraint graph, for a discrete model at least two. *E.g.*, for the four-slider model, the set $A_B(\lambda)$ consists of all four boundary arcs, one of which must be present in the appropriate constraint graph. Note that we can express all six axis-parallel rectangular labelling models we have introduced in Sect. 1 as additional requirements on the constraint graphs.

Lemma 2 (Proof omitted). *Let λ be a label, c be a coordinate assignment respecting $A_L(\lambda)$, $A_P(\lambda)$ and at least d boundary arcs from $A_B(\lambda)$. Then c results in a placement in which λ is represented by a rectangle $r(\lambda)$ of width $w(\lambda)$ and height $h(\lambda)$. The label is placed so that point feature $a(\lambda)$ lies on the boundary of $r(\lambda)$ if $d = 1$ and on a corner of $r(\lambda)$ if $d = 2$.*

Avoiding label overlaps. Until now we have assured that each label is placed correctly with respect to its point feature. It remains to guarantee that the intersection of rectangles is empty. A crucial observation is that it suffices to consider only the pairs of labels that can possibly interact. If there is any overlap, such a pair must be involved.

Consider two different labels λ and μ and their corresponding rectangles $r(\lambda)$ and $r(\mu)$. We call the pair *vertically separated* if $r(\lambda)$ is placed either above or below $r(\mu)$. Similarly, λ and μ are *horizontally separated* if one rectangle is placed left to the other. Two labels overlap if they are neither vertically nor horizontally separated, we can exclude this by introducing one of the following four *label separation arcs* $A_S(\lambda,\mu) = \{(t_\mu,b_\lambda),(t_\lambda,b_\mu),(r_\mu,l_\lambda),(r_\lambda,l_\mu)\}$. Label separation arcs have weight zero.

Let R_λ be the boundary of the region in which label λ can be placed. Note that R_λ is defined by lower left corner $(x(a(\lambda))-w(\lambda), y(a(\lambda))-h(\lambda))$ and upper right corner $(x(a(\lambda))+w(\lambda), y(a(\lambda))+h(\lambda))$. Likewise, we determine R_μ for label μ. If the intersection of R_λ and R_μ is empty, λ and μ can never overlap, and we do not have to add any label separation arcs for this pair. In this case we set $A_S(\lambda,\mu) = \emptyset$.

Consider now the case that the intersection of R_λ and R_μ is not empty, as depicted in Fig. 5. Depending on the position of the corresponding point features $a(\lambda)$ and $a(\mu)$, $A_S(\lambda,\mu)$ contains the following label separation arcs:

1. If $x(a(\mu)) \geq x(a(\lambda))$ we have $(r_\lambda,l_\mu) \in A_S(\lambda,\mu)$.
2. If $x(a(\lambda)) \geq x(a(\mu))$ we have $(r_\mu,l_\lambda) \in A_S(\lambda,\mu)$.
3. If $y(a(\mu)) \geq y(a(\lambda))$ we have $(t_\lambda,b_\mu) \in A_S(\lambda,\mu)$.
4. If $y(a(\lambda)) \geq y(a(\mu))$ we have $(t_\mu,b_\lambda) \in A_S(\lambda,\mu)$.

Note that the only case in which $A_S(\lambda,\mu)$ contains all four label separation arcs occurs if λ and μ label the same point feature, *i.e.*, $a(\lambda) = a(\mu)$.

Lemma 3 (Proof omitted). *Let λ and μ be two labels that can possibly overlap and let c be a coordinate assignment respecting $A_L(\lambda)$, $A_L(\mu)$ and $A \subseteq A_S(\lambda,\mu)$ with $|A| \geq 1$. Then c results in a placement in which the two rectangles $r(\lambda)$ and $r(\mu)$ do not overlap.*

We refer to the boundary and label separation arcs as *potential arcs* $A_{\text{pot}} = \bigcup_{\lambda \in \Lambda} A_B(\lambda) \cup \bigcup_{\lambda,\mu \in \Lambda, \lambda \neq \mu} A_S(\lambda,\mu)$ and state the label number maximisation problem in a combinatorial way. The task is to choose additional arcs from A_{pot} for a maximum number of labels without creating positive directed cycles.

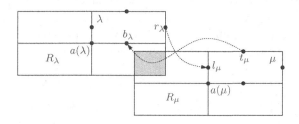

Fig. 5. Label separation arcs between two labels λ and μ

Problem 2 (Constraint Graph Fulfilment problem, CGF). Given an instance of LNM, let (D_x, D_y) be the pair of constraint graphs including only fixed distance arcs, label size arcs and proximity arcs. Let $d = 1$ if in the slider model and $d = 2$ if in the discrete model and let A_x and A_y be the arc sets of D_x and D_y, respectively. Find a set $\Lambda_P \subseteq \Lambda$ of greatest cardinality and an arc set $A \subseteq A_{\text{pot}}$ with the properties:

(F1) $|A \cap A_B(\lambda)| \geq d$ for all $\lambda \in \Lambda_P$.
(F2) $|A \cap A_S(\lambda, \mu)| \geq 1$ for all $\lambda, \mu \in \Lambda_P$, $\lambda \neq \mu$, $R_\lambda \cap R_\mu \neq \emptyset$.
(F3) $A \cap A_x$ and $A \cap A_y$ do not contain a positive cycle.

Theorem 2. *Problems* LNM *and* CGF *are polynomially equivalent.*

Proof. Let A be a solution of CGF. We extend (D_x, D_y) by adding A to the arc sets. Because of (F3) and Theorem 1, there is a coordinate assignment that respects both the horizontal and vertical arc set. Lemmas 1, 2 and 3 ensure that properties (L1), (L2) and (L3) are fulfilled, thus we have a solution for LNM.

For the other direction, we start with the given coordinate assignment c resulting from the placement of labels. We create the set A of additional arcs as follows: For each label λ we add one or two boundary arcs, depending on how $r(\lambda)$ is placed with respect to point feature $a(\lambda)$. Similarly, we add appropriate arcs from $A(\lambda, \mu)$ for pairs of labels λ and μ, depending on the relative position of λ and μ in the labelling. Note that we have chosen the additional arcs so that they are respected by c. Properties (F1) and (F2) follow by construction, property (F3) follows by Theorem 1. $\qquad \square$

3   Integer Linear Programming Formulation

In the previous section we have shown how to transform the label number maximisation problem into the combinatorial optimisation problem CGF. We propose a zero-one integer linear programming formulation and a branch-and-cut algorithm for solving CGF. The goal is to find the set of additional boundary and label separation arcs A and to determine which labels are to be placed.

We introduce two types of binary variables for this task: For each label λ there is a variable $y_\lambda \in \{0, 1\}$, indicating whether λ will be placed or not. Additionally, there are variables $x_a \in \{0, 1\}$ for potential additional arcs $a \in A_{\text{pot}}$. We get

$$y_\lambda = \begin{cases} 1 & \lambda \in \Lambda_P \\ 0 & \lambda \in \Lambda \setminus \Lambda_P \end{cases} \quad \text{and} \quad x_a = \begin{cases} 1 & a \in A \\ 0 & a \in A_{\text{pot}} \setminus A . \end{cases}$$

We present the zero-one integer linear program (ILP) and show that it corresponds to CGF. We define $C_p = A_{\text{pot}} \cap C$.

$$\max \quad \sum_{\lambda \in \Lambda} y_\lambda \qquad \qquad \text{(ILP)}$$

$$\text{subject to} \quad \sum_{a \in A_B(\lambda)} x_a - 2y_\lambda \geq d - 2 \qquad \forall \lambda \in \Lambda \qquad \text{(ILP.1)}$$

$$\sum_{a \in A_S(\lambda,\mu)} x_a - y_\lambda - y_\mu \geq -1 \qquad \forall \lambda, \mu \in \Lambda, \lambda \neq \mu \qquad \text{(ILP.2)}$$

$$\sum_{a \in C_p} x_a \leq |C_p| - 1 \qquad \forall \text{ positive cycles } C \qquad \text{(ILP.3)}$$

$$y_\lambda \in \{0, 1\} \qquad \forall \lambda \in \Lambda \qquad \text{(ILP.4)}$$

$$x_a \in \{0, 1\} \qquad \forall a \in A_{\text{pot}} \qquad \text{(ILP.5)}$$

We refer to (ILP.1) as *boundary inequalities*, to (ILP.2) as *label separation inequalities* and to (ILP.3) as *positive cycle inequalities*.

Theorem 3 (Proof omitted). *Each feasible solution (y, x) to (ILP) corresponds to a feasible solution of* CGF *and vice versa. The value of the objective function equals the cardinality of Λ_P.*

Corollary 1. *An optimal solution of (ILP) corresponds to an optimal labelling.*

Corollary 1 and Theorem 1 suggest an algorithm for attacking practical instances of the label number maximisation problem: In a first step, we solve (ILP) using integer programming techniques. The solution tells us which boundary and label separation arcs should be added to the arc sets of the constraint graphs (D_x, D_y). We use this information in a second step for computing the corresponding coordinate assignment via minimum cost flow (see proof of Theorem 1).

4   First Experiments

We have implemented a simple strategy to solve the integer linear program based on LEDA [6] and the ILP-solver in CPLEX [3]. Due to the fact that the ILP may have an exponential number of positive cycle inequalities, we solve it iteratively, using a cutting plane approach: We start with an integer linear program containing inequalities (ILP.1), (ILP.2), (ILP.4), (ILP.5) and a set of *local cycle inequalities* of type (ILP.3). We determine these inequalities by looking at possible positive cycles involving up to two labels.

Let $\bar{x}$ be a solution of an ILP in the iteration. If the corresponding constraint graphs do not contain any positive cycles, $\bar{x}$ is an optimal solution for CGF. Otherwise, we find a positive cycle and add the corresponding inequality to the new integer linear program. Our separation procedure uses the Bellman-Ford algorithm for detecting negative cycles applied to the constraint graphs after multiplying the arc weights $\omega \in \mathbb{Q}^{|A_x \cup A_y|}$ by -1. Our implementation is based on the one given in [1].

We have tested the implementation on a widely used benchmark generator for randomly creating instances of LNM, following the rules described in [2]: First, we construct a set of n points with random coordinates in the range $\{0, \ldots, 792\}$ for the x- and $\{0, \ldots, 612\}$ for the y-coordinates. To each point feature belongs a label of width 30 and height 7. We have run the algorithm with both the slider and the discrete labelling models for rectilinear map labelling. Figure 2 in Sect. 1 shows an optimal solution for an instance with $n = 700$. In the optimal solution for the four-slider model, 699 labels can be placed, whereas at most 691 can be placed in the four-position model.

Evidently, more freedom in the labelling model results in a higher number of labels that can be placed. Two main factors influence the running time of our implementation for instances of the same size: On the one hand, this is the number of labels that cannot be placed in an optimal solution, $i.e.$, the difference $|\Lambda| - |\Lambda_P|$: To our surprise, we had the longest running times in the one-position model. On the other hand, the tightness of the inequalities seems to have an impact on the running time; the more restrictions on the variables, the faster the algorithms. Both factors, however, interrelate: In the more restricted models we can also place fewer labels.

5   Extensions

As mentioned in Sect. 1, our formulation of LNM allows labels to overlap other, unlabelled point features. In several applications, this may not be allowed. In this case, we take the point features into consideration when introducing the label separation arcs (which then should be called general separation arcs). We then have to check the overlap conditions also for point feature/label pairs; it is worth noting that the boundary arcs arise as a special case of general separation arcs when considering a pair $(\lambda, a(\lambda))$, $i.e.$, a label and the point feature it should be attached to.

In many applications, there are labels of different importance. It is easy to integrate this into our approach: The objective function of the integer linear program changes to $\sum_{\lambda \in \Lambda} z_\lambda$, where z_λ denotes the importance of label λ. The algorithm will then prefer more important labels and remove less important ones more easily. Another practically motivated extension is to model preferable positions of labels: Often, a label should be placed rather at its rightmost and upmost position than at other possible positions. We suggest to define a weight vector for the boundary arcs and incorporate it into the objective function.

Acknowledgements. The authors thank Alexander Wolff for the real-world data and the help with the conversion in our data format, Michael Jünger for spontaneously supporting the development of the negative cycle separator and Andrew V. Goldberg for his negative cycle detection code.

References

[1] B. V. Cherkassky and A. V. Goldberg, *Negative-cycle detection algorithms*, Mathematical Programming **85** (1999), no. 2, 277–311.

[2] J. Christensen, J. Marks, and S. Shieber, *An empirical study of algorithms for point-feature label placement*, ACM Transactions on Graphics **14** (1995), no. 3, 203–232.

[3] ILOG, *CPLEX 6.5 Reference Manual*, 1999.

[4] G. W. Klau and P. Mutzel, *Combining graph labeling and compaction*, Proc. 8th Internat. Symp. on Graph Drawing (GD '99) (Štiřín Castle, Czech Republic) (J. Kratochvíl, ed.), LNCS, no. 1731, Springer–Verlag, 1999, pp. 27–37.

[5] G. W. Klau and P. Mutzel, *Optimal compaction of orthogonal grid drawings*, Integer Programming and Combinatorial Optimization (IPCO '99) (Graz, Austria) (G. P. Cornuéjols, R. E. Burkard, and G. J. Woeginger, eds.), LNCS, no. 1610, Springer–Verlag, 1999, pp. 304–319.

[6] K. Mehlhorn and S. Näher, *LEDA. A platform for combinatorial and geometric computing*, Cambridge University Press, 1999.

[7] M. van Kreveld, T. Strijk, and A. Wolff, *Point labeling with sliding labels*, Computational Geometry: Theory and Applications **13** (1999), 21–47.

[8] B. Verweij and K. Aardal, *An optimisation algorithm for maximum independent set with applications in map labelling*, Proc. 7th Europ. Symp. on Algorithms (ESA '99) (Prague, Czech Republic), LNCS, vol. 1643, Springer–Verlag, 1999, pp. 426–437.

[9] A. Wolff and T. Strijk, *The map labeling bibliography*, http://www.math-inf.uni-greifswald.de/map-labeling/bibliography.

Mappings for Conflict-Free Access of Paths in Elementary Data Structures[*]

Alan A. Bertossi[1] and M. Cristina Pinotti[2]

[1] Department of Mathematics, University of Trento, Trento, Italy
bertossi@science.unitn.it
[2] Istituto di Elaborazione dell' Informazione, CNR, Pisa, ITALY
pinotti@iei.pi.cnr.it

Abstract. Since the divergence between the processor speed and the memory access rate is progressively increasing, an efficient partition of the main memory into multibanks is useful to improve the overall system performance. The effectiveness of the multibank partition can be degraded by *memory conflicts*, that occur when there are many references to the same memory bank while accessing the same memory pattern. Therefore, mapping schemes are needed to distribute data in such a way that data can be retrieved via regular patterns without conflicts. In this paper, the problem of conflict-free access of *arbitrary* paths in bidimensional arrays, circular lists and complete trees is considered for the first time and reduced to variants of graph-coloring problems. Balanced and fast mappings are proposed which require an optimal number of colors (i.e., memory banks). The solution for bidimensional arrays is based on a combinatorial object similar to a Latin Square. The functions that map an array node or a circular list node to a memory bank can be calculated in constant time. As for complete trees, the mapping of a tree node to a memory bank takes time that grows logarithmically with the number of nodes of the tree.

1 Introduction

In recent years, the traditional divergence between the processor speed and the memory access rate is progressively increasing. Thus, an efficient organization of the main memory is important to achieve high-speed computations. For this purpose, the main memory can be equipped with *cache* memories – which have about the same cycle time as the processors – or can be partitioned into *multibanks*. Since the cost of the cache memory is high and its size is limited, the multibank partition has mostly been adopted, especially in shared-memory multiprocessors [3]. However, the effectiveness of such a memory partition can be limited by *memory conflicts*, that occur when there are many references to the same memory bank while accessing the same memory pattern. To exploit to the fullest extent the performance of the multibank partition, mapping schemes

---

[*] This work has been supported by the "Provincia Autonoma di Trento" under a research grant.

D.-Z. Du et al. (Eds.): COCOON 2000, LNCS 1858, pp. 351–361, 2000.

can be employed that avoid or minimize the memory conflicts [10]. Since it is hard to find universal mappings – mappings that minimize conflicts for arbitrary memory access patterns – several specialized mappings, designed for accessing regular patterns in specific data structures, have been proposed in the litera-ture (see [7,2] for a complete list of references). In particular, mappings that provide conflict-free access to complete subtrees, root-to-leaves paths, sublevels, and composite patterns obtained by their combination, have been investigated in [4,5,1,6,9].

In the present paper, optimal, balanced and fast mappings are designed for conflict-free access of paths in bidimensional arrays, circular lists, and complete trees. With respect to the above mentioned papers, paths in bidimensional arrays and circular lists are dealt with for the first time. Moreover, access to any (not only to root-to-leaves) paths in complete trees is provided.

2   Conflict-Free Access

When storing a data structure D, represented in general by a graph, on a memory system consisting of N memory banks, a desirable issue is to map any subset of N arbitrary nodes of D to all the N different banks. This problem can be viewed as a *coloring* problem where the distribution of nodes of D among the banks is done by coloring the nodes with a color from the set $\{0, 1, 2, \ldots, N-1\}$. Since it is hard to solve the problem in general, access of regular patterns, called *templates*, in special data structures – like bidimensional arrays, circular lists, and complete trees – are considered hereafter.

A *template* T is a subgraph of D. The occurrences $\{T_1, T_2, \ldots, T_m\}$ of T in D are the *template instances*. For example, if D is a complete binary tree, then a path of length k can be a template, and all the paths of length k in D are the template instances.

After coloring D, a *conflict* occurs if two nodes of a template instance are assigned to the same memory bank, i.e., they get the same color. An access to a template instance T_i results in c conflicts if $c+1$ nodes of T_i belong to the same memory bank.

Given a memory system with N banks and a template T, the goal is to find a *memory mapping* $U : D \rightarrow N$ that colors the nodes of D in such a way that the number of conflicts for accessing any instance of T is minimal. In fact, the *cost* for T_i colored according to U, $Cost_U(D, T_i, N)$, is defined as the number of conflicts for accessing T_i. The template instance of T with the highest cost determines the overall cost of the mapping U. That is,

$$Cost_U(D, T, N) \stackrel{\text{def}}{=} \max_{T_i \in T} Cost_U(D, T_i, N).$$

A mapping U is *conflict-free* for T if $Cost_U(D, T, N) = 0$.

Among desirable properties for a conflict-free mapping, a mapping should be balanced, fast, and optimal. A mapping U is termed *balanced* if it evenly distributes the nodes of the data structure among the N memory banks. For a balanced mapping, the *memory load* is almost the same in all the banks. A mapping U will be called *fast* if the color of each node can be computed quickly (possibly in constant time) without knowledge of the coloring of the entire data structure. Among all possible conflict-free mappings for a given template of a data structure, the more interesting ones are those that use the minimum possible number of memory banks. These mappings are called *optimal*. It is worth to note that not only the template size but also the overlapping of template instances in the data structure determine a lower bound on the number of memory banks necessary to guarantee a conflict-free access scheme. This fact will be more convincing by the argument below for accessing paths in D.

Let $G_D = (V, E)$ be the graph representing the data structure D. The template P_k is a *path* of length k in D. The template instance $P_k[x, y]$ is the path of length k between two vertices x and y in V, that is, the sequence $x = v_1, v_2, \ldots, v_{k+1} = y$ of vertices such that $(v_h, v_{h+1}) \in E$ for $h = 1, 2, \ldots k$.

The conflicts can be eliminated on $P_k[x, y]$ if $v_1, v_2, \ldots, v_{k+1}$ are assigned to all different memory banks. The conflict-free access to P_k can be reduced to a classical coloring problem on the associated graph G_{DP_k} obtained as follows. The vertex set of G_{DP_k} is the same as the vertex set of G_D, while the edge (r, s) belongs to the edge set of G_{DP_k} iff the distance d_{rs} between the vertices r and s in G_D satisfies $d_{rs} \leq k$, where the distance is the length of the shortest path between r and s. Now, colors must be assigned to the vertices of G_{DP_k} so that every pair of vertices connected by an edge is assigned a couple of different colors and the minimum number of colors is used. Hence, the role of *maximum clique* in G_{DP_k} is apparent for deriving lower bounds on the conflict-free access on paths. A *clique* K for G_{DP_k} is a subset of the vertices of G_{DP_k} such that for each pair of vertices in K there is an edge. By well-known graph theoretical results, a clique of size n in the associated graph G_{DP_k} implies that at least n different colors are needed to color G_{DP_k}. In other words, the size of the largest clique in G_{DP_k} is a lower bound for the number of memory banks required to access paths of length k in D without conflicts.

On the other hand, the conflict-free access to P_k on G_D is equivalent to color the nodes of G_D in such a way that any two nodes which are at distance k or less apart have assigned different colors. Unfortunately, this latter coloring problem is NP-complete [8] for general graphs. In the next three sections, optimal mappings for bidimensional arrays, circular lists and complete binary trees will be derived for conflict-free accessing P_k.

3   Accessing Paths in Bidimensional Arrays

Let a bidimensional array A be the data structure D to be mapped into the multibank memory system. An array $r \times c$ has r rows and c columns, indexed respectively from 0 to $r - 1$ (from top to bottom) and from 0 to $c - 1$ (from left to right), with r and c both greater than 1.

The graph $G_A = (V, E)$ representing A is a mesh, whose vertices correspond to the elements of A and whose arcs correspond to any pair of adjacent elements of A on the same row or on the same column. For the sake of simplicity, A will be used instead of G_A since there is no ambiguity. Thus, a generic node x of A will be denoted by $x = (i, j)$, where i is its row index and j is its column index.

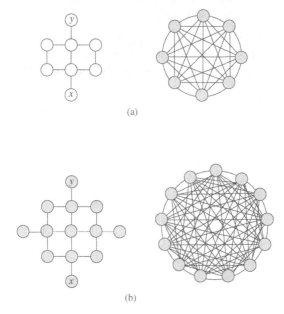

(a)

(b)

Fig. 1. A subset $K_A(x, k)$ of nodes of A that forms a clique in G_{AP_k}: (a) $k = 3$, (b) $k = 4$.

Consider a generic node $x = (i, j)$ of A, and its opposite node at distance k on the same column, i.e., $y = (i - k, j)$. All the nodes of A at distance k or less from both x and y are mutually at distance k or less, as shown in Figure 1. In the associated graph G_{AP_k}, they form a clique, and they must be assigned to different colors. Therefore,

Lemma 1. *At least* $M = \left\lceil \frac{(k+1)^2}{2} \right\rceil$ *memory banks are required for conflict-free accessing* P_k *in* A.

Below, a conflict-free mapping is given to color all the nodes of an array A using as few colors as in Lemma 1. Therefore, the mapping is optimal. From now on, the color assigned to node x is denoted by $\gamma(x)$.

Algorithm Array-Coloring (A, k);

- Set $M = \left\lceil \frac{(k+1)^2}{2} \right\rceil$ and $\Delta = \begin{cases} k + 1 & \text{if } k \text{ is even} \\ k & \text{if } k \text{ is odd} \end{cases}$
- Assign to each node $x = (i, j) \in A$ the color $\gamma(x) = (i\Delta + j) \bmod M$.

Theorem 1. *The Array-Coloring mapping is optimal, fast, and balanced.*

Proof: Intuitively, the above algorithm first covers A with a tessellation of basic sub-arrays of size $M \times M$. Each basic sub-array S is colored conflict-free in a Latin Square fashion as follows:

- the colors in the first row of S appear from left-to-right in the sequence $0, 1, 2, \ldots, M - 1$;
- the color sequence for a generic row is obtained from the sequence at the previous row by a Δ left-cyclic shift.

For $k = 3$, the coloring of A, decomposed into 6 basic sub-arrays of size $M \times M$, is illustrated in Figure 2.

0 1 2 3 4 5 6 7	0 1 2 3 4 5 6 7	0 1 2 3 4 5 6 7
3 4 5 6 7 0 1 2	3 4 5 6 7 0 1 2	3 4 5 6 7 0 1 2
6 7 0 1 2 3 4 5	6 7 0 1 2 3 4 5	6 7 0 1 2 3 4 5
1 2 3 4 5 6 7 0	1 2 3 4 5 6 7 0	1 2 3 4 5 6 7 0
4 5 6 7 0 1 2 3	4 5 6 7 0 1 2 3	4 5 6 7 0 1 2 3
7 0 1 2 3 4 5 6	7 0 1 2 3 4 5 6	7 0 1 2 3 4 5 6
2 3 4 5 6 7 0 1	2 3 4 5 6 7 0 1	2 3 4 5 6 7 0 1
5 6 7 0 1 2 3 4	5 6 7 0 1 2 3 4	5 6 7 0 1 2 3 4
0 1 2 3 4 5 6 7	0 1 2 3 4 5 6 7	0 1 2 3 4 5 6 7
3 4 5 6 7 0 1 2	3 4 5 6 7 0 1 2	3 4 5 6 7 0 1 2
6 7 0 1 2 3 4 5	6 7 0 1 2 3 4 5	6 7 0 1 2 3 4 5
1 2 3 4 5 6 7 0	1 2 3 4 5 6 7 0	1 2 3 4 5 6 7 0
4 5 6 7 0 1 2 3	4 5 6 7 0 1 2 3	4 5 6 7 0 1 2 3
7 0 1 2 3 4 5 6	7 0 1 2 3 4 5 6	7 0 1 2 3 4 5 6
2 3 4 5 6 7 0 1	2 3 4 5 6 7 0 1	2 3 4 5 6 7 0 1
5 6 7 0 1 2 3 4	5 6 7 0 1 2 3 4	5 6 7 0 1 2 3 4

Fig. 2. An array A of size 16×24 with a tessellation of 6 sub-arrays of size 8×8 colored by the Array-Coloring algorithm to conflict-free access P_3.

No conflict arises on the borders of the sub-arrays. In fact, it can be proved that any two nodes colored the same are $k + 1$ apart and their relative positions are depicted in Figure 3.

So, the Array-Coloring Algorithm is conflict-free. Moreover, since it uses the minimum number of colors, the proposed mapping is optimal.

It is easy to see that the time required to color all the $n = rc$ nodes of an array is $O(n)$. Moreover, to color only a single node $x = (i, j)$ of the tree requires only $O(1)$ time, since $\gamma(x) = (i\Delta + j) \bmod M$, and hence the mapping is fast.

In order to prove that the mapping is balanced, observe that each color appears once in each sub-row of size M. Hence, the number m of nodes with the same color verifies $r \lfloor \frac{c}{M} \rfloor \le m \le r \lceil \frac{c}{M} \rceil$.

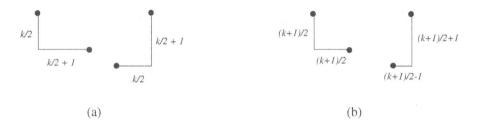

(a)   (b)

Fig. 3. Relative positions in A of two nodes which are assigned to the same color: (a) k even, (b) k odd.

4   Accessing Paths in Circular Lists

Let a circular list C be the data structure D to be mapped into the multibank memory system. A circular list of n nodes, indexed consecutively from 0 to $n-1$, is a sequence of n nodes such that node i is connected to both nodes $(i-1) \bmod n$ and $(i+1) \bmod n$.

The graph $G_C = (V, E)$ representing C is a ring, whose vertices correspond to the elements of C and whose arcs correspond to any pair of adjacent elements of C. For the sake of simplicity, C will be used instead of G_C since there is no ambiguity.

Lemma 2. *Let* $M = \begin{cases} n & \text{if } n < k+1, \\ (k+1) + \left\lceil \dfrac{n \bmod (k+1)}{\left\lfloor \frac{n}{(k+1)} \right\rfloor} \right\rceil & \text{if } n \geq k+1. \end{cases}$

At least M memory banks are required for conflict-free accessing P_k in C.

Proof   For conflict-free accessing P_k in C two nodes with the same color must be at distance at least $k+1$. When $n < k+1$, all the nodes are mutually at distance less than k and must all be colored with different colors. When $n \geq k+1$, each color may appear at most $t = \left\lfloor \frac{n}{(k+1)} \right\rfloor$ times. Therefore, at least $\left\lceil \frac{n}{t} \right\rceil$ colors are needed. Observed that $n = \left\lfloor \frac{n}{(k+1)} \right\rfloor (k+1) + (n \bmod (k+1))$, it follows that at least $M = \left\lceil \frac{n}{t} \right\rceil = (k+1) + \left\lceil \dfrac{n \bmod (k+1)}{\left\lfloor \frac{n}{(k+1)} \right\rfloor} \right\rceil$ memory banks are required.

Below, an optimal conflict-free mapping is provided to color all the nodes of a circular list C using as few colors as in Lemma 2. As before, the color assigned to node x is denoted by $\gamma(x)$.

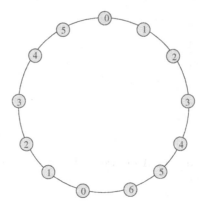

Fig. 4. Conflict-free access to P_4 in a circular list C of 13 nodes colored by the Circular-List-Coloring algorithm with $M = 7$.

Algorithm Circular-List-Coloring (C, k);

- Set $M = \begin{cases} n & \text{if } n < k+1 \\ (k+1) + \left\lceil \dfrac{n \bmod (k+1)}{\left\lfloor \frac{n}{(k+1)} \right\rfloor} \right\rceil & \text{if } n \geq k+1 \end{cases}$

- Set $\theta = sM$ where $s = \begin{cases} n \bmod (M-1), & \text{if } n \bmod M \neq 0 \\ \frac{n}{M}, & \text{if } n \bmod M = 0 \end{cases}$

- Assign to node $x \in C$, the color
  $$\gamma(x) = \begin{cases} x \bmod M & \text{if } x < \theta \\ (x - \theta) \bmod (M-1) & \text{if } x \geq \theta \end{cases}$$

Note that a linear (that is, non circular) list L can be optimally colored to conflict-free access P_k with $M' = k + 1$ colors, which matches the trivial lower bound given by the number of nodes in P_k. In fact, L can be optimally colored by a *naive algorithm* which assigns to node x the color $\gamma(x) = x \bmod M'$. Such a naive algorithm does not work for circular lists.

Theorem 2. *The Circular-List-Coloring mapping is optimal, fast, and balanced.*

5   Accessing Paths in Complete Trees

Let a rooted complete binary tree B be the data structure to be mapped into the multibank memory system. The *level* of node $x \in B$ is defined as the number of edges on the path from x to the root, which is at level 0. The maximum level of the nodes of B is the *height* of B. Let $Lev_B(i)$ be the set of all nodes of B at level $i \geq 0$. A complete binary tree of height H is a rooted tree B in which all the leaves are at the same level and each internal node has exactly 2 children.

Thus, $Lev_B(i)$ contains 2^i nodes. The h-th *ancestor* of the node (i, j) is the node $(i - h, \lfloor \frac{j}{2^h} \rfloor)$, while its children are the nodes $(i+1, 2j)$ and $(i+1, 2j+1)$, in the left-to-right order.

From now on, the generic node x, which is the j-th node of $Lev_B(i)$, with $j \geq 0$ counting from left to right, will be denoted by $x = (i, j)$. Therefore, the generic path instance $P_k[x, y]$ will be denoted by $P_k[(i, j), (r, s)]$, where $x = (i, j)$ and $y = (r, s)$.

Lemma 3. *At least* $M = 2^{\lfloor \frac{k}{2} \rfloor + 1} + 2^{\lceil \frac{k}{2} \rceil} - 2$ *memory banks are required to conflict-free access* P_k *in* B.

Proof   Consider a generic node $x = (i, j)$. All the $2^{\lfloor \frac{k}{2} \rfloor + 1} - 1$ nodes in the subtree S of height $\lfloor \frac{k}{2} \rfloor$ rooted at the $\lfloor \frac{k}{2} \rfloor$-th ancestor of x are mutually at distance not greater than k.

In addition, consider the $\lceil \frac{k}{2} \rceil$ nodes, $\mu_1, \mu_2, \ldots \mu_{\lceil \frac{k}{2} \rceil}$, ancestors of x, on the path I of length $\lceil \frac{k}{2} \rceil$ from the $\lfloor \frac{k}{2} \rfloor$-th ancestor of x up to the k-th ancestor of x. All these nodes are at distance not greater than k from node x, and together with the nodes of S they are at mutual distance not greater than k.

Moreover, for $1 \leq j \leq \lceil \frac{k}{2} \rceil - 1$, consider the $2^{\alpha_j + 1} - 1$ nodes in the complete subtree of height $\alpha_j = k - \lfloor \frac{k}{2} \rfloor - j - 1$, rooted at the μ_j's child which does not belong to I. Such nodes are at distance not greater than k from x. Furthermore, these nodes, along with the nodes of S and I, are all together at mutual distance not greater than k.

Hence, in the associated graph G_{DP_k} there is at least a clique of size

$$2^{\lfloor \frac{k}{2} \rfloor + 1} - 1 + \left\lceil \frac{k}{2} \right\rceil + \sum_{j=1}^{\lceil \frac{k}{2} \rceil - 1} \left(2^{\alpha_j + 1} - 1 \right) = 2^{\lfloor \frac{k}{2} \rfloor + 1} - 1 + \left\lceil \frac{k}{2} \right\rceil + \sum_{h=0}^{\lceil \frac{k}{2} \rceil - 2} \left(2^{h+1} - 1 \right).$$

From that, the claim easily follows. Figure 5 shows a subset $K_B(k)$ of nodes of B which are at pairwise distance not greater than k, for $k = 3$ and 4, and hence forms a clique in the associated graph G_{BP_k}.

An optimal conflict-free mapping to color a complete binary tree B acts as follows.

A basic subtree $K_B(k)$ defined as in the proof of Lemma 3 is identified and colored. Such a tree is then overlaid to B in such a way that the uppermost $\lfloor \frac{k}{2} \rfloor$ levels of B coincide with the lowermost $\lfloor \frac{k}{2} \rfloor$ levels of $K_B(k)$. Then, the complete coloring of B is produced level by level by assigning to each node the same color as an already colored node.

Formally, for a given k, define the binary tree $K_B(k)$ as follows:

- $K_B(k)$ has a leftmost path of $k + 1$ nodes.
- the root of $K_B(k)$ has only the left child;
- a complete subtree of height $i - 1$ is rooted at the right child of the node at level i on the leftmost path of $K_B(k)$.

The $2^{\lfloor \frac{k}{2} \rfloor + 1} + 2^{\lceil \frac{k}{2} \rceil} - 2$ nodes of $K_B(k)$ must be colored with $2^{\lfloor \frac{k}{2} \rfloor + 1} + 2^{\lceil \frac{k}{2} \rceil} - 2$ different colors. Thus, the uppermost $\lfloor \frac{k}{2} \rfloor$ levels of B are already colored.

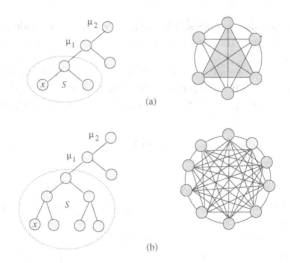

(a)

(b)

Fig. 5. A subset $K_B(k)$ of nodes of B that forms a clique in G_{BP_k}: (a) $k = 3$, (b) $k = 4$.

For the sake of simplicity, to color the remaining part of B, the levels are counted starting from the root of $K_B(k)$. That is, the level of the root of B will be renumbered as level $\lfloor \frac{k}{2} \rfloor + 1$. Now, fixed $x = (0, k)$, the algorithm to color B acts as follows.

Algorithm Binary-Tree-Coloring (B, k);

- Set $M = 2^{\lfloor \frac{k}{2} \rfloor + 1} + 2^{\lceil \frac{k}{2} \rceil} - 2$;
- Color $K_B(k)$ with M colors;
- Visit the tree B in breadth first search, and for each node $x = (i, j)$ of B, with $j \geq k + 1$, do:
  - Set $\pi = j \bmod 2^{\lfloor \frac{k}{2} \rfloor}$, $\alpha = \lceil \log(\pi + 1) \rceil$, $\delta = k - \alpha + 1$ and $\tau = \left(\lfloor \frac{j}{2^{\delta-1}} \rfloor - 1 \right) \bmod 2$;
  - Assign to x the same color as that of the node $y = (r, s)$, where
    $$r = i - \delta + \alpha$$
    and
    $$s = \begin{cases} \lfloor \frac{j}{2^{\delta}} \rfloor & \text{if } \alpha = 0 \\ \lfloor \frac{j}{2^{\delta}} \rfloor 2^{\alpha} + \tau 2^{\alpha-1} + \left(\pi \bmod 2^{\alpha-1} \right) & \text{if } \alpha \neq 0 \end{cases}$$

Examples of colorings to conflict-free access P_3 and P_4 are illustrated in Figure 6.

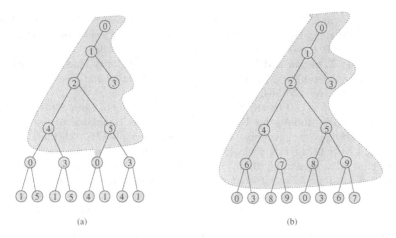

Fig. 6. Coloring of B for conflict-free accessing: (a) P_3, (b) P_4. (Both $K_B(3)$ and $K_B(4)$ are depicted by dash splines.)

Theorem 3. *The Binary-Tree-Coloring mapping is optimal, fast and balanced.*

The results shown for binary trees can be extended to a q-ary tree Q, with $q \geq 2$.

6   Conclusions

In this paper, the problem of conflict-free accessing *arbitrary* paths P_k in particular data structures, such as bidimensional arrays, circular lists and complete trees, has been considered for the first time and reduced to variants of graph-coloring problems. Optimal, fast and balanced mappings have been proposed. Indeed, the memory bank to which a node is assigned is computed in constant time for arrays and circular lists, while it is computed in logarithmic time for complete trees. However, it remains as an open question whether a tree node can be assigned to a memory bank in constant time.

References

1. V. Auletta, S. K. Das, M. C. Pinotti, and V. Scarano, "Toward a Universal Mapping Algorithm for Accessing Trees in Parallel Memory Systems", *Proceedings of IEEE Int'l Parallel Processing Symposium*, Orlando, pp. 447-454, Apr. 1998.
2. V. Auletta, A. De Vivo, V. Scarano, "Multiple Template Access of Trees in Parallel Memory Systems". *Journal of Parallel and Distributed Computing*, Vol. 49, 1998, pp. 22-39.
3. G.E. Blelloch, P.B. Gibbons, Y. Mattias and M. Zagha, "Accounting for Memory Bank Contention and Delay in High-Bandwidth Multiprocessors", *IEEE Trans. on Parallel and Distrib. Systems*, Vol. 8, 1997, pp. 943-958.

4. S. K. Das and F. Sarkar, "Conflict-Free Data Access of Arrays and Trees in Parallel Memory Systems", *Proc. of the Sixth IEEE Symposium on Parallel and Distributed Processing*, Dallas, TX, Oct. 1994, pp. 377-384.

5. S. K. Das, F. Sarkar and M. C. Pinotti, "Parallel Priority Queues in Distributed Memory Hypercubes", *IEEE Transactions on Parallel and Distributed Systems*, Vol. 7, 1996, pp. 555-564.

6. S.K. Das and M.C. Pinotti, "Load Balanced Mapping of Data Structures in Parallel Memory Modules for Fast and Conflict-Free Templates Access" *Proc. 5th Int. Workshop on Algorithms and Data Structures (WADS'97)* Halifax NS, Aug. 1997, LNCS 1272, (Eds. Dehne, Rau-Chaplin, Sack, Tamassia), pp. 272-281.

7. K. Kim, V.K. Prasanna, "Latin Squares for Parallel Array Access", *IEEE Transactions on Parallel and Distributed Systems*, Vol. 4, 1993, pp. 361-370.

8. S.T. McCormick, "Optimal Approximation of Sparse Hessians and its Equivalence to a Graph Coloring Problem", *Mathematical Programming*, Vol. 26, 1983, pp. 153-171.

9. M. C. Pinotti, S. K. Das, and F. Sarkar, "Conflict-Free Template Access in k-ary and Binomial Trees", *Proceedings of ACM-Int'l Conference on Supercomputing*, Wein, Austria, pp. 237-244, July 7-11, 1997.

10. H.D. Shapiro, "Theoretical Limitations on the Efficient Use of Parallel Memories", *IEEE Trans. on Computers*, Vol. 27, 1978, pp. 421-428.

Theory of Trinomial Heaps

Tadao Takaoka

Department of Computer Science, University of Canterbury
Christchurch, New Zealand
tad@cosc.canterbury.ac.nz

Abstract. We design a new data structure, called a trinomial heap, which supports a decrease-key in $O(1)$ time, and an insert operation and delete-min operation in $O(\log n)$ time, both in the worst case, where n is the size of the heap. The merit of the trinomial heap is that it is conceptually simpler and easier to implement than the previously invented relaxed heap. The relaxed heap is based on binary linking, while the trinomial heap is based on ternary linking.

1   Introduction

The Fibonacci heap was invented by Fredman and Tarjan [3] in 1987. Since then, there have been two alternatives that can support n insert and delete-min operations, and m decrease-key operations in $O(m + n \log n)$ time. The relaxed heaps by Driscoll, et. al [2] have the same overall complexity with decrease-key with $O(1)$ worst case time, but are difficult to implement. The other alternative is the 2-3 heap invented by the author [5], which supports the same set of operations with the same time complexity. Although the 2-3 heap is simpler and slightly more efficient than the Fibonacci heap, the $O(1)$ time for decrease-key is in the amortized sense, meaning that the time from one decrease key to the next can not be smooth. Two representative application areas for these operations will be the single source shortest path problem and the minimum cost spanning tree problem. Direct use of these operations in Dijkstra's [1] and Prim's [4] algorithms with those data structures will solve these two problems in $O(m + n \log n)$ time, where n and m are the numbers of vertices and edges of the given graph. Logarithm here is with base 2, unless otherwise specified.

A Fibonacci heap is a generalization of a binomial queue invented by Vuillemin [6]. When the key value of a node v is decreased, the subtree rooted at v is removed and linked to another tree at the root level in the Fibonacci and 2-3 heap. This removal of a subtree may cause a chain effect to keep structural properties of those heaps, resulting in worst case time greater than $O(1)$, although it is $O(1)$ amortized time. As an alternative to the Fibonacci heap, Driscoll, et. al. proposed a data structure called a relaxed heap, whose shape is the same as that of a binomial queue. The requirement of heap order is relaxed in the relaxed heap; a certain number of nodes are allowed to have smaller key values than those of their parents. Those nodes are called bad children in [2] and inconsistent nodes in this paper. On the other hand, the 2-3 heaps is proposed as

D.-Z. Du et al. (Eds.): COCOON 2000, LNCS 1858, pp. 362–372, 2000.

another alternative to the Fibonacci heap. While the Fibonacci heap is based on binary linking, 2-3 heaps are based on ternary linking; we link three roots of three trees in increasing order according to the key values. We call this path of three nodes a trunk. We allow a trunk to shrink by one. If there is requirement of further shrink, we make adjustment by moving a few subtrees from nearby positions. This adjustment may propagate, taking time more than $O(1)$.

In this paper, we combine the ideas of the relaxed heap and 2-3 heap; we use ternary linking and allow a certain number of inconsistent nodes. Ternary linking gives us more flexibility to keep the number of inconsistent nodes under control. The new data structure is called a trinomial heap, and is simpler and easier to implement than the relaxed heap. In the relaxed heap which is based on binary linking, we must keep each bad child at the rightmost branch of its parent, causing difficult book-keeping. The trinomial heap is constructed by ternary linking of trees repeatedly, that is, repeating the process of making the product of a linear tree and a tree of lower dimension. This general description of r-ary trees is given in Section 2. The definition of trinomial heaps and their operations are given in Section 3. In Section 4, we implement decrease-key in $O(1)$ amortized time. In Section 5, we implement it with $O(1)$ worst case time. In Section 6, we give several practical considerations for implementation. Section 7 concludes the paper.

In this paper we analyze the number of key comarisons for computing time as the times for other operations are proportional to it. We mainly deal with decrease-key and delete-min in this paper since the main application areas are the single source shortest path problem and the minimum cost spanning tree problem and we can build up the heap of size n at the beginning. The nodes not adjacent with the source can be inserted with infinite key values at the beginning. As shown in Section 2, the time for building the heap can be $O(n)$. Thus we can concentrate our discussion on decrease-key and delete-min after the heap is built.

2   Polynomial Queues and Their Analysis

We borrow some materials from [5] in this section, as the trinomal heap and the 2-3 heap share the same basic structure. We define algebraic operations on rooted trees as the basis for priority queues. A tree consists of nodes and branches, each branch connecting two nodes. The root of tree T is denoted by $root(T)$. A linear tree of size r is a liner list of r nodes such that its first element is regarded as the root and a branch exists from a node to the next. The linear tree of size r is expressed by bold face $\mathbf{r}$. Thus a single node is denoted by $\mathbf{1}$, which is an identity in our tree algebra. The empty tree is denoted by $\mathbf{0}$, which serves as the zero element. A product of two trees S and T, $P = ST$, is defined in such a way that every node of S is replaced by T and every branch in S connecting two nodes u and v now connects the roots of the trees substituted for u and v in S. Note that $\mathbf{2} * \mathbf{2} \neq \mathbf{4}$, for example, and also that $ST \neq TS$ in general. The symbol " $*$ "

is used to avoid ambiguity. Since the operation of product is associative, we use the notation of $\mathbf{r}^i$ for the products of i $\mathbf{r}$'s.

Let the operation "$\bullet$" be defined by the tree $L = S \bullet T$ for trees S and T. The tree L is made by linking S and T in such a way that $root(T)$ is connected as a child of $root(S)$. Then the product $\mathbf{r}^i = \mathbf{r}\mathbf{r}^{i-1}$ is expressed by

$$\mathbf{r}^i = \mathbf{r}^{i-1} \bullet ... \bullet \mathbf{r}^{i-1} \ (r - 1\bullet\text{'s are evaluated right to left}) \ (1)$$

The whole operation in (1) is to link r trees, called an i-th r-ary linking. The path of length $r - 1$ created by the r-ary linking is called the i-th trunk of the tree $\mathbf{r}^i$, which defines the i-th dimension of the tree in a geometrical sense. The j-th $\mathbf{r}^{i-1}$ in (1) is called the j-th subtree on the trunk counting from $j = 0$. The root of such a j-th subtree is said to be of dimension i. The root of the 0-th subtree is called the head node of the trunk. The dimension of the head node is i or higher, depending on further linking. If v is of dimension i, we write as $dim(v) = i$.

The dimension of tree T, $dim(T)$, is defined by $dim(root(T))$. A sum of two trees S and T, denoted by $S + T$, is just the collection of two trees S and T. Let $\mathbf{a}_i\mathbf{r}^i$ be defined similarly to (1) by linking a_i trees of $\mathbf{r}^i$. An r-ary polynomial of trees of degree $k - 1$, P, is defined by

$$P = \mathbf{a}_{k-1}\mathbf{r}^{k-1} + ... + \mathbf{a}_1\mathbf{r} + \mathbf{a}_0 \ (2)$$

where $\mathbf{a}_i$ is a linear tree of size a_i and called a coefficient in the polynomial. Let $|P|$ be the number of nodes in P and $|\mathbf{a}_i| = a_i$. Then we have $|P| = a_{k-1}r^{k-1} + ... + a_1r + a_0$. We choose a_i to be $0 \le a_i \le r - 1$, so that n nodes can be expressed by the above polynomial of trees uniquely, as the k digit radix-r expression of n is unique with $k = \lceil \log_r(n + 1) \rceil$. The term $\mathbf{a}_i\mathbf{r}^i$ is called the i-th term. We call $\mathbf{r}^i$ the complete tree of dimension i. The path created by linking a_i trees of $\mathbf{r}^i$ is called the main trunk of the tree corresponding to this term. Each term $\mathbf{a}_i\mathbf{r}^i$ in form (2) is a tree of dimension $i + 1$ if $a_i > 1$, and i if $a_i = 1$. A polynomial of trees is regarded as a collection of trees of dimensions up to k.

We next define a polynomial queue. An r-nomial queue is an r-ary polynomial of trees with a label $label(v)$ attached to each node v such that if u is a parent of v, $label(u) \le label(v)$. A binomial queue is a 2-nomial queue. We use "label" and "key" interchangeably.

Example 1. A polynomial queue with an underlying polynomial of trees $P = 2 * 3^2 + 2 * 3 + 2$ is given in Fig. 1.

The merging of two linear trees $\mathbf{r}$ and $\mathbf{s}$ is to merge the two lists by their labels. The result is denoted by the sum $\mathbf{r} + \mathbf{s}$. The merging of two terms $\mathbf{a}_i\mathbf{r}^i$ and $\mathbf{a}'_i\mathbf{r}^i$ is to merge the main trunks of the two trees by their labels. When the roots are merged, the trees underneath are moved accordingly. If $a_i + a'_i < r$, we have the merged tree with coefficient $\mathbf{a}_i + \mathbf{a}'_i$. Otherwise we have a carry tree $\mathbf{r}^{i+1}$ and the remaining tree with the main trunk of length $a_i + a'_i - r$. The sum of two polynomial queues P and Q is made by merging two polynomial queues in a very similar way to the addition of two radix-r numbers. We start from

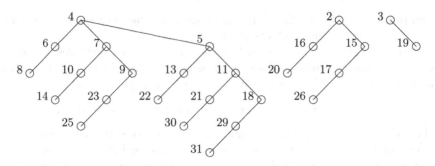

Fig. 1. Polynomial of trees with $r = 3$

the 0-th term. Two i-th terms from both queues are merged, causing a possible carry to the $(i + 1)$-th terms. Then we proceed to the $(i + 1)$-th terms with the possible carry.

An insertion of a key into a polynomial queue is to merge a single node with the label of the key into the 0-th term, taking $O(r \log_r n)$ time for possible propagation of carries to higher terms. The time for n insertions to form a polynomial queue P looks like taking $O(nr \log_r n)$ time. Actually the time is $O(rn)$, as shown below. After the heap is made, we can take n successive minima from the queue by deleting the minimum in some tree T, adding the resulting polynomial queue Q to $P - T$, and repeating this process. This will sort the n numbers in $O(nr \log_r n)$ time after the queue is made. Thus the total time for sorting is $O(nr \log_r n)$. In the sorting process, we do not change key values. If the labels are updated frequently, however, this structure of polynomial queue is not flexible.

Amortized analysis for insert: Let the deficit function Φ_i after the i-th insert be defined by $r - 1$ times the number of trees in the heap. We use the word deficit rather than potential because the number of roots gives a negative aspect of the heap. Let t_i and a_i be the actual time and the amortized time for the i-th insert operation. Then we have $a_i = t_i + \Phi_i - \Phi_{i-1}$. When we insert a single node into the heap, we spend r-1 comparisons by scanning the main trunk, whenever we make a carry to the next position. We decrease the number of trees by one as a result. We terminate the carry propagation either by making a new complete tree or inserting a carry into a main trunk. Thus we have $a_i \leq r - 1$. Note that $\Sigma a_i = \Sigma t_i + \Phi_n - \Phi_0$, $\Phi_0 = 0$, and $\Phi_n = O(\log n)$.

3   Trinomial Heaps

We linked r trees in heap order in form (1). We relax this condition in the following way. A node is said to be inconsistent if its key value is smaller than that of its head node. Otherwise the node is said to be consistent. An active node is either an inconsistent node or a node which was once inconsistent. The latter case occurs when a node becomes inconsistent and then the key value of its head node is decreased, and the node becomes consistent again. Since it is expensive to check whether the descendants of a node become consistent when the key value of the node is decreased, we keep the children intact. Note that we need to look at the roots and active nodes to find the minimum key in the heap. If the number of active nodes in the given r-nomial queue is bounded by t, the queue is said to be an r-nomial heap with tolerance t. We further require that the nodes except for the head node on each trunk are sorted in non-decreasing order of their key values, and that there are no active nodes in each main trunk, that is, main trunks are sorted. Note that an r-nomial queue in the previous section is an r-nomial heap with tolerance 0.

When $r = 3$, we call it a trinomial heap. In a trinomial heap we refer to the three nodes on a trunk of dimension i as the head node, the first child and the second child. Note that each node of dimension $i(> 0)$ is connected downwards with first children of dimension 1, ..., $i - 1$. It is further connected downward and/or upward with trees of dimension i or higher. We refer to the trees in (2) and their roots as those at the top level, as we often deal with subtrees at lower levels and need to distinguish them from the top level trees. The sum $P + Q$ of the two trinomial heaps P and Q is defined similarly to that for polynomial queues. Note that those sum operations involve computational process based on merging. We set $t = \lceil \log_3(n + 1) \rceil - 1$ as the default tolerance.

Example 2. A trinomial heap with tolerance 3 with an underlying polynomial of trees $P = 2 * 3^2 + 2 * 3 + 2$ is given below. Active nodes are shown by black circles. We use $t = 3$ just for explanation purposes, although the default is 2.

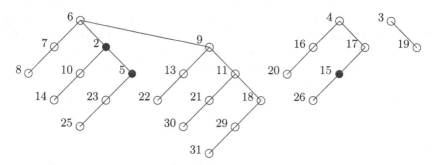

Fig. 2. Trinomial heap with $t = 3$

We describe delete-min, insertion, and decrease-key operations for a trinomial heap. If we make active nodes non-active by spending key comparisons, we say we clean them up.

Let the dimension of node v be i, that is, the trunk of the highest dimension on which node v stands is the i-th. Let $tree_i(v)$ be the complete tree 3^i rooted at v. To prepare for delete-min, we first define the break-up operation. A similar break-up operation for a binomial queue is given in [6]

Break-up: The break-up operation caused by deleting node v is defined in the following way. Let v_p, v_{p-1}, ..., v_1 be the head nodes on the path from the root $(= v_p)$ to the head node $(= v_1)$ of $v = (v_0)$ in the tree $\mathbf{a_i 3^i}$ in which v exists. We cut off all subtrees $tree_{j_l}(v_l)$ where $j_l = dim(v_{l-1})$ and the corresponding links to their first children of $dim(v_{l-1}) + 1$, ..., $dim(v_l)$ for $l = p, p-1, ..., 1$ in this order. Then we delete v, and cut the links to all the first children of v. The trunks from which subtrees are removed are shortened. This operation will create i trees of the form $\mathbf{2*3^j}$ for $j = 0, ..., i-1$, and $\mathbf{3^i}$ if $\mathbf{a_i = 2}$. See Example 3.

Delete-min: Perform break-up after deleting the node with the minimum key. Swap the two subtrees on the main trunk of each resulting tree to have sorted order if necessary. Then merge the trees with the remaining trees at the top level. The time taken is obviously $O(\log n)$. The nodes on the main trunks of the new trees are all cleaned up, and thus the number of active nodes is decreased accordingly.

Reordering: When we move a node v on a trunk of dimension i in the following, we mean we move $tree_{i-1}(v)$. Let v and w be the first and second children and active nodes on the same trunk of dimension i. Then reordering is to re-order the positions of u, v and w after comparing $label(u)$ with $label(w)$. If $label(u) \le label(w)$, we make w non-active, and do no more. Otherwise we move u to the bottom. When we move u, we move $tree_{i-1}(u)$, although $dim(u)$ may be higher. See Fig. 3. We can decrease the number of active nodes. Note that this operation may cause an effect equivalent to decrease-key at dimension $i+1$.

Rearrangement: Let v and w be active first children of dimension i on different trunks. Suppose v' and w' are the non-active second children on the same trunks. Now we compare keys of v' and w', and v and w, spending two comparisons. Without loss of generality assume $label(v') \le label(w')$ and $label(v) \ge label(w)$. Then arrange v' and w', and w and v respectively to be the first and second children on each trunk. We call this operation rearrangement. Note that this operation is done within the same dimension of v. See Fig. 4. Since w and v are active on the same trunk, we call reordering described above.

Decrease-key: Suppose we decreased the key value of node v such that $dim(v) = i$. Let its head node and the other child on the same trunk be u and w. If v is the second child and $label(w) \le label(v)$, do nothing. Otherwise swap w and v and go to the next step. If v is a first child and $label(u) \le label(v)$, clean v if v was active. We have spent one or two comparisons. If we did not increase active nodes, we can finish decrease-key. Otherwise go to the next step. Assume at this stage v is the first child, $label(u) > label(v)$, and v has been turned active. We perform rearrangement and/or reordering a certain number of times to keep the number of active nodes under control. There are two control mechanisms as shown later.

Insert: To insert one node, merge a tree of a single node with the right most tree of the heap. A possible carry may propagate to the left. Thus the worst case time is $O(\log n)$. Note that main trunks are sorted and there are no active nodes on them, meaning that there is no substantial difference between polynomial queues with $r=3$ and trinomial heaps. As shown in Section 2, the amortized time for insert is $O(1)$ with $r = 3$.

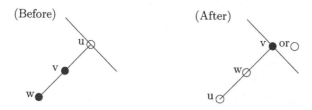

Fig. 3. reordering

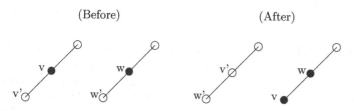

Fig. 4. Rearrangement

Example 3. Let us perform delete-min on the trinomial heap in Fig. 2. Then the biggest tree is broken up as in Fig. 5. After the subtrees rooted at nodes with key 6 and 5 are swapped, the resulting trees and the trees at the top level will be merged.

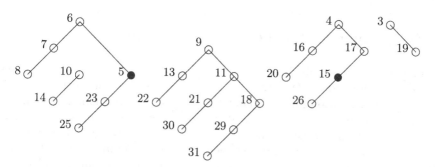

Fig. 5. Trinomial heap with $k = 3$

Example 4. We name a node with key x by $node(x)$. Suppose key 21 is decreased to 1 in Fig. 2. We connect $node(30)$ to $node(26)$, and $node(15)$ to $node(1)$. The result is shown in Fig. 6.

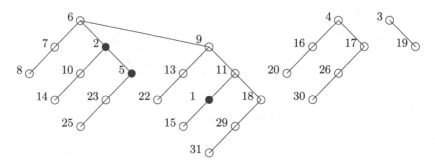

Fig. 6. Trinomial heap with $k = 3$

There are two ways to keep the number of active nodes within tolerance. The first is to allow at most one active node for each dimension. If v becomes active after we perform decrease-key on v, and there is another active w of the same dimension, we can perform rearrangement and/or reordering. The reordering may cause two active nodes at the next higher dimension, causing a possible chain effect of adjustments.

The second is to keep a counter for the number of active nodes. If the counter exceeds the tolerance bound t, we can perform rearrangement and/or reordering on two active nodes of the same dimension, and decrease the counter by one. Note that if the number of active nodes exceeds the tolerance, there must be two active nodes of the same dimension by the pigeon hole principle. There is no chain effect in this case thanks to the global information given by the counter. We allow more than one active nodes of the same dimension, even on the same trunk.

4 $O(1)$ Amortized Time for Decrease-Key

We prepare an array, called *active*, of size $t(= k - 1)$, whose elements are active nodes of dimension i ($i = 1, ..., k - 1$). That is, we have at most one active node for each dimension greater than 0. If there is no active node of dimension i, we denote $active[i] = \phi$. Suppose we perform decrease-key on node v of dimension i, which is now active. If $active[i] = \phi$, we set $active[i] = v$ and finish. If $active[i] = u$, we perform rearrangement and/or reordering on u and v, and adjust *active* accordingly. If we perform reordering, rearrangement and/or reordering will proceed to higher dimensions. This is like carry propagation over array *active*, so the worst case time is $O(\log n)$. Similarly to the insert operation, we can easily show that the amortized time for decrease-key is $O(1)$. As mentioned in Introduction, we can perform m decrease-key operations, and n insert and delete-min operations in $O(m + n \log n)$ time. This implementation is more efficient for applications to shortest paths, etc. than the implementation in the next section, and can be on a par with Fibonacci and 2-3 heaps.

Amortized analysis:. After the key value of a node is decreased, several operations are performed, including checking inconsistencies. The total maximum number of key comparisons leading to a reordering is five. Thus if we define a deficit function Ψ_i after the i-th decrease-key by five times the number of active nodes, we can show that the amortized time for one decrease-key is $O(1)$. If we have inserts in a series of operations, we need to define the deficit function after the i-th operation by $\Phi_i + \Psi_i$.

5 $O(1)$ Worst Case Time for Decrease-Key

In this section, we implement trinomial heaps with $O(1)$ worst case time for decrease-key. We prepare a one-dimensional array *point* of pointers to the list of active nodes for each dimension, an array *count*, and *counter*. The variable *counter* is to count the total number of active nodes. The size of the arrays is bounded by the number of dimensions in the heap. The list of nodes for dimension i contains the current active nodes of dimension i. The element *count*[i] shows the size of the i-th list. To avoid scanning arrays *point* and *count*, we prepare another linked list called *candidates*, which maintains a first-in-first-out structure for candidate dimensions for the clean-up operation. With these additional data structures, we can perform decrease-key with a constant number of operations. Note that we need to modify these data structures on delete-min as well, details of which are omitted.

Example 5. In this example, we have $t = 7$. If we create an active node of dimension 1, we go beyond the tolerance bound, and we perform clean-up using nodes a and b. Suppose this is a rearrangement not followed by reordering. Then a or b is removed from the 3rd list and the first item in *candidates* is removed. Next let us perform reordering using e and f. First we remove e and f from the 6-th list. Suppose $label(e) \leq label(f)$, and e becomes active in the 7-th dimension. Then we append e to the 7-th list, and remove 6 from and append 7 to the list of *candidates*. See Fig. 7. This is a simplified picture. To remove an active node from a list of active nodes in $O(1)$ time, those lists need to be doubly linked.

6   Practical Considerations

For the data structure of a trinomial heap, we need to implement it by pointers. For the top level trees, we prepare an array of pointers of size $d = \lceil \log_3(n+1) \rceil$, each of which points to the tree of the corresponding term. Let the dimension of node v be i. The data structure for node v consists of integer variables *key*

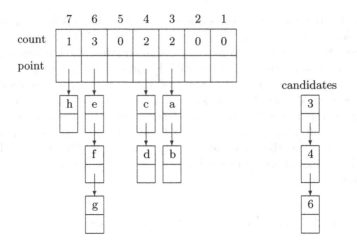

Fig. 7. Auxiliary data structure for $O(1)$ worst case time

and $dim = i$, a pointer to the head node of the i-th trunk, and an array of size d, whose elements are pairs $(first, second)$. The $first$ of the j-th element of the array points to the first node on the j-th trunk of v. The $second$ is to the second node. If we prepare the fixed size d for all the nodes, we would need $O(n \log n)$ space. By implementing arrays by pointers, we can implement our algorithm with $O(n)$ space, although this version will be less time-efficient.

7   Concluding Remarks

Measuring the complexity of n delete-min's, n insert's, and m decrease-key's by the number of comparisons, we showed it to be $O(m + n \log n)$ in a trinomial heap by two implementations. The first implementation performs decrease-key in $O(1)$ amortized time, whereas the second achieves $O(1)$ worst case time for decrease-key. Although due to less overhead time the first is more efficient for applications such as shortest paths where the total time is concerned, the second has an advantage when the data structure is used on-line to respond to updates and queries from outside, as it can give smoother responses.

The clean-up mechanism is not unique. For example, we could make a node active without key comparison after decrease-key. We will need experiments to see if this is better.

References

1. Dijkstra, E.W., A note on two problems in connexion with graphs, Numer. Math. 1 (1959) 269-271.

2. Driscoll, J.R., H.N. Gabow, R. Shrairman, and R.E. Tarjan, An alternative to Fibonacci heaps with application to parallel computation, Comm. ACM, 31(11) (1988) 1343-1345.
3. Fredman, M.L. and R,E, Tarjan, Fibonacci heaps and their uses in inproved network optimization algorithms, Jour. ACM 34 (1987) 596-615
4. Prim, R.C., Shortest connection networks and some generalizations, Bell Sys. Tech. Jour. 36 (1957) 1389-1401.
5. Takaoka, T., Theory of 2-3 Heaps, COCOON 99, Lecture Notes of Computer Science (1999) 41-50
6. Vuillemin, J., A data structure for manipulating priority queues, Comm. ACM 21 (1978) 309-314.

Polyhedral Aspects of the Consecutive Ones Problem

Marcus Oswald and Gerhard Reinelt

Institut für Angewandte Mathematik, Universität Heidelberg, Im Neuenheimer
Feld 368, D-69120 Heidelberg, Germany

Abstract. A 0/1-matrix has the *consecutive ones property for rows* if
its columns can be permuted such that in every row all ones appear
consecutively. Whereas it is easy to decide whether a given matrix has
the consecutive ones property, it is difficult to find for a given 0/1-matrix
B a consecutive ones matrix A that resembles B as closely as possible
(in a sense to be specified). In this paper we study the latter problem
from a polyhedral point of view and discuss an integer programming
formulation that can serve as a basis for a branch-and-cut algorithm
and give separation algorithms. **Key words:** consecutive ones, polytope,
branch-and-cut

1. Introduction

A 0/1-matrix $A \in \{0,1\}^{m \times n}$ has the *consecutive ones property (for rows)* if the
columns of A can be permuted so that the ones in each row appear consecutively.
For convenience we just say that A is C1P.

Consecutive ones matrices play an important role in computational biology.
For the purposes of this paper we do not discuss possible applications, but address the general problem.

Let some 0/1-matrix B be given. For a 0/1-matrix $A \in \{0,1\}^{m \times n}$ let n_o^A
(n_z^A) denote the number of 1's (of 0's) that have to be switched to transform
A into B. For given nonnegative numbers c_o and c_z, we define the *weighted
consecutive ones problem (WC1P)* as the task to find a matrix A which is C1P
and minimizes $c_o n_o^A + c_z n_z^A$. This problem is known to be NP-hard ([Boo75]).
This general version of the problem as well as some specialized variants have to
be solved in computational biology. For a particular variant, where the first and
last one in each row are prescribed, see [COR98].

Note that, if A is C1P, then all column permutations which transform A such
that all ones appear consecutively in every row can be found in time $O(n + m)$
by the so called *PQ-tree algorithm* ([BL76]).

In this paper we study WC1P from a polyhedral point of view. Section 2
discusses a first IP formulation based on the characterization of C1P matrices
by Tucker. This formulation is improved in section 3 based on the study of the
so-called consecutive ones polytope. The separation problem to be solved in a
branch-and-cut approach based on this formulation is addressed in section 4.

D.-Z. Du et al. (Eds.): COCOON 2000, LNCS 1858, pp. 373-382, 2000.

2. Tucker's Characterization and an IP Formulation

Tucker ([T72]) has characterized the C1P matrices by exhibiting five types of matrices M_{1_k}, M_{2_k}, M_{3_k}, M_4 and M_5 which must not occur as submatrices.

M_{1_k}, M_{2_k} and M_{3_k} are $(k+2, k+2)$-, $(k+3, k+3)$- and $(k+2, k+3)$-matrices with $k \geq 1$, respectively. M_4 and M_5 are $(4,6)$- and $(4,5)$-matrices. The five Tucker matrices are displayed in Figure 1.

$$
\begin{pmatrix} 1 & 1 & & & \\ & 1 & 1 & & \\ & & \ddots & \ddots & \\ & & & 1 & 1 \\ 1 & 0 & \cdots & 0 & 1 \end{pmatrix}
\begin{pmatrix} 1 & 1 & & & & 0 \\ & 1 & 1 & & & 0 \\ & & \ddots & \ddots & & \vdots \\ & & & 1 & 1 & 0 \\ 0 & 1 & \cdots & 1 & 1 & 1 \\ 1 & 1 & \cdots & 1 & 0 & 1 \end{pmatrix}
\begin{pmatrix} 1 & 1 & & & & 0 \\ & 1 & 1 & & & 0 \\ & & \ddots & \ddots & & \vdots \\ & & & 1 & 1 & 0 \\ 0 & 1 & \cdots & 1 & 0 & 1 \end{pmatrix}
$$

$$
\begin{pmatrix} 1 & 1 & 0 & 0 & 0 & 0 \\ 0 & 0 & 1 & 1 & 0 & 0 \\ 0 & 0 & 0 & 0 & 1 & 1 \\ 0 & 1 & 0 & 1 & 0 & 1 \end{pmatrix}
\qquad
\begin{pmatrix} 1 & 1 & 0 & 0 & 0 \\ 1 & 1 & 1 & 1 & 0 \\ 0 & 0 & 1 & 1 & 0 \\ 1 & 0 & 0 & 1 & 1 \end{pmatrix}
$$

Figure 1: Tucker matrices M_{1_k}, M_{2_k}, M_{3_k}, M_4 and M_5

Tucker has shown that a 0/1-matrix A is C1P if and only if it is not possible to permute the rows and columns of A such that any of the five forbidden matrices occurs as a submatrix.

Based on tis result we can obtain an integer programming formulation of WC1P. We will first introduce appropriate notations.

For a given WC1P we are looking for an (m, n)-matrix with 0/1 entries which has the consecutive ones property for rows and minimizes a certain objective function. This matrix will be represented by variables x_{ij}, $i = 1, \ldots, m$, $j = 1, \ldots, n$, where x_{ij} represents the matrix entry of row i and column j. We will interpret $x = (x_{11}, \ldots, x_{1n}, \ldots, x_{m1}, \ldots, x_{mn})$ as vector or as matrix whatever is more appropriate. In the following we deal with inequalities that have to be satisfied by x. In most cases, the coefficients of these inequalities are specified by a matrix. Let A be an (l, k)-matrix of coefficients. For an ordered l-tuple $I = (r_1, \ldots, r_l)$ with pairwise distinct entries $r_i \in \{1, \ldots, m\}$ and an ordered k-tuple $J = (c_1, \ldots, c_k)$ with pairwise distinct entries $c_j \in \{1, \ldots, n\}$ we define

$$
A \circ x_{IJ} = \sum_{i=1}^{l} \sum_{j=1}^{k} a_{ij} x_{r_i c_j}.
$$

For simplicity we will just say, for example, "$A \circ x_{IJ} \leq a_0$ for all (l, k)-tuples (I, J)", meaning that all l-tuples $I = (r_1, \ldots, r_l)$ and all k-tuples $J = (c_1, \ldots, c_k)$ are allowed for mapping A to x. Whenever we use $A \circ x$ we assume that $I = (1, \ldots, m)$ and $J = (1, \ldots, n)$.

Now let a 0/1-matrix B be given as input. We are looking for a C1P matrix x that resembles B as closely as possible. If x contains a 1 where B contains a 0, we add a penalty c_o, if x contains a 0, where 1 would be preferred, we add a penalty c_z. Therefore the following objective function value (penalty) is associated with x:

$$c_o \sum_{\substack{i,j \\ b_{ij}=0}} x_{ij} + c_z \sum_{\substack{i,j \\ b_{ij}=1}} (1 - x_{ij})$$

$$= c_o \sum_{\substack{i,j \\ b_{ij}=0}} x_{ij} - c_z \sum_{\substack{i,j \\ b_{ij}=1}} x_{ij} + \sum_{\substack{i,j \\ b_{ij}=1}} c_z$$

For M_{1_k}, M_{2_k}, M_{3_k}, M_4 and M_5 let $\mathcal{M}_{1_k}$, $\mathcal{M}_{2_k}$, $\mathcal{M}_{3_k}$, $\mathcal{M}_4$ and $\mathcal{M}_5$ denote the corresponding matrices where all zero entries are replaced by -1.

If we set

$$c_{ij} = \begin{cases} -c_z & \text{if } b_{ij} = 1, \\ c_o & \text{if } b_{ij} = 0, \end{cases}$$

then an integer programming formulation of WC1P can be given as

$$\min \sum_{i,j} c_{ij} x_{ij}$$

$$\mathcal{M}_{1_k} \circ x_{IJ} \le 2k + 3 \quad \text{for all } (k+2, k+2)\text{-tuples } (I, J) \text{ and } k \ge 1,$$

$$\mathcal{M}_{2_k} \circ x_{IJ} \le 4k + 5 \quad \text{for all } (k+3, k+3)\text{-tuples } (I, J) \text{ and } k \ge 1,$$

$$\mathcal{M}_{3_k} \circ x_{IJ} \le 3k + 2 \quad \text{for all } (k+2, k+3)\text{-tuples } (I, J) \text{ and } k \ge 1,$$

$$\mathcal{M}_4 \circ x_{IJ} \le 8 \quad \text{for all } (4, 6)\text{-tuples } (I, J),$$

$$\mathcal{M}_5 \circ x_{IJ} \le 10 \quad \text{for all } (4, 5)\text{-tuples } (I, J),$$

$$x_{ij} \in \{0, 1\} \quad \text{for all } i = 1, \ldots, m, \ j = 1, \ldots, n.$$

The validity of this formulation is clear, since for a 0/1-matrix x all Tucker matrices are forbidden as submatrices by this set of inequalities.

3.   An IP Formulation with Facets

We will now show that the IP formulation given above is not a suitable one in the sense that stronger inequalities can be derived. To this end we will associate a polytope with the consecutive ones problem. For background in polyhedral combinatorics we refer to [P95].

We define the *consecutive ones polytope* as

$$P_{C1}^{m,n} = \text{conv}\{M \mid M \text{ is an } (m, n)\text{-matrix with C1P}\}.$$

It is easy to see that $P_{C1}^{m,n}$ has full dimension $m \cdot n$. Namely, the zero matrix is C1P and, for every $1 \le i \le m$ and $1 \le j \le n$, the matrix consisting of zeroes only except for a one in position ij is C1P. This gives a set of $n \cdot m + 1$ affinely independent C1P matrices.

Let $a^T x \leq a_0$ be a valid inequality for $P_{C1}^{m,n}$ and let $m' \geq m$ and $n' \geq n$. We say that the inequality $\bar{a}^T x \leq a_0$ for $P_{C1}^{m',n'}$ is obtained from $a^T x \leq a_0$ by *trivial lifting* if

$$\bar{a}_{ij} = \begin{cases} a_{ij} & \text{if } i \leq m \text{ and } j \leq n, \\ 0 & \text{otherwise.} \end{cases}$$

Theorem 1

Let $a^T x \leq a_0$ be a facet-defining inequality for $P_{C1}^{m,n}$ and let $m' \geq m$ and $n' \geq n$. If $a^T x \leq a_0$ is trivially lifted then the resulting inequality defines a facet of $P_{C1}^{m',n'}$.

Proof. Denote the lifted inequality by $\bar{a}^T x \leq a_0$, and let $\bar{x}$ be a matrix satisfying $a^T \bar{x} = a_0$.

First consider the case $m' > m$ and $n' = n$, where without loss of generality $m' = m + 1$.

We add one row to $\bar{x}$ and form the matrix $\bar{x}_0$ by adding a zero row and matrices $\bar{x}_j$ by adding a row of all zeroes except for a 1 in column j, $j = 1, \ldots, n$. Obviously all generated matrices are C1P, and we have $\bar{a}^T \bar{x}_j = a_0$, for all $j = 0, 1, \ldots, n$.

Therefore, if we have $m \cdot n$ affinely independent matrices satisfying $a^T x = a_0$ we can obtain by the above construction $(m+1) \cdot n$ affinely independent matrices satisfying $\bar{a}^T x = a_0$, thus proving that the trivially lifted inequality is also facet-defining.

The case $m' = m$ and $n' > n$, where without loss of generality $n' = n + 1$ follows along a similar line. We add one column to $\bar{x}$ and form the matrix $\bar{x}_0$ by adding a zero column and matrices $\bar{x}_i$ by adding a column of all zeroes except for a 1 in row i, $i = 1, \ldots, m$. Again all generated matrices are C1P since column $n + 1$ can be moved to an appropriate position, and we have $\bar{a}^T \bar{x}_i = a_0$, for all $i = 0, 1, \ldots, m$. From $m \cdot n$ affinely independent matrices satisfying $a^T x = a_0$ we can obtain this way $m \cdot (n+1)$ affinely independent matrices satisfying $\bar{a}^T x = a_0$.

The general result follows from these two constructions. $\qquad\square$

If trivial lifting is possible, this means, that larger polytopes inherit all facets of smaller polytopes.

Inequalities that are obviously valid for $P_{C1}^{m,n}$ are the *trivial inequalities* $x_{ij} \geq 0$ and $x_{ij} \leq 1$, for all $1 \leq i \leq m$, $1 \leq j \leq n$. It is easily seen that they also define facets.

Theorem 2

For all $m \geq 1$, $n \geq 1$, $1 \leq i \leq m$, $1 \leq j \leq n$, the inequalities $x_{ij} \geq 0$ and $x_{ij} \leq 1$ define facets of $P_{C1}^{m,n}$.

Proof. We have $P_{C1}^{1,1} = \text{conv}\{0, 1\}$ and therefore $x_{11} \geq 0$ and $x_{11} \leq 1$ are facet-defining for $P_{C1}^{1,1}$. The general result holds due to the trivial lifting property. $\quad\square$

We are interested in getting more insight into the facet structure of $P_{C1}^{m,n}$. In this section we will desribe four types of facet-defining inequalities that can then be used to replace the IP formulation above with a stronger formulation.

The first two inequalities are based on two classes of matrices $\mathcal{F}_{1_k}$ and $\mathcal{F}_{2_k}$ for $k \geq 1$. These matrices are $(k+2, k+2)$-, respectively $(k+2, k+3)$-matrices and have entries -1, 0 and $+1$, where for convenience we write "$-$" ("$+$") instead of "-1" ("$+1$"). The matrices are shown in Figure 2.

$$
\begin{pmatrix}
+ & + & & & & - \\
 & + & + & & & - \\
 & & \ddots & \ddots & & \vdots \\
 & & & + & + & - \\
- & 0 & \cdots & 0 & + & + \\
+ & 0 & \cdots & 0 & - & +
\end{pmatrix}
\qquad
\begin{pmatrix}
+ & + & & & & - & - \\
 & + & + & & & - & - \\
 & & \ddots & \ddots & & \vdots & \vdots \\
 & & & + & + & - & - \\
- & 0 & \cdots & 0 & + & + & - \\
- & 0 & \cdots & 0 & + & - & +
\end{pmatrix}
$$

Figure 2: Matrices $\mathcal{F}_{1_k}$ and $\mathcal{F}_{2_k}$

We show that these matrices lead to facet-defining inqualities.

Theorem 3

The inequalities $\mathcal{F}_{1_k} \circ x_{IJ} \leq 2k + 3$, $k \geq 1$, for all $(k + 2, k + 2)$-index sets, are facet-defining for $P_{C1}^{m,n}$ for all $m \geq k + 2$ and $n \geq k + 2$.

Proof. Due to Theorem 1, we only need to show that this class of inequalities is facet-defining for $P_{C1}^{m,n}$ with $m = n = k + 2$. Moreover, we only need to consider the canonical ordered index sets $I = \{1, 2, \ldots, m\}$ and $J = \{1, 2, \ldots, n\}$. Let $a^T x \leq a_0$ denote this inequality and let $F = \{x \mid a^T x = a_0\} \cap P_{C1}^{m,n}$ denote the induced face.

We first show that the inequality is valid. Let x be a 0/1-matrix such that $\mathcal{F}_{1_k} \circ x > 2k + 3$. Then x must have 0 entries at the "$-$"-positions of $\mathcal{F}_{1_k}$ and 1 entries at the "$+$"-positions. We now show that we cannot obtain a C1P matrix no matter how the remaining entries of x are assigned and how the columns of x are permuted. We call a row *bad* if and only if there is a forced 0 between two forced 1 entries. Note that x has exactly one bad row. Now consider any sequence of permutations of adjacent columns of x. The status of a row changes if and only if the entries of this row in the two columns are a forced 0 and a forced 1. For each pair of columns of x there is an even number (0 or 2) of rows with this property. Thus for any column permutation of x the number of bad rows remains odd and therefore at least one. But a matrix with a bad row cannot be C1P and validity of the inequality follows.

Since $P_{C1}^{m,n}$ is full-dimensional, facet-defining inequalities defining the same facet only differ by multiplication with a positive scalar. Now let $b^T x \leq b_0$ be a facet-defining inequality for $P_{C1}^{m,n}$ such that $F \subseteq \{x \mid b^T x = b_0\} \cap P_{C1}^{m,n}$. If we can show that $b = \lambda a$, for some $\lambda > 0$, then is is proven that $a^T x \leq a_0$

is facet-defining. We will show this in three steps. Let $\beta = b_{11}$. We call C1P matrices x that satisfy $a^T x = a_0$ *solutions*.

Every 0/1-matrix with $2k+3$ 1's in the "+" positions and 0's otherwise is a solution. We call such matrices *standard solutions* in the following. Let x^1 and x^2 be two standard solutions where $x^1_{11} = 0$ and $x^2_{12} = 0$. Then we have

$$0 = b^T x^1 - b^T x^2 = b_{12} - b_{11}$$

and therefore $b_{12} = b_{11} = \beta$. By using appropriate matrices we can thus show that $b_{ij} = \beta$ for all "+"-positions ij.

Consider the standard solution x^1 with $x_{k+2,k+2} = 0$. Let x^2 be a matrix which is identical to x^1 except for an additional 1 in a "0"-position next to an "+"-position in any row i. Let this position be il. Then x^2 is a solution and we obtain

$$0 = b^T x^1 - b^T x^2 = -b_{il},$$

i.e., $b_{il} = 0$. Extending the chain of 1's to the next "0"-positions eventually shows that $b_{is} = 0$ for all "0"-positions is of row i. Since this holds for every row i we can show that $b_{ij} = 0$ for all "0"-positions ij.

Finally let x^1 be a standard solution with $x_{11} = 0$. Construct x^2 from x^1 as follows. Set $x^2_{11} = 1$ and $x^2_{1,k+2} = 1$ and insert column $k+2$ after column 1. This matrix is not C1P, but if all 0-entries of row $k+1$ are changed to 1, then we obtain a C1P matrix. Now

$$0 = b^T x^1 - b^T x^2 = -b_{11} - b_{1,k+2},$$

and therefore $b_{1,k+2} = -\beta$. Using similar arguments we eventually obtain $b_{ij} = -\beta$ for all "−"-positions ij.

Thus we have shown that $b = \beta a$. It is clear that $\beta > 0$ since if we change a 1-entry in a standard solution to 0 then we obtain another C1P matrix which would violate the inequality if $\beta < 0$. $\qquad\square$

Theorem 4

The inequalities $\mathcal{F}_{2_k} \circ x_{IJ} \leq 2k+3$, $k \geq 1$, for all $(k+2, k+3)$-index sets, are facet-defining for $P^{m,n}_{C1}$ for all $m \geq k+2$ and $n \geq k+3$.

Proof. The proof is similar to the previous one. Here rows of x are called bad if and only if there is forced 0 between two forced 1 entries and a forced 1 between two forced 0 entries. Then the verification of validity and the proof that the inequalities are facet-defining follows along the same lines as for $\mathcal{F}_{1_k}$. $\qquad\square$

Corresponding to the two single Tucker matrices M_4 and M_5 we have the $(4,6)$-matrix $\mathcal{F}_3$ and the $(4,5)$-matrix $\mathcal{F}_4$ shown in Figure 3.

Theorem 5

(a) The inequalities $\mathcal{F}_3 \circ x_{IJ} \leq 8$, for all $(4,6)$-index sets, define facets of $P^{m,n}_{C1}$ for all $m \geq 4$ and $n \geq 6$.

$$\begin{pmatrix} + & + & - & 0 & - & 0 \\ - & 0 & + & + & - & 0 \\ - & 0 & - & 0 & + & + \\ - & + & - & + & - & + \end{pmatrix} \qquad \begin{pmatrix} + & + & 0 & + & - \\ + & 0 & 0 & + & - \\ - & 0 & + & + & - \\ + & - & - & + & + \end{pmatrix}$$

Figure 3: Matrices $\mathcal{F}_3$ and $\mathcal{F}_4$

(b) The inequalities $\mathcal{F}_4 \circ x_{IJ} \leq 8$, for all $(4,5)$-index sets, define facets of $P_{C1}^{m,n}$ for all $m \geq 4$ and $n \geq 5$.

Proof. Using the software PORTA ([CL98]) we could verify that $\mathcal{F}_3 \circ x \leq 8$ is facet-defining for $P_{C1}^{4,6}$ and $\mathcal{F}_4 \circ x \leq 8$ is facet-defining for $P_{C1}^{4,5}$. The result then follows from trivial lifting.   □

If $m \leq 2$ or $n \leq 2$, all (m,n)-matrices are C1P and therefore the trivial inequalities completely describe $P_{C1}^{m,n}$. The facets discussed above give a complete description if $n = 3$ and $m = 3, 4, 5$, or 6 or if $m = 3$ and $n = 4$ or 5. For $m = 3$ and $n > 5$ or for $m > 3$ and $n > 3$ further facet-defining inequalities are needed.

$$x_{ij} \geq 0 \qquad x_{ij} \leq 1 \qquad \begin{pmatrix} 1 & 1 & -1 \\ 1 & -1 & 1 \\ -1 & 1 & 1 \end{pmatrix} \circ x \leq 5$$

$$\begin{pmatrix} 1 & 1 & -1 & -1 \\ 1 & -1 & 1 & -1 \\ 1 & -1 & -1 & 1 \end{pmatrix} \circ x \leq 5 \qquad \begin{pmatrix} 1 & 1 & 0 & -1 \\ 1 & 0 & 1 & -1 \\ 0 & 1 & -1 & 1 \\ 0 & -1 & 1 & 1 \end{pmatrix} \circ x \leq 7$$

$$\begin{pmatrix} 1 & 1 & 0 & -1 \\ 1 & 1 & -1 & 0 \\ 1 & -1 & 1 & -1 \\ 1 & -1 & -1 & 1 \end{pmatrix} \circ x \leq 7 \qquad \begin{pmatrix} 1 & 1 & 0 & -1 \\ 1 & 0 & -1 & 1 \\ -1 & 1 & 1 & -1 \\ -1 & 1 & 0 & 1 \end{pmatrix} \circ x \leq 7$$

$$\begin{pmatrix} 1 & 1 & 1 & -1 \\ 1 & -1 & -1 & 1 \\ -1 & 1 & -1 & 1 \\ -1 & -1 & 1 & 1 \end{pmatrix} \circ x \leq 7 \qquad \begin{pmatrix} 1 & 1 & 1 & -1 \\ 1 & 0 & -1 & 1 \\ 0 & -1 & 1 & 1 \\ -1 & 1 & 0 & 1 \end{pmatrix} \circ x \leq 8$$

$$\begin{pmatrix} 1 & 1 & 1 & -1 \\ 1 & 1 & -1 & 1 \\ 1 & -1 & 1 & 1 \\ -1 & 1 & 1 & 1 \end{pmatrix} \circ x \leq 10 \qquad \begin{pmatrix} 2 & 2 & -1 & -2 \\ 2 & -1 & 2 & -2 \\ 2 & -2 & -2 & 2 \\ -2 & 1 & 1 & 2 \end{pmatrix} \circ x \leq 13$$

$$\begin{pmatrix} 2 & 2 & 2 & -2 \\ 1 & 1 & -2 & 2 \\ 1 & -2 & 1 & 2 \\ -2 & 1 & 1 & 2 \end{pmatrix} \circ x \leq 15 \qquad \begin{pmatrix} 2 & 2 & 2 & -2 \\ 2 & -1 & -1 & 2 \\ -1 & 2 & -1 & 2 \\ -1 & -1 & 2 & 2 \end{pmatrix} \circ x \leq 15$$

Figure 4: Complete description of $P_{C1}^{4,4}$

In Figure 4 we give the complete description of $P_{C1}^{4,4}$ obtained using PORTA.

The complete description of $P_{C1}^{4,4}$ thus requires already 9 additional classes of facet-defining inequalities. The total number of facets of $P_{C1}^{4,4}$ is 1880.

Theorem 6

The facet-defining inequalities from Theorems 3, 4 and 5 cut off all Tucker matrices.

Proof. It is easily verified that
  i) $\mathcal{F}_{1_k} \circ M_{1_k} = 2k + 4$,
 ii) $\mathcal{F}_{1_{k+1}} \circ M_{2_k} = 2k + 6$,
iii) $\mathcal{F}_{2_k} \circ M_{3_k} = 2k + 4$,
 iv) $\mathcal{F}_3 \circ M_4 = 9$,
  v) $\mathcal{F}_4 \circ M_5 = 9$.

Therefore every Tucker matrix violates one of these inequalities.   □

Based on the above results we obtain the following integer programming formulation of WC1P. It consists only of inequalities which are facet-defining for $P_{C1}^{m,n}$.

$$\min \sum_{i,j} c_{ij} x_{ij}$$

$$\mathcal{F}_{1_k} \circ x_{IJ} \leq 2k + 3 \quad \text{for all } (k+2, k+2)\text{-tuples } (I, J) \text{ and } k \geq 1,$$

$$\mathcal{F}_{2_k} \circ x_{IJ} \leq 2k + 3 \quad \text{for all } (k+2, k+3)\text{-tuples } (I, J) \text{ and } k \geq 1,$$

$$\mathcal{F}_3 \circ x_{IJ} \leq 8 \quad \text{for all } (4, 6)\text{-tuples } (I, J),$$

$$\mathcal{F}_4 \circ x_{IJ} \leq 8 \quad \text{for all } (4, 5)\text{-tuples } (I, J),$$

$$x_{ij} \in \{0, 1\} \quad \text{for all } i = 1, \ldots, m, \ j = 1, \ldots, n.$$

4. Separation

In this section we address how to solve the LP relaxation obtained from the IP formulation by dropping the integrality requirement. To this end we have to give separation algorithms for the given classes of inequalities. We will only discuss exact separation of the $\mathcal{F}_{1_k}$-inequalities in detail. Separation for the $\mathcal{F}_{2_k}$-inequalities works along similar lines, $\mathcal{F}_3$- and $\mathcal{F}_4$-inequalites can be separated using quadratic assignment heuristics.

Actually, in the case of $\mathcal{F}_{1_k}$-inequalities we will separate a more general class of inequalities. These inequalities can be obtained by observing that the "-1" entry in the last row can be moved to any position changing the first and last column in an appropriate way. The corresponding $\overline{F}_{1_k}$-inequalities can also be shown to be facet-defining for $P_{C1}^{m,n}$ and can be visualized as follows where the left hand side matrix $\overline{\mathcal{F}}_{1_k}$ is a $(k+2, k+2)$-matrix:

$$
\begin{array}{c}
\\
\\
\\
\\
r_1 \\
\vdots \\
r_{d-1} \\
r_d \\
l
\end{array}
\begin{array}{c}
\begin{array}{cccccccccc}
i & & & & j & c_2 & \cdots & c_d & h \\
\end{array} \\
\left(
\begin{array}{ccccccccc}
+ & + & & & & & & & - \\
0 & + & + & & & & & & - \\
\vdots & & \ddots & \ddots & & & & & \vdots \\
0 & & & + & + & & & & - \\
- & & & + & + & & & & 0 \\
\vdots & & & & \ddots & \ddots & & & \vdots \\
- & & & & & + & + & & 0 \\
- & & & & & & + & + & \\
+ & 0 & \cdots & 0 & - & 0 & \cdots & 0 & + \\
\end{array}
\right)
\end{array}
\circ x \le 2k+3
$$

$$\text{Figure 5: } \overline{\mathcal{F}}_{1_k}\text{-inequality}$$

We obtain the original F_{1_k}-inequalities when there are no columns $c_2, \ldots, c_d$, i.e., when $d = 1$.

The main task of the separation algorithm is to identify the row l and the columns i, j, and h and to sum appropriate coefficients for rows and columns in between.

We proceed as follows. For every column $i = 1, \ldots, n$ we create the complete undirected bipartite graph G^i with the n columns and m rows as the two node sets. With every edge cr we associate the weight $w_{cr}^i = 1 - x_{rc} + \frac{1}{2} \cdot x_{ri}$ where x is the given LP solution to be cut off.

In every weighted graph G^i we now compute for every pair $j, j \ne i$ and $h, h \ne i, h \ne j$ of columns a shortest path between j and h with respect to the assigned edge weights. This way we obtain shortest lengths p_{jh}^i.

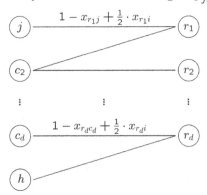

$$\text{Figure 6: Path between columns } j \text{ and } h$$

For every quadrupel i, j, h, l of columns i, j, h and rows l we evaluate

$$
p_{jh}^i + p_{ji}^h - x_{li} + x_{lj} - x_{lh} = 2k+2 - \overline{\mathcal{F}}_{1_k} \circ x.
$$

For every expression that has value less than -1 we can construct a violated $\overline{F}_{1_k}$-inequality using the shortest paths computed above to include columns $c_2, \ldots, c_d$ and rows $r_1, \ldots, r_d$. If none of these values is less than -1, then no violated $\overline{F}_{1_k}$-inequality and thus no violated F_{1_k}-inequality exists.

The running time is dominated by the all-pairs shortest path computation for every column, and we obtain time complexity $O(n^3(n+m))$. Therefore the F_{1_k}-inequalities can be separated in polynomial time.

We do not elaborate on the similar separation procedure for F_{2_k}-inequalities. The time complexity for the respective procedure turns out to be $O(n^4m^2)$.

In total we have shown that all inequalities constituting the canonical LP relaxation of our IP formulation can be separated in polynomial time. Note that for this result we can just enumerate all possible F_3- and F_4-inequalities. Thus the LP relaxation can be solved in polynomial time.

It is an interesting future research project to employ these separation algorithms in a branch-and-cut algorithm for solving the weighted consecutive ones problem.

References

[Boo75] K.S. Booth 1975 PQ-Tree Algorithms PhD Thesis, University of California, Berkeley

[BL76] K.S. Booth and G. Lueker 1976x Testing for the Consecutive Ones Property, Interval Graphs, and Graph Planarity Using PQ-Tree Algorithms *Journal of Computer and System Sciences*, 13, 335–379

[CL98] T. Christof, A. Loebel 1998 PORTA - A Polyhedron Representation Algorithm www.iwr.uni-heidelberg.de/iwr/comopt/software/PORTA/

[COR98] T. Christof, M. Oswald and G. Reinelt 1998 Consecutive Ones and A Betweenness Problem in Computational Biology *Proceedings of the 6th IPCO Conference, Houston*, 1998, 213–228

[P95] M. Padberg 1995 Linear Optimization and Extensions Springer, Berlin

[T72] A. Tucker 1972 A Structure Theorem for the Consecutive 1's Property *Journal of Combinatorial Theory B*, 12, 153–162

The Complexity of Physical Mapping with Strict Chimerism

Stephan Weis and Rüdiger Reischuk

Institut für Theoretische Informatik, MU Lübeck
Wallstraße 40, 23560 Lübeck, Germany
reischuk@tcs.mu-luebeck.de
http://www.tcs.mu-luebeck.de/

Abstract. We analyze the algorithmic complexity of physical mapping by hybridization in situations of restricted forms of chimeric errors, which is motivated by typical experimental conditions. The constituents of a chimeric probe always occur in pure form in the data base, too. This problem can be modelled by a variant of the k-consecutive ones problem. We show that even under this restriction the corresponding decision problem is $\mathcal{NP}$-complete. Considering the most important situation of strict 2-chimerism, for the related optimization problem a complete separation between efficiently solvable and $\mathcal{NP}$-hard cases is given based on the sparseness parameters of the clone library. For the favourable case we present a fast algorithm and a data structure that provides an effective description of all optimal solutions to the problem.

1 Introduction

Physical mapping by hybridization plays a major role in exploring the genomic landscape (for biological background, see e.g. [11]). When drawing a physical map for a strand of DNA, the problem arises to locate a number of "landmarks", called *probes*, on the sequence. Since the DNA is often too big to handle, a *clone library* is compiled whose elements, called *fragments*, are overlapping stretches of the strand. Associated to each such fragment is a *fingerprint*, which is defined as the set of probes that are located on the fragment. This information is gathered by a series of *hybridization*-experiments. The outcomes of the hybridization experiments are stored in a binary matrix, such that each row represents the fingerprint of the fragment associated to that row.

In order to locate the landmarks we have to find the relative order of the probes. If every fragment of the clone library represents an interval on the strand, this task has been called the *consecutive ones problem* for the hybridization matrix; it can be solved in linear time ([5],[4],[15],[12]). However, during the cutting procedure of the DNA two or more stretches that stem from distant regions may stick together, forming a *chimeric fragment*. In practice, a large fraction of elements of a clone library is chimeric. A permutation of the columns of the hybridization matrix, which has to represent the correct probe order for

D.-Z. Du et al. (Eds.): COCOON 2000, LNCS 1858, pp. 383–395, 2000.
© Springer-Verlag Berlin Heidelberg 2000

such a library, will reveal chimeric rows as noncontiguous blocks of consecutive ones.

Furthermore, the hybridization experiment itself can be faulty. This may result in a noisy hybridization matrix, i.e. some entries provide wrong values (*false positives or false negatives*). The presence of any of the three types of errors leads to intractable problems in the search for the correct probe order. In [6] several variants of the physical mapping problem are formulated as $\mathcal{NP}$-complete interval graph problems. Even when there are no false entries in the matrix, to test the *k-consecutive ones property,* that is the question whether no chimeric fragment falls apart into more than k pieces, is $\mathcal{NP}$-complete [3].

Previous attempts to solve these problems approximatively have mainly led to heuristic methods (see [2],[7],[8],[13],[14]). In [1] sophisticated methods have been developed that lead to good solutions even for relatively high error rates. We consider a restriction of chimeric errors that takes into account properties of real data. One can assume that the pieces of a chimeric fragment appear as single fragments of the clone library, too, since not all copies of such a piece glue to other stretches of the DNA strand. That means, some will be visible in pure form. We show that even in this favourable situation the physical mapping problem remains intractable in general. This already holds when no other errors occur.

Secondly, we consider the optimization problem to determine the minimum number of chimeric rows such that a given matrix obtains the consecutive ones property. With the help of two characteristic parameters of the clone library, namely the maximal length of a fragment and the maximal degree of overlap, we can exactly separate the efficiently solvable from the intractable cases of the optimization problem. For the tractable cases we present an algorithm, which has the important feature that it reveals the complete set of optimal solutions. This is necessary, because usually a great number of combinatorial equally legitimate solutions exist, but only very few of them are biologically meaningful.

The paper is organized as follows. After giving a formal definition of the algorithmic problem in the next section, section 3 analyses the problem in the special situation of overlap degree 2. In section 4 we prove the $\mathcal{NP}$-completeness of the decision problem. The optimization problem will be investigated in the two final sections: we first show $\mathcal{NP}$-hardness for the cases overlap degree 3 with fragment length 3, resp. overlap degree 4 with fragment length 2. For the only remaing case overlap degree 3 with fragment length 2 we then present an efficient solution.

2 Basic Definitions and Notation

Let M be a binary matrix with r rows and c columns. The i-th row (resp. j-th column) in M is denoted by $m_{i,\bullet}$ (resp. $m_{\bullet,j}$). The number of '1's in $m_{i,\bullet}$ (resp. $m_{\bullet,j}$) is given by $\|m_{i,\bullet}\|$ (resp. $\|m_{\bullet,j}\|$). $[a,b]$ denotes the interval of natural numbers i with $a \leq i \leq b$. The set of columns given by a set $S \subseteq [1,c]$ is denoted by $m_{\bullet,S}$; it defines a submatrix of M containing exactly the columns of the set $m_{\bullet,S}$ with retention of the previous order. A row will also be identified by the set of column indices containing a '1', i.e. $m_{i,\bullet} \equiv \{j \mid m_{i,j} = 1\}$.

Definition 1 *A **clone library** is a set F of elements $f_1, \ldots, f_r$ called **fragments**. Each fragment f is characterized by a set called its **fingerprint**, a subset of the set $P = \{p_1, \ldots, p_c\}$ of probes.*

A clone library can be represented by a binary matrix M such that the fragments are associated to its rows and the probes to its columns. An entry of M equals 1 iff the corresponding fragment and probe hybridize. We call $||m_{i,\bullet}||$, the number of probes that stick to fragment f_i, the **length** of f_i, and $||m_{\bullet,j}||$, the number of fragments to which p_j sticks, the **degree of overlap** of the j-th probe.

Definition 2 *Let M be a binary $r \times c$-matrix and k a natural number. Row $m_{i,\bullet}$ is **k-chimeric** iff $m_{i,\bullet} = [s_1, t_1] \,\dot{\cup}\, [s_2, t_2] \,\dot{\cup}\, \cdots \,\dot{\cup}\, [s_k, t_k]$ for some $s_1, t_1, s_2, t_2, \ldots, s_k, t_k \in [1, c]$ with $s_1 \leq t_1 < s_2 - 1$, $s_2 \leq t_2 < s_3 - 1$, $\ldots$, $s_k \leq t_k$. This means that $m_{i,\bullet}$ consists of k blocks of '1's, which are separated by one ore more '0's.*
*$m_{i,\bullet}$ is **strict k-chimeric** iff it is k-chimeric and for each of its k blocks $[s_a, t_a]$ there exists another row $m_{i_a,\bullet} = [s_a, t_a]$. These $m_{i_1,\bullet}, \ldots, m_{i_k,\bullet}$ are called the **isolated fragments** of the chimeric row $m_{i,\bullet}$.*
*M obeys the **(strict) k-consecutive ones order** iff each of its rows $m_{i,\bullet}$ is (strict) k'-chimeric for some $k' \leq k$.*

The strictness property requires that each of the pieces that stick together to form a chimeric fragment must itself be a non-chimeric fragment of the clone library. Figure 1 illustrates the difference between chimerism and strict chimerism. This scenario can be extended to require strict k-chimeric fragments, for arbitrary $k > 3$. The special cases k-chimeric for $k = 0$ and $k = 1$, that means the row is either identical 0 or has only a single block of '1's, we will call also **non-chimeric**. In the following we will assume that the binary matrices considered do not have rows or columns that are identical 0 since such fragments or probes do not provide any information about the relative order of the others.

Definition 3 *For $k \geq 2$, row $m_{i,\bullet}$ of a matrix M is called **potentially k-chimeric** iff there exists nonempty rows $p_1, \ldots, p_k \in [1, \ell_1] \setminus \{i\}$ such that $m_{i,\bullet} = m_{p_1,\bullet} \,\dot{\cup}\, \ldots \,\dot{\cup}\, m_{p_k,\bullet}$. It is **potentially chimeric** iff there exists a $k > 1$ such that $m_{i,\bullet}$ is potentially k-chimeric.*

The problem EXACT COVER BY 3-SETS, which is known to be $\mathcal{NP}$-complete [9], can be reduced to the question whether a row is potentially k-chimeric. We therefore get for the case that k is part of the input:

Lemma 1 *It is $\mathcal{NP}$-complete to decide whether a specific row of a matrix is potentially k-chimeric.*

For constant k, however, this question can easily be solved in time polynomial in the size of the matrix.

Definition 4 *M has the **(strict) k-consecutive ones property** iff by per-muting its columns it can be reordered to a matrix M' that is in (strict) k-consecutive ones order.*

1-consecutive equals the consecutive ones property as defined in [5, 4]. Figure 1 also shows that it is necessary to know the largest k, for which a row is potentially k-chimeric, in order to estimate the degree of chimerism for a strict ordering: In the example it does not suffice to know that the two highlighted fragments are potentially 2-chimeric.

Fig. 1. A 2-chimeric order for a matrix that requires a 3-chimeric fragment to obtain a strict ordering.

When conducting hybridization experiments one first has no information about the relative order of probes, that means the columns of the result matrix M are in an arbitrary order. The '1's of a fragment may be spread all over. The algorithmic task now is to find a permutation such that the '1's in each row generate as few blocks as possible.

Definition 5 *The problem **(u|l)-Physical-Mapping with strict k-chimeric errors** is the following: Given a hybridization matrix M with maximal fragment length l and maximal degree of overlap u as input, decide whether M has the strict k-consecutive ones property. If there is no bound on l we write (u|−), and for unbounded u the notation (−|l) will be used. The optimization version of **(u|l)-Physical-Mapping with strict k-chimeric errors** requires to com-pute a permutation M' of M that obeys the strict k-consecutive ones order and that minimizes the number of rows remaining in M' that are not non-chimeric.*

In the following we will analyse the complexity of PHYSICAL-MAPPING for the basic and in practice most important case of strict 2-chimeric errors. Consider a clone library where some fragments are missing and which therefore does not overlap the complete strand of DNA. This means that the strand falls apart into regions, which are completely overlapped by the library, separated by gaps. Such contiguous regions are called **islands** or **contigs**. A formal definition of this concept focuses on the two most important structural properties of islands in our context, namely the impossibility to determine a relative order of the islands, and a certain guarantee to obtain a precise ordering inside each island.

This can be achieved to some extent when we ignore those fragments that might destroy the consecutive ones property, i.e. the potentially chimeric fragments.

Definition 6 *The* **overlap graph** *of a hybridization matrix M is given by $G = (P, E)$ with $E := \{ \{p_j, p_{j'}\} \mid \exists f_i, f_{i'} \text{ with } p_j \in m_{f_i, \bullet}, p_{j'} \in m_{f_{i'}, \bullet} \text{ and } m_{f_i, \bullet} \cap m_{f_{i'}, \bullet} \neq \emptyset \}$. Let M be a matrix and M' the submatrix that contains exactly those rows of M that are not potentially chimeric. Let $G = (P, E)$ be the overlap graph of M' and $G_1 = (P_1, E_1), \ldots, G_z = (P_z, E_z)$ its connected components. The vertex sets P_x are called* **islands** *of M.*

3 Two Simple Cases

The $(-|1)$–PHYSICAL-MAPPING problem is trivial because any such clone library is in consecutive ones order by definition.

Theorem 1 *Let M be an instance for $(2|-)$–PHYSICAL-MAPPING with strict 2-chimeric errors. If M has the 2-consecutive ones property then it has the 1-consecutive ones property, too.*

Proof: Assume that M does not have the 1-consecutive ones property. Let M' be a permutation of M with minimum number of strict 2-chimeric rows. A strict 2-chimeric row $m'_{i, \bullet}$ consisting of 1-blocks A and B has to be accompanied by rows $m'_{j_1, \bullet} = A$ and $m'_{j_2, \bullet} = B$. Since the overlap is at most 2 there cannot be any other row m_x with $m_{x, \bullet} \cap m_{i, \bullet} \neq \emptyset$. Thus, rearranging the columns of M' such that the ones in A are followed directly by the ones in B row i becomes non-chimeric without breaking the 1-block of any other row.   ∎

Corollary 1 $(2|-)$–PHYSICAL-MAPPING *with strict chimeric errors can be decided in linear time. The corresponding optimization problem is trivial.*

4 The Complexity of the Decision Problem

We will show that the decision problem already becomes intractable for sparse clone libraries. Compared to the results about potential chimerism the following result goes a step further and shows that in general the strictness property does not help to reduce the complexity of PHYSICAL-MAPPING (cf. [3]).

Theorem 2 $(11|5)$–PHYSICAL-MAPPING *with strict 2-chimeric errors is $\mathcal{NP}$-complete.*

Proof: By reduction of the following version of the Hamiltonian path problem [10],[16]: Given an undirected graph $G = (\{v_1, \ldots, v_n\}, E)$ with two marked nodes v_1 and v_n of degree 2 and all other nodes of degree 3, does there exist a Hamiltonian path between v_1 and v_n?

Note that n has to be even because of the degree properties. Let $m := |E| = 3n/2 - 1$. We construct a matrix M with n islands $I_1, \ldots, I_n$. I_1 consists of

6 probes, I_n of 4, and all others of 5 probes. M has the strict 2-consecutive ones property iff G has a Hamiltonian path with endpoints v_1 and v_n. If so any 2-consecutive ones order for M places the islands such that the sequence of their indices is the sequence of node indices of a Hamiltonian path. Let

$$M_1 \;=\; \begin{pmatrix} 100011 \\ 111110 \end{pmatrix} \;,\quad M_i = (11111)\,,\quad \text{and}\quad M_n = (1111)\,.$$

These are the building blocks of an $(n+1) \times (5n)$-matrix M_I that forms a diagonal structure to define the n islands. As an example, figure 2 displays the complete matrix M for a graph with 6 nodes. For every island I, the first column of submatrix $m_{\bullet,I}$ of M that corresponds to I will be called the island's *pole*, the second column its *null-column*. Outside the above mentioned diagonal structure the null-columns do not contain any '1'. The construction of M will be such that the following holds.

Property 1. Every row of M that is not in M_I is either potentially chimeric or contains only a single '1'.

Therefore the set $I_1, \ldots, I_n$ will indeed be the set of islands of M. The three columns (resp. two columns) of $m_{\bullet,I}$ for $I \in I_1, \ldots, I_{n-1}$ (resp. $I = I_n$) that follow the null-column are called the island's λ-*columns*. We proceed with the row-wise construction of M by adding m rows with four '1's each as follows. Let G be an instance for the Hamiltonian path problem described above with $E = \{e_1, \ldots, e_m\}$. Let $e_i = \{v_a, v_b\}$, $i \in [1, m]$. The i-th row gets a '1' each in the pole of I_a and the pole of I_b. The row gets another '1' each in I_a and I_b which is placed in one of each island's λ-columns with the following goal. After adding all m rows every λ-column has got *no more* than one new '1'. Because of the degree constraints for G this goal can indeed be achieved. We call this submatrix M_E.

Property 2. A Hamiltonian path between v_1 and v_n defines a sequence of the islands that can be refined to a strict 2-consecutive ones order for M by local permutations of the columns of each island such that the following holds.

1.) The pole becomes the third column of the first island and the second column in all other islands of the sequence.

2.) A row in M_E that corresponds to an edge used in the path falls apart into two fragments after permutation: One is a single pole-'1' of one island and the other has a λ-'1' in its center surrounded by a pole-'1' of the 'next' island and a λ-'1' of the 'previous' island. After permutation all other rows of M_E fall apart into two fragments, each of length 2.

To conclude the construction of M, we add another $23n/2 - 7$ rows with the following properties:

3.) These rows generate the isolated fragments for the rows in M_E and for the chimeric rows that will occur within themselves.

4.) They prevent M from having the strict 2-consecutive ones property without revealing a Hamiltonian path.

These fragments are defined as follows by the matrices M_{1-1st}, M_{V-1st}, M_{E-1st}, M_{1-2nd} and M_{E-2nd}:

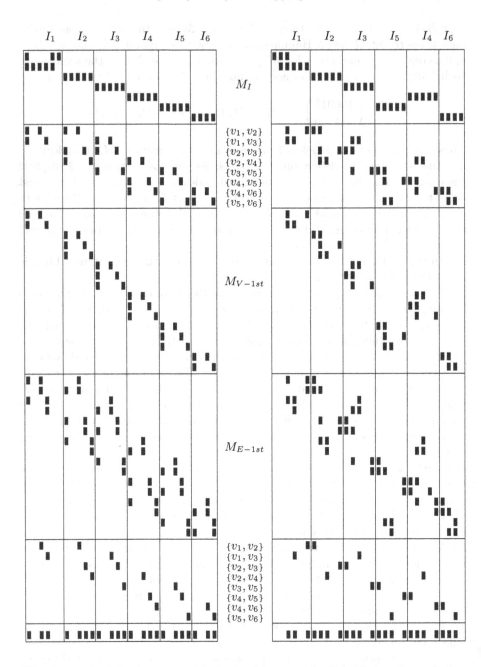

Fig. 2. On the left: the matrix M obtained for a graph G (with 6 nodes and edges as listed in the middle) according to the reduction in the proof of Theorem 2. Rows containing only a single '1' – that is, the rows of M_{1-1st} and M_{1-2nd} – are projected to a single row at the bottom of the diagram, i.e. every '1' there means that M has a row which consists of exactly this '1'.
On the right: an optimal permutation of M.

- M_{1-1st} and M_{1-2nd} have rows of single '1's, one row for every pole- and λ-column, except for the third λ-column column of I_1, for which no such row exists.

- M_{V-1st} has – with the same exception – one row for every λ-column with one '1' in that λ-column and one '1' in the pole of the island of that λ-column.

- It remains to define M_{E-1st} and M_{E-2nd}. Consider an arbitrary row $m_{i,\bullet}$ in M_E and let $m_{i,\bullet} = \{p_1, \lambda_1, p_2, \lambda_2\}$ such that p_1 and p_2 are the poles where $m_{i,\bullet}$ has a '1' each. Add the two rows $\{p_1, \lambda_1, \lambda_2\}$ and $\{\lambda_1, p_2, \lambda_2\}$ to M_{E-1st} and the row $\{\lambda_1, \lambda_2\}$ to M_{E-2nd}.

This concludes the construction of M. It remains to show the existence of a permutation that fulfills the strictness property in case that a Hamiltonian path exists in G. This can be done using similar ideas as in [3]. This finishes the proof sketch. A complete elaboration is given in [16].   ∎

5   The Complexity of the (4|2)- and (3|3)-Optimization Problems

The two cases (4|2) and (3|3) turn out to be the frontiers of intractability. Their reductions are related.

Theorem 3 *The optimization problem* (4|2)–PHYSICAL-MAPPING *with strict 2-chimeric errors is* $\mathcal{NP}$ *-hard.*

Proof: By reduction from the following version of the directed Hamiltonian path problem, which has been established as $\mathcal{NP}$ -complete [17]:

> Given a directed graph $G = (\{v_1, \ldots, v_n\}, E)$ with two marked nodes v_1 and v_n of indegree, resp. outdegree $\delta_{in}(v_1) = \delta_{out}(v_n) = 0$, $\delta_{out}(v_1) = \delta_{in}(v_n) = 2$ and each other node either of indegree 2 and outdegree 1 or vice versa, does there exist a Hamiltonian path from v_1 to v_n?

The matrix M to be constructed consists of n islands. Let $E = \{e_1, \ldots, e_m\}$, with $m := |E| = 3n/2 - 1$. G has a Hamiltonian path from v_1 to v_n iff M can be permuted to a strict 2-consecutive ones order that has no more than $m - n + 1 = n/2$ strict chimeric rows. Let

$$M_d := \begin{pmatrix} 110 \\ 011 \end{pmatrix} \quad \text{and} \quad M_{iso} := \begin{pmatrix} 100 \\ 001 \end{pmatrix}.$$

M is divided vertically into three parts. The first part is a $(2n \times 3n)$-matrix with n occurrences of M_d on its diagonal, all other entries are '0'. This structure defines the islands of M. The second part is constructed similarly, with the diagonal submatrices M_{iso}. The third part consists of $m = 3n/2 - 1$ rows, such that the i-th row has a '1' in the first column of the a-th island and one in the third column of the b-th island iff $e_i = (v_a, v_b)$. M has the strict 2-consecutive ones property and is a valid instance for the $(4|2)$-problem. The left of Fig. 3 shows M for a graph of 6 vertices and edge set as listed in the middle.

Any permutation of M has $n - 1$ transitions between islands. Each such transition allows to remove chimerism in no more than one row, and only when the two islands are in proper orientation to form an edge in G. To obtain a solution with $n/2$ chimeric rows, the islands associated to v_1 and v_n have to be placed at the borders of the matrix. Using this observation, one can easily deduce the correctness of the reduction. ∎

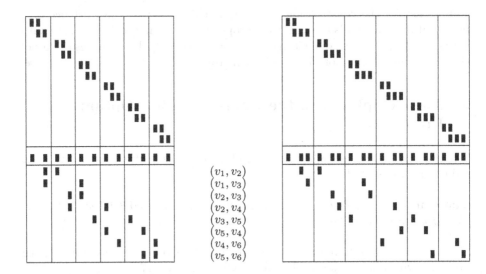

$$\begin{pmatrix} v_1, v_2 \\ v_1, v_3 \\ v_2, v_3 \\ v_2, v_4 \\ v_3, v_5 \\ v_5, v_4 \\ v_4, v_6 \\ v_5, v_6 \end{pmatrix}$$

Fig. 3. The $(4|2)$-matrix and the $(3|3)$-matrix for a 6-node graph with edge set as listed. The submatrix that is spread out by the diagonal structure M_{iso} of isolated fragments is projected to a single row in the middle.

Theorem 4 *The optimization problem* $(3|3)$–PHYSICAL–MAPPING *with strict 2-chimeric errors is* $\mathcal{NP}$ *-hard.*

Proof: We modify the construction with the goal that the third part of the matrix has no two '1's in the same column. This decreases the degree of overlap in the clone library, but we need fragments of length 3. Let

$$M_d := \begin{pmatrix} 1100 \\ 0111 \end{pmatrix} \quad \text{and} \quad M_{iso} := \begin{pmatrix} 1000 \\ 0010 \\ 0001 \end{pmatrix} .$$

In the third part, the two '1's of a row corresponding to edge $e = (v_a, v_b)$ are placed as follows. If v_a (resp. v_b) is incident with another edge that has the same direction as e relative to it, place a '1' in one of the two right columns of the a-th (resp. b-th) island, thereby select a column that does not yet have a '1' in the third part of the matrix. Otherwise put a '1' into the left column of the a-th (resp. b-th) island. See the right part of figure 3 for an example. The correctness follows similarly as in the proof of Theorem 3. ∎

6   An Efficient Algorithm for (3|2)-Optimization

In this section we consider the only efficiently solvable case where a skillful ordering of the probes leads to a substantial reduction of chimerism. Usually there exists a large set of solutions to the optimization problem. However, these are equally valid merely from a combinatorial point of view, whereas in the underlying biological context only very few of them might be meaningful. Therefore, it is desirable to generate all solutions out of which the biologicaly meaningful ones have to be selected with the aid of other criteria. Below we will develop a method to describe the complete set of optimal solutions in a compact form.

Given a hybridization matrix M, let $G_M := (F \dot{\cup} P, E)$ be the corresponding bipartite graph with $F = \{f_1, \ldots, f_r\}$, $P = \{p_1, \ldots, p_c\}$, and $\{f_i, p_j\} \in E$ iff $m_{i,j} = 1$. Let n denote the length of a compact coding of this sparse graph, and $\Gamma(f) \subseteq P$ the neighbors of f, and similarly $\Gamma(p) \subseteq F$.

Theorem 5 *The optimization problem* (3|2)–PHYSICAL-MAPPING *with strict 2-chimeric errors can be solved in time* $O(n \log n)$.

Proof: For short, we call a permutation of M *allowed* if it turns M into strict 2-consecutive ones order. Let M' be the submatrix of M without the potentially chimeric rows. These rows can easily be identified with the help of G_M, since they exactly correspond to all nodes $f \in F$ in G_M with the property

$$\delta(f) = 2 \ \wedge \ \forall p \in \Gamma(f) \ \exists \ f' \in \Gamma(p) \ : \ \delta(f') = 1 \ . \tag{1}$$

Of course, M' has the consecutive ones property. For M' we construct $G_{M'}$, which can be derived from G_M by deleting all edges incident to a fragment f having property (1). Let $\mathcal{I}_{M'}$ be the set of islands of M'. Such an island corresponds to a set of probes that form a connected component in the graph $G_{M'}$.

Consider an island which consists of more than two probes. Because the length of a fragment is bounded by 2 (and by the definition of an island) no two probes overlap with exactly the same set of fragments. This forces a unique order of the start and end points of all fragments that belong to the island. Therefore, a consecutive ones order of the columns of an island in M' is unique except for inversion. It remains to add the potentially chimeric rows. While doing so, we construct an ordering of the islands together with the orientation of each island that leaves as few as possible rows 2-chimeric.

Let us now examine how an potentially chimeric row $m_{i,\bullet}$ added can distribute its two '1's to the islands. Since the inner columns of a consecutive ones ordering of an island in $\mathcal{I}_{M'}$ have at least two '1's, there can be no additional '1' forming a potentially chimeric row. Thus, every potentially chimeric row has its '1's at the coasts of the islands in the final permutation. Let $m_{i,\bullet} = \{\varphi_1, \varphi_2\}$ and distinguish four cases. For an illustration see figure 4.

Case 1. $m_{\bullet,\varphi_1}$ and $m_{\bullet,\varphi_2}$ belong to the same island I. Observe that I consists of at least three columns. Neither $m_{\bullet,\varphi_1}$ nor $m_{\bullet,\varphi_2}$ overlap with a potentially chimeric row different from $m_{i,\bullet}$. Any allowed permutation of M leaves $m_{i,\bullet}$ chimeric.

Case 2. $m_{\bullet,\varphi_1} \in I_1$, $m_{\bullet,\varphi_2} \in I_2$ with $I_1 \neq I_2$, and both I_j are more than one column wide. Again, no other potentially chimeric row can overlap with $m_{\bullet,\varphi_1}$ or $m_{\bullet,\varphi_2}$.

Case 3. $m_{\bullet,\varphi_1} = I_1$ and $m_{\bullet,\varphi_2} = I_2$. Similar to case 2, except that there might exist an $m_{i',\bullet} = m_{i,\bullet}$, $i' \neq i$.

Case 4. Either I_1 or I_2 is one column wide, say $I_1 = m_{\bullet,\varphi_1}$. Only $m_{\bullet,\varphi_1}$ might overlap with another potentially chimeric row $m_{i',\bullet}$, $i' \neq i$.

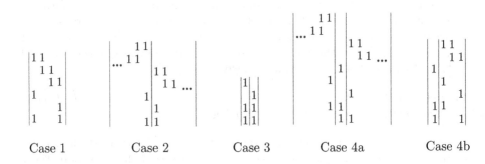

Case 1   Case 2   Case 3   Case 4a   Case 4b

Fig. 4. Characteristic situations for the four cases.

The last three cases allow M' plus $m_{i,\bullet}$ to be turned into consecutive ones order. We will now define a graph G_{pc} that can be regarded as an implicit representation of all optimal solutions. Let $\mathcal{I}_{M'} = \{I_1, \ldots, I_z\}$, $z := |\mathcal{I}_{M'}|$.

Definition 7 Let $G_{pc} := (V_{pc}, E_{pc})$ be an undirected weighted multi-graph with $w(\{v_i, v_j\}) = g$ for $\{v_i, v_j\} \in E_{pc}$ iff M contains g identical potentially chimeric rows whose associated two isolated fragments distribute to the coasts of I_i and I_j. If $i = j$ both fragments are on opposite coasts of one island.

Figure 5 shows how a potentially chimeric row defines the local structure of G_{pc} at the edge corresponding to that row. A dotted line means that there can be one edge, depending on the distribution of potentially chimeric rows. Other edges are not allowed.

Obviously G_{pc} is degree-2 bounded. Each potentially chimeric row appears as one unit in the weight of one edge of G_{pc}. One such row can be non-chimeric if its isolated fragments belong to different islands I_a and I_b with $\{v_a, v_b\} \in E_{pc}$. Two different rows can both be non-chimeric if the two corresponding edges

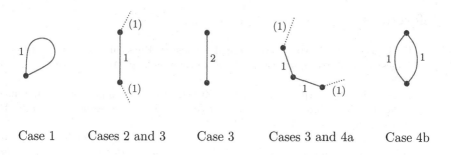

| Case 1 | Cases 2 and 3 | Case 3 | Cases 3 and 4a | Case 4b |

Fig. 5. Local structure of G_{pc}.

do not form a circle. By induction, t different rows can be non-chimeric if the t corresponding edges do not form a circle. Therefore, in order to achieve a consecutive ones order for the probes one has to delete at least as much rows as G_{pc} has circles and self-loops. In other words, the minimum number of rows that remain chimeric in an optimal solution is equal to the number of circles and self-loops, because edges with weight 1 exist only there.

Hence, an optimal solution can be obtained by the following procedure.

1. Delete one edge each from every self-loop and circle.
2. Put every island of M' into consecutive ones order.
3. For each remaining edge e in G_{pc} make the corresponding row (resp. the two rows, if $w(e) = 2$) non-chimeric, but retain the partial order that has been obtained up to now; possibly one of the ordered areas has to be inverted.
4. Arrange the ordered areas in arbitrary sequence and orientation. (i.e. forward or reverse).

Each of the four steps contains a nondeterministic choice. The algorithm terminates with all optimal solutions revealed. Using an efficient representation it can be shown that the whole procedure requires only $O(n \log n)$ steps.   ∎

7   Conclusions

We have shown that the strictness property for chimeric errors does not reduce the complexity of PHYSICAL-MAPPING – this problem remains $\mathcal{NP}$-hard in general. This even holds for sparse clone libraries, for which potential chimerism can be detected fast. A tight complexity estimate for computing an optimal ordering has been obtained based on the sparseness parameters *maximum fragment length* and *maximum degree of probe overlap*.

For very sparse cases, the problem can be solved in linear or almost linear time, where the algorithmically interesting case turns out to be overlap 3 and fragment length 2. Any relaxation of sparsity beyound this point makes PHYSICAL-MAPPING intractable.

References

[1] F. Alizadeh, R. Karp, L. Newberg, D. Weisser, Physical mapping of chromosomes: a combinatorial problem in molecular biology, *Algorithmica* 13, 52–76, 1995.

[2] F. Alizadeh, R. Karp, D. Weisser, G. Zweig, Physical mapping of chromosomes using unique probes, *Proc. 5th Annual ACM-SIAM Symposium on Discrete Algorithms* SODA'94, 489–500, 1994.

[3] J. Atkins, M. Middendorf, On physical mapping and the consecutive ones property for sparse matrices, *DAMATH: Discrete Applied Mathematics and Combinatorial Operations Research and Computer Science* 71, 1996.

[4] K. Booth, G. Lueker, Testing for the consecutive ones property, interval graphs, and graph planarity using PQ-tree algorithms, *J. Computer and System Sciences* 13, 335–379, 1976.

[5] D. Fulkerson, O. Gross, Incidence matrices and interval graphs, *Pacific Journal of Mathematics* 15, 835–856, 1965.

[6] P. Goldberg, M. Golumbic, H. Kaplan, R. Shamir, Four strikes against physical mapping of DNA, *J. Computational Biology* 2, 139–152, 1995.

[7] D. Greenberg, S. Istrail, The chimeric mapping problem: Algorithmic strategies and performance evaluation on synthetic genomic data, *Computers and Chemistry* 18, 207–220, 1994.

[8] D. Greenberg, S. Istrail, Physical Mapping by STS Hybridization: Algorithmic Strategies and the challenge of Software Evaluation, *J. Comp. Biology* 2, 219-273, 1995.

[9] M. Garey, D. Johnson, *Computers and Intractability: A Guide to NP-Completeness*, Freeman, 1979.

[10] M. Garey, D. Johnson, R. Tarjan, The planar Hamiltonian circuit problem is NP-complete, *SIAM J. Computing* 5, 704–714, 1976.

[11] *To Know Ourselves*. Human Genome Program, U.S. Department of Energy, 1996.

[12] W. Hsu. A simple test for the consecutive ones property, *Proc. 3rd Int. Symposium on Algorithms and Computation* ISAAC'92, LNCS 650, 459–468, 1992.

[13] W. Hsu, On physical mapping algorithms: an error tolerant test for the consecutive ones property, *Proc. 3rd Int. Conf. Computing and Combinatorics*, COCOON'97, LNCS 1267, 242–250, 1997.

[14] H. Kaplan, R. Shamir, R. Tarjan. Tractability of parameterized completion problems on chordal and interval graphs: Minimum fill-in and physical mapping, *Proc. 35th Symp. on Foundations of Computer Science* FOCS'94, 780–793, 1994.

[15] N. Korte, R. Möhring. An incremental linear-time algorithm for recognizing interval graphs, *SIAM J. Computing* 18, 68–81, 1989.

[16] S. Weis, Das Entscheidungsproblem Physikalische Kartierung mit starkchimerischen Fehlern, Technical Report A-99-05, Med. Uni. Lübeck, Institut für Theoretische Informatik, 1999.

[17] S. Weis, Zur algorithmischen Komplexität des Optimierungsproblems Physikalische Kartierung mit starkchimerischen Fehlern, Technical Report A-99-06, Med. Uni. Lübeck, Institut für Theoretische Informatik, 1999.

Logical Analysis of Data with Decomposable Structures

Hirotaka Ono[1], Kazuhisa Makino[2], and Toshihide Ibaraki[1]

[1] Department of Applied Mathematics and Physics, Graduate School of Informatics,
Kyoto University, Kyoto 606-8501, Japan.
({htono,ibaraki}@amp.i.kyoto-u.ac.jp)
[2] Department of Systems and Human Science, Graduate School of Engineering
Science, Osaka University, Toyonaka, Osaka, 560-8531, Japan.
(makino@sys.es.osaka-u.ac.jp)

Abstract. In such areas as knowledge discovery, data mining and logical analysis of data, methodologies to find relations among attributes are considered important. In this paper, given a data set (T, F) of a phenomenon, where $T \subseteq \{0,1\}^n$ denotes a set of positive examples and $F \subseteq \{0,1\}^n$ denotes a set of negative examples, we propose a method to identify decomposable structures among the attributes of the data. Such information will reveal hierarchical structure of the phenomenon under consideration. We first study computational complexity of the problem of finding decomposable Boolean extensions. Since the problem turns out to be intractable (i.e., NP-complete), we propose a heuristic algorithm in the second half of the paper. Our method searches a decomposable partition of the set of all attributes, by using the error sizes of almost-fit decomposable extensions as a guiding measure, and then finds structural relations among the attributes in the obtained partition. The results of numerical experiment on synthetically generated data sets are also reported.

1 Introduction

Extracting knowledge from given data sets has been intensively studied in such fields as knowledge discovery, knowledge engineering, data mining, logical analysis of data, artificial intelligence and database theory (e.g., [4,3,5]). We assume in this paper that the data set is given by a pair of a set T of true vectors (positive examples) and a set F of false vectors (negative examples), where $T, F \subseteq \{0,1\}^n$ and $T \cap F = \emptyset$ are assumed. We denote by $S = \{1, 2, \ldots, n\}$ the set of attributes. We are interested in finding a decomposable structure; i.e., given a family $\mathcal{S} = \{S_0, S_1, \ldots, S_k\}$ of subsets of S, we want to establish the existence of Boolean functions $g, h_1, h_2, \cdots, h_k$, if any, so that $f(S) = g(S_0, h_1(S_1), \ldots, h_k(S_k))$ is true (resp., false) in every given true (resp., false) vector in T (resp., F).

In the above scheme, the sets S_i may represent intermediate groups of attributes, and we ask whether these groups define new "meta-attributes", which can completely specify the positive or negative character of the examples. This

D.-Z. Du et al. (Eds.): COCOON 2000, LNCS 1858, pp. 396–406, 2000.

problem of structure identification is in fact one form of knowledge discovery. As an example, assume that $f(x)$ describes whether the species are primates or not; e.g., $f(x) = 1$ for $x = (1100\cdots)$, denotes that the chimpanzee $(= x)$, which has characteristics of viviparous $(x_1 = 1)$, vertebrate $(x_2 = 1)$, do not fly $(x_3 = 0)$, do not have claw $(x_4 = 0)$, and so on, is a primate. In the case of the hawk, on the other hand, we shall have $f(0111\cdots) = 0$. In this example, we can group properties "viviparity" and "vertebrate" as the property of the mammals, and the chimpanzee is a mammal. That is, $f(x)$ can be represented as $g(S_0, h(S_1))$, where $S_1 = \{x_1, x_2\}$, $S_0 = S \setminus S_1$, and h describes whether species are the mammal or not. This "mammal" is a meta-attribute, and we can recognize primates by regarding $S_1 = \{x_1, x_2\}$ as one attribute $h(S_1)$. In this sense, finding a family of attribute sets $\mathcal{S} = \{S_0, S_1, \ldots, S_k\}$, which satisfy the above decomposition property, can be understood as computing essential relations among the original attributes [2,7].

In this paper, we concentrate on finding a basic decomposition structure for a partition into two sets $(S_0, S_1(= S \setminus S_0))$. We first show that determining the existence of such a partition (S_0, S_1) is intractable (i.e., NP-complete). Since finding decomposable structures is a very important problem, we then propose a heuristic algorithm. Our method searches a decomposable partition (S_0, S_1) of the attribute set S, by using the error sizes of almost-fit decomposable extensions as a guiding measure. In order to compute the error size of an almost-fit decomposable extension, in the above algorithm, we also propose a fast heuristic algorithm.

We then apply our method to synthetically generated data sets. Experimental results show that our method has good performance to identify such decomposable structures.

2   Preliminaries

2.1   Extensions and Best-Fit Extensions

A *Boolean function*, or a *function* in short, is a mapping $f : \{0,1\}^n \to \{0,1\}$, where $x \in \{0,1\}^n$ is called a *Boolean vector* (a *vector* in short). If $f(x) = 1$ (resp., 0), then x is called a *true* (resp., *false*) vector of f. The set of all true vectors (false vectors) is denoted by $T(f)$ $(F(f))$. Let, for a vector $v \in \{0,1\}^n$, $ON(v) = \{j \mid v_j = 1, j = 1, 2, \ldots, n\}$ and $OFF(v) = \{j \mid v_j = 0, j = 1, 2, \ldots, n\}$. For two sets A and B, we write $A \subseteq B$ if A is a subset of B, and $A \subset B$ if A is a proper subset of B. For two functions f and g on the same set of attributes, we write $f \leq g$ if $f(x) = 1$ implies $g(x) = 1$ for all $x \in \{0,1\}^n$, and we write $f < g$ if $f \leq g$ and $f \neq g$.

A function f is *positive* if $x \leq y$ (i.e., $x_i \leq y_i$ for all $i \in \{1, 2, \ldots, n\}$) always implies $f(x) \leq f(y)$. The attributes $x_1, x_2, \ldots, x_n$ and their complements $\bar{x}_1, \bar{x}_2, \ldots, \bar{x}_n$ are called *literals*. A *term* is a conjunction of literals such that at most one of x_i and $\bar{x}_i$ appears in it for each i. A *disjunctive normal form* (*DNF*) is a disjunction of terms. Clearly, a DNF defines a function, and it is well-known that every function can be represented by a DNF (however, such a representation may not be unique).

A *partially defined Boolean function* (*pdBf*) is defined by a pair of sets (T, F) of Boolean vectors of n attributes, where T denotes a set of true vectors (or positive examples) and F denotes a set of false vectors (or negative examples). A function f is called an *extension* of the pdBf (T, F) if $T(f) \supseteq T$ and $F(f) \supseteq F$.

Evidently, the disjointness of the sets T and F is a necessary and sufficient condition for the existence of an extension, if it is considered in the class of all Boolean functions. It may not be evident, however, to find out whether a given pdBf has a extension in $\mathcal{C}$, where $\mathcal{C}$ is a subclass of Boolean functions, such as the class of positive functions. Therefore, we define the following problem [3]:

PROBLEM EXTENSION($\mathcal{C}$)

Input: A pdBf (T, F), where $T, F \subseteq \{0, 1\}^n$.
Question: Is there an extension $f \in \mathcal{C}$ of (T, F)?

Let us note that this problem is also called *consistency problem* in computational learning theory [1].

For a set A, $|A|$ denotes the cardinarity of A, and we define the *error size* of a function f with respect to (T, F) by $e(f; (T, F)) = |\{a \in T \mid f(a) = 0\}| + |\{b \in F \mid f(b) = 1\}|$. As EXTENSION($\mathcal{C}$) may not always have answer "yes", we introduce the following problem [3]:

PROBLEM BEST-FIT($\mathcal{C}$)

Input: A pdBf (T, F), where $T, F \subseteq \{0, 1\}^n$.
Output: $f \in \mathcal{C}$ that realizes $\min_{f \in \mathcal{C}} e(f; (T, F))$.

Clearly, problem EXTENSION is a special case of problem BEST-FIT, since EXTENSION has a solution f if and only if BEST-FIT has a solution f with $e(f; (T, F)) = 0$.

For a pdBf (T, F), an extension f is called *almost-fit* if it has a *small* error size $e(f; (T, F))$, while a best-fit extension f^* gives the smallest error size $e(f^*; (T, F))$. Thus the definition of almost-fit extension is not exact in the mathematical sense, since "small" is only vaguely defined. In the cases where best-fit extensions are difficult to compute, we use almost-fit extensions instead.

2.2   Decomposable Functions

For a vector $a \in \{0, 1\}^n$ and a subset $S^* \subseteq S$, let $a[S^*]$ denote the *projection* of a on S^*, and let $\{0, 1\}^{S^*}$ denote the vector space defined by an attribute set S^*. For example, if $a = (1011100)$, $b = (0010011)$ and $S^* = \{2, 3, 5\}$, then $a[S^*] = (011)$, $b[S^*] = (010)$, and $a[S^*], b[S^*] \in \{0, 1\}^{S^*}$. Furthermore, for a subset of vectors $U \subseteq \{0, 1\}^n$, we use notation $U[S^*] = \{a[S^*] \mid a \in U\}$. For a Boolean function h that depends only on $S^* \subseteq S$, we write $h(S^*)$ to denote $h(v[S^*])$ for

$v \in \{0,1\}^n$. Given subsets $S_0, S_1 \subseteq S$, a function f is called $F_0(S_0, F_1(S_1))$-decomposable [2,11] if there exist Boolean functions h_1, and g satisfying the following conditions, where F_i stands for the class of all Boolean functions:

(i) $f(v) = g(v[S_0], h_1(v[S_1]))$, for all $v \in \{0,1\}^n$,
(ii) $h_1 : \{0,1\}^{S_1} \to \{0,1\}$,
(iii) $g: \{0,1\}^{S'} \to \{0,1\}$, where $S' = S_0 \cup \{h_1\}$.

We denote by $\mathcal{C}_{F_0(S_0, F_1(S_1))}$ the class of $F_0(S_0, F_1(S_1))$-decomposable functions. In some cases, we further assume that a pair (S_0, S_1) is a *partition*, i.e., $S_0 \cap S_1 = \emptyset$ and $S = S_0 \cup S_1$.

It is known that, given a pair of subsets (S_0, S_1), problem EXTENSION($\mathcal{C}$ $_{F_0(S_0, F_1(S_1))}$) can be solved in polynomial time if one exists [2] (see Proposition 1), while problem BEST-FIT($\mathcal{C}_{F_0(S_0, F_1(S_1))}$) is NP-hard [3].

Given a pdBf (T, F) and a pair of sets of attributes S_0 and S_1, we define its *conflict graph* $G^*_{(T,F)} = (V, E)$ by $V = \{v[S_1] \mid v \in T \cup F\}$, $E = \{(a[S_1], b[S_1]) \mid a \in T, b \in F, a[S_0] = b[S_0]\}$. For example, (T, F) in Fig. 1, and Consider the pdBf given in the truth table of Fig.1, $S_0 = \{x_1, x_2, x_3\}$ and $S_1 = \{x_4, x_5, x_6\}$ have the conflict graph G^* of Fig. 1. This is not $F_0(S_0, F_1(S_1))$-decomposable by the next proposition [2].

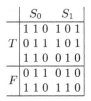
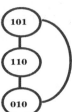

	S_0	S_1
	1 1 0	1 0 1
T	0 1 1	1 0 1
	1 1 0	0 1 0
F	0 1 1	0 1 0
	1 1 0	1 1 0

Fig. 1. A pdBf (T, F) and its conflict graph.

Proposition 1. A pdBf (T, F) has an extension $f = g(S_0, h_1(S_1))$ for a given pair of subsets (S_0, S_1) (not necessarily a partition) if and only if its conflict graph $G^*_{(T,F)}$ is bipartite (this condition can be tested in polynomial time). □

3 Computational Complexity of Decomposability

3.1 NP-Hardness of Finding a (Positive) Decomposable Structure

By Proposition 1, it can be decided in polynomial time whether a pdBf (T, F) has an $F_0(S_0, F_1(S_1))$-decomposable extension for a given (S_0, S_1). However, we sometimes want to know the existence of subsets S_0 and S_1 such that a pdBf (T, F) is $F_0(S_0, F_1(S_1))$-decomposable, where (S_0, S_1) is restricted to be a partition.

PROBLEM $F_0(S_0, F_1(S_1))$-DECOMPOSABILITY

Input: A pdBf (T, F), where $T, F \subseteq \{0,1\}^n$.

Question: Is there a partition (S_0, S_1) of S with $|S_0| \geq 3$ and $|S_1| \geq 2$, such that pdBf (T, F) is $F_0(S_0, F_1(S_1))$-decomposable?

Theorem 1. Problem $F_0(S_0, F_1(S_1))$-DECOMPOSABILITY is NP-complete.
□

Moreover, even if F_0 and F_1 are positive functions, this problem remains NP-complete. Since these theorems can be proved similarly, we only outline the proof of Theorem 1.

Outline of the proof of Theorem 1. We reduce the following NP-complete problem [8] to $F_0(S_0, F_1(S_1))$-DECOMPOSABILITY.

PROBLEM ONE-IN-THREE 3SAT

Input: Set U of attributes, collection $\mathcal{C}$ of clauses over U such that each clause $C \in \mathcal{C}$ consists of 3 positive literals.

Question: Is there a truth assignment for U such that each clause in C has exactly one literal assigned to 1?

Given an instance of ONE-IN-THREE 3SAT, let E_p denote the set of attributes corresponding to a clause $C_p \in \mathcal{C}$ (e.g., if $C_p = x_1 \vee x_2 \vee x_3$, then $E_p = \{x_1, x_2, x_3\}$), and let $\mathcal{E}$ denote the family of all E_p. The instance has answer "yes" if there exists a partition (U_0, U_1) of U such that $|U_0 \cap E_p| = 2$ and $|U_1 \cap E_p| = 1$ hold for all $E_p \in \mathcal{E}$. We assume without the loss of generality that $|U_0| \geq 3$ and $|U_1| \geq 2$.

Corresponding to the above instance of ONE-IN-THREE 3SAT, we now construct the instance (i.e., pdBf (T, F)) of $F_0(S_0, F_1(S_1))$-DECOMPOSABILITY as follows: $S = U$, $T = T_1 \cup T_3$, $F = F_2 \cup F_4 \cup F_5$, where

$$T_1 = \{v \in \{0,1\}^S \mid |ON(v)| = 1\},$$
$$T_3 = \{v \in \{0,1\}^S \mid ON(v) = E_p \text{ and } E_p \in \mathcal{E}\},$$
$$F_2 = \{v \in \{0,1\}^S \mid |ON(v)| = 2 \text{ and } ON(v) \subset E_p, \text{ for some } E_p \in \mathcal{E}\},$$
$$F_4 = \{v \in \{0,1\}^S \mid |ON(v)| = 4 \text{ and } ON(v) \supset E_p \text{ for some } E_p \in \mathcal{E}\},$$
$$F_5 = \{v \in \{0,1\}^S \mid |ON(v)| = 5 \text{ and } ON(v) \supset E_p \text{ for some } E_p \in \mathcal{E}\},$$

For example, if there is a clause $C = x_1 \vee x_2 \vee x_3$, we construct the true vectors $\{(100000 \cdots), (010000 \cdots), (001000 \cdots), (111000 \cdots)\}$ and the false vectors $\{(110000 \cdots), (101000 \cdots), (011000 \cdots), (111100 \cdots), (111010 \cdots), (111001 \cdots), \ldots, (111110 \cdots), (111101 \cdots), (111011 \cdots), \ldots\}$.

Then, using Proposition 1, we can show that there is a partition (S_0, S_1) with $|S_0| \geq 3$ and $|S_1| \geq 2$ such that pdBf (T, F) is $F_0(S_0, F_1(S_1))$-decomposable if and only if a given collection of clauses $\mathcal{C}$ has a partition (U_0, U_1) with $|U_0| \geq 3$ and $|U_1| \geq 2$ such that $|U_0 \cap E_p| = 2$ and $|U_1 \cap E_p| = 1$ hold for all $E_p \in \mathcal{E}$. We however omit the details for the sake of space.
□

4 Heuristic Algorithm to Find a Decomposable Structure

By Theorem 1, it is in general difficult to find a decomposable structure of a given pdBf (T, F). As the problem of finding a decomposable structure is very important in applications, however, we propose a heuristic algorithm DECOMP, which finds a partition (S_0, S_1) for which the given pdBf (T, F) has an $F_0(S_0, F_1(S_1))$-decomposable extension f with a small error size $e(f; (T, F))$. Allowing errors is essential in practice because the real-world data often contain errors, such as measurement and classification errors. Due to these errors, the given pdBf (T, F) may not be $F_0(S_1, F_1(S_1))$-decomposable for any partition (S_0, S_1), although the underlying phenomenon has a decomposable structure. By allowing errors of small size, it is expected that this kind of misjudgement is prevented.

Our algorithm uses the error size of each partition (S_0, S_1) as a measure to evaluate the distance from (S_0, S_1) to (S_0^*, S_1^*), where pdBf (T, F) is assumed to be $F_0(S_0^*, F_1(S_1^*))$-decomposable. For this purpose, we consider how to evaluate the error size efficiently in the next subsection.

4.1 Finding Almost-Fit Decomposable Extensions

Since the problem of computing a best-fit decomposable extension for a given (S_0, S_1) is NP-hard [3], we propose two heuristic algorithms in this subsection. The first one is theoretically interesting as it has a guaranteed approximation ratio, while the latter is more practical. In our experiment in Section 5, the latter algorithm is employed.

Given a pdBf (T, F) and a partition (S_0, S_1), we construct an almost-fit $F_0(S_0, F_1(S_1))$-decomposable extension f with error size $e(f; (T, F))$. To make the error size small, we introduce an *extended conflict graph* G' such that the error vectors in the extension f correspond one-to-one to the vertices in G', whose deletion makes the resulting G' bipartite.

Given a pdBf (T, F) and a pair of sets of attributes (S_0, S_1) (not necessarily a partition), we define its extended conflict graph $G'_{(T,F)} = (V, E)$, where $V = V_1 \cup V_2$ is given by $V_1 = \{a \mid a \in T\} \cup \{b \mid b \in F\}, V_2 = \{a[S_1] \mid a \in T \cup F\}$ and $E = E_1 \cup E_2$ is given by $E_1 = \{(a, a[S_1]) \mid a \in T \cup F\}$, $E_2 = \{(a, b) \mid a \in T, b \in F, a[S_0] = b[S_0]\}$. The vertices in G' have positive weights ω defined by $\omega(a) = 1$ if $z = a \in V_1$, $+\infty$ otherwise.

Now let $M \subseteq V_1$ be a set of vertices whose removal makes $G'_{(T,F)}$ bipartite. It can be proved that deleting the corresponding sets $T' \subseteq T$ and $F' \subseteq F$ from (T, F) gives the pdBf $(T \setminus T', F \setminus F')$ which has an $F(S_0, F_1(S_1))$-decomposable extension. The converse direction also holds. Therefore, we have the next lemma.

Lemma 1. Let a pdBf (T, F) and a pair of attribute sets (S_0, S_1) be given, and let $V^* \subseteq V$ be a minimum weighted set of vertices whose removal makes $G'_{(T,F)}$ bipartite. Then we have $\min_{f \in C_{F(S_0, F_1(S_1))}} e(f; (T, F)) = |V^*|$. □

The problem of minimizing the number of vertices in a given graph $G = (V, E)$, whose deletion make G bipartite is called VERTEX-DELETION-BIPARTIZA-

TION, or VDB for short. VDB is known to be MAX SNP-hard, but has an approximation algorithm with approximation ratio $O(\log |V|)$ [9,10,6].

Theorem 2. Given a pdBf (T, F) and a pair of attribute sets (S_0, S_1), there is an approximate algorithm for BEST-FIT$(\mathcal{C}_{F(S_0, F_1(S_1))})$, which attains the approximation ratio of

$$\frac{e(f; (T, F))}{e(f^*; (T, F))} = O(\log(|T| + |F|)), \tag{1}$$

where f^* is a best-fit $F(S_0, F_1(S_1))$-decomposable extension of (T, F).

However, the approximation algorithm in [10] for VDB may require too much computational time for practical purposes. Since our heuristic algorithm in Section 4.2 uses the error size as a guiding measure, it has to be computed many times. Therefore, we propose a faster algorithm ALMOST-FIT even though its approximation ratio is not guaranteed theoretically.

ALMOST-FIT first constructs the conflict graph $G^* = G^*_{(T,F)}$. Next, it traverses G^* in the depth-first manner, and, whenever a cycle of odd length is found, deletes a vector in pdBf (T, F) so that the most recently visited edge in the cycle is eliminated. Then, after reconstructing the resulting conflict graph G^*, it resumes the depth-first search in the new G^*.

In order to eliminate an edge in the detected cycle of odd length, we only have to delete one of the two vectors in $T \cup F$ that correspond to the two end vertices of the edge. We now explain which vector to delete. By the property of depth-first search (in short, DFS), we find a cycle only when we traverse a so-called back edge [13]. Denote the back edge by (v_1, v_2), where DFS traverses this edge from v_1 to v_2, and let $a_1, a_2 \in T \cup F$ satisfy $v_1 = a[S_1]$ and $v_2 = a[S_1]$, where one of the a_1 and a_2 belongs to T and the other belongs to F. In this case, we say that a_1 (resp., a_2) corresponds to v_1 (resp., v_2), and delete a_1. This deletion does not change the set of ancestor vertices of v_1 on the current DFS tree and hence DFS can be resumed from v_1. (If we delete a_2, on the other hand, the structure of DFS tree changes substantially).

ALGORITHM ALMOST-FIT

Input: A pdBf (T, F) and a pair of attribute sets (S_0, S_1), where $T, F \subseteq \{0, 1\}^S$ and $S_0, S_1 \subseteq S$.

Output: The error size $e(f; (T, F))$, where f is an almost-fit $F_0(S_0, F_1(S_1))$-decomposable extension for (T, F).

Step 1. Construct the conflict graph $G^* = G^*_{(T,F)}$ for pdBf (T, F) and (S_0, S_1). Set $e := 0$

Step 2. Apply the DFS to the conflict graph G^*. If a cycle of odd length is found, then delete one vector from (T, F) in the above manner, reconstruct the conflict graph G^*, set $e := e + 1$ and return to Step 2. If there is no odd cycle, output e as $e(f; (T, F))$ and halt. (f is an extension of the resulting pdBf (T, F).)   □

4.2   Finding Decomposable Structures

In this section, we propose a local search algorithm to find decomposable partitions (S_0, S_1), by making use of the error size obtained by the heuristic algorithm ALMOST-FIT of section 4.1. A local search algorithm is defined by specifying the neighbourhood of the current solution (S_0, S_1). In our algorithm, $N(S_0, S_1) = \{(S_0 \cup \{j\} \setminus \{i\}, S_1 \cup \{i\} \setminus \{j\}) \mid i \in S_0, j \in S_1\}$. However, as all partitions $(S_0', S_1') \in N(S_0, S_1)$ satisfy $|S_0'| = |S_0|$ and $|S_1'| = |S_1|$, we apply local search to initial solutions (S_0, S_1) with $|S_0| = k$, separately, for $k = 2, 3, \ldots, |S| - 1$.

ALGORITHM DECOMP

Input: A pdBf (T, F), where $T, F \subseteq \{0,1\}^S$ and S is the attribute set.
Output: Partitions (S_0, S_1) with $|S_1| = k$, for $k = 2, 3, \ldots, |S| - 1$, having almost-fit $F_0(S_0, F_1(S_1))$-decomposable extensions of small error sizes.

Step 0: $k := 2$.

Step 1: Choose a partition (S_0, S_1) with $|S_1| = k$ randomly, and compute the error size $\hat{e}(S_0, S_1)$ of an almost-fit $F_0(S_0, F_1(S_1))$-decomposable extension f by algorithm ALMOST-FIT. Set $\epsilon := \hat{e}(S_0, S_1)$.

Step 2: For each $(S_0', S_1') \in N(S_0, S_1)$, apply ALMOST-FIT. If there is a partition (S_0', S_1') with a smaller error size than ϵ, then set $S_0 := S_0'$, $S_1 := S_1'$, $\epsilon := \hat{e}(S_0', S_1')$, and return to Step 2. Otherwise, output partition (S_0, S_1) and its error size ϵ for the current k.

Step 3: If $k := |S| - 1$, then halt. Otherwise, let $k := k + 1$ and return to Step 1.   □

5   Numerical Experiments

We apply algorithm DECOMP described in Section 4 to synthetically generated pdBfs (T, F), in order to evaluate their power of finding decomposable structures.

5.1   Generation of (T, F)

For our experiment, we generate random data sets (T, F), $T, F \subseteq \{0,1\}^n$, which are $F_0(S_0, F_1(S_1))$-decomposable. To ensure a decomposable structure, we first specify a partition $(\tilde{S}_0, \tilde{S}_1)$ and then construct a function $\tilde{f} : \{0,1\}^n \to \{0,1\}$ such that $\tilde{f} = \tilde{g}(\tilde{S}_0, \tilde{h}(\tilde{S}_1))$, $\tilde{h} : \{0,1\}^{\tilde{S}_1} \to \{0,1\}$, $\tilde{g} : \{0,1\}^{\tilde{S}_0 \cup \{\tilde{h}\}} \times \{0,1\} \to \{0,1\}$. Then for a given rate r, where $0 \leq r \leq 1$, we randomly draw a set Q_r of $r \cdot 2^n$ vectors from $\{0,1\}^n$ as samples, and classify all $u \in Q_r$ into T (resp., F) according to whether $\tilde{f}(u) = 1$ (resp., $\tilde{f}(u) = 0$) holds. Obviously $T \cap F = \emptyset$ hold. In our experiment, we use $n = 18$, and test $r = 0.01, 0.02, \ldots, 0.09, 0.1, 0.2, \ldots, 0.5$. The original partitions $(\tilde{S}_0, \tilde{S}_1)$ satisfy $|\tilde{S}_0| = |\tilde{S}_1| = 9$. For the hidden function $\tilde{f}$, we used the following three types.

(i) Randomly assigned functions: $\tilde{f}$ is constructed by assigning the value $\tilde{f}(a) = 0$ or 1 randomly with equal probability for each $a \in \{0, 1\}^n$. Any Boolean function can be generated in this way. However, most of the randomly assigned functions may not appear as functions in real-world.

(ii) DNF functions: DNF of $\tilde{f}$ is defined by a set of terms, as follows: A term is randomly generated by specifying each variable to be a positive literal, a negative literal or not used with equal probability, and taking the conjunction of the generated literals. We generate $|\tilde{S}_0| + 1$ terms for $\tilde{g}$ and $|\tilde{S}_1|$ terms as $\tilde{h}$. Any Boolean function can be represented by a DNF, and those functions representing real-world phenomena are considered to have short DNFs.

(iii) Threshold functions: A threshold function $\tilde{f}$ is defined by weights w_i, $i = 1, 2, \ldots, n$ and a threshold θ, as $f(x) = 1$ if $\sum_i w_i x_i \geq \theta, 0$ otherwise [12]. We generate a threshold function by choosing all weights w_i from $(0, 1]$ randomly and by setting $\theta = \sum_i w_i / n$.

5.2 Computational Results

We generate 5 pdBfs for each type, and execute DECOMP 10 times for each pdBf (i.e., 50 times in total). Let the success rate denote the ratio with which DECOMP could find decomposable partitions and the cpu time denote the average of total cpu time of 10 time executions of DECOMP for one function. We show the results of success rate and cpu time for the randomly assigned functions in Fig. 2. The success rates for DNF functions and threshold functions are given in Fig. 3. We omit the results of cpu time for other two types, since they have similar tendency with random assigned functions.

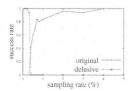 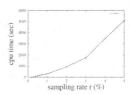

Fig. 2. (i) Success rate (left) and (ii) average cpu time (right) for randomly assigned functions

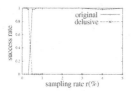 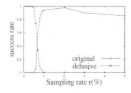

Fig. 3. (i) Success rate for DNF functions (left) and (ii) for threshold functions (right)

Fig. 2 (i) and Fig. 3 show the success rate of finding the original partition $(\tilde{S}_0, \tilde{S}_1)$ (solid curves) as well as that of finding delusive partitions (broken curves), when sampling rate r (%) is changed from 0.01 to 0.5.

The results for randomly assigned functions (Fig. 2) say that DECOMP can almost always find the original $(\tilde{S}_0, \tilde{S}_1)$ if sampling rate satisfies $r \geq 0.5\%$. For DNF functions and threshold functions (Fig. 3), similar tendency is also observed. If sampling rate r is lower than 0.5%, each pdBf (T, F) has only a small number of edges in its conflict graph, and many partitions $(S_0, S_1) \neq (S_0^*, S_1^*)$ also turn out to be decomposable. These partitions are called *delusive*, and also depicted in Fig. 2 (i) and Fig. 3. These indicate that a sampling rate larger than a certain threshold value is necessary to ensure the discovery of the original decomposable structure $(\tilde{S}_0, \tilde{S}_1)$.

6   Conclusion

In this paper, we used the concept of decomposability to extract the structural information from the given data (T, F). Such information can explain the hierarchical relations that exist among the attributes. We first clarified the computational complexity of the problem of finding a decomposable structure. Since it is NP-complete, we proposed a heuristic algorithm, which is based on local search method. As a guiding measure in this local search, we used the error sizes of almost-fit $F_0(S_0, F_1(S_1))$-decomposable extensions, for which a fast heuristic algorithm is developed.

We then performed numerical experiments on three types of synthetically generated data sets. Judging from the experimental results, our approach appears to be able to detect decomposition structures reasonably effectively. However, its performance critically depends on the number of data vectors in the data sets; it appears essential to have enough number of data vectors for our method to be effective. The number of necessary data vectors depends on the types of data sets, and the sizes of the variable sets. More theoretical and experimental studies appear necessary in order to know the effect of these parameters more accurately.

Acknowledgement

This work was partially supported by the Scientific Grant-in-Aid by the Ministry of Education, Science, Sports and Culture of Japan.

References

1. M. Anthony and N. Biggs, *Computational Learning Theory* (Cambridge University Press, 1992).
2. E.Boros, V.Gurvich, P.L.Hammer, T.Ibaraki and A.Kogan, Decompositions of partially defined Boolean functions, *Discrete Applied Mathematics*, 62 (1995) 51-75.

3. E. Boros, T. Ibaraki and K. Makino, Error-free and best-fit extensions of a partially defined Boolean function, *Information and Computation*, 140 (1998) 254-283.
4. E. Boros, P. L. Hammer, T. Ibaraki, A. Kogan, E. Mayoraz and I. Muchnik, An implementation of logical analysis of data, RUTCOR Research Report RRR 22-96, Rutgers University, 1996; to appear in IEEE Trans. on Data Engineering.
5. Y. Crama, P. L. Hammer and T. Ibaraki, Cause-effect relationships and partially defined Boolean functions, *Annals of Operations Research*, 16 (1988) 299-325.
6. P. Crescenzi and V. Kann, A compendium of NP optimization problems, http://www.nada.kth.se/ viggo/index-en.html.
7. U. M. Fayyad, G. Piatetsky-Shapiro, P. Smyth, and R. Uthurusamy(eds.), *Advances in Knowledge Discovery and Data Mining*, 1996, AAAI Press.
8. M. R. Garey and D. S. Johnson, *Computers and Intractability*, Freeman, New York, 1979.
9. N. Garg, V. V. Vazirani and M. Yannakakis, Approximate max-flow min-(multi)cut theorems and their applications, *SIAM J. on Computing*, 25 (1996) 235-251.
10. N. Garg, V. V. Vazirani and M. Yannakakis, Multiway cuts in directed and node weighted graphs, *ICALP'94*, LNCS 820 (1994) 487-498.
11. K. Makino, K. Yano and T. Ibaraki, Positive and Horn decomposability of partially defined Boolean functions, *Discrete Applied Mathematics*, 74 (1997) 251-274.
12. S. Muroga, *Threshold Logic and Its Applications*, Wiley-Interscience, 1971.
13. R.C Read and R.E. Tarjan, Bounds on backtrack algorithms for listing cycles, paths, and spanning tress, *Networks*, 5 (1975) 237-252.

Learning from Approximate Data

Shirley Cheung H.C.

Department of Mathematics, City University of Hong Kong
83 Tat Chee Avenue, Kowloon Tong, HONG KONG
s5111127@math.cityu.edu.hk

Abstract. We give an algorithm to PAC-learn the coefficients of a multivariate polynomial from the signs of its values, over a sample of real points which are only known approximately. While there are several papers dealing with PAC-learning polynomials (e.g. [3,11]), they mainly only consider variables over finite fields or real variables with no round-off error. In particular, to the best of our knowledge, the only other work considering rounded-off real data is that of Dennis Cheung [6]. There, multivariate polynomials are learned under the assumption that the coefficients are independent, eventually leading to a linear programming problem. In this paper we consider the other extreme: namely, we consider the case where the coefficients of the polynomial are (polynomial) functions of a single parameter.

1 Introduction

In the PAC model of learning one often finds concepts parameterized by (exact) real numbers. Examples of such concepts appear in the first pages of well-known textbooks such as [9]. The algorithmics for learning such concepts follows the same pattern as that for learning concepts parameterized by Boolean values. One randomly selects some elements $x^{(1)}, \ldots, x^{(m)}$ in the instance space X. Then, with the help of an oracle, one decides which of them satisfy the target concept c^*. Finally, one computes an hypothesis c^h which is consistent with the sample, i.e., a concept c^h which is satisfied by exactly those $x^{(i)}$ which satisfy c^*.

A main result from Blumer et al. [5] provides a bound for the sample size m which guarantees that the error of c^h is less than ε with probability at least $1 - \delta$, namely

$$m \geq C_0 \left(\frac{1}{\varepsilon} \log \left(\frac{1}{\delta} \right) + \frac{\text{VCD}(\mathcal{C})}{\varepsilon} \log \left(\frac{\delta}{\varepsilon} \right) \right), \tag{1}$$

where C_0 is a universal constant and $\text{VCD}(\mathcal{C})$ is the Vapnik-Chervonenkis dimension of the concept class at hand. This result is especially useful when the concepts are not discrete entities (i.e., not representable using words over a finite alphabet), since in this case one can bound the size of the sample without using the VC dimension.

A particularly important case of concepts parameterized by real numbers is where the membership test of an instance $x \in X$ to a concept c in the concept

D.-Z. Du et al. (Eds.): COCOON 2000, LNCS 1858, pp. 407–415, 2000.

class $\mathcal{C}$ can be expressed by a quantifier-free first-order formula. In this case, concepts in $\mathcal{C}_{n,N}$ are parameterized by elements in $\mathbb{R}^n$, the instance space X is the Euclidean space $\mathbb{R}^N$, and the membership of $x \in X$ to $c \in \mathcal{C}$ is given by the truth of $\Psi_{n,N}(x, c)$ where $\Psi_{n,N}$ is a quantifier-free first-order formula (over the theory of the reals) with $n + N$ free variables. In this case, a result of Goldberg and Jerrum [8] bounds the VC-dimension of $\mathcal{C}_{n,N}$ by

$$\mathrm{VCD}(\mathcal{C}_{n,N}) \leq 2n \log(8eds). \tag{2}$$

Here d is a bound for the degrees of the polynomials appearing in $\Psi_{n,N}$ and s is a bound for the number of distinct atomic predicates in $\Psi_{n,N}$.

One may say that at this stage the problem of PAC learning a concept $c^* \in \mathcal{C}_{n,N}$ is solved. Given $\varepsilon, \delta > 0$ we simply compute m satisfying (1) with the VC-dimension replaced by the bound in (2). We then randomly draw $x^{(1)}, \ldots, x^{(m)} \in X$ and finally compute a hypothesis $c^h \in \mathcal{C}_{n,N}$ consistent with the membership of $x^{(i)}$ to c^*, $i = 1, \ldots, m$ (which we obtain from some oracle). To obtain c^h we may use any of the algorithms proposed recently to solve the first-order theory of the reals (cf. [10,2]).

It is at this stage, however, that our research begins: from a practical viewpoint, we can not read the elements $x^{(i)}$ exactly. Instead, we typically obtain rational approximations $\tilde{x}^{(i)}$. The membership of $x^{(i)}$ to c^* depends nevertheless on $x^{(i)}$ and not on $\tilde{x}^{(i)}$. Our problem thus becomes that of learning c^* from approximate data. A key assumption is that we know the precision ρ of these approximations and that we can actually modify ρ in our algorithm to obtain better approximations.

In this paper, we will not deal with general concepts in $\mathcal{C}_{n,N}$ as defined above but with a subclass where the membership test is given by applying the sign function to a single polynomial. We will design an algorithm PAC-learning a concept in this class from approximate data. In studying the complexity of this algorithm we will naturally deal with a classical theme in numerical analysis — conditioning — and we will find yet another instance of the dependence of running time on the condition number of the input (cf. [7]).

2   The Problem

Consider a polynomial $\mathcal{F} \in \mathbb{R}[c, \bar{x}]$ where $\bar{x} = (x_1, \ldots, x_N)$. Let d be the total degree of $\mathcal{F}$. Then we can write

$$\mathcal{F} = \sum_{i=0}^{d} g_i(\bar{x}) c^i$$

where $g_i(\bar{x})$ is a polynomial of degree at most $d - i$ in $x_1, \ldots, x_N$.

Replacing the parameter c for a given $c \in \mathbb{R}$, $\mathcal{F}$ becomes a polynomial in $x_1, \ldots, x_N$ which we will denote by $\mathcal{F}_c$. Let $\mathcal{M} \in \mathbb{R}$, $\mathcal{M} > 0$. When c varies in $[-\mathcal{M}, \mathcal{M}]$, $\mathcal{F}_c$ describes a class of polynomials $\mathcal{C}(\mathcal{F})$. For an instance $x \in X$, we say that x satisfies c^* when $\mathcal{F}_{c^*}(x) \geq 0$. This makes $\mathcal{C}(\mathcal{F})$ into a concept class by associating to each $f \in \mathcal{C}(\mathcal{F})$ the concept set $\{x \in X \mid f(x) \geq 0\}$.

Our goal is to PAC learn a target polynomial $\mathcal{F}_{c^*} \in \mathcal{C}(\mathcal{F})$ with parameter $c^* \in [-\mathcal{M}, \mathcal{M}]$. The instance space X is a subset of $\mathbb{R}^N$ and we assume a probability distribution $\mathcal{D}$ over it. The error of a hypothesis $\mathcal{F}_{c^h}$ is given by

$$\text{Error}(c^h) = \text{Prob}\left(\text{sign}(\mathcal{F}_{c^*}(x)) \neq \text{sign}(\mathcal{F}_{c^h}(x))\right)$$

where the probability is taken according to $\mathcal{D}$ and the sign function is defined by

$$\text{sign}(z) = \begin{cases} 1 \text{ if } z \geq 0 \\ 0 \text{ otherwise} \end{cases}$$

As usual, we will suppose that an oracle $\text{EX}_{c^*} : X \to \{0, 1\}$ is available computing $\text{EX}_{c^*}(x) = \text{sign}(\mathcal{F}_{c^*}(x))$. We finally recall that a randomized algorithm PAC learns $\mathcal{F}_{c^*}$ with error ε and confidence δ when it returns a concept c^h satisfying $\text{Error}(c^h) \leq \varepsilon$ with probability at least $1 - \delta$.

The goal of this paper is to PAC-learn $\mathcal{F}_{c^*}$ from approximate data. In the next section we briefly explain how to do so from exact data since some of the ideas used in the exact case will come handy in the approximate one.

3   Learning $\mathcal{F}_{c^*}$ under Infinite Precision

Should we be able to deal with arbitrary real numbers, the following algorithm would PAC learn $\mathcal{F}_{c^*}$.

Algorithm 1
1. **Input** $(\mathcal{F}, \varepsilon, \delta, \mathcal{M})$
2. Compute m using (1) and (2) with $s = 1$
3. Draw m random points $x^{(i)} \in \mathbb{R}^N$
4. Use the function EX_{c^*} to obtain $\text{sign}(\mathcal{F}_{c^*}(x^{(i)}))$ for $i = 1, \ldots, m$
5. From step 4, we obtain a number of polynomial inequalities in c
   $$f_1(c) \geq 0, f_2(c) < 0, f_3(c) < 0, \ldots, f_m(c) \geq 0$$
6. Find any real $c^h \in [-\mathcal{M}, \mathcal{M}]$ satisfying the system in step 5
7. Output: c^h

Here $f_i(c) = \mathcal{F}(c, x^{(i)}) \in \mathbb{R}[c]$ and the sign (≥ 0 or < 0) is given by $\text{EX}_{c^*}(x^{(i)})$. Denote by φ the system of inequalities in step 5 above.

Note that, to execute step 6, we don't need the general algorithms for solving the first-order theory over the reals mentioned in the preceding section but only an algorithm to find a point satisfying a polynomial system of inequalities in one variable (whose solution set is non-empty since c^* belongs to it). We next briefly sketch how this may be done.

First, we need to know how to isolate the roots of a single polynomial $f \in \mathbb{R}[c]$.

Definition 1. *Suppose $\xi_1 < \xi_2 < \cdots < \xi_r$ are all the real roots of f in $[-\mathcal{M}, \mathcal{M}]$. Then a finite union of intervals $\mathcal{J} = \mathcal{I}_1 \cup \cdots \cup \mathcal{I}_\ell$ is called an* **interval set for** f *iff the following conditions hold:*

1. *Every $\mathcal{I}_i$ is an interval either open or closed.*
2. *For every $i \leq r$ there exists $j \leq \ell$ such that $\xi_i \in \mathcal{I}_j$.*
3. *For every $j \leq \ell$ there exists $i \leq r$ such that $\xi_i \in \mathcal{I}_j$.*
4. *$\mathcal{I}_i \cap \mathcal{I}_j = \emptyset$ for all $j \neq i$.*

If in addition $r = \ell$ then we say that $\mathcal{J}$ is an **isolating set for** f.

To find an isolating set for f we use Sturm sequences [1] to count the number of real roots in $[-\mathcal{M}, \mathcal{M}]$ and then bisection to isolate each root of f in an interval. (There are of course more intermediate uses of Sturm sequences to ensure that each interval indeed contains exactly one root.) We can thus easily obtain an isolating set via classical means.

Refinements of isolating sets are then defined in the most natural way.

Definition 2. *Let $\mathcal{J} = \mathcal{I}_1 \cup \cdots \cup \mathcal{I}_r$ and $\mathcal{J}' = \mathcal{I}'_1 \cup \cdots \cup \mathcal{I}'_r$ be isolating sets for f, where r is the number of real roots of f in $[-\mathcal{M}, \mathcal{M}]$ and every $\mathcal{I}_i$ and $\mathcal{I}'_i$ is an interval either open or closed. The set $\mathcal{J}'$ is then said to be a* **refinement** *of $\mathcal{J}$ iff $\mathcal{I}'_i \subseteq \mathcal{I}_i$ for all i.*

By using bisection to further refine several isolating sets at once, we can solve our original semi-algebraic system φ.

More precisely, we construct isolating sets $\mathcal{J}_1, \ldots, \mathcal{J}_m$ (one set for each f_i), which are sufficiently refined to lead to a solution of our original system φ.

Definition 3. *Following the notation above, we say that $\mathcal{L}_i$ is a* **basic feasible set for** f_i *iff $f_i(c) > 0$ for all $c \in \mathcal{L}_i$. We then say that $\mathcal{L}$ is a* **basic feasible set for** φ *iff $\mathcal{L}$ is a basic feasible set for f_i for all $i \in \{1, \ldots, m\}$.*

We can construct a basic feasible set $\mathcal{L}$ for φ from a collection of basic feasible sets $\mathcal{L}_1, \ldots, \mathcal{L}_m$ for $f_1, \ldots, f_m$ simply by setting $\mathcal{L} := \mathcal{L}_1 \cap \cdots \cap \mathcal{L}_m$. If the intersection is empty, then we can try to find larger basic feasible sets so that the intersection $\mathcal{L}$ becomes nonempty.

To find a particular $\mathcal{L}_i$ we can first find an isolating set $\mathcal{J}_i$ for f_i, and then do the following: Take $\mathcal{L}_i$ to be the union of those intervals I, lying in the complement of $\mathcal{J}_i$, such that $f_i > 0$. Thus, as we refine the $\mathcal{J}_i$, the intervals making up the $\mathcal{L}_i$ become larger, and, if the solution set of φ has non-empty interior, we eventually arrive at a nonempty $\mathcal{L}$.

Let us now consider the case where we don't actually know φ for certain.

4   Learning $\mathcal{F}_{c^*}$ from Approximate Data

In the previous section, we assumed that exact data instances were used by our algorithm. Now, we assume that we can select the value of a parameter ρ such that, for $x \in \mathbb{R}^N$, each coordinate x_i of x is read with relative precision ρ. That is, instead of x_i, we read $\tilde{x}_i$ satisfying

$$|x_i - \tilde{x}_i| \leq \rho |x_i|.$$

Denote by $\tilde{f}$ the polynomial obtained from evaluating $\mathcal{F}$ at $\tilde{x}$ instead of x.

If we get $\tilde{f}_i$ instead of f_i Algorithm 1 will find a set $\mathcal{L}$ whose points all satisfy $\tilde{\varphi}$. Our question now is how to ensure that $\varphi(c)$ is true as well, when $c \in \mathcal{L}$ and $\tilde{\varphi}(c)$ is true. The rest of this paper is devoted to solve this question. Through it, we will search for a solution c^h which is not in the boundary of the solution set of φ. This makes sense since solutions in the boundary of this set are unstable under arbitrarily small perturbations. Thus, replacing f_i by $-f_i$ if necessary, we will assume that our system φ has the form

$$f_1 > 0, \dots, f_m > 0.$$

This assumption simplifies notation in the rest of this paper.

Consider $\mathcal{F} = \sum_{i\eta} a_{i\eta} c^i x^\eta$, $f \in \mathbb{Z}[c, x_1, \dots, x_N]$. Recall that we write $\mathcal{F} = \sum_i g_i(\tilde{x}) c^i$, i.e., $g_i(\tilde{x}) = \sum_\eta a_{i\eta} \tilde{x}^\eta$. Here $\eta = (\eta_1, \dots, \eta_N) \in \mathbb{N}^N$ denotes a multiindex and $|\eta| = \eta_1 + \cdots + \eta_N \leq d - i$.

Define $\sigma = (1 + \rho)^d - 1$ and assume in the sequel that $\rho \leq (3/2)^{\frac{1}{d}} - 1$ (and thus, that $\sigma \leq 1/2$).

We now propose a new algorithm for learning f_{c^*} when instances are read only approximately. Its main loop can be written as follows.

Algorithm 2
1. **Input** $(\mathcal{F}, \varepsilon, \delta, \mathcal{M})$
2. **Compute** m using (1) and (2) with $s = 1$
3. Pick $x^{(1)}, \dots, x^{(m)} \in X$ randomly and use the oracle EX_{c^*} to get $\mathrm{sign}(\mathcal{F}(c^*, x^{(i)}))$ for all $i \in \{1, \dots, m\}$
4. Set $\rho := (3/2)^{\frac{1}{d}} - 1$ and $\alpha_0 = 1/8$
5. Let $\sigma := (1 + \rho)^d - 1$
   For all $i \in \{1, \dots, m\}$
   a. For $j = 1, \dots, N$ read $x_j^{(i)}$ with precision ρ, i.e.
      such that $|\tilde{x}_j^{(i)} - x_j^{(i)}| \leq \rho |x_j^{(i)}|$
   b. **Compute** $\tilde{f}_i(c)$
6. **Call** procedure SEARCH($\tilde{f}_1, \dots, \tilde{f}_m, \sigma, \mathcal{M}$), possibly halting here.
7. **Set** $\rho := \rho^2$ and **goto** step 5

Procedure SEARCH, which will be described in Subsection 4.2, attempts to find a solution of φ knowing $\tilde{\varphi}$ and σ. If SEARCH finds a solution of φ then it halts the algorithm and returns a solution.

Let $x \in X$ and $f \in \mathbb{R}[c]$ given by

$$f = \sum_i g_i(x)c^i = \sum_{i\eta} a_{i\eta} x^\eta c^i.$$

We define $\|f\|_* = \sum_{i\eta} |a_{i\eta} x^\eta|$. Note that if $\mathcal{S} = \max_{j \leq N} |x_j|$ then $\|f\|_* \leq \|\mathcal{F}\|_1 \mathcal{S}^d$.

4.1   Newton's Method

Recall that **Newton's method for** f, starting at the point $c_0 \in \mathbb{R}$, consists of the following recurrence:

$$c_{i+1} = N_f(c_i), \quad \text{where} \quad N_f(c_i) := c_i - f(c_i)/f'(c_i).$$

We call c_i the $i^{\underline{th}}$ **Newton iterate of** c_0, and we call the sequence $(c_i)_{i=1}^\infty$ the **Newton iterates of** c_0.

Definition 4. *Consider a univariate polynomial* $f \in \mathbb{R}[x]$. *A point* $b \in \mathbb{R}$ *is an* **approximate zero of** f *iff the Newton iterates of* b *are all well-defined, and there exists* $\xi \in \mathbb{R}$ *(the* **associated zero of** b*) such that* $f(\xi) = 0$ *and*

$$|b_i - \xi| \leq \left(\frac{1}{2}\right)^{2^i - 1} |b - \xi|, \quad \text{for all} \quad i \geq 0.$$

Definition 5. *Suppose* f *is analytic and* $f'(c) \neq 0$. *Define*

$$\alpha(f, c) = \beta(f, c)\gamma(f, c)$$

where

$$\beta(f, c) = |N_f(c) - c| = \left|\frac{f(c)}{f'(c)}\right| \quad \text{and} \quad \gamma(f, c) = \sup_{k \geq 2} \left|\frac{f^{(k)}(c)}{f'(c)k!}\right|^{\frac{1}{k-1}}.$$

By definition, $\beta(f, c)$ is the length of the Newton's step for f at c. The next result provides insight on the nature of $\alpha(f, c)$.

Theorem 1. *[4] There exists a universal constant* $\alpha_0 \leq 1/8$ *such that* $\alpha(f, c) \leq \alpha_0 \Longrightarrow c$ *is an approximate zero of* f. *Also, if* ξ *is the associated zero of* c, *then* $|c - \xi| \leq 2\beta(f, c)$.

Note that Theorem 1 gives us a sufficient condition to guarrantee the existence of a root of f near a point c. In addition, it provides a bound for the distance from c to that root.

4.2   The Procedure SEARCH

We now describe the procedure SEARCH mentioned above.

Procedure SEARCH
Input $(\tilde{f}_1, \ldots, \tilde{f}_m, \sigma, \mathcal{M})$
For $i = 1$ **to** m **do**
    Compute an interval set $\mathcal{L}_i$ for $\tilde{f}_i$ such that
    all the intervals in $\mathcal{L}_i$ have length at most $2\sigma\mathcal{M}^d$
        # The above can be done via bisection and Sturm sequences #
    For $j = 1$ **to** size$(\mathcal{L}_i)$ **do**
    if, for either $z^* = a$ or $z^* = b$, $(\alpha(\tilde{f}_i, z^*) \leq \alpha_0/2$
        and $\beta(\tilde{f}_i, z^*) \leq 16d\sigma\mathcal{M}^{2d-1}\mu_*^2(\tilde{f}_i, z^*)$ **and** $420\sigma(d\mathcal{M}^{2d-1}\mu_*^3(\tilde{f}_i, z^*) \leq \alpha_0/2)$
        and $|\tilde{f}_i'(z^*)| \geq 2d\mathcal{M}^{d-2}\|\tilde{f}_i\|_*(\sigma\mathcal{M} + 64d(d-1)\sigma\mathcal{M}^{2d-1}\mu_*^2(\tilde{f}_i, z^*))$
    then
        set $h := 16d\sigma\mathcal{M}^{2d-1}\mu_*^2(\tilde{f}_i, z^*)$
        set $(c_1, c_2) := (z^* - 4h, z^* + 4h)$
            # An isolating interval for a real root of f_i #
        if $\tilde{f}_i'(z^*) < 0$ **then set** $c^h := c_1$
            else set $c^h := c_2$
        set $k := 1$
        while $k < m$ **and** $(k \neq i$ **or** $\tilde{f}_k(c^h) > 2\sigma\mathcal{M}^d\|\tilde{f}_k\|_*)$ **do**
            $k := k + 1$
        endwhile
        if $k = i$ **or** $\tilde{f}_k(c^h) > 2\sigma\mathcal{M}^d\|\tilde{f}_k\|_*$ **then** HALT **and Return** c^h
    endfor
endfor

5   Correctness and Complexity of Algorithm 2

Theorem 2. *If procedure* SEARCH *halts and returns a point* c^h, *then* c^h *is a solution of* φ.

We now consider the time complexity of Algorithm 2. A key parameter to describe this complexity is the condition of φ (see [7] for complexity and conditioning).

Consider a set of inequalities φ. The set of its solutions is a union of open intervals

$$\text{Sol}(\varphi) = \bigcup_{j=1}^{s} (\xi_{2j-1}, \xi_{2j})$$

with $-\mathcal{M} \leq \xi_1 < \xi_2 < \ldots < \xi_{2s} \leq \mathcal{M}$.

Let ξ be an extremity of one of these intervals and $i \leq m$ such that $f_i(\xi) = 0$. We define

$$K(\xi) = \min_{\substack{k \leq m \\ k \neq i}} \left\{ 1, \frac{f_k(\xi)}{\|f_k\|_*}, \frac{|f_i'(\xi)|}{\|f_i\|_*} \right\}.$$

Finally, let the *condition number of φ* be

$$\mathcal{C}(\varphi) = \min_{j \leq 2s} \frac{1}{\mathcal{K}(\xi_i)}.$$

Theorem 3. *If $\mathcal{C}(\varphi) < \infty$ then Algorithm 2 halts and returns a solution c^h. The number of iterations it performs (i.e. the number of times the procedure* SEARCH *is executed) is bounded by*

$$\left\lceil \log_2 \left(\frac{\log_2([1 + \sigma_{\min}]^{\frac{1}{d}} - 1)}{\log_2(\rho_0)} \right) \right\rceil.$$

where $\sigma_{\min} = \dfrac{\alpha_0}{840d^3 \mathcal{M}^{4d-2}[5\mathcal{C}(\varphi)]^3}$ and ρ_0 is the initial value set in the beginning of the algorithm.

The arithmetic complexity (number of arithmetic operations performed by the algorithm) is bounded by

$$\mathcal{O}[d^3 m(d \log \mathcal{M} + \log(\mathcal{C}(\varphi))) \log(d \log \mathcal{M} + \log(\mathcal{C}(\varphi)))].$$

Remark. In the previous discussion we have not considered the situation where $\mathcal{M}$ (or $-\mathcal{M}$) belongs to $\mathrm{Sol}(\varphi)$ but it is not a root of f_i for $i = 1, \ldots, m$. A simple modification of procedure SEARCH would deal with that situation. The definition of $\mathcal{C}(\varphi)$ needs to be slightly modified as well. Theorem 3 remains unchanged.

References

1. A. Akritas. *Elements of Computer Algebra with Applications*. John Wiley & Sons, 1989.
2. S. Basu, R. Pollack, and M.-F. Roy. On the combinatorial and algebraic complexity of quantifier elimination. In *35th annual IEEE Symp. on Foundations of Computer Science*, pages 632–641, 1994.
3. F. Bergadano, N.H. Bshouty, and S. Varrichio. Learning multivariate polynomials from substitutions and equivalence queries. Preprint, 1996.
4. L. Blum, F. Cucker, M. Shub, and S. Smale. *Complexity and Real Computation*. Springer-Verlag, 1998.
5. A. Blumer, A. Ehrenfeucht, D. Haussler, and M.K. Warmuth. Learnability and the Vapnik-Chervonenkis dimension. *Journal of the ACM*, 36:929–965, 1989.
6. D. Cheung. Learning real polynomials with a Turing machine. In O. Watanabe, and T. Yokomori, editor, *ALT'99*, volume 1720 of *Lect. Notes in Artificial Intelligence*, pages 231–240. Springer-Verlag, 1999.
7. F. Cucker. Real computations with fake numbers. In J. Wiedermann, P. van Emde Boas, and M. Nielsen, editors, *ICALP'99*, volume 1644 of *Lect. Notes in Comp. Sci.*, pages 55–73. Springer-Verlag, 1999.
8. P.W. Goldberg and M.R. Jerrum. Bounding the Vapnik-Chervonenkis dimension of concept classes parameterized by real numbers. *Machine Learning*, 18:131–148, 1995.

9. M.J. Kearns and U.V. Vazirani. *An Introduction to Computational Learning Theory*. The MIT Press, 1994.

10. J. Renegar. On the computational complexity and geometry of the first-order theory of the reals. Part I. *Journal of Symbolic Computation*, 13:255–299, 1992.

11. R.E. Shapire and L.M. Sellie. Learning sparse multivariate polynomials over a field with queries and counterexamples. In *6th annual ACM Symp. on Computational Learning Theory*, pages 17–26, 1993.

A Combinatorial Approach to Asymmetric Traitor Tracing

Reihaneh Safavi-Naini and Yejing Wang

School of IT and CS, University of Wollongong,
Wollongong 2522, Australia
[rei/yw17]@uow.edu.au

Abstract. To protect against illegal copying and distribution of digital objects, such as images, videos and software products, merchants can 'fingerprint' objects by embedding a distinct codeword in each copy of the object, hence allowing unique identification of the buyer. The buyer does not know where the codeword is embedded and so cannot tamper with it. However a group of dishonest buyers can compare their copies of the object, find some of the embedded bits and change them to create a pirate copy. A c-traceability scheme can identify at least one of the colluders if up to c colluders have generated a pirate copy.
In this paper we assume the merchant is not trusted and may attempt to 'frame' a buyer by embedding the buyer's codeword in a second copy of the object. We introduce a third party called the 'arbiter' who is trusted and can arbitrate between the buyer and the merchant if a dispute occurs. We describe the system as a set system and give two constructions, one based on polynomials over finite fields and the other based on orthogonal arrays, that provide protection in the above scenario.

Keywords:

Digital fingerprint, traceability schemes, cryptography, information security

1   Introduction

Traceability schemes are cryptographic systems that provide protection against illegal copying and redistribution of digital data. When a digital object is sold, the merchant embeds a distinct fingerprint in each copy hence allowing identification of the buyer. A fingerprint is a sequence of *marks* embedded in the object which is not distinguishable from the original content. The buyer does not know where the fingerprint is inserted and so cannot remove or tamper with it. The collection of all mark strings used by the merchant is the *fingerprinting code*. A copy of the object with an embedded fingerprint belonging to this code is called a *legal copy*.

   If two or more buyers collude they can compare their copies which will be identical on non-mark places and different on all mark positions that at least

D.-Z. Du et al. (Eds.): COCOON 2000, LNCS 1858, pp. 416–425, 2000.

two colluders have differing marks. Colluders then may change the marks to construct an *illegal copy*. In an illegal copy the data is intact but the mark sequence is not in the fingerprinting code.

The aim of the fingerprinting code are the following.

1. to ensure that a collusion of buyers cannot remove or tamper with the fingerprint and produce a legal copy other than their own copies.
2. to allow identification of at least one member of a colluding group who has produced an illegal copy.

We use the term *piracy* to refer to the generation of illegal copies of digital objects. In this paper we only consider binary marks.

1.1 Previous Works

Fingerprinting schemes have been widely studied in recent years. The main aim of these schemes is to discourage illegal copying of digital objects such as softwares, by allowing the merchant to identify the original buyer of a copy.

Frameproof codes are introduced by Boneh and Shaw [1], [7], and provide a *general method* of fingerprinting digital objects. The code is a collection of *binary marks* and protects against generation of illegal copies. They also defined *c*-secure codes that allow *deterministic* identification of at least one colluder if an illegal copy of the object is constructed by a collusion of up to *c* colluders. They showed [1] that *c*-secure codes for $c \geq 2$ can not exist, and proposed a code which determines the colluders with a very high probability.

Traceability schemes are introduced by Chor, Fiat and Naor [2] in the context of broadcast encryption schemes used for pay television. In their proposed scheme each authorised user has a decoder with a set of keys that uniquely determines the owner and allows him to decrypt the broadcast. Colluders attempt to construct a decoder that decodes the broadcast by combining the key sets in their decoders. A subset of keys given to an authorised user defines a fingerprinting codeword and the set of all such codewords defines the code — this is also called *key fingerprinting*. Traceability schemes allow detection of at least one of the colluders in the above scenario. Stinson and Wei in [7] defined traitor tracing schemes as set systems, studied their combinatorial properties, and showed constructions using combinatorial designs. Kurosawa and Desmedt gave a construction of one-time traceability scheme for broadcast encryption in which the decryption key will be revealed once it is used. It was shown [8] that this construction is not secure and colluders can construct a decoder that can decrypt the broadcast and is not identical to the decoder set of any of the colluders.

1.2 Asymmetric Tracing

In all the above schemes the merchant is assumed to be honest and so his knowledge of the fingerprints of all the buyers is not a security threat. This however is not always true and in particular in open environments such as the Internet assuming full trust on the merchant is not realistic. If the merchant is dishonest he may *frame* a buyer by inserting his unique code in another copy. In this case there is no way for the buyer to prove his innocence.

Asymmetric traceability schemes proposed by Pfitzmann and Schunter [5] remove honesty assumption of the merchant and in the case of piracy allow the merchant to identify the original buyer, and be able to construct a 'proof' for a judge. The fingerprinted data can only be seen by the buyer but if a merchant can capture such an object he can 'prove' the identity of the buyer to the judge. Their system requires four protocols, *key-gen*, *fing*, *identify* and *dispute* and are all polynomial-time in the security parameter k. The security requirements are: (i) a buyer should obtain a fingerprinted copy if all protocol participants honestly execute the protocols; (ii) the merchant is protected from colluding buyers who would like to generate an illegal copy of the data; and (iii) buyers must be protected from the merchant and other buyers. Although security of the merchant is against collusions of maximum size c, a buyer must remain secure against any size collusion. The proposed general scheme has computational security and uses a general 2-party protocol, a signature scheme and a one-way function.

Pfitzmann and Waidner's [6] extended this work by giving constructions that are efficient for larger collusions. In these schemes a symmetric fingerprinting scheme with unconditional security is combined with a number of computationally secure primitives such as signature schemes and one-way functions, and so the final security is computational.

Kurosawa and Desmedt [4] also constructed a computationally secure asymmetric scheme in which there are ℓ agents that cooperatively generate the key. The agents are not required in the tracing of a traitor, or providing a proof for the judge (this is done by the merchant). They also proposed the first unconditionally secure scheme which uses polynomials over finite fields. However in this scheme merchant cannot trace the traitor and the tracing is performed by the arbiter. Moreover the scheme can be subjected to an extension of the attack pointed out by Stinson and Wei [8]. Details of this attack is omitted because of space limitation.

In our model of asymmetric traceability there is a set of publicly known codewords that is used for fingerprinting objects. The merchant chooses the codeword to be inserted into an object, but only inserts part of it. The scheme consists of three procedures.

- *Fingerprinting* is performed in two steps, starting with the merchant and then completed by the arbiter.

  First, the merchant chooses a fingerprint B and then inserts his partial fingerprint $B|_M$ in the object $\mathcal{O}$ to produce $\mathcal{O}_{B|_M}$. Merchant then gives $(\mathcal{O}_{B|_M}, M, B|_M, B)$ to the arbiter, who verifies that the object $\mathcal{O}_{B|_M}$ has fingerprint $B|_M$ inserted in positions given by M. Then he inserts the rest of the fingerprint to produce $\mathcal{O}_B$ which is the fingerprinted object which will be given directly to the buyer without merchant being able to access it. Both the merchant and the arbiter store buyer B identification information in a secure database.

- *Tracing* is performed by the merchant. If the merchant finds a pirate copy of the object, it extracts his embedded fingerprint $F|_M$ and identifies one of the traitors using his information of partial fingerprints.

- *Arbitration* is required if merchant's accusation is not acceptable by the accused buyer. In this case the merchant presents $\mathcal{O}_F$ and the partial fingerprint $B_j|_M$ that identifies the buyer B_j. The accused buyer shows his ID. The arbiter will use his arbitration rule which either verifies the merchant's claims, if merchant had honestly followed his tracing algorithm and the accused buyer is a correct output of this algorithm, or reject his accusation.

In this paper we describe a model for unconditionally secure asymmetric traitor tracing schemes. We describe asymmetric traceability as a set system and give two constructions based on polynomials over finite fields and orthogonal arrays.

The paper organisation is as follows. In Section 2 we introduce the model. In Section 3 we give a necessary and sufficient condition for existence of asymmetric traceability scheme. In section 4 we show two constructions and in Section 5 conclude the paper.

2   Model

Suppose there are n buyers labelled by $B_1, B_2, \cdots, B_n$. *Hamming weight* of a binary vector is the number of non-zero components of the vector. A buyer B_i has a *fingerprint* which is a binary vector of length ℓ and fixed Hamming weight w, $B_i = (x_{i1}, x_{i2}, \cdots, x_{i\ell})$ that will be embedded in ℓ positions $P = \{p_1, p_2, \cdots, p_\ell\}$ of the object $\mathcal{O}$. The collection of all such vectors define the *fingerprint code*. The code is public but the assignment of the codewords to the buyers is secret. A *legal copy* of the object $\mathcal{O}$ has a fingerprint belonging to the fingerprint code. The position set is fixed, and for a user B_i the mark x_{ij} will be inserted in position p_j of the object. By *restriction* of a vector B to $M \subset P$, denoted by $B|_M$, we mean the vector obtained by deleting all components of B, except those in M. By *intersection* of two vectors B_i and B_j, denoted by $B_i \cap B_j$, we mean the binary ℓ-vector having a 1 in all positions where B_i and B_j are both 1, and 0 in all other positions. We use $|B|$ to denote the Hamming weight (number of ones) of B.

The full set of embedding positions P is only known to the arbiter. There is a subset $M \subset P$ of the positions known by the merchant. A buyer's fingerprint B includes two parts $B|_M$ and $B|_A$. We assume that $|B|_M| = w_1$ for all buyers. For a buyer, the merchant chooses the fingerprint B to be inserted into the object.

For a subset $C = \{i_1, i_2, \cdots, i_c\} \subseteq \{1, 2, \cdots, n\}$ a position p_j is said to be *undetectable* if $i_1, i_2, \cdots, i_c$ have the same mark on that position. Let U be the set of undetectable mark positions for C, then we define the *feasible set of C* as

$$F(C) = \{B \in \{0,1\}^\ell : B|_U = B_{i_k}|_U, \text{ for some } B_{i_k} \in C\}.$$

Let F denote a pirate fingerprint. We say F is *produced by a collusion* C if $F \in F(C)$.

Attacks

1. *Colluding groups of buyers creating an illegal object:* The colluding users C use their fingerprinted objects $\mathcal{O}_{B_{i_1}}, \cdots, \mathcal{O}_{B_{i_c}}$ to create a *pirate object* $\mathcal{O}_F$ where $F \in F(C)$.
2. *Merchant's framing of a user:* the merchant uses his knowledge of the system, that is the code and the codeword he has allocated to the buyer B , to construct another object fingerprinted with buyer B's fingerprint.

It is important to note that we do not allow merchant to use a pirate copy, $\mathcal{O}_F$ in the framing attack. Such an attack would imply collusion of merchant and buyers which is not considered in our attack model.

We will use the following tracing algorithm and arbitration rule for the merchant and the arbiter, respectively.

Merchant's tracing algorithm

If the merchant captures a fingerprinted object $\mathcal{O}_F$, he can obtain $F|_M$ which is the part of fingerprint embedded by him. Then for all i, $1 \leq i \leq n$, merchant calculates $|(F \cap B_i)_M|$, and *accuses* B_j whose allocated fingerprint restricted to M, has the largest intersection with F. That is it satisfies:

$$|(F \cap B_j)|_M = max_{1 \leq i \leq n}|(F \cap B_i)|_M$$

Arbitration rule

If an accused user B_j does not accept the accusation, the arbiter will be called.

The merchant will provide the captured object, $\mathcal{O}_F$, the accused buyer's fingerprint $B_j|_M$, and the accused buyer's identity. The arbiter will (i) extract the fingerprint F from $\mathcal{O}_F$; (ii) find the number a_F; and (iii) accepts the accusation if $a_F - |F \cap B_j| \leq s$, and rejects otherwise.

Here the number s is a fixed predetermined threshold parameter of the system and a_F is a number depending on the pirate word F.

Now we can formally define an asymmetric c-traceability scheme.

Definition 1. *An asymmetric c-traceability scheme is a binary fingerprinting code which has the following properties.*

1. *The merchant tracing algorithm can always find a traitor in any collusion group of up to c colluders. That is the accused buyer B_j by the merchant's tracing algorithm is a member of the collusion C whenever a pirate fingerprint F is produced by C with size $|C| \leq c$;*
2. *If the merchant correctly follows the tracing algorithm, his accusation will always be accepted by the arbiter; and*
3. *If the merchant accuses an innocent buyer, the arbiter always reject his accusation.*

The above definition implies that if a merchant accuses a traitor only based on his suspicion and not by following the algorithm, the arbiter may accept or reject his accusation. That is merchant's correct application of his tracing algorithm will serve as a proof for the arbiter.

3   A Combinatorial Approach

Stinson and Wei gave a combinatorial description of symmetric traceability schemes. In symmetric schemes, the merchant is assumed honest and there is no need for arbitration. The merchant embeds the fingerprints (codewords) in each object and can trace a traitor if a pirate copy is found. The tracing algorithm is similar to the one in section 2 but now the algorithm uses the whole codeword and not the part restricted to the merchant. An asymmetric scheme restricted to positions M will reduce to a symmetric scheme.

For a pirate word F, B_j is called an *exposed user* if

$$|(F \cap B_j)|_M = max_{1 \leq i \leq n} |(F \cap B_i)|_M.$$

Stinson and Wei (see [7], Definition 1.2) defined a symmetric c-traceability scheme as follows.

Consider a set $X = \{x_1, x_2, \cdots, x_\ell\}$, and a family $\mathcal{B}$ of subsets of X. A set system (X, B) is a c-traceability if (i) $|B| = w$ and (ii) whenever a pirate decoder F is produced by C and $|C| \leq c$, then any exposed user U is a member of the coalition C.

The following theorem characterises symmetric c-traceability schemes as set systems satisfying special properties.

Theorem 1. *([7]) There exists a c-traceability scheme if and only if there exists a set system $(X, \mathcal{B})$ such that $|B| = w$ for every $B \in \mathcal{B}$, with the following property*

$$
\boxed{
\begin{array}{l}
\textit{for any } d \leq c \textit{ blocks } B_1, B_2, \cdots, B_d \in \mathcal{B} \textit{ and for any } w\textit{-subset} \\
F \subseteq \bigcup_{j=1}^{d} B_j, \textit{ there does not exist a block} \\
B \in \mathcal{B} \setminus \{B_1, B_2, \cdots, B_d\} \textit{ such that } |F \cap B_j| \leq |F \cap B| \\
\textit{for } 1 \leq j \leq d.
\end{array}
}
\tag{1}
$$

Note that in this theorem it is assumed that $|F| = w$ for every pirate decoder F. It is not difficult to show that this theorem is also true after relaxing the assumption on F.

Theorem 2. *There exists a (symmetric) c-traceability scheme if and only if there exists a set system $(X, \mathcal{B})$ such that $|B| = w$ for every $B \in \mathcal{B}$, with the following property*

$$
\boxed{
\begin{array}{l}
\textit{for any } d \leq c \textit{ blocks } B_1, B_2, \cdots, B_d \in \mathcal{B} \textit{ and for any subset } F \\
\textit{such that } F \subseteq \bigcup_{j=1}^{d} B_j, \textit{ there does not exist a block} \\
B \in \mathcal{B} \setminus \{B_1, B_2, \cdots, B_d\} \textit{ such that } |F \cap B_j| \leq |F \cap B| \\
\textit{for } 1 \leq j \leq d.
\end{array}
}
\tag{2}
$$

We present the following theorem which characterises asymmetric schemes in terms of set systems with special properties. The proof is omitted due to space limitation.

Theorem 3. *There exists an asymmetric c-traceability scheme if and only if there exists a set system $(X, \mathcal{B})$ with $|B| = w$ for all $B \in \mathcal{B}$, and a number s that satisfy the following properties:*

1. *There exists a non-emptyset $M \subset X$ such that $0 < |B \cap M| = w_1 < w$ for all $B \in \mathcal{B}$, and for the set system $(M, \mathcal{B}_M)$, obtained by restricting $(X, \mathcal{B})$ to M, property (2) in theorem 2 holds;*
2. *For every F, where $\bigcap_{i=1}^{d} B_i \subseteq F \subseteq \bigcup_{i=1}^{d} B_i$ for some choice of $B_1, \cdots, B_d \in \mathcal{B}$ and $d \leq c$, there exists a number $a_F > 0$ such that*

$$|B \cap F| < a_F - s \leq |B_j \cap F|, \quad \text{for all } B \in \mathcal{B} \setminus \{B_1, B_2, \cdots, B_d\}, \qquad (3)$$

where

$$|F \cap B_j \cap M| = \max_{B \in \mathcal{B}} |F \cap B \cap M|.$$

¿From theorem 3 it follows that both the fingerprinting code and the sub-code inserted by the merchant are (symmetric) c-traceability schemes, and buyers' collusions are tracable.

4 Constructions

In this section we will construct asymmetric traceability schemes assuming the weight of any pirate fingerprint is *at least* w. In symmetric traceability schemes in [7], the weight of pirate fingerprint is exactly w.

First we give sufficient conditions for a set system to satisfy conditions of theorem 3 and hence can be used to construct an asymmetric c-traceability scheme.

Theorem 4. *Suppose there exists a set system $(X, \mathcal{B})$ and three positive integers c, μ and s such that,*

1. *For any $B \in \mathcal{B}$, $|B| = w > c^2\mu + 2cs$;*
2. *For any pair $B_i, B_j \in \mathcal{B}$, $|B_i \cap B_j| \leq \mu$;*
3. *There exists a subset $X_1 \subset X$ such that $0 < |X_1 \cap B| = s$ for any $B \in \mathcal{B}$.*

Then there exists an asymmetric c-traceability scheme.

Proof. Let $M = X \setminus X_1$. Since $|B| = w$ and $|X_1 \cap B| = s$, then for any $B \in \mathcal{B}$,

$$|B \cap M| = w - s > c^2\mu + (2c - 1)s > c^2\mu.$$

Also for any pair $B_i, B_j \in \mathcal{B}$,

$$|B_i \cap B_j \cap M| \leq |B_i \cap B_j| \leq \mu.$$

Therefore using Lemma 61 of [8] the system $(M, \mathcal{B}_M)$ forms a symmetric c-traceability scheme. Hence condition 1 of theorem 3 is satisfied.

Now suppose $F \subseteq B_1 \cup B_2 \cup \cdots \cup B_d$ and $d \le c$. Assume $|F| \ge w$. Let $x_i = |B_i \cap F|$. Then $\sum_i x_i \ge w$ and so $\max_i\{x_i\} \ge w/c = c\mu + 2s$. Let $a_F = \max_i\{x_i\}$. For any $B \in \mathcal{B} \setminus \{B_1, B_2, \cdots, B_d\}$ we have

$$a_F - |B \cap F| > c\mu + 2s - |B \cap (\cup_{i=1}^d B_i)| = c\mu + 2s - |\cup_{i=1}^d (B \cap B_i)|$$
$$\ge c\mu + 2s - c\mu = 2s. \tag{4}$$

Now let $B_j \in \mathcal{B}$ be such that

$$|(B_j \cap F)_M| = \max_{B \in \mathcal{B}} |(B \cap F)_M|.$$

Noting that

$$|B_{j_0} \cap F| - |B_j \cap F| = (|(B_j \cap F)_M| - |(B_{j_0} \cap F)_M|) + (|(B_{j_0} \cap F)_{X_1}| - |(B_j \cap F)_{X_1}|),$$

and

$$|(B_j \cap F)_M| - |(B_{j_0} \cap F)_M| \le |(B_{j_0} \cap F)_{X_1}| - |(B_j \cap F)_{X_1}| \le s.$$

We have

$$a_F - |B_j \cap F| \le 2s. \tag{5}$$

From (4) and (5) we know that for any $B \in \mathcal{B} \setminus \{B_1, B_2, \cdots, B_d\}$,

$$|B \cap F| < a_F - 2s \le |B_j \cap F|,$$

and so condition 2 of theorem 3 is satisfied.

In the following we will construct two set systems satisfying conditions of theorem 4, and hence resulting in asymmetric traceability schemes.

4.1   Construction 1

Theorem 5. *There exists an asymmetric c-traceability scheme with*

$$c = \lfloor \frac{-(q - t) + \sqrt{(q - t)^2 + tq}}{t} \rfloor.$$

Proof. Let F_q be a field of q elements, $t < q - 1$ be a positive integer, and X be the set of all points $\{(x, y) : x, y \in F_q\}$. For a polynomial $a_t x^t + \cdots + a_1 x + a_0$, $a_i \in F_q$, $0 \le i \le t$ and $a_t \ne 0$, define a block $B \subset X$ as

$$B = \{(x, y) \in X \ : \ y = a_t x^t + \cdots + a_1 x + a_0\}.$$

Let $\mathcal{B}$ be the family of all blocks defined as above. Then for any $B \in \mathcal{B}$, $|B| = q$. Since any set of $t + 1$ points $(x_1, y_1), (x_2, y_2), \cdots, (x_{t+1}, y_{t+1}) \in X$ is in at most one block, therefore $|B \cap B'| \le t$ for every pair of $B, B' \in \mathcal{B}$.

Now fix s elements $x_1, x_2, \cdots, x_s \in F_q$, where $s \geq q - t$. Let $X_1 = \{(x_i, y) : 1 \leq i \leq s, y \in F_q\}$. For any block $B \in \mathcal{B}$ we have $|X_1 \cap B| = s$. So if $q > c^2 t + 2cs$, $(X, \mathcal{B})$ satisfies conditions of theorem 4. Since $q > c^2 t + 2cs$ implies $q > c^2 t + 2c(q - t)$, we have

$$c = \lfloor \frac{-2(q-t) + \sqrt{4(q-t)^2 + 4tq}}{2t} \rfloor = \lfloor \frac{-(q-t) + \sqrt{(q-t)^2 + tq}}{t} \rfloor,$$

and the theorem is proved.

4.2   Construction 2

The second construction is based on orthogonal arrays. The following definition and theorem on the existence of orthogonal arrays can be found in [3].

Definition 2. *An orthogonal array $OA(t, k, v)$ is a $k \times v^t$ array with entries from a set of $v \geq 2$ symbols, such that in any t rows, every $t \times 1$ column vector appears exactly once.*

Theorem 6. *Let q be a prime power, and $t < q$ be an integer. Then there exists an $OA(t, q + 1, q)$.*

The following theorem gives the second construction.

Theorem 7. *Let q be a prime power, t, s be integers such that $1 < s < t < q$. Then there exists an asymmetric c-traceability scheme with $c = \lfloor \frac{-s + \sqrt{s^2 + (t-1)(q+1)}}{t-1} \rfloor$.*

Proof. Let A be an $OA(t, q + 1, q)$. Let $X = \{(x, y) : 1 \leq x \leq k, 1 \leq y \leq v\}$, $\mathcal{B}$ be the family of all $B = \{(1, y_1), (2, y_2), \cdots, (k, y_k)\}$ where $(y_1, y_2, \cdots, y_k)^T$ is a column of A. Now consider the set system $(X, \mathcal{B})$. We have $|B| = k$ for every $B \in \mathcal{B}$, and $|B_i \cap B_j| \leq t - 1$ for every pair of $B_i, B_j \in \mathcal{B}$. This latter is true because every t-tuple $(y_1, y_2, \cdots, y_t)^T$ appears once in a given t rows of the orthogonal array. Let s be an integer such that $1 < s < t$. Define $X_1 = \{(x, y) : k - s + 1 \leq x \leq k, 1 \leq y \leq v\}$. Then $|X_1 \cap B| > 0$ for every $B \in \mathcal{B}$. Also $|X_1 \cap B| = s$ for every $B \in \mathcal{B}$. If $q + 1 > c^2(t - 1) + 2cs$, we obtain a set system $(X, \mathcal{B})$, satisfying conditions of theorem 4. Therefore there exists an asymmetric c-traceability scheme with $c = \lfloor \frac{-s + \sqrt{s^2 + (t-1)(q+1)}}{t-1} \rfloor$.

5   Conclusions

In this paper we propose a model for unconditionally secure asymmetric traceability scheme in which the merchant can independently trace a traitor and as long as he correctly follows the tracing algorithm, his decision will also be approved by the arbiter. The system ensures that an innocent buyer can neither be framed nor accused unlawfully by the merchant. In the latter case merchant's accusation will always be rejected by the arbiter. We gave a characterisation of asymmetric traceable system in terms of set systems and constructed two classes of such systems.

References

1. D. Boneh and J. Shaw. Collusion-secure fingerprinting for digital data. In "Advanced in Cryptology - CRYPTO'95, Lecture Notes in Computer Science, " volume 963, pages 453-465. Springer-Velag, Berlin, Heidelberg, New York, 1995

2. B. Chor, A. Fiat, and M Naor. Tracing traitors In "Advanced in Cryptology - CRYPTO'94, Lecture Notes in Computer Science, " volume 839, pages 257-270. Springer-Velag, Berlin, Heidelberg, New York, 1994

3. C. J. Collbourn and J. H. Dinitz Eds. CRC "Handbook of Combinatorial Designs". CRC Press, 1996.

4. K. Kurosawa and Y. Desmedt. Optimum traitor tracing and asymmetric schemes. In "Advanced in Cryptology - EUROCRYPT'98, Lecture Notes in Computer Science, " volume 1462, pages 502-517. Springer-Velag, Berlin, Heidelberg, New York, 1998

5. B. Pfitzmann and M. Schunter. Asymmetric fingerprinting. In "Advanced in Cryptology - EUROCRYPT'96, Lecture Notes in Computer Science, " volume 1070, pages 84-95. Springer-Velag, Berlin, Heidelberg, New York, 1996

6. B. Pfitzmann and M. Waidner. Asymmetric fingerprinting for large collusions. In "proceedings of 4th ACM conference on computer and communications security," pages 151-160, 1997.

7. D. Stinson and R. Wei. Combinatorial properties and constructions of traceability schemes and framproof codes. "SIAM Journal on Discrete Mathematics," 11:41-53, 1998.

8. D. Stinson and R. Wei. Key preassigning traceability schemes for broadcast encryption. In "Proceedings of SAC'98, Lecture Notes in Computer Science, " volume 1556, pages 144-156. Springer-Velag, Berlin, Heidelberg, New York, 1999

Removing Complexity Assumptions from Concurrent Zero-Knowledge Proofs⋆
(Extended Abstract)

Giovanni Di Crescenzo

Telcordia Technologies Inc., 445 South Street, Morristown, NJ, 07960.
giovanni@research.telcordia.com

Abstract. Zero-knowledge proofs are a powerful tool for the construction of several types of cryptographic protocols. Recently, motivated by practical considerations, such protocols have been investigated in a concurrent and asynchronous distributed model, where protocols have been proposed relying on various synchronization assumptions and unproven complexity assumptions.

In this paper we present the first constructions of proof systems that are concurrent zero-knowledge without relying on unproven complexity assumptions. Our techniques transform a non-concurrent zero-knowledge protocol into a concurrent zero-knowledge one. They apply to large classes of languages and preserve the type of zero-knowledge: if the original protocol is computational, statistical or perfect zero-knowledge, then so is the transformed one.

1 Introduction

The notion of zero-knowledge (zk) proof systems was introduced in the seminal paper of Goldwasser, Micali and Rackoff [13]. Using a zk proof system, a prover can prove to a verifier that a certain string x is in a language L without revealing any additional information that the verifier could not compute alone. Since their introduction, zk proofs have proven to be very useful as a building block in the construction of several cryptographic protocols, as identification schemes, private function evaluation, electronic cash schemes and election schemes. Due to their importance, considerable attention has been given to the study of which adversarial settings and complexity assumptions are necessary for implementing zk protocols, the ultimate goal being that of achieving the most adversarial possible setting together with the minimal possible assumptions or none at all.

Complexity assumptions for zk proofs. Computational zk proofs were shown to be constructible first for all languages in NP [12] and then for all languages having an interactive proof system [1,14]. These proofs have interesting generality features but rely on unproven complexity assumptions, such as the existence of one-way functions. Perfect zk proofs, instead, require no unproven

---

⋆ Copyright 2000, Telcordia Technologies, Inc. All Rights Reserved.

D.-Z. Du et al. (Eds.): COCOON 2000, LNCS 1858, pp. 426–435, 2000.
© Springer-Verlag Berlin Heidelberg 2000

complexity assumption; it is known that they cannot be given for all languages in NP, unless the polynomial time hierarchy collapses [4,11]. These proofs seem to capture the intrinsic notion of zk and are notoriously harder to achieve; only a class of random self-reducible languages [13,12,17] and formula composition over them [6] are known to have perfect zk proofs. We also note that perfect zk proofs are the ones used in practical applications, because of their efficiency.

Settings for zk proofs: concurrent zk. Recently, a lot of attention has been paid to the setting where many concurrent executions of the same protocol take place, capturing practical scenarios as, for instance, the Internet. Here the zk property is much harder to achieve since an adversary can corrupt many verifiers that are executing protocols with different provers. Constructions of concurrent zk protocols in such a setting have been given in [2,9,10,16,7], relying on unproven complexity assumptions and various synchronization assumptions. A negative result on the possibility of obtaining small round complexity in this setting was showed in [15].

Our results. In this paper we present the first protocols (i.e., proof systems or arguments) that are proved to be concurrent zk *without* using unproven complexity assumptions. Our techniques transform a non-concurrent zk protocol to a concurrent one by preserving the type of zk: if the original protocol is computational or perfect zk then so is the resulting protocol. We present two techniques, the first being applicable to a larger class of languages but making stronger synchronization assumptions than the second. We note that the synchronization assumptions of our protocols are probably stronger than those made previously in the literature. However we believe that our techniques provide a crucial step towards the construction of concurrent zk proofs in the asynchronous model without complexity assumption, an important open question, and probably the ultimate in the area.

In the asynchronous model, we show some relationship between the problem of obtaining concurrent zk proofs and the well studied problem of reducing the soundness error of non-concurrent zk proofs. This relationship allows to derive, under some assumptions, lower bounds on the round complexity of concurrent zk proofs, and gives some evidence of round-optimality of our techniques.

Formal descriptions of protocols and proofs are omitted for lack of space.

2 Definitions

We present all definitions of interest: interactive protocols and interactive proof systems (introduced in [13]), zk proof systems in the two-party model (introduced in [13]), and in various multi-party models, in order of increasing generality: a so-called strongly-synchronous model (firstly considered in this paper), a so-called weakly-synchronous model (firstly considered in this paper), and the asynchronous model (introduced in [9]).

Interactive protocols and proof systems. An *interactive Turing machine* is a Turing machine with a public input tape, a public communication tape, a private random tape and a private work tape. An *interactive protocol* (A,B) is a pair of interactive Turing machines A,B sharing their public input tape and communication tape. An interactive protocol $\pi = (A, B)$ is an *interactive proof system* with soundness parameter k for the language L if B runs in polynomial time and the following two requirements are satisfied: *completeness*, stating that for any input $x \in L$, at the end of the interaction between A and B, B accepts with probability at least $1 - 2^{-|x|}$, and *soundness*, stating that for any input $x \notin L$, and any algorithm A$'$, the probability that, at the end of the interaction between A$'$ and B, B accepts is at most 2^{-k}. We say that an interactive proof system is *public-coin* if the messages of the verifier only consist of uniformly chosen bits.

Zk proof systems. We define zk proof systems in various settings, under a unified framework. Informally, a zk proof system for a language L in a certain setting is an interactive proof system for L such that any efficient adversary cannot use his power in this setting to gain any information from his view that he did not know before running the protocol. An efficient adversary is modeled as a probabilistic polynomial time algorithm. The concept of gaining no information not known before the protocol is modeled using the simulation notion put forward in [13]. The specification of the power of the adversary and of his view, given below, depend on the specific setting. In the two-party setting a prover and a verifier are intended to run a protocol for proving that $x \in L$, for some language L and some common input x. In the multi-party settings we consider a set of provers $\mathcal{P} = \{P_1, \ldots, P_q\}$ and a set of verifiers $\mathcal{V} = \{V_1, \ldots, V_q\}$, such that for $i = 1, \ldots, q$, prover P_i and verifier V_i are intended to run a protocol (for simplicity, the same protocol) for proving that $x_i \in L$, for some language L and some input x_i common to P_i and V_i.

Two-party setting. In this setting the adversary's power consists of corrupting a single verifier; namely, the adversary can impersonate such verifier and therefore use his randomness and send and receive messages on his behalf.

Multi-party strongly-synchronous setting. Here the adversary's power consists of corrupting up to all verifiers $V_1, \ldots, V_q$; namely, the adversary can impersonate all such verifiers and therefore use their randomness and send and receive messages on their behalf. In fact, we can assume wlog that he always corrupts all of them. In this setting there exists a global clock and the time measured by such clock is available to all provers, to all verifiers and to the adversary. No matter what the adversary does, however, each message takes at most a certain fixed amount of time, denoted as d, in order to arrive to its receiver, where the value d is known to all parties. The adversary is allowed to start any of the q executions at any time (i.e., not necessarily the same time); however, the number q of pairs of parties is known to all parties and cannot be changed by the adversary.

Multi-party weakly-synchronous setting. As in the previous setting, in this setting there is a global clock, the adversary can corrupt up to all verifiers and he is allowed to start any of the q executions at any time (i.e., not necessarily the same time). Contrarily to the previous setting, here there is no fixed amount of time delay d that a message can take to arrive to its destination (for instance, d

could depend on the number of messages currently sent in the system); however, we assume that at any time d is the same for all messages. Here, for instance, the adversary could at any time modify d to his advantage. Also, the absolute time measured by the global clock does not need to be available to all parties. Still, each party can measure relative time according to such clock (i.e., can measure whether d seconds have passed since the end of some other event).

Multi-party asynchronous setting. As in the previous two settings, in this setting the adversary can corrupt up to all verifiers; namely, the adversary can impersonate all such verifiers, use their randomness and send and receive messages on their behalf; moreover, the adversary is allowed to start any of the q executions at any time (i.e., not necessarily the same time). As in the weakly-synchronous setting, the global clock is not available to all parties in the system, and there is no fixed bound on the amount of time that each message takes in order to arrive to its receiver. However, contrarily to the previous two settings, the local clock of each party can measure time and at a rate possibly different from that of another party. This can be used by the adversary to arbitrarily delay the messages sent by the various verifiers that he is impersonating, and eventually create arbitrary interleavings between the concurrent executions of a certain protocol. Moreover, in this setting the number q of pairs of parties is not fixed in advance; therefore, the adversary can generate any polynomial number (on the size of the inputs) of executions of the same protocol.

We call the adversary in model X the X *model adversary* and his view, including all public inputs and all verifiers' random and communication tapes, the X *model adversary's view*. We are now ready to define zk proof systems in each of the above settings. We note that there exist three notions of zk: computational, statistical and perfect, in order of increasing strength. Here, we define the third one (although our results hold for all three).

Definition 1. Let $X \in \{$ two-party, multi-party strongly-synchronous, multi-party weakly-synchronous, and asynchronous $\}$, let 1^n be a security parameter, let q be a polynomial and let $\pi = (A, B)$ be an interactive proof system with soundness parameter k for the language L. We say that π is q-*concurrent perfect zk* in model X if for all X model probabilistic polynomial time adversaries $\mathcal{A}$, there exists an efficient algorithm $S_{\mathcal{A}}$, called the *simulator*, such that for all $x_1, \ldots, x_{q(n)} \in L$, where $|x_1| = \cdots = |x_{q(n)}| = n$, the two distributions $S_{\mathcal{A}}(x_1, \ldots, x_{q(n)})$ and $View_{\mathcal{A}}(x_1, \ldots, x_{q(n)})$ are equal, where $View_{\mathcal{A}}(x_1, \ldots, x_{q(n)})$ is the X model adversary's view.

Remarks. Variants of synchronous and asynchronous distributed models are being widely investigated in the research field of Distributed Computing; our approach of investigating these models was partially inspired by the research in this field as well. Among all these models, the most adversarial scenario is in the asynchronous model; the models considered in [2,9,10,16,7] all make some synchronization assumptions, our weakly-synchronous and strongly-synchronous models make probably more restricting synchronization assumptions (the first being less restricting than the second). Note that a protocol that is zk in a certain model is also zk in a model with stronger synchronization assumptions.

3   Concurrent Zk vs. Soundness Error Reduction

We show a relationship between the following two problems: (1) transforming a two-party zk proof system into a concurrent zk proof system in the strongly-synchronous model, and (2) reducing the soundness error of a two-party zk proof system. The latter problem, widely studied in the theory of interactive proofs and zk proofs, asks whether, given a zk proof system with soundness parameter k_1, it is possible to construct a zk proof system for the same language with soundness parameter k_2, for $k_2 > k_1$ (so that the error 2^{-k_1} decreases to 2^{-k_2}). Known techniques for such reductions are: sequential and parallel composition (namely, repeating the original protocol several times in series or in parallel, respectively). Note that these techniques use no unproven complexity assumption. Other techniques in the literature are for specific languages and use algebraic properties of such languages, or unproven complexity assumptions. Informally, our result says that a solution to problem (1) provides a solution to problem (2), and relates the round complexity of the transformations.

Theorem 1. Let k, n, q be integers > 1 and let π be a q-concurrent zk proof system for language L in the asynchronous model and with soundness parameter k. Then there exists (constructively) a two-party zk proof system π' for L and with soundness parameter k such that the following holds. If the number of rounds of π is $r(n, k, q)$, where n is the size of the common inputs, then the number of rounds of π' is $r' = r(n, 1, k)$.

In the proof, protocol π' runs k parallel executions of π, each using the same input x, and each using 1 as soundness parameter. One application of this result is that any procedure for transforming a proof system that is two-party zk into one that is zk even in the strongly-synchronous model gives a procedure for reducing the soundness error of two-party zk proof systems. Another application is that if a round-optimal procedure for reducing the soundness error of a two-party zk proof systems is known then one obtains a lower bound on the round complexity of any zk proof system in the asynchronous model for the same language. For a large class of protocols [6,8], the most round-efficient technique (assuming no unproven complexity assumptions or algebraic properties of the language are used) to reduce the error from 1 to k consists of $\Theta(k/\log k)$ sequential repetitions of $\Theta(\log k)$ parallel repetition of the atomic protocol with soundness parameter 1. As a corollary of Theorem 1, we obtain that by further assuming that such procedure is round-optimal then for such languages we obtain that any proof system with soundness parameter 1, and that is q-concurrent zk (even) in the strongly-synchronous model, must have round-complexity at least $\Omega(q/\log q)$. Conversely, a more round-efficient zk protocol in the asynchronous model would give a better soundness error reduction procedure in the two-party model. These arguments can be extended so that they apply to the two concurrent zk protocols that we will present in Section 4 and 5.

4   A Protocol for the Strongly Synchronous Model

In this section we present a technique for constructing concurrent zk proof systems in the strongly synchronous model. The technique consists of transforming, in this setting, a protocol that is two-party zk and satisfies a certain special property into a protocol that is concurrent zk. We stress that this result does not use any unproven complexity assumption and preserves the type of zk of the original proof system. We define CL1 as the class of languages having a two-party zk proof system π for which there exists an efficient simulator S running in time $c^{q(n)} \cdot t(n)$, for some constant c, some polynomial t, and on inputs of length n. Note that a language in CL1 is $(\log n)$-concurrent zk but not necessarily poly(n)-concurrent zk. We remark that class CL1 contains essentially all languages of interest. In particular, we do not know of a language that has been given a two-party zk proof system in the literature and does not belong to it.

Theorem 2.   *Let L be a language belonging to class CL1 and let π be the two-party zk proof system associated with it. Also, let q be a polynomial. Then L has a $(q(n))$-concurrent zk proof system π' for L in the strongly-synchronous model, where n is the length of the inputs. Moreover, if π is computational, statistical, or perfect zk then so is π'.*

The rest of this section is devoted to the proof of Theorem 2.

Description of the technique. Assume for simplicity that the protocol π' starts at time 0. Let $\pi = (A,B)$ be the two-party zk proof system associated with language L, let r be the number of rounds of π, let q, k be the desired concurrency and soundness parameters for π', respectively, and let $z = \lceil q(n)/(\log q(n) + \log k) \rceil$. Also, let d be the bound on the time that a message takes to arrive to its recipient, and let tm_i be time marks, such that $tm_i = i \cdot (2dr)$, for all positive integers i. We construct a protocol $\pi' = $(P,V), as follows. On input the n-bit string x, P uniformly chooses an integer $w \in \{1, \ldots, z\}$ and starts executing π, by running A's program, on input x, at the time step corresponding to the w-th time mark following the current time (i.e., if the current time is t, P will start protocol π when the current time is $(\lfloor t/2dr \rfloor + w) \cdot (2dr)$). V checks that the execution of protocol π on input x was accepting by running B's program.

Properties of protocol π'. We stress that protocol π' assumes that party P knows the current time measured by the global clock (for instance, P needs to check the value of the current time). Here we need the assumption that the global time is available to all parties, and this is one reason for which this technique does not seem to directly extend to the weakly synchronous model. The completeness and soundness of the protocol π' directly follow from the analogue properties of π. The simulator S' for π' consists of running P's program when choosing the time marks, and then simulator S for π when simulating all concurrent proofs within each time mark. It is not hard to show that the quality of the simulation of S is preserved by S'. In order to show that S' runs in expected polynomial time, we observe that the execution of S' is analogous to a variant of a classical occupancy game of randomly tossing balls into bins, and analyze this game.

A balls-into-bins game. Let n be an integer, q be a polynomial, and $wt : \mathcal{N} \to \mathcal{N}$ be a penalty function; also, assume an adversary sequentially scans a possibly infinite sequence of bins. The adversary is able to choose $q(n)$ bins $sbin_1, \ldots, sbin_{q(n)}$ in an adaptive way (namely, the choice being possibly dependent on the history of the game so far). Each time bin $sbin_i$ is chosen by the adversary, a ball is tossed into one out of the z bins immediately following bin $sbin_i$, which bin being chosen uniformly and independently from all previous random choices. Now, denoting by $nball_j$ be the number of balls in bin j, we have that the goal of the adversary is to maximize the value $val = \sum_{j \geq 1} wt(nball_j)$, while our goal is to show that for any adversary, val is poly(n). In our analysis, we will only need to consider the case in which wt is set equal to a specific function obtained from our simulation process.

Equivalence between game and simulation. The equivalence between the above described game and the simulation process comes from the following associations: each bin is a time area between any two time marks; each ball is a concurrent execution of protocol π; the action of tossing a ball into a uniformly chosen bin is equivalent to P's action of setting a begin time for the next execution of protocol π; the penalty function wt, when evaluated on $nball_j$, corresponds to the expected running time taken by S' in order to simulate the $nball_j$ concurrent executions of π starting at the j-th time mark. Moreover, we observe that by our construction all executions of π have to start at some time mark; and that by our choice of the time distance between any two consecutive time marks, each execution of π is concluded within the same time area (namely, before the next time mark). Therefore the entire simulation process can be seen as a sequence of steps, each simulating a single time area. This implies that the expected simulation time can be written as the sum of the values of the penalty function at values $nball_j$, for all j.

Analysis of the game. From the definition of function wt, we derive that: (1) $wt(0) = 0$; (2) $wt(d) \leq c^d \cdot t(n)$, for any d, for some constant c, and some polynomial t. Specifically, (1) follows from the fact that no time is spent to simulate a time area containing no execution of π to be simulated and (2) follows from our original assumptions on protocol π. Note that during the duration of the game at most $q(n)$ balls are ever tossed, and each of them can land in one out of at most $z = \lceil q(n)/(\log q(n) + \log k) \rceil$ bins. This, together with (1) and the fact that q is a polynomial, implies that there are at most a polynomial number of terms $wt(d)$ contributing a non-zero factor to the value val. Therefore, in order to show that val is bounded by a polynomial in n, it is enough to show that for each $nball_j \neq 0$, it holds that $wt(nball_j)$ is bounded by a polynomial in n. To this purpose, let us set $z_0 = 1 + 3c(\lceil \log q(n) + \log k \rceil)$, fix a bin number j and denote by E_d, for any natural number d, the event that $nball_j = d$, and, finally, by E the event that one of $E_1, \ldots, E_{z_0-1}$ happens. Then it holds that val is

$$\leq q(n)t(n) \cdot \left(\text{Prob}\,[E] \cdot (k \cdot q(n))^{3c^2} + \sum_{d=z_0}^{z} \text{Prob}\,[E_d] \cdot c^d \right)$$

$$\leq \text{poly}(kn) + \sum_{d=z_0}^{z} \binom{q(n)}{d} \left(\frac{\log q(n) + \log k}{q(n)} \right)^d \left(1 - \frac{\log q(n) + \log k}{q(n)} \right)^{q(n)-d} \cdot c^d \cdot t(n)$$

$$\leq \text{poly}(kn) + \sum_{d=z_0}^{q(n)} \left(\frac{3q(n)}{d}\right)^d \left(\frac{\log q(n) + \log k}{q(n)}\right)^d \cdot c^d \cdot t(n)$$

$$\leq \text{poly}(kn) + t(n) \cdot q(n) = \text{poly}(kn),$$

where the first inequality follows from (1) and (2) above, and the second from the expression of the distribution of the number of balls into a bin.

5   A Protocol for a Weakly Synchronous Model

In this section we present a concurrent zk proof system in a weakly synchronous model for a large class of languages. This result does not make any unproven complexity assumption and preserves the type of zk of the original proof system. We define CL2 as the class of languages having a two-party zk proof system π for which completeness holds with probability 1, soundness with probability at most $1/2 + \epsilon(n)$, for a negligible function ϵ, and with the following property. There exists a prover A', called the *lucky prover*, such that: (1) A' runs in polynomial time; (2) A' makes B accept with probability exactly equal to $1/2$ for any input x (i.e., in L or not); (3) for any probabilistic polynomial time B', the transcript generated by the original prover A when interacting with B' and the transcript generated by the lucky prover A' when interacting with B', are equally distributed whenever the latter transcript is accepting and $x \in L$. We note that all languages having a 3-round public-coin two-party perfect zk proof system belong to CL2. Also, all languages in NP (using the computational zk proof system in [3]) and all languages having an interactive proof system (using the computational zk proof system in [14]) belong to CL2.

Theorem 3. *Let L be a language belonging to class CL2 and let π be the two-party zk proof system π associated with L. Also, let q be a polynomial. Then L has a $(q(n))$-concurrent zk proof system π' in the weakly-synchronous model, where n is the length of the inputs. Moreover, if π is computational, statistical, or perfect zk then so is π'.*

Note that the result in Theorem 3 is uncomparable with the result in Theorem 2. While in the former the assumptions on the synchronicity of the model is weaker, in the latter the class of languages to which it applies is larger. The rest of this section is devoted to the proof of Theorem 3.

Description of the technique. A first step in the construction of our proof system $\pi' =$(P,V) consists of performing some modified sequential composition of the proof system $\pi =$(A,B); specifically, a composition in which P runs the program of the associated lucky prover A' and V uses algorithm B to verify that only one of the proofs is accepting (rather than all of them). More precisely, we would like to repeat these executions until P finally gives a proof (using the lucky prover) that is accepting. However note that we cannot repeat these executions too many times otherwise a cheating prover could have a too high chance of completing at least one of the several executions in which he was running the lucky prover. Therefore, we need to guarantee a final execution of π on which P cannot delay his accepting proof any further. Specifically, the final execution will be the $(\log q(n))$-th execution, where $q(n)$ is the max number of concurrent executions of the protocol that the adversary can generate, and k is the desired soundness parameter. In this final execution P runs algorithm A and V runs algorithm B. Let us call π_0' the protocol thus obtained; protocol π' is obtained by running $kq(n)$ sequential executions of π_0'.

Properties of protocol π'. Contrarily to the protocol in Section 4, the prover and the verifier of the protocol π' in this section do not use any information about the current time or about the value of the delay that a message takes to go from its sender to its receiver. For the proof of the zk requirement, however, we assume that at any time such delay is the same for all messages. We observe that the completeness property of π' directly follows from the analogue property of π. Moreover, the soundness property is proved as follows. Since π has soundness parameter 1, the probability that a prover is able to compute an accepting proof in π_0' is $\leq 1 - 2^{-\log q(n)} \leq 1 - 1/q(n)$. Since the number of independent executions of π_0' in π' is $kq(n)$, for any prover, the verifier accepts with probability $\leq (1 - 1/q(n))^{kq(n)} \leq e^{-k}$, and therefore the soundness parameter of π' is at least k. In the rest of the section we prove that the requirement of concurrent zk is satisfied by π'. The construction of the simulator S' consists of 'almost always' running the same program that is run by the prover. Specifically, in order to simulate the first $\log q(n) - 1$ executions of π in each iterations of π', the simulator S' can just run the program of the prover P, since it requires only polynomial time (recall that the prover P is running the lucky prover associated with π, which runs in polynomial time). The only difficulty arises when the simulator has to simulate the $(\log q(n))$-th execution of π inside a certain iteration of π'. Note that in this case in a real execution of π' the prover always provides an accepting proof. Here the strategy of the simulator S' will be that of repeatedly rewinding the adversary $\mathcal{A}$ up to the time right before the beginning of this particular execution, until, using the lucky prover again, S' is able to produce an accepting proof for all concurrent executions of π. Note that properties (1) and (3) in the definition of lucky prover imply that the simulation is perfect. In order to prove that S' runs in expected polynomial time, first observe that the only executions that may ever cause S' to rewind are those corresponding to a final execution of a certain iteration; then, observe that the expected number of concurrent executions of one iteration of π' that will ever have to run the final execution of π is, in fact, very small. Therefore, since the lucky prover can be run in polynomial time, we only need to show that the expected time of the simulation of all final executions in a single iteration is polynomial. To see this, observe that this expected running time is

$$\leq \sum_{d=1}^{q(n)} \binom{q(n)}{d} \left(\frac{1}{2^{\log q(n)-1}} \right)^d \left(1 - \frac{1}{2^{\log q(n)-1}} \right)^{q(n)-d} \cdot 2^d$$

$$\leq \sum_{d=1}^{q(n)} \left(\frac{3q(n)}{d} \right)^d \left(\frac{1}{2^{\log q(n)-1}} \right)^d \cdot 2^d \leq \Theta(q(n)),$$

where the first inequality is obtained as the probability that in d executions of π' a final execution of π is run, times the expected time 2^d to simulate all final executions using the lucky prover, and summing over all $d = 1, \ldots, q(n)$. We note that the number of rounds of π' is $kq(n) \log q(n)$ but can be reduced to $kq(n)$ by replacing the 'sequential repetitions' of π_0' with a 'sequential pipelined repetition' (details omitted).

Acknowledgements. Many thanks to Alfredo De Santis, Russell Impagliazzo, Rafail Ostrovsky, Giuseppe Persiano and Moti Yung for interesting discussions.

References

1. M. Ben-Or, J. Hastad, J. Kilian, O. Goldreich, S. Goldwasser, S. Micali, and P. Rogaway, *Everything Provable is Provable in Zero-Knowledge*, CRYPTO 88.
2. T. Beth and Y. Desmedt, *Identification Tokens - or: Solving the Chess Grandmaster Problem*, in Proc. of CRYPTO 90.
3. M. Blum, *How to Prove a Theorem So No One Else Can Claim It*, in Proc. of the International Congress of Mathematicians, 1986.
4. R. Boppana, J. Håstad, and S. Zachos, *Does co-NP have Short Interactive Proofs ?*, in Inf. Proc. Lett., vol. 25, May 1987.
5. I. Damgaard, *Interactive Hashing Simplifies Zero-Knowledge Protocol Design Without Complexity Assumptions*, in Proc. of CRYPTO 93.
6. A. De Santis, G. Di Crescenzo, G. Persiano and M. Yung, *On Monotone Formula Closure of SZK*, in Proc. of FOCS 94.
7. G. Di Crescenzo and R. Ostrovsky, *On Concurrent Zero-Knowledge with Pre-Processing*, in Proc. of CRYPTO 99.
8. G. Di Crescenzo, K. Sakurai and M. Yung, *On Zero-Knowledge Proofs: "From Membership to Decision"*, in Proc. of STOC 2000.
9. C. Dwork, M. Naor, and A. Sahai, *Concurrent Zero-Knowledge*, STOC 98.
10. C. Dwork and A. Sahai, *Concurrent Zero-Knowledge: Reducing the Need for Timing Constraints*, in Proc. of CRYPTO 98.
11. L. Fortnow, *The Complexity of Perfect Zero-Knowledge*, in Proc. of STOC 87.
12. O. Goldreich, S. Micali, and A. Wigderson, *Proofs that Yield Nothing but their Validity or All Languages in NP Have Zero-Knowledge Proof Systems*, in Journal of the ACM, vol. 38, n. 1, 1991.
13. S. Goldwasser, S. Micali, and C. Rackoff, *The Knowledge Complexity of Interactive Proof-Systems*, in SIAM Journal on Computing, vol. 18, n. 1, February 1989.
14. R. Impagliazzo and M. Yung, *Direct Minimum-Knowledge Computation*, in Proc. of CRYPTO 87.
15. J. Kilian, E. Petrank, and C. Rackoff, *Lower Bounds for Zero-Knowledge on the Internet*, in Proc. of FOCS 98.
16. R. Richardson and J. Kilian, *On the Concurrent Composition of Zero-Knowledge Proofs*, in Proc. of EUROCRYPT 99.
17. M. Tompa and H. Woll, *Random Self-Reducibility and Zero-Knowledge Interactive Proofs of Possession of Information*, in Proc. of FOCS 87.

One-Way Probabilistic Reversible and Quantum One-Counter Automata

Tomohiro Yamasaki, Hirotada Kobayashi, Yuuki Tokunaga, and Hiroshi Imai

Department of Information Science, Faculty of Science, University of Tokyo
7-3-1 Hongo, Bunkyo-ku, Tokyo 113-0033, Japan
{yamasaki, hirotada, tokunaga, imai}@is.s.u-tokyo.ac.jp

Abstract. Kravtsev introduced 1-way quantum 1-counter automata (1Q1CAs), and showed that several non-context-free languages can be recognized by bounded error 1Q1CAs. In this paper, we first show that all of these non-context-free languages can be also recognized by bounded error 1PR1CAs (and so 1Q1CAs). Moreover, the accepting probability of each of these 1PR1CAs is strictly greater than, or at least equal to, that of corresponding Kravtsev's original 1Q1CA. Second, we show that there exists a bounded error 1PR1CA (and so 1Q1CA) which recognizes $\{a_1^n a_2^n \cdots a_k^n\}$, for each $k \geq 2$. We also show that, in a quantum case, we can improve the accepting probability in a strict sense by using quantum interference. Third, we state the relation between 1-way deterministic 1-counter automata (1D1CAs) and 1Q1CAs. On one hand, all of above mentioned languages cannot be recognized by 1D1CAs because they are non-context-free. On the other hand, we show that a regular language $\{\{a, b\}^* a\}$ cannot be recognized by bounded error 1Q1CAs.

1 Introduction

It has been widely considered that quantum mechanism gives new great power for computation after Shor [8] showed the existence of quantum polynomial time algorithm for integer factoring problem. However, it has been still unclear why quantum computers are so powerful. In this context, it is worth considering simpler models such as finite automata.

Quantum finite automata were introduced by Moore and Crutchfield [6] and Kondacs and Watrous [3], independently. The latter showed that the class of languages recognizable by bounded error 1-way quantum finite automata (1QFAs) is properly contained in the class of regular languages. This means that 1QFAs are strictly less powerful than classical 1-way deterministic finite automata. This weakness comes from the restriction of reversibility. Since any quantum computation is performed by unitary operators and unitary operators are reversible, any transition function of quantum computation must be reversible. Ambainis and Freivalds [2] studied the characterizations of 1QFAs in more detail by comparing 1QFAs with 1-way probabilistic reversible finite automata (1PRFAs), for 1PRFAs are clearly special cases of 1QFAs. They showed that there exist languages, such as $\{a^* b^*\}$, which can be recognized by bounded error 1QFAs but

D.-Z. Du et al. (Eds.): COCOON 2000, LNCS 1858, pp. 436–446, 2000.

not by bounded error 1PRFAs. However, as we show in this paper, this situation seems different in case of automata with one counter.

Kravtsev [4] introduced 1-way quantum 1-counter automata (1Q1CAs), and showed that several non-context-free languages such as $L_1 = \{a^i ba^j ba^k \mid i=j=k\}$, $L_2 = \{a^i ba^j ba^k \mid k=i\neq j \vee k=j\neq i\}$, and $L_3 = \{a^i ba^j ba^k \mid$ exactly 2 of i,j,k are equal$\}$, can be recognized by bounded error 1Q1CAs. No clear comparisons with other automata such as 1-way deterministic 1-counter automata (1D1CAs) or 1-way probabilistic reversible 1-counter automata (1PR1CAs) were done in [4]. In this paper, we investigate the power of 1Q1CAs in comparison with 1PR1CAs and 1D1CAs.

We first show that all of these non-context-free languages can be also recognized by bounded error 1PR1CAs (and so 1Q1CAs). Moreover, the accepting probability of each of these 1PR1CAs is strictly greater than, or at least equal to, that of corresponding Kravtsev's original 1Q1CA.

Second, we show that there exists a bounded error 1PR1CA (and so 1Q1CA) which recognizes $L_{k,4} = \{a_1^* a_2^* \cdots a_k^*\}$, for each $k \geq 2$. This result is in contrast to the case of no counter shown by Ambainis and Freivalds [2]. We extend this result by showing that there exists a bounded error 1PR1CA (and so 1Q1CA) which recognizes $L_{k,5} = \{a_1^n a_2^n \cdots a_k^n\}$, for each $k \geq 2$. We also show that, in a quantum case, we can improve the accepting probability in a strict sense by using quantum interference.

Third, we state the relation between 1D1CAs and 1Q1CAs. On one hand, all of above mentioned languages cannot be recognized by 1D1CAs because they are non-context-free. On the other hand, we show that a regular language $\{\{a,b\}^* a\}$ cannot be recognized by bounded error 1Q1CAs.

2   Definitions

Definition 1 *A 1-way deterministic 1-counter automaton (1D1CA) is defined by $M = (Q, \Sigma, \delta, q_0, Q_{\mathrm{acc}}, Q_{\mathrm{rej}})$, where Q is a finite set of states, Σ is a finite input alphabet, q_0 is the initial state, $Q_{\mathrm{acc}} \subset Q$ is a set of accepting states, $Q_{\mathrm{rej}} \subset Q$ is a set of rejecting states, and $\delta : Q \times \Gamma \times S \rightarrow Q \times \{-1, 0, +1\}$ is a transition function, where $\Gamma = \Sigma \cup \{\mathrm{\cent}, \$\}$, symbol $\mathrm{\cent} \notin \Sigma$ is the left end-marker, symbol $\$ \notin \Sigma$ is the right end-marker, and $S = \{0, 1\}$.*

We assume that each 1D1CA has a counter which can contain an arbitrary integer and the counter value is 0 at the start of computation. According to the second element of δ, $-1, 0, +1$ respectively, corresponds to decrease of the counter value by 1, retainment the same and increase by 1.

Let $s = \mathrm{sign}(k)$, where k is the counter value and $\mathrm{sign}(k) = 0$ if $k = 0$, otherwise 1. We also assume that all inputs are started by $\mathrm{\cent}$ and terminated by $\$$.

The automaton starts in q_0 and reads an input w from left to right. At the ith step, it reads a symbol w_i in the state q, checks whether the counter value is 0 or not (i.e. checks s) and finds an appropriate transition $\delta(q, w_i, s) = (q', d)$. Then it updates its state to q' and the counter value according to d. The automaton accepts w if it enters the final state in Q_{acc} and rejects if it enters the final state in Q_{rej}.

Definition 2 *A 1-way reversible 1-counter automaton (1R1CA) is defined as a 1D1CA such that, for any $q \in Q$, $\sigma \in \Gamma$ and $s \in \{0,1\}$, there is at most one state $q' \in Q$ such that $\delta(q', \sigma, s) = (q, d)$.*

Definition 3 *A 1-way probabilistic 1-counter automaton (1P1CA) is defined by $M = (Q, \Sigma, \delta, q_0, Q_{\text{acc}}, Q_{\text{rej}})$, where Q, Σ, q_0, Q_{acc}, and Q_{rej} are the same as for 1D1CAs. A transition function δ is defined as $Q \times \Gamma \times S \times Q \times \{-1, 0, +1\} \to \mathbb{R}^+$, where $\Gamma, \cent, \$, $ and S are the same as for 1D1CAs. For any $q, q' \in Q, \sigma \in \Gamma, s \in \{0,1\}, d \in \{-1, 0, +1\}$, δ satisfies the following condition:*

$$\sum_{q',d} \delta(q, \sigma, s, q', d) = 1.$$

The definition of a counter remains the same as for 1D1CAs.

A language L is said recognizable by a 1P1CA with probability p if there exists a 1P1CA which accepts any input $x \in L$ with probability at least $p > 1/2$ and rejects any input $x \notin L$ with probability at least p. We may use the term "accepting probability" for denoting this probability p.

Definition 4 *A 1-way probabilistic reversible 1-counter automaton (1PR1CA) is defined as a 1P1CA such that, for any $q \in Q$, $\sigma \in \Gamma$ and $s \in \{0,1\}$, there is at most one state $q' \in Q$ such that $\delta(q', \sigma, s, q, d)$ is non-zero.*

Definition 5 *A 1-way quantum 1-counter automaton (1Q1CA) is defined by $M = (Q, \Sigma, \delta, q_0, Q_{\text{acc}}, Q_{\text{rej}})$, where Q, Σ, q_0, Q_{acc}, and Q_{rej} are the same as for 1D1CAs. A transition function δ is defined as $Q \times \Gamma \times S \times Q \times \{-1, 0, +1\} \to \mathbb{C}$, where $\Gamma, \cent, \$, $ and S are the same as for 1D1CAs. For any $q, q' \in Q, \sigma \in \Gamma, s \in \{0,1\}, d \in \{-1, 0, +1\}$, δ satisfies the following conditions:*

$$\sum_{q',d} \delta^\dagger(q_1, \sigma, s_1, q', d)\delta(q_2, \sigma, s_2, q', d) = \begin{cases} 1 & (q_1 = q_2) \\ 0 & (q_1 \neq q_2) \end{cases},$$

$$\sum_{q',d} \delta^\dagger(q_1, \sigma, s_1, q', +1)\delta(q_2, \sigma, s_2, q', 0) + \delta^\dagger(q_1, \sigma, s_1, q', 0)\delta(q_2, \sigma, s_2, q', -1) = 0,$$

$$\sum_{q',d} \delta^\dagger(q_1, \sigma, s_1, q', +1)\delta(q_2, \sigma, s_2, q', -1) = 0.$$

The definition of a counter remains the same as for 1D1CAs.

The number of configurations of a 1Q1CA on any input x of length n is precisely $(2n + 1)|Q|$, since there are $2n + 1$ possible counter value and $|Q|$ internal states. For a fixed M, let C_n denote this set of configurations.

A computation on an input x of length n corresponds to a unitary evolution in the Hilbert space $\mathcal{H}_n = l_2(C_n)$. For each $(q, k) \in C_n, q \in Q, k \in [-n, n]$, let $|q, k\rangle$ denote the basis vector in $l_2(C_n)$. A superposition of a 1Q1CA corresponds to a unit vector $\sum_{q,k} \alpha_{q,k}|q, k\rangle$, where $\alpha_{q,k} \in \mathbb{C}$ is the amplitude of $|q, k\rangle$.

A unitary operator U_σ^δ for a symbol σ on $\mathcal{H}_n$ is defined as follows:

$$U_\sigma^\delta|q,k\rangle = \sum_{q',d} \delta(q,\sigma,\text{sign}(k),q',d)|q',k+d\rangle.$$

After each transition, a state of a 1Q1CA is observed. The computational observable O corresponds to the orthogonal decomposition $l_2(C_n) = E_{\text{acc}} \oplus E_{\text{rej}} \oplus E_{\text{non}}$. The outcome of any observation will be either "accept"(E_{acc}) or "reject"(E_{rej}) or "non-halting"(E_{non}). The probability of the acceptance, rejection and non-halting at each step is equal to the sum of the squared amplitude of each basis state in new state for the corresponding subspace.

The definition of the recognizability remains the same as for 1P1CAs.

To describe concrete automata easily, we use the concept of simple 1Q1CAs. A 1Q1CA is said simple if, for any $\sigma \in \Gamma, s \in \{0,1\}$, there is a unitary operator $V_{\sigma,s}$ on $l_2(Q)$ and a counter function $D : Q \times \Gamma \to \{-1,0,+1\}$ such that

$$\delta(q,\sigma,s,q',d) = \begin{cases} \langle q'|V_{\sigma,s}|q\rangle & \text{if } D(q',\sigma) = d \\ 0 & \text{otherwise} \end{cases},$$

where $\langle q'|V_{\sigma,s}|q\rangle$ is the coefficient of $|q\rangle \in V_{\sigma,s}|q\rangle$. We also use this representation for 1D1CA, 1R1CA, and 1PR1CA.

3   Recognizability of L_1, L_2, and L_3

Kravtsev [4] showed that several non-context-free languages such as $L_1 = \{a^i ba^j ba^k \mid i=j=k\}$, $L_2 = \{a^i ba^j ba^k \mid k=i\neq j \vee k=j\neq i\}$, and $L_3 = \{a^i ba^j ba^k \mid \text{exactly 2 of } i,j,k \text{ are equal}\}$, can be recognized by bounded error 1Q1CAs. In this section, we show that all of these languages can be also recognized by bounded error 1PR1CAs. Moreover, the accepting probability of each of these 1PR1CAs is strictly greater than, or at least equal to, that of corresponding Kravtsev's original 1Q1CAs. This also indicates the existence of a 1Q1CA for each of these languages whose accepting probability is strictly greater than, or at least equal to, that of corresponding Kravtsev's original one, since a 1PR1CA is regarded as a special case of a 1Q1CA.

Let $L_{i=j} = \{a^i ba^j ba^k \mid i=j\}$ and $L_{i=(j+k)/2} = \{a^i ba^j ba^k \mid i=(j+k)/2\}$. The existence of a 1R1CA for each of these can be shown easily.

Lemma 1 *There exist 1R1CAs $M_R(L_{i=j})$, $M_R(L_{j=k})$, $M_R(L_{k=i})$ for $L_{i=j}$, $L_{j=k}$, $L_{k=i}$, respectively.*

Lemma 2 *There exist 1R1CAs $M_R(L_{i=(j+k)/2})$, $M_R(L_{j=(k+i)/2})$, $M_R(L_{k=(i+j)/2})$ for $L_{i=(j+k)/2}$, $L_{j=(k+i)/2}$, $L_{k=(i+j)/2}$, respectively.*

Proof: We show the case of $L_{i=(j+k)/2}$. Other cases of $L_{j=(k+i)/2}, L_{k=(i+j)/2}$ can be shown in similar ways.

Let the state set $Q = \{q_0, q_1, q_2, q_3, q_4, q_5, q_{\text{acc}}, q_{\text{rej}1}, q_{\text{rej}2}, q_{\text{rej}3}, q_{\text{rej}4}, q_{\text{rej}5}\}$, where q_0 is an initial state, q_{acc} is an accepting state, and $q_{\text{rej}1}, q_{\text{rej}2}, q_{\text{rej}3}, q_{\text{rej}4}$,

q_{rej5} are rejecting states. Define the transition matrices $V_{\sigma,s}$ and the counter function D of $M_R(L_{i=(j+k)/2})$ as follows:

$$V_{\math{c},0}|q_0\rangle = |q_1\rangle, \quad V_{\$,0}|q_1\rangle = |q_{\text{rej1}}\rangle, \quad V_{a,0}|q_1\rangle = |q_1\rangle, \quad V_{b,0}|q_1\rangle = |q_2\rangle,$$
$$V_{\$,0}|q_2\rangle = |q_{\text{rej2}}\rangle, \quad V_{a,0}|q_2\rangle = |q_{\text{rej2}}\rangle, \quad V_{b,0}|q_2\rangle = |q_4\rangle,$$
$$V_{\$,0}|q_4\rangle = |q_{\text{acc}}\rangle, \quad V_{a,0}|q_4\rangle = |q_{\text{rej4}}\rangle, \quad V_{b,0}|q_4\rangle = |q_{\text{rej4}}\rangle,$$

$$D(q_1, a) = +1, \quad V_{\$,1}|q_1\rangle = |q_{\text{rej1}}\rangle, \quad V_{a,1}|q_1\rangle = |q_1\rangle, \quad V_{b,1}|q_1\rangle = |q_2\rangle,$$
$$D(q_2, a) = -1, \quad V_{\$,1}|q_2\rangle = |q_{\text{rej2}}\rangle, \quad V_{a,1}|q_2\rangle = |q_3\rangle, \quad V_{b,1}|q_2\rangle = |q_4\rangle,$$
$$D(q_4, a) = -1, \quad V_{\$,1}|q_3\rangle = |q_{\text{rej3}}\rangle, \quad V_{a,1}|q_3\rangle = |q_2\rangle, \quad V_{b,1}|q_3\rangle = |q_5\rangle,$$
$$D(q, \sigma) = 0, \quad V_{\$,1}|q_4\rangle = |q_{\text{rej4}}\rangle, \quad V_{a,1}|q_4\rangle = |q_5\rangle, \quad V_{b,1}|q_4\rangle = |q_{\text{rej4}}\rangle,$$
$$\text{otherwise}, \quad V_{\$,1}|q_5\rangle = |q_{\text{rej5}}\rangle, \quad V_{a,1}|q_5\rangle = |q_4\rangle, \quad V_{b,1}|q_5\rangle = |q_{\text{rej5}}\rangle.$$

Reversibility of this automaton can be checked easily. □

Kravtsev [4] showed the recognizability of $L_1 = \{a^i b a^j b a^k \mid i = j = k\}$ with probability $1 - 1/c$ for arbitrary chosen $c \geq 3$ by a 1P1CA and a 1Q1CA. This 1P1CA for L_1 is clearly reversible, and so, L_1 is recognized by 1PR1CA with probability $1 - 1/c$.

Here we show the recognizability of $L_2 = \{a^i b a^j b a^k \mid k = i \neq j \vee k = j \neq i\}$.

Theorem 1 *There exists a 1PR1CA $M_{\text{PR}}(L_2)$ which recognizes L_2 with probability $3/5$.*

Proof: After reading the left end-marker $\math{c}$, $M_{\text{PR}}(L_2)$ enters one of the following three paths, path-1, path-2, path-3, with probability $1/4, 1/4, 1/2$, respectively.

In path-1(path-2), $M_{\text{PR}}(L_2)$ checks whether $j = k(k = i)$ or not, by behaving in the same way as $M_R(L_{j=k})(M_R(L_{k=i}))$ except for the treatment of acceptance and rejection. The input is accepted with probability $4/5$ if $j = k(k = i)$ is satisfied, while it is always rejected if $j \neq k(k \neq i)$.

In path-3, $M_{\text{PR}}(L_2)$ checks whether $i \neq (j + k)/2$ or not, by behaving in the same way as $M_R(L_{i=(j+k)/2})$ except for the treatment of acceptance and rejection. The input is accepted with probability $4/5$ if $i \neq (j + k)/2$ is satisfied, while it is always rejected if $i = (j + k)/2$.

Then the input $x \in L_2$ always satisfies the condition of path-3 and exactly one of the conditions of first two paths. Hence, $M_{\text{PR}}(L_2)$ accepts it with probability $3/5$. On the other hand, $M_{\text{PR}}(L_2)$ rejects any input $x \notin L_2$ with probability at least $3/5$. Indeed, when the input satisfies $i = j = k$, the conditions of path-1 and path-2 are satisfied while the condition of path-3 is not satisfied, hence, $M_{\text{PR}}(L_2)$ rejects it with probability $3/5$. Next, when i, j, k differ from one another, none of the conditions of path-1 and path-2 are satisfied, hence $M_{\text{PR}}(L_2)$ rejects it with probability at least $3/5$. Finally, when the input is not in the form of $a^i b a^j b a^k$, it is always rejected, obviously.

Reversibility of this automaton is clear by its construction. □

Corollary 1 *There exists a 1Q1CA $M_Q(L_2)$ which recognizes L_2 with probability $3/5$.*

Note that the accepting probability $3/5$ of this 1Q1CA $M_Q(L_2)$ for L_2 is greater than the original Kravtsev's $4/7$.

Next we show that $L_3 = \{a^i b a^j b a^k \mid$ exactly 2 of i, j, k are equal$\}$ can be recognized by 1PR1CA with bounded error.

Theorem 2 *There exists a 1PR1CA $M_{\mathrm{PR}}(L_3)$ which recognizes L_3 with probability 4/7.*

Proof: After reading the left end-marker $\cent$, $M_{\mathrm{PR}}(L_3)$ enters one of the following four paths, path-1, path-2, path-3, path-4, with probability $1/6, 1/6, 1/6, 1/2$, respectively.

In path-1(path-2)[path-3], $M_{\mathrm{PR}}(L_3)$ checks whether $i = j(j = k)[k = i]$ or not, by behaving in the same way as $M_{\mathrm{R}}(L_{i=j})(M_{\mathrm{R}}(L_{j=k}))[M_{\mathrm{R}}(L_{k=i})]$ except for the treatment of acceptance and rejection. The input is accepted with probability $6/7$ if $i = j(j = k)[k = i]$ is satisfied, while it is always rejected if $i \neq j(j \neq k)[k \neq i]$.

In path-4, $M_{\mathrm{PR}}(L_3)$ checks whether $i \neq (j + k)/2$ or not, by behaving in the same way as $M_{\mathrm{R}}(L_{i=(j+k)/2})$ except for the treatment of acceptance and rejection. The input is accepted with probability $6/7$ if $i \neq (j + k)/2$ is satisfied, while it is always rejected if $i = (j + k)/2$.

Then the input $x \in L_3$ always satisfies the condition of path-4 and exactly one of the conditions of first three paths. Hence, $M_{\mathrm{PR}}(L_3)$ accepts it with probability $4/7$. On the other hand, $M_{\mathrm{PR}}(L_3)$ rejects any input $x \notin L_3$ with probability at least $4/7$. Indeed, when the input satisfies $i = j = k$, the conditions of path-1, path-2, and path-3 are satisfied while the condition of path-4 is not satisfied, hence, $M_{\mathrm{PR}}(L_3)$ rejects it with probability at least $4/7$. Next, when i, k, j differ from one another, none of the conditions of first three paths are satisfied, hence, $M_{\mathrm{PR}}(L_3)$ rejects it with probability at least $4/7$. Finally, when the input is not in the form of $a^i b a^j b a^k$, it is always rejected, obviously.

Reversibility of this automaton is clear by its construction.   □

Corollary 2 *There exists a 1Q1CA $M_{\mathrm{Q}}(L_3)$ which recognizes L_3 with probability 4/7.*

Note that the accepting probability $4/7$ of this 1Q1CA $M_{\mathrm{Q}}(L_3)$ for L_3 is greater than the original Kravtsev's $1/2 + \epsilon$.

4   Recognizability of $L_{k,5} = \{a_1^n a_2^n \cdots a_k^n\}$

Here we show that another family of non-context-free languages $L_{k,5} = \{a_1^n a_2^n \cdots a_k^n\}$ for each fixed $k \geq 2$, is also recognizable by bounded error 1PR1CAs.

First we show that $L_{k,4} = \{a_1^* a_2^* \cdots a_k^*\}$, for each fixed $k \geq 2$, is recognizable by a 1PR1CA with bounded error.

For each $k \geq 2$, let $L_{k,i|i+1} = \{\{a_1, \ldots, a_i\}^* \{a_{i+1}, \ldots, a_k\}^*\}$ for each i, $1 \leq i \leq k - 1$.

Lemma 3 *For each $k \geq 2$, there exists a 1R1CA $M_{\mathrm{R}}(L_{k,i|i+1})$ for each $L_{k,i|i+1}$, $1 \leq i \leq k - 1$.*

Proof: Let the state set $Q = \{q_0, q_1, q_{\text{acc}}, q_{\text{rej}}\}$, where q_0 is an initial state, q_{acc} is an accepting state, and q_{rej} is a rejecting state. Define the transition matrices $V_{\sigma,s}$ and the counter function D of $M_R(L_{k,i|i+1})$ as follows:

$$V_{\mathrm{\cent},0}|q_0\rangle = |q_1\rangle, \quad V_{a_j,0}|q_1\rangle = |q_1\rangle, \ 1 \le j \le i \qquad D(q_1, a_j) = +1, \ i+1 \le j \le k$$
$$V_{a_j,1}|q_1\rangle = |q_{\text{rej}}\rangle, \ 1 \le j \le i$$
$$V_{\$,0}|q_1\rangle = |q_{\text{acc}}\rangle, \qquad\qquad\qquad\qquad D(q, \sigma) = 0, \qquad \text{otherwise.}$$
$$V_{\$,1}|q_1\rangle = |q_{\text{acc}}\rangle, \quad V_{a_j,0}|q_1\rangle = |q_1\rangle, \ i+1 \le j \le k$$
$$V_{a_j,1}|q_1\rangle = |q_1\rangle, \ i+1 \le j \le k$$

Reversibility of this automaton can be checked easily.   □

Theorem 3 *For each $k \ge 2$, there exists a 1PR1CA $M_{\text{PR}}(L_{k,4})$ for $L_{k,4}$ with probability $1/2 + 1/(4k - 6)$.*

Proof: After reading the left end-marker $\mathrm{\cent}$, one of the following $k - 1$ paths is chosen with the same probability $1/(k - 1)$.

In the ith path, $M_{\text{PR}}(L_{k,4})$ checks whether the input is in $L_{k,i|i+1}$ or not, utilizing $M_R(L_{k,i|i+1})$, for $1 \le i \le k - 1$. If the input is in $L_{k,i|i+1}$, $M_{\text{PR}}(L_{k,4})$ accepts it with probability p, while if the input is not in $L_{k,i|i+1}$, $M_{\text{PR}}(L_{k,4})$ always rejects it.

Since the input $x \in L_{k,4}$ satisfies the condition in any path, $M_{\text{PR}}(L_{k,4})$ accepts it with probability p. On the other hand, for any input $x \notin L_{k,4}$, there exists at least one path whose condition is not satisfied. Thus, the probability $M_{\text{PR}}(L_{k,4})$ is at most $p \cdot (k - 2)/(k - 1)$. Hence, if we take p such that $p \cdot (k - 2)/(k - 1) < 1/2 < p$, $M_{\text{PR}}(L_{k,4})$ recognizes $L_{k,4}$ with bounded error. To maximize the accepting probability, we solve $1 - p \cdot (k - 2)/(k - 1) = p$, which gives $p = 1/2 + 1/(4k - 6)$.

Reversibility of this automaton is clear by its construction.   □

Corollary 3 *For each $k \ge 2$, there exists a 1Q1CA $M_Q(L_{k,4})$ for $L_{k,4}$ with probability $1/2 + 1/(4k - 6)$.*

It has been known that, while there exists a 1QFA which recognizes $L_{k,4}$ with bounded error, any 1PRFA cannot recognize $L_{k,4}$ with bounded error [2]. In this point, Theorem 3 gives a contrastive result between no-counter and one-counter cases.

Before showing the recognizability of $L_{k,5}$, we prove one more lemma. Let each $L_{k,\#a_i=\#a_{i+1}} = \{x \mid (\# \text{ of } a_i \text{ in } x)=(\# \text{ of } a_{i+1} \text{ in } x)\}$ for $1 \le i \le k - 1$.

Lemma 4 *For each $k \ge 2$, there exists a 1R1CA $M_R(L_{k,\#a_i=\#a_{i+1}})$ for each $L_{k,\#a_i=\#a_{i+1}}, 1 \le i \le k - 1$.*

Proof: Let the state set $Q = \{q_0, q_1, q_{\text{acc}}, q_{\text{rej}}\}$, where q_0 is an initial state, q_{acc} is an accepting state, and q_{rej} is a rejecting state. Define the transition matrices $V_{\sigma,s}$ and the counter function D of $M_R(L_{k,\#a_i=\#a_{i+1}})$ as follows:

$$V_{\mathvarphi,0}|q_0\rangle = |q_1\rangle, \quad V_{a_l,0}|q_1\rangle = |q_1\rangle, \quad 1 \le l \le k \quad D(q_1, a_i) = +1,$$
$$V_{a_l,1}|q_1\rangle = |q_{\mathrm{rej}}\rangle, \quad 1 \le l \le k \quad D(q_1, a_{i+1}) = -1,$$
$$V_{\$,0}|q_1\rangle = |q_{\mathrm{acc}}\rangle,$$
$$V_{\$,1}|q_1\rangle = |q_{\mathrm{acc}}\rangle, \qquad\qquad\qquad D(q, \sigma) = 0, \text{ otherwise.}$$

Reversibility of this automaton can be checked easily.   □

Now we show the recognizability of $L_{k,5} = \{a_1^n a_2^n \cdots a_k^n\}$.

Theorem 4 *For each $k \ge 2$, there exists a 1PR1CA $M_{\mathrm{PR}}(L_{k,5})$ which recognizes $L_{k,5}$ with probability $1/2 + 1/(8k - 10)$.*

Proof: After reading the left end-marker ¢, one of the following $2(k-1)$ paths, path-1-1, ..., path-1-$(k-1)$, path-2-1, ..., path-2-$(k-1)$, is chosen with the same probability $1/2(k-1)$.

In each path-1-i, $M_{\mathrm{PR}}(L_{k,5})$ checks whether the input is in $L_{k,i|i+1}$ or not, utilizing $M_{\mathrm{R}}(L_{k,i|i+1})$, for $1 \le i \le k-1$. If the input is in $L_{k,i|i+1}$, $M_{\mathrm{PR}}(L_{k,5})$ accepts it with probability p, while if the input is not in $L_{k,i|i+1}$, $M_{\mathrm{PR}}(L_{k,5})$ always rejects it.

In each path-2-i, $M_{\mathrm{PR}}(L_{k,5})$ checks whether the input is in $L_{k,\#a_i=\#a_{i+1}}$ or not, utilizing $M_{\mathrm{R}}(L_{k,\#a_i=\#a_{i+1}})$, for $1 \le i \le k-1$. If the input is in $L_{k,\#a_i=\#a_{i+1}}$, $M_{\mathrm{PR}}(L_{k,5})$ accepts it with probability p, while if the input is not in $L_{k,\#a_i=\#a_{i+1}}$, $M_{\mathrm{PR}}(L_{k,5})$ always rejects it.

Since the input $x \in L_{k,5}$ satisfies the condition in any path, $M_{\mathrm{PR}}(L_{k,5})$ accepts it with probability p. On the other hand, for any input $x \notin L_{k,5}$, there exists at least one path whose condition is not satisfied. Thus, the probability $M_{\mathrm{PR}}(L_{k,5})$ accepts it is at most $p \cdot (2k-3)/(2k-2)$. Hence, if we take p such that $p \cdot (2k-3)/(2k-2) < 1/2 < p$, $M_{\mathrm{PR}}(L_{k,5})$ recognizes $L_{k,5}$ with bounded error. To maximize the accepting probability, we solve $1 - p \cdot (2k-3)/(2k-2) = p$, which gives $p = 1/2 + 1/(8k - 10)$.

Reversibility of this automaton is clear by its construction.   □

Corollary 4 *For each $k \ge 2$, there exists a 1Q1CA $M_{\mathrm{Q}}(L_{k,5})$ which recognizes $L_{k,5}$ with probability $1/2 + 1/(8k - 10)$.*

5   Improving the Accepting Probability of 1Q1CA for $L_{k,5}$

In the previous subsection, we showed that $L_{k,5} = \{a_1^n a_2^n \cdots a_k^n\}$ is recognizable by a bounded error 1PR1CA. In this section, we also show that, in a quantum case, we can improve the accepting probability in a strict sense by using quantum interference. We utilize the following result.

Theorem 5 (Ambainis et. al. [1]) $L_{k,4} = \{a_1^* a_2^* \cdots a_k^*\}$ *can be recognized by a 1QFA $M_{\mathrm{1QFA}}(L_{k,4})$ with probability p, where p is the root of $p^{(k+1)/(k-1)} + p = 1$ in the interval of $(1/2, 1)$.*

By using $M_{\mathrm{1QFA}}(L_{k,4})$, we prove the existence of a 1Q1CA which recognizes $L_{k,4}$. The following two lemmas can be shown easily.

Lemma 5 *For each $k \geq 3$, if $p^{(k+1)/(k-1)} + p = 1$, then $1/2 < p < 2/3$.*

Lemma 6 *For arbitrary $m \times m$ unitary matrices U_1, U_2, there exists an 2×2 block unitary matrix $N(U_1, U_2)$ such that*

$$N(U_1, U_2) = \frac{1}{\sqrt{2}} \underbrace{\begin{pmatrix} U_1 & * \\ U_2 & * \end{pmatrix}}_{\text{2blocks}},$$

where the blocks indicated by $*$ are determined to hold unitary of N.

Now, we prove the main theorem.

Theorem 6 *For each $k \geq 2$, $L_{k,5}$ can be recognized by a 1Q1CA with probability p, where p is the root of $p^{(k+1)/(k-1)} + p = 1$ in the interval of $(1/2, 1)$.*

Proof: By using $M_{1QFA}(L_{k,4})$, we can construct a 1Q1CA $M = (Q, \Sigma, \delta, q_1^1, Q_{acc}, Q_{rej})$ as follows. Let $Q = \{q_i^m \mid 1 \leq i \leq 3k, m = 1, 2\}$, $\Sigma = \{a_i \mid 1 \leq i \leq k\}$, $Q_{acc} = \{q_{2k}^m \mid m = 1, 2\}$, and $Q_{rej} = \{q_i^m \mid k+1 \leq i \leq 2k-1, 2k+1 \leq i \leq 3k, m = 1, 2\}$. For each $\sigma \in \Gamma$, we define the transition matrices $\{W_{\sigma,s}\}$ and the counter function D as follows:

$$W_{\mathbb{c},0} = N\left(\begin{pmatrix} V_{\mathbb{c}} & O \\ O & I_k \end{pmatrix}, \begin{pmatrix} V_{\mathbb{c}} & O \\ O & I_k \end{pmatrix}\right), \text{ for } k \geq 3, \quad W_{\mathbb{c},0} = \begin{pmatrix} V_{\mathbb{c}} & O \\ O & I_k \end{pmatrix} \oplus \begin{pmatrix} I_k & O \\ O & I_k \end{pmatrix}, \text{ for } k = 2,$$

$$W_{a_{2i-1},0} = \begin{pmatrix} V_{a_{2i-1}} & O \\ O & I_k \end{pmatrix} \oplus \begin{pmatrix} V_{a_{2i-1}} & O \\ O & I_k \end{pmatrix}, \quad W_{a_{2i-1},1} = \begin{pmatrix} O & I_{2k} \\ I_k & O \end{pmatrix} \oplus \begin{pmatrix} V_{a_{2i-1}} & O \\ O & I_k \end{pmatrix},$$

$$W_{a_{2i},0} = \begin{pmatrix} V_{a_{2i}} & O \\ O & I_k \end{pmatrix} \oplus \begin{pmatrix} V_{a_{2i}} & O \\ O & I_k \end{pmatrix}, \quad W_{a_{2i},1} = \begin{pmatrix} V_{a_{2i}} & O \\ O & I_k \end{pmatrix} \oplus \begin{pmatrix} O & I_{2k} \\ I_k & O \end{pmatrix},$$

$$W_{\$,0} = \begin{pmatrix} V_{\$} & O \\ O & I_k \end{pmatrix} \oplus \begin{pmatrix} V_{\$} & O \\ O & I_k \end{pmatrix}, \quad W_{\$,1} = \begin{pmatrix} O & I_{2k} \\ I_k & O \end{pmatrix} \oplus \begin{pmatrix} O & I_{2k} \\ I_k & O \end{pmatrix},$$

$$D(q_j^1, a_{2i-1}) = +1, \text{ for } 1 \leq j \leq k, 1 \leq i \leq \lfloor k/2 \rfloor,$$
$$D(q_j^1, a_{2i}) = -1, \text{ for } 1 \leq j \leq k, 1 \leq i \leq \lfloor k/2 \rfloor,$$
$$D(q_j^1, a_k) = 0, \text{ for } 1 \leq j \leq k, k \text{ is odd},$$
$$D(q_j^2, a_1) = 0, \text{ for } 1 \leq j \leq k,$$
$$D(q_j^2, a_{2i}) = +1, \text{ for } 1 \leq j \leq k, 1 \leq i \leq \lfloor (k-1)/2 \rfloor,$$
$$D(q_j^2, a_{2i+1}) = -1, \text{ for } 1 \leq j \leq k, 1 \leq i \leq \lfloor (k-1)/2 \rfloor,$$
$$D(q_j^2, a_k) = 0, \text{ for } 1 \leq j \leq k, k \text{ is even},$$

where each V_σ is the transition matrix of $M_{1QFA}(L_{k,4})$ and the columns of the transition matrices correspond to the states in order of $q_1^1, q_2^1, \ldots, q_k^1, q_1^2, q_2^2, \ldots, q_k^2$ (i.e. the order of the basis states is $q_1^1, q_2^1, \ldots, q_k^1, q_1^2, q_2^2, \ldots, q_k^2$). Let δ be defined in the manner described in Section 2.

If the input string is of the form $a_1^n a_2^n \ldots a_k^n$, in each of two paths, the input is accepted. Thus, the probability of accepting is $(p/2) \cdot 2 = p$.

If $k = 2$, the input string is of the form $a_1^{m_1} a_2^{m_2}$, and $m_1 \neq m_2$, in the first path, the input string is rejected and the states in the second path are never entered. Thus, the input is always rejected.

If $k \geq 3$, the input string is of the form $a_1^{m_1} a_2^{m_2} \ldots a_k^{m_k}$, and there exists at least one pair of (i, j) such that $m_i \neq m_j$, in at least one of two paths, the counter value is not 0 upon reading the right end-marker. Thus, the probability of accepting is at most $(p/2) \cdot 1 = p/2$. By Lemma 5, the probability of rejecting is at least $1 - p/2 > 1 - (2/3) \cdot (1/2) = 2/3 > p$.

Finally, if the input string is not of the form $a_1^* a_2^* \ldots a_k^*$, in each of two paths, the input string is rejected with probability at least p, since each path is equivalent to $M_{1\mathrm{QFA}}(L_{k,4})$ when the counter is left out of consideration. Therefore, the probability of rejecting is at least p. □

Proposition 1 *The accepting probability p of M is greater than $1/2 + 1/(8k - 10)$, the accepting probability of $M_Q(L_{k,5})$.*

Proof: Omitted. □

6   Relation between 1D1CAs and 1Q1CAs

As we have seen in Section 3, 4, and 5, some non-context-free languages can be recognized by bounded error 1Q1CAs. It is clear that 1D1CAs cannot recognize any non-context-free languages, since 1D1CAs are special cases of 1-way pushdown automata. This indicates the strength of 1Q1CAs. Conversely, we present the weakness of 1Q1CAs by showing that there is a regular language which can be recognized by a 1D1CA but not by a 1Q1CA with bounded error.

Theorem 7 *The language $\{\{a, b\}^* a\}$ cannot be recognized by a 1Q1CA with bounded error.*

Proof: Nayak [7] showed that, for each fixed $n \geq 0$, any general 1-way QFA recognizing $\{wa \mid w \in \{a, b\}^*, |w| \leq n\}$ must have $2^{\Omega(n)}$ basis states. Thus a 1Q1CA for $\{\{a, b\}^* a\}$ should have at least $2^{\Omega(n)}$ quantum basis states if the input length is n. However, the number of basis states of a 1Q1CA for a language of length n has precisely $(2n + 1)|Q|$. Since $(2n + 1)|Q| < 2^{\Omega(n)}$ for sufficiently large n, it proves the theorem. □

7   Conclusions and Open Problems

In this paper, we proved that there are non-context-free languages which can be recognized by 1PR1CAs and 1Q1CAs, but cannot be recognized by 1D1CAs. We also showed that there is a regular language which can be recognized by a 1D1CA, but cannot be recognized by a 1Q1CA.

One interesting question is what languages are recognizable by 1Q1CAs but not by 1PR1CAs. Similarly, what are the languages recognizable by 1Q1CAs but not by 1P1CAs?

Another question is concerning to a 2-counter case. It is known that a 2-way deterministic 2-counter automaton can simulate a deterministic Turing machine [5]. How about the power of 2-way quantum 2-counter automata, or 2-way quantum 1-counter automata?

References

[1] A. Ambainis, R. Bonner, R. Freivalds, and A. Ķikusts. Probabilities to accept languages by quantum finite automata. In *Proceedings of the 5th Annual International Conference on Computing and Combinatorics (COCOON'99), Lecture Notes in Computer Science*, volume 1627, pages 174–183, 1999.

[2] A. Ambainis and R. Freivalds. 1-way quantum finite automata: Strengths, weakness and generalizations. In *Proceedings of the 39th Annual Symposium on Foundation of Computer Science*, pages 332–341, 1998.

[3] A. Kondacs and J. Watrous. On the Power of Quantum Finite State Automata. In *Proceedings of the 38th Annual Symposium on Foundation of Computer Science*, pages 66–75, 1997.

[4] M. Kravtsev. Quantum finite one-counter automata. In *Proceedings of the 26th Conference on Current Trends in Theory and Practice of Informatics (SOFSEM'99), Lecture Notes in Computer Science*, volume 1725, pages 431–440, 1999.

[5] M. L. Minsky. Recursive unsolvability of post's problem of 'tag' and other topics in the theory of turing machines. *Annals of Math.*, 74:3:437–455, 1961.

[6] C. Moore and J. Crutchfield. Quantum automata and quantum grammars. Technical Report 97-07-02, Santa-Fe Institute Working Paper, 1997. Also available at http://xxx.lanl.gov/archive/quant-ph/9707031.

[7] A. Nayak. Optimal lower bounds for quantum automata and random access codes. In *Proceedings of the 40th Annual Symposium on Foundation of Computer Science*, pages 369–376, 1999.

[8] P. Shor. Algorithms for quantum computation: Discrete log and factoring. In *Proceedings of the 35th Annual Symposium on Foundation of Computer Science*, pages 56–65, 1994.

Similarity Enrichment in Image Compression through Weighted Finite Automata

Zhuhan Jiang[1], Bruce Litow[2], and Olivier de Vel[3]

[1] School of Mathematical and Computer Sciences, University of New England,
Armidale NSW 2351, Australia
zhuhan@mcs.une.edu.au
[2] School of Information Technology, James Cook University,
Townsville, QLD 4811, Australia
bruce@cs.jcu.edu.au
[3] Information Technology Division, DSTO, PO Box 1500, Salisbury SA 5108,
Australia
olivier.devel@dsto.defence.gov.au

Abstract. We propose and study in details a similarity enrichment scheme for the application to the image compression through the extension of the weighted finite automata (WFA). We then develop a mechanism with which rich families of legitimate similarity images can be systematically created so as to reduce the overall WFA size, leading to an eventual better WFA-based compression performance. A number of desirable properties, including WFA of minimum states, have been established for a class of packed WFA. Moreover, a codec based on a special extended WFA is implemented to exemplify explicitly the performance gain due to extended WFA under otherwise the same conditions.

1 Introduction

There are many different compression schemes each of which concentrates on often specific image or compression characteristics. For instance, a wavelet based compression scheme makes use of the multiresolution analysis, see e.g. [1] and the references there, while the JPEG standard hinges on the signal superposition of the Fourier harmonics. Alternatively, similarities among given images and their subimages may be explored to decorrelate the image components for the perceived compression purpose. In this regard, weighted finite automata (WFA) have been successfully applied to both image and video compressions, see e.g. [2,3,4]. These WFA are basically composed of subimages as the states, and the linear expansions of the state images in terms of themselves and the rest of state images. For a given image under consideration, a typical inferred WFA will select a collection of subimages to form the WFA states in such a way that the given original image can be reconstructed, often as a close approximation, from the inferred WFA. Such a WFA, when further encoded, may then be regarded as a compressed image. The main objective of this work is therefore to propose a similarity enrichment scheme that will reduce the overall size of an inferred

D.-Z. Du et al. (Eds.): COCOON 2000, LNCS 1858, pp. 447–456, 2000.
© Springer-Verlag Berlin Heidelberg 2000

WFA, to set up the basic structure and develop interrelated properties of the extended WFA, and to establish a mechanism that determines systematically what type of similarity enrichment is theoretically legitimate for a WFA based compression. In fact we shall also devise a simplified yet still fully-fledged image compression codec based on a newly extended WFA, and demonstrate explicitly that, under otherwise the same conditions, a codec based on similarity enriched WFA will outperform those without such an enrichment.

Throughout this work we shall follow the convention and assume that all images are square images, and that all WFA based compressors are lossy compressors. We recall that an image $\mathbf{x}$ of the resolution of $n \times n$ pixels is mathematically an $n \times n$ matrix whose (i, j)-th element $x_{i,j}$ takes a real value. For any image of $2^d \times 2^d$ pixels, we define its *multiresolution representation* $\rho(\mathbf{x})$ by $\rho(\mathbf{x}) = \{\rho_D(\mathbf{x}) : D \in \mathbb{N}\}$, where $\mathbb{N} = \{0, 1, 2, 3, ...\}$ and $\rho_D(\mathbf{x})$ denotes the same image at the resolution of $2^D \times 2^D$ pixels. In fact we will identify any image $\mathbf{x}$ with its multiresolution representation $\rho(\mathbf{x})$ unless otherwise stated.

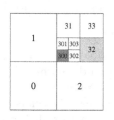

In order to address pixels of all resolutions in $\rho(\mathbf{x})$, we follow the convention in [2] and divide $\mathbf{x}$ into 4 equal-sized quadrants, numbering them clockwise from the top left by 1, 3, 2 and 0 respectively. These numbers are the *addresses* of the corresponding quadrant subimages. Let ϵ be an empty string and $Q = \{0, 1, 2, 3\}$ be the set of alphabet. Then $\sigma_a(\mathbf{x})$ with $a \in Q$ will denote all four quadrant subimages of $\mathbf{x}$. Furthermore the subimage defined by

Fig. 1. Subimages addressable via a quadtree

$$\sigma_{a_1 a_2 \cdots a_k}(\mathbf{x}) = \sigma_{a_k}(\sigma_{a_{k-1}}(\cdots \sigma_{a_2}(\sigma_{a_1}(\mathbf{x})) \cdots)) \quad (1)$$

for $a_1, a_2, ..., a_k \in Q$ will thus represent a_k-th quadrant of a_{k-1}-th quadrant of $\cdots$ of a_2-th quadrant of a_1-th quadrant of $\mathbf{x}$. For instance, $\sigma_{32}(\mathbf{x})$ represents the larger shaded block in Fig. 1 and $\sigma_{300}(\mathbf{x})$, the smaller shaded block.

Let the set of quadrant addresses be given by Q^* where $Q^* = \bigcup_{m=0}^{\infty} Q^m$ with $Q^0 = \{\epsilon\}$ and $Q^m = \{a_1 \cdots a_m : a_1, ..., a_m \in Q\}$ for $m \geq 1$. If we define a *multiresolution function* Γ as a mapping $\Gamma : Q^* \to \mathbb{R}$ satisfying the area-preserving condition $\Gamma(\omega) = \frac{1}{|Q|} \sum_{a \in Q} \Gamma(\omega a)$, where $|Q|$ is the cardinality of Q and $\mathbb{R}$ denotes the set of all real numbers, then each multiresolution image $\rho(\mathbf{x})$ corresponds to a unique multiresolution function $\Gamma_{\mathbf{x}} : Q^* \to \mathbb{R}$ defined by $\Gamma_{\mathbf{x}}(\omega) = \sigma_{\omega}(\mathbf{x})$. In fact all images of finite resolution will be automatically promoted to multiresolution images. For simplicity we shall always denote $\sigma_{\omega}(\mathbf{x})$ for any $\omega \in Q^*$ by $\mathbf{x}^{\omega}$ when no confusion is to occur, typically when $\mathbf{x}$ is in bold face, and denote by $\mu(\mathbf{x})$ the average intensity of the image $\mathbf{x}$.

There exist so far two forms of WFA that have been applied to image compression. The first was adopted by Culik et al [2,5] while the second was due to Hafner [3,4]. For convenience, Culik et al's WFA will be referred to as a *linear* WFA while Hafner's (m)-WFA will be termed a *hierarchic* WFA due to its hierarchic nature. We note that a linear WFA may also be regarded as a generalised stochastic automaton [6], and can be used to represent real functions [7] and wavelet coefficients [8], to process images [9], as well as to recognise speech [10].

In a WFA based image compressor, the design of an *inference algorithm* plays an important role. An inference algorithm will typically partition a given image into a collection of *range blocks*, i.e. subimages of the original given image, then find proper *domain blocks* to best approximate the corresponding range blocks. In general, the availability of a large number of domain blocks will result in fewer number states when encoding a given image. This will in turn lead to better compression performance. Some initial but inconclusive attempts to engage more domain blocks have already been made in [11] for bi-level images. The extra blocks there were in the form of certain transformations such as some special rotations, or in the form of some unions of such domain blocks. Due to the lack of theoretical justification there, it is not obvious whether and what extra image blocks can be *legitimately* added to the pool of domain images. Hence one of the main objectives of this work is to provide a sound theoretical background on which one may readily decide how *additional* domain images may be artificially generated by the existing ones, resulting in a similarity enriched pool of domain images.

This paper is organised as follows. We first in the next section propose and study two types of important mappings that are indispensible to a proper extension of WFA. They are the resolution-wise and resolution-driven mappings, and are formulated in practically the broadest possible sense. Section 3 then sets out to the actual extension of WFA, its associated graph representation as well as the unification of the existing different forms of WFA. A number of relevant properties and methods, such as the determination of minimum states for an extended WFA, have also been established. Finally in section 4, the benefits of the proposed similarity enrichment are demonstrated on the compression of the Lena image through a complete codec based on a specific extended WFA.

We note that the important results in this work are often summarised in the form of theorems. Although all proofs are fairly short, they are not explicitly presented here due to the lack of space. Interested readers however may consult [12] for further details.

2   Resolution-wise and Resolution-Driven Mappings

A fundamental question related to the generalisation of WFA is to find what kind of similarity enrichment will preserve the original WFA's crucial features that underpin their usefulness in the image compression. The resolution-driven and resolution-wise mappings are therefore specifically designed for this purpose and for their broadness. A *resolution-driven mapping* f is a mapping that maps a finite sequence of multiresolution images $\{\mathbf{x}_i\}$ into another multiresolution image $\mathbf{x}$, satisfying the following invariance condition: for any $k \in \mathbb{N}$, if $\mathbf{x}_i$ and $\mathbf{y}_i$ are indistinguishable at the resolution of $2^k \times 2^k$ pixels for all i, then image $\mathbf{x} = f(\{\mathbf{x}_i\})$ and image $\mathbf{y} = f(\{\mathbf{y}_i\})$ are also indistinguishable at the same resolution. A resolution-driven mapping is said to be a *resolution-wise* mapping if the resulting image $\mathbf{x}$ at any given resolution is completely determined through f by the images $\{\mathbf{x}_i\}$ at the *same* resolution.

Let $\mathfrak{F}$ and $\mathfrak{G}$ be the set of resolution-wise mappings and the set of resolution-driven mappings respectively, and let $\mathfrak{I}$ be the set of all *invertible* operators in $\mathfrak{F}$. We will always use the addressing scheme (1) associated with $Q = \{0, 1, 2, 3\}$ unless otherwise stated. Then the following theorem gives a rich family of resolution-wise mappings.

Theorem 1. *Suppose a mapping $\mathfrak{h}$ permutes pixels at all resolutions. That is, for any multiresolution image $\mathbf{x}$, the multiresolution image $\mathfrak{h}(\mathbf{x})$ at any resolution is a pixel permutation of $\mathbf{x}$ at the same resolution. If every addressible block in the form of $\mathbf{x}^\omega$ at any resolution will always be mapped onto an addressible block $\mathfrak{h}(\mathbf{x})^{\omega'}$ for some $\omega' \in Q^*$ in the mapped image, then the mapping $\mathfrak{h}$ is an invertible resolution-wise mapping, i.e. $\mathfrak{h} \in \mathfrak{I}$.*

A straightforward corollary is that $\mathfrak{I}$ contains the horizontal mirror mapping $\mathfrak{m}$, the diagonal mirror mapping $\mathfrak{d}$, the anticlockwise rotation $\mathfrak{r}$ by $90°$, as well as the permutation of the $|Q|$ quadrants. Obviously $\mathfrak{I} \subset \mathfrak{F} \subset \mathfrak{G}$, identity mapping $\mathbf{1} \in \mathfrak{I}$, and $\mathfrak{I}$ is a group with respect to the "$\cdot$" product, i.e. the composition of functions. Moreover $\mathfrak{F}$ and $\mathfrak{G}$ are both linear spaces. We also note that, for any $\omega \in Q^*$, the zoom-in operator $\mathfrak{z}^\omega$ defined by $\mathfrak{z}^\omega(\mathbf{x}) = \mathbf{x}^\omega$ for all $\mathbf{x}$ belongs to $\mathfrak{G}$.

To understand more about $\mathfrak{I}$, we first introduce $\mathbb{P}$, the group of all permutations of Q's elements. A typical permutation $p \in \mathbb{P}$ can be written as $p = \begin{pmatrix} 0 & 1 & 2 & 3 \\ n_0 & n_1 & n_2 & n_3 \end{pmatrix}$ which means symbol $i \in Q$ will be permuted to symbol $n_i \in Q$. As a convention, every permutation of k elements can be represented by a product of some disjoint cycles of the form $(n_1, n_2, ..., n_m)$, implying that n_1 will be permuted to n_2 and n_2 will be permuted to n_3 and so on and that n_m will be permuted back to n_1. Any elements that are not in any cycles will thus not be permuted at all. Hence for instance the permutation $\begin{pmatrix} 0 & 1 & 2 & 3 & 4 & 5 \\ 4 & 3 & 2 & 1 & 5 & 0 \end{pmatrix}$ can be written as $(1, 3)(0, 4, 5)$, or equivalently $(0, 4, 5)(1, 3)(2)$ because the "(2)" there is redundant.

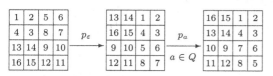

Fig. 2. A resolution-wise mapping in action

Let $\mathfrak{p} = \{p_\omega\}_{\omega \in Q^*}$ be a sequence of permutations $p_\omega \in \mathbb{P}$, then $\mathfrak{p}$ represents an invertible resolution-wise mapping in the following sense: for any multiresolution image $\mathbf{x}$, the mapped multiresolution image $\mathfrak{p}(\mathbf{x})$ at any resolution of $2^k \times 2^k$ pixels is defined by, for $m = 0$ to $k - 1$ in sequence, permuting the $|Q|$ quadrants of the subimage $\mathbf{x}^\omega$ according to p_ω for all $\omega \in Q^m$. The horizontal mirror mapping $\mathfrak{m}$ thus corresponds to $\{p_\omega\}$ with $p_\omega = (0, 2)(1, 3)$ for *all* $\omega \in Q^*$. As another example, let $\mathfrak{p} = \{p_\omega\}$ with $p_\varepsilon = (0, 1, 3, 2)$, $p_0 = (0, 2)(1, 3)$, $p_1 = (0, 1)(2, 3)$, $p_2 = (0, 3)$, $p_3 = (1, 2)$ and $p_\omega = (0)$ for all $|\omega| > 1$. Then the mapping process of $\mathfrak{p}$ is shown in Fig. 2 in which the leftmost image is gradually transformed to the rightmost resulting image. Additional properties are summarised in the next theorem.

Theorem 2. *Let* $\mathfrak{P} = \bigcup_{m=0}^{\infty} \mathfrak{P}_m$ *and* $\mathfrak{P}_m = \{\, \mathfrak{p} \in \mathfrak{J} : \mathfrak{p} = \{p_\omega\}_{\omega \in Q^*} \,$ *such that* $p_{\omega'} = p_{\omega''} \;\; \forall \omega', \omega'' \in Q^{m'}$ *with* $m' \geq m \,\}$, *and let* m *be any non-negative integer. Then*

(i) $\mathfrak{P}_m \subset \mathfrak{P}_{m+1} \subset \mathfrak{P} \subset \mathfrak{J} \subset \mathfrak{F} \subset \mathfrak{G}$.

(ii)   *For any* $\mathfrak{p} \in \mathfrak{P}_m$, *its inverse,* $\mathfrak{p}^{-1}$, *is also in* $\mathfrak{P}_m$.

(iii)   *For any* $\mathfrak{p}, \mathfrak{q} \in \mathfrak{P}_m$, *the composition* $\mathfrak{p} \cdot \mathfrak{q} \equiv \mathfrak{p}(\mathfrak{q})$ *is again in* $\mathfrak{P}_m$.

(iv) $\mathfrak{P}_m$ *is a finite group under the mapping composition.*

(v)   *For any* $\mathfrak{p} = \{p_\omega\}_{\omega \in Q^*} \in \mathfrak{P}_m$ *and any zoom-in operator* $\mathfrak{z}^a$ *with* $a \in Q$, *we set* $a' = p_\varepsilon(a) \in Q$, $\mathfrak{q} = \{p_{a'\omega}\}_{\omega \in Q^*}$, *and* $\mathfrak{P}_{-1} \equiv \mathfrak{P}_0$ *in the case of* $m = 0$. *Then* $\mathfrak{q} \in \mathfrak{P}_{m-1}$ *and* $\mathfrak{z}^{a'}\mathfrak{p} = \mathfrak{q}\mathfrak{z}^a$. *Moreover the mapping* $a \to a'$ *on* Q *is one-to-one and onto.*

For easy reference, a resolution-wise mapping will be called *permutative* if it belongs to $\mathfrak{P}$. Fig. 3 exemplifies some of the typical operators in $\mathfrak{G}$: $\mathfrak{m}$ is the horizontal mirror, $\mathfrak{d}$ is the diagonal mirror, $\mathfrak{r}$ is an anticlockwise rotation, $(0,3)$ is a quadrant permutation/exchange and $\mathfrak{z}^3$ is a zoom-in operation into quadrant 3. In fact, $\mathfrak{m}$, $\mathfrak{d}$ and $\mathfrak{r}$ all belong to $\mathfrak{P}_0$. Moreover we can conclude from the properties of $\mathfrak{G}$ that the image negation is for instance also a resolution-wise mapping. We note that the function space $\mathfrak{F}$ is of great importance as is manifested in the following theorem.

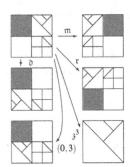

Fig. 3.   Some typical resolution-driven mappings

Theorem 3. *Let* $f_i^a \in \mathfrak{F}$ *and* $\beta_i \in \mathbb{R}$ *for all* $i \in \mathcal{I}$ *and* $a \in Q$. *If*

$$\sum_{q \in Q} f_i^a(\{\beta_j\}) = |Q| \cdot \beta_i, \qquad i \in \mathcal{I}, \tag{2}$$

then the images $\mathbf{x}_i^{(k)}$ *of* $2^k \times 2^k$ *pixel resolution determined iteratively by*

$$\mathbf{x}_i^{(0)} = \beta_i, \qquad \left[\mathbf{x}_i^{(1)}\right]^a = f_i^a(\{\mathbf{x}_j^{(0)}\}), \qquad \cdots$$

$$\left[\mathbf{x}_i^{(m)}\right]^a = f_i^a(\{\mathbf{x}_j^{(m-1)}\}), \qquad \left[\mathbf{x}_i^{(m+1)}\right]^a = f_i^a(\{\mathbf{x}_j^{(m)}\}), \qquad \cdots \tag{3}$$

uniquely define for each $i \in \mathcal{I}$ *a multiresolution image* $\mathbf{x}_i$ *of intensity* $\mu(\mathbf{x}_i) = \beta_i$. *The representation of* $\mathbf{x}_i$ *at the resolution of* $2^k \times 2^k$ *pixels is given by* $\mathbf{x}_i^{(k)}$ *precisely for all* $i \in \mathcal{I}$. *Moreover the* $\mathbf{x}_i$ *will satisfy*

$$\mathbf{x}_i^a = f_i^a(\{\mathbf{x}_j\}), \qquad f_i^a \in \mathfrak{F}, \quad i \in \mathcal{I}. \tag{4}$$

Conversely if (4) holds for all $i \in \mathcal{I}$, *then (2) must be valid for* $\beta_i = \mu(\mathbf{x}_i)$.

In the case that corresponds eventually to a linear WFA, the results can be represented in a very concise matrix, see [12] for details. To conclude this section, we note that the symbols such as $\mathfrak{P}_m$, $\mathfrak{J}$, $\mathfrak{F}$, $\mathfrak{G}$ and $\mathfrak{z}^a$ will always carry their current definitions, and that resolution-driven mappings can be regarded as the similarity generating or enriching operators.

3   Extended Weighted Finite Automata

The objective here is to show how resolution-wise or resolution-driven mappings come into play for the representation, and hence the compression, of images. We shall first propose an extended form of WFA and its graph representation, and then devise a compression scheme based on such an extended WFA. In fact we shall establish a broad and unified theory on the WFA for the purpose of image compression, including the creation of extended WFA of minimum number of states.

3.1   Definition and Graph Representation of Extended WFA

Suppose a finite sequence of multiresolution images $\{\mathbf{x}_k\}_{1 \leq k \leq n}$ satisfy

$$\mathbf{x}_i^a = f_i^a(\mathbf{x}_1, ..., \mathbf{x}_r), \quad \mathbf{x}_j = g_j(\mathbf{x}_1, ..., \mathbf{x}_{j-1}), \quad f_i^a \in \mathfrak{F}, \ g_j \in \mathfrak{G},$$

for $1 \leq i \leq r$ and $r < j \leq n$. Then all images $\mathbf{x}_k$ can be constructed at any resolution from the image intensities $\beta_k = \mu(\mathbf{x}_k)$. For all practical purposes we however consider the following canonical form

$$\mathbf{x}_i^a = \sum_{k=1}^{r} f_{i,k}^a(\mathbf{x}_k), \quad \mathbf{x}_j = \sum_{k=1}^{j-1} g_{j,k}(\mathbf{x}_k), \quad a \in Q, \ 0 \leq q < i \leq r, \ r < j \leq n. \quad (5)$$

We first represent (5) by an extended WFA of n states. This extended WFA will consist of (i) a finite set of states, $S = \{1, 2, ..., n\}$; (ii) a finite alphabet Q; (iii) a target state; (iv) a final distribution $\beta_1, ..., \beta_n \in \mathbb{R}$; (v) a set of weight functions $f_{i,k}^a \in \mathfrak{F}$ and $g_{j,\ell} \in \mathfrak{G}$ for $a \in Q$, $q < k \leq r$, $r < j \leq n$ and $1 \leq \ell < j$. The first q states will be called the *pump states*, and the next $r - q$ states and the last $n - r$ states will be called the *generic states* and the *relay states* respectively. For each state $k \in S$, image $\mathbf{x}_k$ is the corresponding *state image* although a more convenient but slightly abusive approach is to identify a state simply with the state image. The *target image* is the state image corresponding to the target state. In general the pump states refer to some preselected base images, and these base images at any resoltions may be generated, or pumped out, independent of the other state images. As an example, we consider the following extended WFA with $Q = \{0, 1, 2, 3\}$, $r = n = 5$, $q = 0$ and

$$\mathbf{x}_1^0 = \mathbf{x}_3, \quad \mathbf{x}_2^0 = \mathbf{x}_4, \quad \mathbf{x}_3^0 = \mathbf{x}_5, \quad \mathbf{x}_4^0 = \mathbf{x}_4 + \eth(\mathbf{x}_4),$$

$$\mathbf{x}_1^1 = \mathbf{x}_2, \quad \mathbf{x}_2^1 = \mathbf{x}_2, \quad \mathbf{x}_3^1 = \frac{1}{2}\mathbf{x}_3 + \frac{1}{2}\mathbf{x}_4, \mathbf{x}_4^1 = \mathbf{x}_4,$$

$$\mathbf{x}_1^2 = \mathbf{x}_1, \quad \mathbf{x}_2^2 = \mathbf{x}_2, \quad \mathbf{x}_3^2 = \frac{1}{2}\mathbf{x}_3, \quad \mathbf{x}_4^2 = \mathbf{x}_4,$$

$$\mathbf{x}_1^3 = \eth(\mathbf{x}_1), \quad \mathbf{x}_2^3 = 0, \quad \mathbf{x}_3^3 = 0, \quad \mathbf{x}_4^3 = 0,$$

$$\mathbf{x}_5^0 = \frac{1}{2}\mathbf{x}_5, \quad \mathbf{x}_5^1 = \frac{1}{2}\mathbf{x}_5 + \frac{1}{4}\mathbf{x}_4 + \frac{1}{4}\eth(\mathbf{x}_4),$$

$$\mathbf{x}_5^2 = \frac{1}{2}\mathbf{x}_5, \quad \mathbf{x}_5^3 = \frac{1}{2}\mathbf{x}_5 + \frac{1}{4}\mathbf{x}_4 + \frac{1}{4}\eth(\mathbf{x}_4), \quad (6)$$

where $\eth$ denotes the diagonal reflection, see Fig. 3. Let $\beta_i = \mu(\mathbf{x}_i)$ for $i = 1, ..., 5$. Then (6) defines a valid WFA (5) if, due to (2), $(\beta_1, \beta_2, \beta_3, \beta_4, \beta_5) = (\frac{5}{12}, \frac{1}{2}, \frac{1}{3}, 1, \frac{1}{2})\xi$ for some $\xi \in \mathbb{R}$. The diagram for the WFA, with $\mathbf{x}_1$ as the target image, is drawn in Fig. 4. It is not difficult to see that the WFA will generate the target image $\mathbf{x}_1$ given in Fig. 5. We note that, in a graph representation,

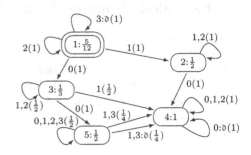

Fig. 4. WFA of 5 states for (6)

$i : \beta_i$ inside a circle represents state i with intensity β_i, and label $a(w)$ or $a : \mathfrak{h}(w)$ on an arrow from state i to state j implies $\mathbf{x}_i^a$ contains linearly $w\mathbf{x}_j$ or $w\mathfrak{h}(\mathbf{x}_j)$ respectively, where w is the weight. Moreover the doubly circled state denotes the target state.

We note that the target image determined by the extended WFA Fig. 4 can also be generated by a linear WFA, at the cost of having more than doubled number of states. The linear WFA is given in Fig. 5 in which each state is now labelled by the corresponding state image and the image intensities $\beta_i = \mu(\mathbf{x}_i)$ are given by $\beta_1 = \beta_7 = \frac{5}{12}$, $\beta_2 = \beta_5 = \beta_8 = \beta_{11} = \frac{1}{2}$, $\beta_3 = \beta_9 = \frac{1}{3}$, $\beta_4 = \beta_{10} = 1$ and $\beta_6 = 2$. Incidentally, if we keep just the first 6 states in Fig. 5 and assign $\beta^T = (5/36, 1/4, 1/6, 1/2, 1/4, 1)\xi$ to the final distribution β in Theorem 3, then the equations in (4) again define a valid WFA with 6 states and 23 edges. The target im-

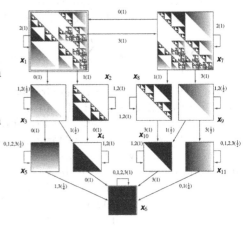

Fig. 5. Transition diagram of a WFA for the target image determined by (6)

age $\mathbf{x}_1$ in this case reduces to the diminishing triangles by Culik and Kari [5], which is just the image $\mathbf{x}_1$ in Fig. 5 without the details above the diagonal.

3.2   Packed WFA and the Associated Properties

For the practicality of applications, it is of interest to restrict ourselves to the following special yet sufficiently general form of WFA relations

$$\mathbf{x}_i^a = \sum_{k=1}^{n} \sum_{\tau \in \mathfrak{R}} W_{i,k}^{a,\tau} \tau(\mathbf{x}_k), \quad \mathbf{x} = \sum_{k=1}^{n} \sum_{\tau \in \mathfrak{R}} \alpha_k^{\tau} \tau(\mathbf{x}_k), \, a \in Q, \, 0 \le q < i \le n, \quad (7)$$

where $W_{i,k}^{a,\tau}, \alpha_k^{\tau} \in \mathbb{R}$ and $\mathfrak{R} \subset \mathfrak{F}$ is a nonempty finite set of resolution-wise mappings. The target image $\mathbf{x}$ corresponds to the only relay state, implicitly the state $n + 1$. However **x** may be absent if the target state is just one of the generic states. An extended WFA defined through (7) will be called a *packed* WFA, and the WFA is also said to be *packed under* $\mathfrak{R}$.

To analyse properly the properties of packed WFA, we first introduce a feature that refines a set of resolution-driven mappings. More specifically, a set $\mathfrak{R} \subset \mathfrak{I}$ is said to be *quadrantly-supportive* if all mappings in $\mathfrak{R}$ are linear and that for all $\tau \in \mathfrak{R}$ and $a \in Q$, there exist $\tau' \in \mathfrak{R}$ and $a' \in Q$ such that $\mathfrak{z}^a \tau = \tau' \mathfrak{z}^{a'}$, i.e. $\left[\tau(\mathbf{x})\right]^a = \tau'(\mathbf{x}^{a'})$ holds for all multiresolution images **x**. As an example, $\mathfrak{P}$ and $\mathfrak{P}_m$ for $m \geq 0$ are all quadrantly-supportive. With these preparations, we now summarise some additional finer properties for packed WFA in the next two theorems.

Theorem 4. *A packed WFA under a finite group $\mathfrak{R}$ can always be expanded into a linear WFA if $\mathfrak{R}$ is also quadrantly-supportive. In particular, a WFA packed under a finite subset of $\mathfrak{P}$ can always be expanded into a linear WFA.*

Alternatively, a packed WFA can be regarded as a linear WFA reduced or packed under the similarities or symmetries dictated by $\mathfrak{R}$. We note that symmetries or similarities are immensely useful in various applications, including for example the group theoretical analysis by Segman and Zeevi in [13], the image processing and understanding in [14], and the solutions and classification of differential and difference systems in [15,16].

Given a linear WFA and a set $\mathfrak{R}$ of resolution-wise mappings, we may reduce the linear WFA to a packed WFA of fewer number of states. For this purpose we say a set of images $\{\mathbf{x}_i\}$ are $\mathfrak{R}$-*dependent* if there exists an i_0 such that $\mathbf{x}_{i_0} = \sum_{i \neq i_0}^{finite} \sum_{\tau \in \mathfrak{R}}^{finite} c_i^\tau \tau(\mathbf{x}_i)$ for some constants $c_i^\tau \in \mathbb{R}$. It is easy to see that $\mathfrak{R}$-independent images are also linearly independent if $\mathbf{1} \in \mathfrak{R}$. Moreover one can show for a WFA packed under a finite group $\mathfrak{R} \subset \mathfrak{I}$, if the generic state images are $\mathfrak{R}$-dependent, then the WFA can be shrunk by at least 1 generic state.

Naturally it is desirable to have a WFA with as few states as possible while not incurring too many additional edges. For any given image there is a lower bound on the number of states a packed WFA can have if the WFA is to regenerate the given image.

Theorem 5. *Suppose $\mathfrak{R} \subset \mathfrak{I}$ is a finite group and is quadrantly-supportive, and that a packed WFA $\mathfrak{A}$ under $\mathfrak{R}$ has a target image **x** and has no pump states. Let $\{\mathbf{x}_i\}_{i=1}^n$ be all the state images of the WFA. Then $\mathfrak{A}$ has the minimum number of states, the minimum required to reconstruct the target image **x**, if and only if*

(i)   all the state images $\mathbf{x}_i$ are $\mathfrak{R}$-independent;

*(ii)   all state images are combinations of addressible subimages of **x**, i.e. all state images are of the form*

$$\sum_{\tau \in \mathfrak{R}} \sum_{\omega \in Q^*}^{finite} c_{\tau,\omega} \tau(\mathbf{x}^\omega), \quad c_{\tau,\omega} \in \mathbb{R};$$

*(iii) target image **x** corresponds to a generic state.*

To conclude this section, we note that the set $\mathfrak{R}$ practically never exceeds the scope of $\mathfrak{P}$ in the application to image compression. This means that all assumptions for the above two theorems are in fact automatically satisfied if $\mathfrak{R}$ is just a finite group. As far as WFA based image compression is concerned, $\mathfrak{R}$ doesn't even have to be a group: it suffices to be just a finite subset of $\mathfrak{F}$. We also remark that the two apparently different formulations of WFA due to Culik et al and Hafner respectively can be easily unified: for each hierarchic WFA due to Hafner we can struct explicitly a linear WFA adopted by Culik et al, but not vice versa. The details are again available in [12].

4   Implementation of a Packed WFA and the Resulting Compression Improvement

The purpose here is to show, through an implemented image compressor based on the theory developed in the previous sections, that similarity enriched WFA are indeed more desirable and beneficial to the

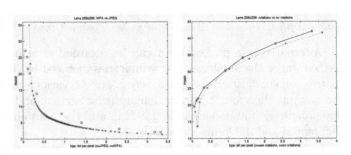

Fig. 6. Extended WFA vs JPEG and vs linear WFA

image compression. For clarity and simplicity of the actual implementation, we do not consider any pump states here and will use only three resolution-wise operators. They are respectively the horizontal, vertical and diagonal reflections.

Hence our compressor is in fact built on the simplistic treatment of the auxiliary techniques associated with the compression scheme based on an extended WFA. For instance no entropy coding is performed after the quantisation of the WFA weights. Despite these simplifications, the compression performance is still on a par with the JPEG standard on natural images: slightly better than JPEG at very low bitrates but otherwise marginally worse than JPEG in the present simplistic implementation, see the l.h.s. figure in Fig. 6 for the compression of the standard Lena image at 256×256 pixels in terms of RMSE. The little circles there correspond to the WFA based compressor.

We now compare the compression performance on the Lena image for the linear WFA and that for the packed WFA. First we note that in the encoding of a linear WFA, the storage spaces for the resolution-wise mappings are completely eliminated. Despite this apparent equal footing, a codec based on extended WFA demonstrates a consistently superior compression ratio. In the case of compressing the standard Lena image at 256×256 pixels, for instance, the r.h.s. figure in Fig. 6 illustrates explicitly the better compression performance by the codec based on an extended WFA in terms of PSNR. The curve traced out there by the little circles corresponds to our currently implemented codec based on a packed WFA. The other curve corresponds to the same codec when the similarity enrichment is turned off. We finally note that the above benchmark codec, for simplicity and speed, made use of only a topdown non-recursive inference algorithm. Hence a sizable improvement is naturally expected if a more computation-intensive *recursive* inference algorithm, akin to that adopted in [5] for the *linear* WFA, is instead implemented to the codec. These will however be left to our future pursuit.

References

1. Jiang Z, Wavelet based image compression under minimum norms and vanishing moments, in Yan H, Eades P, Feng D D and Jin J (edrs), Proceedings of PSA Workshop on Visual Information Processing, (1999)1–5.
2. Culik II K and Kari J, Image compression using weighted finite state automata, Computer Graphics **17** (1993)305–313.
3. Hafner U, Refining image compression with weighted finite automata, Proceedings of Data Compression Conference , edrs. Storer J and Cohn M, (1996)359–368.
4. Hafner U, Weighted finite automata for video compression, IEEE Journal on Selected Areas in Communications , edrs. Enami K, Krikelis A and Reed T R, (1998)108–119.
5. Culik II K and Kari J, Inference algorithm for WFA and image compression, in Fisher Y (edr), Fractal Image Compression: theory and application, Springer, New York 1995.
6. Litow B and De Vel O, A recursive GSA acquisition algorithm for image compression, TR **97/2**, Department of Computer Science, James Cook University 1997.
7. Culik II K and Karhumaki, Finite automata computing real functions, SIAM Journal of Computations **23** (1994)789–814.
8. Culik II K, Dube S and Rajcani P, Effective compression of wavelet coefficients for smooth and fractal-like data, Proceedings of Data Compression Conference , edrs. Storer J and Cohn M, (1993)234–243.
9. Culik II K and Fris I, Weighted finite transducers in image processing, Discrete Applied Mathematics **58** (1995)223–237.
10. Pereira F C and Riley M D, Speech recognition by composition of weighted finite automata, At&T Labs, Murray Hill 1996.
11. Culik II K and Valenta V, Finite automata based compression of bi-level and simple color images, presented at the Data Compression Conference, Snowbird, Utah 1996.
12. Jiang Z, Litow B and De Vel O, Theory of packed finite state automata, Technical Report 99-173, University of New England, June 1999.
13. Laine A (edr), Wavelet Theory and Application: A Special Issue of the Journal of Mathematical Imaging and Vision, Kluwer, Boston 1993.
14. Lenz R, Group Theoretical Methods in Image Processing, Lecture Notes in Computer Science **413**, Springer, Berlin 1990.
15. Jiang Z, Lie symmetries and their local determinancy for a class of differential-difference euqations, Physics Letters A **240** (1998)137–143.
16. Jiang Z, The intrinsicality of Lie symmetries of $u_n^{(k)}(t) = F_n(t, u_{n+a}, ..., u_{n+b})$, Journal of Mathematical Analysis and Applications **227** (1998)396–419.

On the Power of Input-Synchronized Alternating Finite Automata

Hiroaki Yamamoto

Department of Information Engineering, Shinshu University,
4-17-1 Wakasato, Nagano-shi, 380-8553 Japan.
yamamoto@cs.shinshu-u.ac.jp

Abstract. In this paper, we introduce a new model of automata, called input-synchronized alternating finite automata (i-SAFAs), and study the power of i-SAFAs. We here consider two types of i-SAFAs, one-way i-SAFAs and two-way i-SAFAs. Then we show that (1) the class of languages accepted by one-way i-SAFAs is equal to the class of regular languages, (2) two-way i-SAFAs are more powerful than one-way i-SAFAs, that is, there exists a language L such that L is accepted by a two-way i-SAFA but not accepted by any one-way i-SAFAs. In addition, we show that the class of languages accepted by $f(n)$ parallel-bounded two-way i-SAFAs is included in Nspace($S(n)$), where $S(n) = \min\{n \log n, f(n) \log n\}$.

1 Introduction

Studying the power of synchronization in parallel computations is one of important problems in the field of computer science, and several researchers [2,5,6,7,9] have studied the power of synchronization by using alternating machines. They have shown that synchronization increases the power of alternating machines. This paper is concerned with alternating finite automata with a new kind of synchronization, called *an input-synchronized alternating finite automata* (i-SAFAs), which have a different synchronization from the existing synchronized alternating machines. Yamamoto [11] has shown an efficient recognition algorithm for semi-extended regular expressions by using a variant of i-SAFAs. Thus the input-synchronization is an important concept in the practical point of view as well as the theoretical point of view. We show an interesting result that i-SAFAs have a different characteristic from the existing models. In this paper, two types of i-SAFAs are discussed; one is one-way i-SAFAs whose input head is permitted only to move right or to remain stationary and the other is two-way i-SAFAs whose input head is permitted to move left, to move right or to remain stationary.

Chandra et al. [1] introduced alternating finite automata (AFAs for short) as a generalization of NFAs and showed that one-way AFAs also exactly accept regular languages. Ladner et al. [8] studied two-way AFAs and showed that two-way AFAs also accept only regular languages. Thus one-way AFAs and two-way AFAs have the same power.

D.-Z. Du et al. (Eds.): COCOON 2000, LNCS 1858, pp. 457–466, 2000.

The concept of synchronization was first introduced by Hromkovic. Slobodova [9] studied the power of synchronized alternating finite automata (SAFAs for short), which are an extension of AFAs, and showed that SAFAs can accept a wider class of languages than the class of regular languages. Dassow et al. [3] showed that two-way SAFAs accept exactly context-sensitive languages, that is, languages in Nspace(n). Here, for any function $S(n)$ from natural numbers to real numbers, by Nspace($S(n)$) we denote the class of languages accepted by nondeterministic Turing machines (NTMs) in $O(S(n))$ space. Hromkovic et al. [5] improved this result and showed that the class of languages accepted by one-way SAFAs is exactly equal to the class of context-sensitive languages. Thus SAFAs are more powerful than AFAs, but one-way SAFAs and two-way SAFAs also have the same power.

We will here show that one-way i-SAFAs accept only regular languages, but two-way i-SAFAs accept a wider class than the class of regular languages. In addition, we also show an upper bound of the power of two-way i-SAFAs.

The difference between i-SAFAs and SAFAs is in the synchronization system. SAFAs have a set of synchronizing symbols, and each time one of the processes working in parallel enters a state with a synchronizing symbol, it must wait until all other processes working in parallel enter accepting states or states with the same synchronizing symbol. Thus all processes of an SAFA are synchronized by the same synchronizing symbol. Other models of synchronized automata also adopt a similar definition. On the other hand, i-SAFAs have a family $\mathcal{S} = \{S_1, \ldots, S_t\}$ of synchronizing sets such that each S_i is a set of synchronizing states, and processes of an i-SAFA are synchronized by the position of the input head when entering a synchronizing state. That is, each time several processes working in parallel enter synchronizing states belonging to the same synchronizing set S_i, they all must start their computations from the same position of the input. Note that it is not necessary that all processes of an i-SAFA working in parallel enter synchronizing states in the same synchronizing set. For example, it is possible that a process enters a synchronizing state in a synchronizing set S_i and other process enters a synchronizing state in a different synchronizing set S_j. This time, these processes do not need to synchronize. Clearly, i-SAFAs are also a generalization of AFAs because i-SAFAs without any synchronizing states are AFAs.

In this paper, we will show the following results:

1. The class of languages accepted by one-way i-SAFAs is equal to the class of regular languages,
2. There exists a language L such that L is accepted by a two-way i-SAFA but not accepted by any one-way i-SAFAs. In fact, we show that there is a two-way i-SAFA accepting non-regular language $L = \{w\#w\# \mid w \in \{0,1\}^*\}$. In addition, we show that the class of languages accepted by $f(n)$ parallel-bounded two-way i-SAFAs is included in Nspace($S(n)$), where $S(n) = \min\{n \log n, f(n) \log n\}$. We say that a two-way i-SAFA M is $f(n)$ parallel-bounded if, for any accepted input w of length n, the number of universal

configurations on the accepting computation tree of M on w is at most $O(f(n))$.

2   One-Way i-SAFAs

2.1   Definitions

We first define one-way i-SAFAs.

Definition 1. *A one-way i-SAFA M is a seven-tuple $M = (Q, \mathcal{S}, \Sigma, \delta, q_0, \mu, F)$, where*

- *Q is a finite set of states,*
- *$\mathcal{S} = \{S_1, \ldots, S_t\}$ is a family of synchronizing sets such that each S_i ($\subseteq Q$) is a set of synchronizing states and $S_i \cap S_j = \emptyset$ for any $1 \leq i, j \leq t$,*
- *Σ is a finite input alphabet,*
- *q_0 ($\in Q$) is the initial state,*
- *μ is a function mapping Q to $\{\vee, \wedge\}$,*
- *F ($\subseteq Q$) is a set of final states,*
- *δ is a transition function mapping $Q \times (\Sigma \cup \{\epsilon\})$ to 2^Q, where ϵ denotes the empty string.*

If $\mu(q) = \wedge$ ($\vee$, respectively), then q is called *a universal* (*existential*, respectively) *state*. A configuration of M is defined to be a pair (q, pos), where q is a state and pos is a position of the input head. If q is a universal (existential, respectively) state, then (q, pos) is called *a universal* (*existential*, respectively) *configuration*. Among these configurations, if a state q is in F, then the configuration is also called *an accepting configuration*. The configuration $(q_0, 1)$ is called *an initial configuration*. The interpretation of $\delta(q, a) = \{q_1, \ldots, q_l\}$ is that M reads the input symbol a and changes the state from q to $q_1, \ldots, q_l$. This time, if $a \neq \epsilon$, then M advances the input head one symbol right, and if $a = \epsilon$, then M does not advance the input head (this is called *an ϵ-move*). We give a precise definition of acceptance for one-way i-SAFAs.

Definition 2. *A full computation tree of a one-way i-SAFA M on an input $w = a_1 \cdots a_n$ is a labelled tree T such that*

- *each node of T is labelled by a configuration of M,*
- *each edge of T is labelled by a symbol in $\{a_1, \ldots, a_n, \epsilon\}$,*
- *the root of T is labelled by $(q_0, 1)$,*
- *if a node v of T is labelled by (q, pos) and $\delta(q, a) = \{q_1, \ldots, q_k\}$ for a symbol $a \in \{\epsilon, a_{pos}\}$ is defined, then v has k children $v_1, \ldots, v_k$ such that each v_i is labelled by (q_i, pos') and every edge e_i from v to v_i is labelled by the symbol a. Furthermore, if $a = a_{pos}$, then $pos' = pos + 1$, and if $a = \epsilon$, then $pos' = pos$.*

Definition 3. *Let T be a full computation tree of a one-way i-SAFA M on an input w of length n and let v_0 be the root of T. For a node v of T, let $\alpha = (p_1, b_1) \cdots (p_u, b_u)$ be the maximum sequence of labels on the path from v_0 to v satisfying the following: (1) $b_1 \leq b_2 \leq \cdots \leq b_u \leq n$, and (2) for any i $(1 \leq i \leq u)$, p_i is a synchronizing state. For the sequence α, we make a subsequence $(p_{i_1}, b_{i_1}) \cdots (p_{i_e}, b_{i_e})(p_u, b_u)$ by taking out all labels (p_i, b_i) satisfying $b_i < b_{i+1}$ for $1 \leq i \leq u-1$ and (p_u, b_u) in the order. Then the sequence $(p_{i_1}, b_{i_1}) \cdots (p_{i_e}, b_{i_e})(p_u, b_u)$ is called a synchronizing sequence of v. This sequence means that only the last synchronizing state is considered when M travels synchronizing states by consecutive ϵ-moves. In addition, for any synchronizing set S_s $(\in \mathcal{S})$, let us consider a sequence $(s_1, b_{i_1}) \cdots (s_l, b_{i_l})$ made by picking up all labels (p_i', b_i') with $p_i' \in S_s$ from a synchronizing sequence $(p_1', b_1') \cdots (p_l', b_l')$ of v. Then we call this subsequence an S_s-synchronizing sequence of v.*

Definition 4. *A computation tree T' of a one-way i-SAFA M on an input w of length n is a subtree of a full computation tree T such that*

- *if v is labelled by a universal configuration, then v has the same children as in T,*
- *if v is labelled by an existential configuration, then v has at most one child,*
- *let v_1 and v_2 be arbitrary nodes. For any synchronizing set S_s $(\in \mathcal{S})$, let $(s_1, b_1) \cdots (s_{l_1}, b_{l_1})$ and $(s_1', c_1) \cdots (s_{l_2}', c_{l_2})$ $(l_1 \leq l_2)$ be S_s-synchronizing sequences of v_1 and v_2, respectively. Then the sequence $b_1 \cdots b_{l_1}$ is an initial subsequence of the sequence $c_1 \cdots c_{l_2}$. This condition ensures that processes working in parallel read an input symbol at the same position if they enter synchronizing states belonging to S_s. We call this* an input-synchronization.

We now associate a computation tree T' with a boolean function by regarding states of M as variables as follows. Let $(p_1, b_1), \ldots, (p_t, b_t)$ be leaves of T'. Then the boolean function f denoted by T' is defined to be $p_1 \wedge \cdots \wedge p_t$. Furthermore, by $eval(f)$, we denote the value of f evaluated by assigning 1 to variables $p \in F$ and 0 to variables $p \notin F$.

Definition 5. *An accepting computation tree of a one-way i-SAFA M on an input w of length n is a finite computation tree of M on w such that*

- *the root is labelled by the initial configuration,*
- *each leaf is labelled by an accepting configuration with the input head position $n+1$, that is, labelled by a label $(q, n+1)$ with $q \in F$.*

We say that a one-way i-SAFA M accepts an input w if there exists an accepting computation tree on w. We denote the language accepted by M by $L(M)$. Clearly, all states occurring in the boolean function denoted by an accepting computation tree are in F.

2.2 ϵ-Free One-Way i-SAFAs

We will here define ϵ-free one-way i-SAFAs, and show an algorithm to construct an ϵ-free one-way i-SAFA from any given one-way i-SAFA. The only difference from one-way i-SAFAs is that ϵ-free one-way i-SAFAs have no ϵ-moves. Thus an ϵ-free one-way i-SAFA on an input w of length n always halts in n steps. Unlike one-way i-SAFAs, we define the transition function of an ϵ-free one-way i-SAFA to be a mapping from a pair $(state, input\_symbol)$ to a monotone boolean function in order to make the construction easier. Below is the formal definition.

Definition 6. *An ϵ-free one-way i-SAFA M is a six-tuple $M=(Q, \mathcal{S}, \Sigma, \delta, q_0, F)$, where*

- *Q is a finite set of states,*
- *$\mathcal{S} = \{S_1, \ldots, S_t\}$ is a family of synchronizing sets such that each S_i ($\subseteq Q$) is a set of synchronizing sates and $S_i \cap S_j = \emptyset$ for any $1 \leq i, j \leq t$,*
- *Σ is a finite input alphabet,*
- *q_0 ($\in Q$) is the initial state,*
- *F ($\subseteq Q$) is a set of final states,*
- *δ is a transition function mapping $Q \times \Sigma$ to $\mathcal{F}$,*
  where $\mathcal{F}$ is a set of monotone boolean functions mapping $\{0, 1\}^{|Q|}$ to $\{0, 1\}$.

Definition 7. *The full computation tree of an ϵ-free one-way i-SAFA M on an input word $w = a_1 \cdots a_n$ is a labelled tree T such that*

- *each node of T is labelled by a configuration of M,*
- *the root of T is labelled by the initial configuration $(q_0, 1)$,*
- *each edge of T is labelled by an input symbol $a \in \{a_1, \ldots, a_n\}$,*
- *if a node v of T is labelled by (q, pos) and $\delta(q, a_{pos}) = g(q_1, \ldots, q_k)$, then v has k children, $v_1, \ldots, v_k$, such that each v_i is labelled by $(q_i, pos + 1)$ and each edge e_i from v to v_i is labelled by a_{pos}.*

A synchronizing sequence and an S_s-synchronizing sequence are defined in the same way as one-way i-SAFAs. Now we define a language accepted by an ϵ-free one-way i-SAFA.

Definition 8. *Let M be an ϵ-free one-way i-SAFA. Then a computation tree T_w^{ϵ} of M on an input w is a subtree of the full computation tree T satisfying the following:*

- *For any node v, let $V = \{v' \mid v' \text{ is a child of } v \text{ in } T\}$. Then, for a subset $V_1 \subseteq V$, v has only the nodes of V_1 as the children,*
- *let v_1 and v_2 be arbitrary nodes other than leaves. For any synchronizing set S_s ($\in \mathcal{S}$), let $(s_1, b_1) \cdots (s_{l_1}, b_{l_1})$ and $(s_1, c_1) \cdots (s_{l_2}, c_{l_2})$ ($l_1 \leq l_2$) be S_s-synchronizing sequences of v_1 and v_2, respectively. Then the sequence $b_1 \cdots b_{l_1}$ is an initial subsequence of the sequence $c_1 \cdots c_{l_2}$.*

In the following, we associate a computation tree of an ϵ-free one-way i-SAFA with a boolean function by regarding a configuration (q, i) as a variable.

Definition 9. *Let* $M = (Q, \mathcal{S}, \Sigma, \delta, q_0, F)$ *be an ϵ-free one-way i-SAFA. By* $T^\epsilon_{w,q}$, *we mean a computation tree of M on an input $w = a_1 \cdots a_n$ such that it is computed by starting from the state q. Then we define a boolean function denoted by each node v of $T^\epsilon_{w,q}$ as follows. Let the label of v be (q', pos), where* $q' \in Q$.

1. *If v is a leaf, then the boolean function denoted by v is (q', pos),*
2. *if v is not a leaf and $\delta(q', a_{pos}) = g(q_1, \ldots, q_k)$, then the boolean function f denoted by v is defined as follows; let $v'_1, \ldots, v'_{k_1}$ be children of v having labels $(q'_1, i), \ldots, (q'_{k_1}, i)$, respectively, and let $f_1, \ldots, f_{k_1}$ be boolean functions denoted by $v'_1, \ldots, v'_{k_1}$, respectively. Then $f = g(g_1, \ldots, g_k)$, where, for any $1 \le i \le k$, $g_i = f_j$ if there is q'_j such that $q_i = q'_j$; otherwise $g_i = 0$.*

Let r be the root of $T^\epsilon_{w,q}$. Then, by the boolean function denoted by $T^\epsilon_{w,q}$, we mean the boolean function f_w denoted by r, and say that $T^\epsilon_{w,q}$ has a boolean function f_w. Such a boolean function f_w is evaluated by the following assignment to each variable, and its value is denoted by $value(f_w)$. The assignment is done as follows. Here note that $|w| = n$.

For each variable (q, i) occurring in f_w,

1. if $i = n + 1$, then

$$(q, i) = \begin{cases} 1 & \text{if } q \in F, \\ 0 & \text{if } q \notin F, \end{cases}$$

2. if $i < n + 1$, then $(q, i) = 0$.

Let $L(M)$ be the language accepted by an ϵ-free one-way i-SAFA M. Then

$$L(M) = \{w \mid \text{there exists a computation tree } T^\epsilon_{w,q_0} \text{ with a boolean function } f_w$$
$$\text{such that } q_0 \text{ is the initial state and } value(f_w) = 1\}.$$

2.3 Conversion from a One-Way i-SAFA to an ϵ-Free One-Way i-SAFA

Since an image of the transition function of an ϵ-free one-way i-SAFA is a boolean function, it is convenient to associate a subtree of a computation tree of a one-way i-SAFA on ϵ^i for the construction of an ϵ-free one-way i-SAFA. We introduce two types of such computation trees.

Definition 10. *Let* $M = (Q, \mathcal{S}, \Sigma, \delta, q_0, \mu, F)$ *be a one-way i-SAFA. Then,*

1. *by $T^1_{\epsilon^i, q}$, we mean a computation tree of M on an input ϵ^i such that (1) it is computed by starting from the state q and (2) all the nodes other than the root are labelled by a non-synchronizing state,*
2. *by $T^2_{\epsilon^i, q}$, we mean a computation tree of M on an input ϵ^i such that (1) it is computed by starting from the state q and (2) each leaf is labelled by either a synchronizing state or a non-synchronizing state, and if it is labelled by a non-synchronizing state, then there is not any synchronizing states on the path from the root to the leaf (hence note that q must be also non-synchronizing state in this case).*

Now, we extend the definition of ϵ-CLOSURE defined for an NFA with ϵ-moves (see [4]) to a one-way i-SAFA.

Definition 11. *Let* $M = (Q, \mathcal{S}, \Sigma, \delta, q_0, \mu, F)$ *be a one-way i-SAFA. Then, for any state* $q \in Q$, *two sets* ϵ-$CL_1(q)$ *and* ϵ-$CL_2(q)$ *are defined as follows.*

ϵ-$CL_1(q) = \{\ f \mid$ *for* $\exists i \geq 0$, f *is the boolean function denoted by a computation tree* $T^1_{\epsilon^i, q}$ *of* $M\}$.

ϵ-$CL_2(q) = \{\ f \mid$ *for* $\exists i \geq 0$, f *is the boolean function denoted by a computation tree* $T^2_{\epsilon^i, q}$ *of* $M\}$.

Let $M = (Q, \mathcal{S}, \Sigma, \delta, q_0, \mu, F)$ with $Q = \{q_0, \ldots, q_l\}$ be a one-way i-SAFA. Then the boolean function denoted by a computation tree of M has $l + 1$ variables. We will construct an ϵ-free one-way i-SAFA $M_1 = (Q \cup \{\tilde{q}\}, \mathcal{S}, \Sigma \cup \{\#\}, \delta_1, \tilde{q}, F_1)$ from M, where $\tilde{q}$ is a new state and $\#$ is a new symbol. The differences between M and M_1 are just in the transition functions and the sets of final states. Let q be any state in Q. Then we define δ_1 as follows:

$$\delta_1(q, a) = \sum_{h \in H} h(q_0, q_1, \ldots, q_l), \tag{1}$$

where

$$H = \bigcup_{g \in G} \{g(g_0, g_1, \ldots, g_l) \mid g_0 \in \epsilon\text{-}CL_2(q_0), \ldots, g_l \in \epsilon\text{-}CL_2(q_l)\}, \tag{2}$$

and

$$G = \bigcup_{f \in \epsilon\text{-}CL_1(q)} \{g \mid g = f(f_0, f_1, \ldots, f_l)\}, \tag{3}$$

where $\delta(q_i, a) = \{q_{i_1}, \ldots, q_{i_j}\}$ and if $\mu(q_i) = \vee$ then $f_i = q_{i_1} \vee \cdots \vee q_{i_j}$, if $\mu(q_i) = \wedge$ then $f_i = q_{i_1} \wedge \cdots \wedge q_{i_j}$. Note that $\sum$ means the repeated boolean OR. For the initial state $\tilde{q}$, we have $\delta_1(\tilde{q}, \#) = \sum_{f \in \epsilon\text{-}CL_2(q_0)} f$ and a set F_1 is defined as follows. For any $q_i \in Q$, let $g_i = \sum_{f \in \epsilon\text{-}CL_1(q_i)} f$. Then $F_1 = F \cup \{q_i \mid eval(g_i) = 1\}$. Clearly, from the above construction, the following property holds for M and M_1. Here, for any boolean functions f_1 and f_2, f_1 and f_2 are said to be equivalent, denoted by $f_1 \equiv f_2$, if and only if the value of f_1 is equal to the value of f_2 for any assignment to variables.

Lemma 12. *Let* M *and* M_1 *be a one-way i-SAFA and an* ϵ-*free i-SAFA described above, respectively. Then, for any* $q \in Q$ *and* $a \in \Sigma$, *if* h *is in* H, *then there exists a computation tree* $T_{a,q}$ *of* M *on the symbol* a *with a boolean function* f_a *such that* $h \equiv f_a$, *and conversely, if* $T_{a,q}$ *is a computation tree of* M *on the symbol* a *with a boolean function* f_a, *then there exists* $h \in H$ *such that* $h \equiv f_a$. *Furthermore, if* M *enters a synchronizing state after reading* a, *then* M_1 *also enters the same synchronizing state and vice versa.*

By this lemma, we immediately have the following theorem.

Theorem 13. *Let M be a one-way i-SAFA with m states. Then we can construct an ϵ-free one-way i-SAFA M_1 such that M_1 has $m+1$ states and $L(M_1) = \{\#w \mid w \in L(M)\}$.*

2.4   The Power of One-Way i-SAFAs

In the previous section, we have shown that one-way i-SAFAs can be converted to ϵ-free one-way i-SAFAs. We here show that ϵ-free one-way i-SAFAs can be simulated by AFAs, and thus one-way i-SAFAs accept exactly the class of regular languages by Theorem 13.

Lemma 14. *Let M be an ϵ-free one-way i-SAFA with m states including t synchronizing states. Then there exists an AFA N with $m \times 2^t$ states such that $L(M) = L(N)$.*

Theorem 15. *The class of languages accepted by one-way i-SAFAs is equal to the class of regular languages.*

3   Two-Way i-SAFAs

In this section, we show two results; one is that a two-way i-SAFAs can accept a non-regular language, the other is that the class of languages accepted by $f(n)$ parallel-bounded two-way i-SAFAs is included in $\text{Nspace}(S(n))$, where $S(n) = \min\{n \log n, f(n) \log n\}$. Thus two-way i-SAFAs are more powerful than one-way i-SAFAs.

3.1   Definitions

Definition 16. *A two-way i-SAFA M is a seven-tuple $M = (Q, \mathcal{S}, \Sigma, \delta, q_0, \mu, F)$, where*

- Q *is a finite set of states,*
- $\mathcal{S} = \{S_1, \ldots, S_t\}$ *is a family of synchronizing sets such that each S_i $(\subseteq Q)$ is a set of synchronizing sates and $S_i \cap S_j = \emptyset$ for any $1 \leq i, j \leq t$,*
- Σ *is a finite input alphabet,*
- q_0 $(\in Q)$ *is the initial state,*
- μ *is a function mapping Q to $\{\vee, \wedge\}$,*
- F $(\subseteq Q)$ *is a set of final states,*
- δ *is a transition function mapping $Q \times \Sigma$ to $2^{Q \times \{L, R, N\}}$.*

The interpretation of $\delta(q, a) = \{(q_1, D_1), \ldots, (q_l, D_l)\}$ is that M reads the input symbol a , changes the state from q to q_i and moves the input head one cell depending on D_i. If $D_i = L$ then M moves the input head one cell to the left, if $D_i = R$ then M moves the input head one cell to the right, if $D_i = N$ then M does not move the input head. We give a precise definition of acceptance of an input for two-way i-SAFAs.

Definition 17. *The full computation tree of a two-way i-SAFA M on an input* $w = a_1 \cdots a_n$ *is a labelled tree T such that*

- *each node of T is labelled by a triple* $(q, pos, depth)$ *of a state of M, a head position of M and the depth of the node,*
- *the root of T is labelled by* $(q_0, 1, 0)$,
- *if a node v of T is labelled by* $(q, pos, depth)$ *and* $\delta(q, a_{pos}) = \{(q_1, D_1), \ldots, (q_k, D_k)\}$, *then v has k children* $v_1, \ldots, v_k$ *such that each* v_i *is labelled by* $(q_i, pos_i, depth + 1)$. *This time, if* $D_i = L$ *then* $pos_i = pos - 1$, *if* $D_i = R$ *then* $pos_i = pos + 1$, *if* $D_i = N$ *then* $pos_i = pos$.

Definition 18. *Let T be the full computation tree of a two-way i-SAFA M on an input w of length n and let* v_0 *be the root of T. For a node v of T, let* $(p_1, b_1, d_1) \cdots$ (p_u, b_u, d_u) *be the maximum sequence of labels on the path from* v_0 *to v such that for any i* $(1 \leq i \leq u)$, p_i *is a synchronizing state and* $b_i \leq n$. *Then this sequence is called* a synchronizing sequence of v. *In addition, for any synchronizing set* S_s $(\in S)$, *let us consider the sequence* $(p_{i_1}, b_{i_1}, d_{i_1}) \cdots (p_{i_l}, b_{i_l}, d_{i_l})$ *made by picking up all labels* (p_i, b_i, d_i) *with* $p_i \in S_s$ *from a synchronizing sequence* $(p_1, b_1, d_1) \cdots (p_u, b_u, d_u)$ *of v. Then, we call this sequence an* S_s-synchronizing sequence of v.

Definition 19. *A computation tree* T' *of a two-way i-SAFA M on an input w of length n is a subtree of the full computation tree T such that*

- *if v is labelled by a universal configuration, then v has the same children as in T,*
- *if v is labelled by an existential configuration, then v has at most one child,*
- *let* v_1 *and* v_2 *be arbitrary nodes. For any synchronizing set* S_s $(\in S)$, *let* $(s_1, b_1, d_1) \cdots (s_{l_1}, b_{l_1}, d_{l_1})$ *and* $(s'_1, c_1, d'_1) \cdots (s'_{l_2}, c_{l_2}, d'_{l_2})$ $(l_1 \leq l_2)$ *be* S_s-synchronizing sequences of v_1 and v_2, *respectively. Then these two sequences must satisfy the following: (1) for any i* $(1 \leq i \leq l_1 - 1)$, $d_i \leq d'_i \leq d_{i+1}$ *or* $d'_i \leq d_i \leq d'_{i+1}$ *holds, (2) the sequence* $b_1 \cdots b_{l_1}$ *is an initial subsequence of the sequence* $c_1 \cdots c_{l_2}$.

Suppose that several processes enter synchronizing states in a synchronizing set S_s. Then the condition (1) means that a process cannot enter successively synchronizing states in S_s before other processes enter a synchronizing state in S_s. An accepting computation tree of a two-way i-SAFA M is defined in same way as Definition 5. We say that M accepts an input w if there exists an accepting computation tree on w. Let $f(n)$ be a function from natural numbers to real numbers. Then it is said that M is $f(n)$ parallel-bounded if, for any accepted input w of length n, the number of universal configurations on an accepting computation tree of M on w is at most $O(f(n))$.

3.2   The Power of Two-Way i-SAFAs

We show that two-way i-SAFAs are more powerful than one-way i-SAFAs, and then show an upper bound for the power of two-way i-SAFAs.

Theorem 20. *There exists a language L such that L can be accepted by a two-way i-SAFA but not accepted by any one-way i-SAFAs.*

Let us consider $L = \{w\#w\# \mid w \in \{0,1\}^*\}$. Since the language L is not regular, it is clear that any one-way i-SAFAs cannot accept L because one-way i-SAFAs accept only regular languages. Hence the following lemma leads to the theorem.

Lemma 21. *The language $L = \{w\#w\# \mid w \in \{0,1\}^*\}$ can be accepted by a two-way i-SAFA.*

The following theorem states an upper bound.

Theorem 22. *The class of languages accepted by $f(n)$ parallel-bounded two-way i-SAFAs is included in Nspace($S(n)$), where $S(n) = \min\{n \log n, f(n) \log n\}$.*

References

1. A.K. Chandra, D.C. Kozen and L.J. Stockmeyer, Alternation, J. Assoc. Comput. Mach. 28,1, 114-133, 1981.
2. O.H. Ibarra and N.Q. Tran, On communication-bounded synchronized alternating finite automata, Acta Informatica, 31, 4, 315-327, 1994.
3. J.Dassow, J.Hromkovic, J.Karhuaki, B.Rovan and A. Slobodova, On the power of synchronization in parallel computation, In Proc. 14th MFCS'89, LNCS 379,196-206, 1989.
4. J.E. Hopcroft and J.D. Ullman, Introduction to automata theory language and computation, Addison Wesley, Reading Mass, 1979.
5. J. Hromkovic, K. Inoue, B. Rovan, A. Slobodova, I. Takanami and K.W. Wagner, On the power of one-way synchronized alternating machines with small space, International Journal of Foundations of Computer Science, 3, 1, 65-79, 1992.
6. J. Hromkovic and K. Inoue, A note on realtime one-way synchronized alternating one-counter automata, Theoret. Comput. Sci., 108, 2, 393-400, 1993.
7. J. Hromkovic, B. Rovan and A. Slobodova, Deterministic versus nondeterministic space in terms of synchronized alternating machines, Theoret. Comput. Sci., 132, 2, 319-336, 1994.
8. R.E. Ladner, R.J. Ripton and L.J. Stockmeyer, Alternating pushdown and stack automata, SIAM J. Computing, 13, 1, 1984.
9. A. Slobodova, On the power of communication in alternating machines, In Proc. 13th MFCS'88, LNCS 324,518-528, 1988.
10. M.Y. Vardi, Alternating automata and program verification, In Computer Science Today-Recent Trends and Developments, LNCS 1000, 471-485, 1996.
11. H. Yamamoto, An automata-based recognition algorithm for semi-extended regular expressions, manuscript, 2000.

Ordered Quantum Branching Programs Are More Powerful than Ordered Probabilistic Branching Programs under a Bounded-Width Restriction

Masaki Nakanishi, Kiyoharu Hamaguchi, and Toshinobu Kashiwabara

Graduate School of Engineering Science, Osaka University,
Toyonaka, Osaka 560-8531, Japan

Abstract. One of important questions on quantum computing is whether there is a computational gap between the model that may use quantum effects and the model that may not. Researchers have shown that some quantum automaton models are more powerful than classical ones. As one of classical computational models, branching programs have been studied intensively as well as automaton models, and several types of branching programs are introduced including read-once branching programs and bounded-width branching programs. In this paper, we introduce a new quantum computational model, a quantum branching program, as an extension of a classical probabilistic branching program, and make comparison of the power of these two models. We show that, under a bounded-width restriction, ordered quantum branching programs can compute some function that ordered probabilistic branching programs cannot compute.

1 Introduction

There are many results that quantum computers might be more powerful than classical computers [2,5], it is unclear whether there is a computational gap between the model that may use quantum effects and the model that may not. Researchers have shown that some quantum automaton models are more powerful than classical ones [1,3]. It would give hints on the power of quantum computation to study about other computational models to see whether quantum computational models can be more powerful than classical ones.

As one of classical computational models, branching programs have been studied intensively as well as automaton models, and several types of branching programs are introduced including read-once branching programs and bounded-width branching programs [4].

In this paper, we introduce a new quantum computational model, a quantum branching program, as an extension of a classical probabilistic branching program, and make comparison of the power of these two models. We show that, under a bounded-width restriction, ordered quantum branching programs can

D.-Z. Du et al. (Eds.): COCOON 2000, LNCS 1858, pp. 467–476, 2000.
© Springer-Verlag Berlin Heidelberg 2000

compute some function that ordered probabilistic branching programs cannot compute.

The remainder of this paper has the following organization. In Sect. 2, we define several types of quantum branching programs and probabilistic branching programs. In Sect. 3, we show that, under a bounded-width restriction, ordered quantum branching programs can compute some function that ordered probabilistic branching programs cannot compute. We conclude with future work in Sect. 4.

2   Preliminaries

We define technical terms.

Definition 1. *Probabilistic Branching Programs*

A probabilistic branching program (a PBP) is a directed acyclic graph that has two terminal nodes, which are labeled by 0 and 1, and internal nodes, which are labeled by boolean variables taken from a set $X = \{x_1, \ldots, x_n\}$. There is a distinguished node, called source, which has in-degree 0. Each internal node has two types of outgoing edges, called the 0-edges and the 1-edges respectively. Each edge e has a weight $w(e)$ $(0 \leq w(e) \leq 1)$. Let $E_0(v)$ and $E_1(v)$ be the set of 0-edges and the set of 1-edges of a node v respectively. The sum of the weights of the edges in $E_0(v)$ and $E_1(v)$ is 1. That is,

$$\sum_{e \in E_0(v)} w(e) = 1, \quad \sum_{e \in E_1(v)} w(e) = 1 .$$

A PBP reads n inputs and returns a boolean value as follows: Starting at the source, the value of the labeled variable of the node is tested. If this is 0 (1), an edge in $E_0(v)$ $(E_1(v))$ is chosen according to the probability distribution given as the weights of the edges. The next node that will be tested is the node pointed by the chosen edge. Arriving at the terminal node, the labeled boolean value is returned.

We say that a PBP P computes a function f (with error rate $1/2 - \delta$) if P returns the correct value of f for any inputs with probability at least $1/2 + \delta$ $(\delta > 0)$. □

We show examples of PBPs in Fig. 1.

Definition 2. *Quantum Branching Programs*

A quantum branching program (a QBP) is an extension of a probabilistic branching program, and its form is same as a probabilistic branching program except for edge weights. In a QBP, the weight of each edge has a complex number $w(e)$ $(0 \leq ||w(e)|| \leq 1)$. The sum of the squared magnitude of the weights of the edges in $E_0(v)$ and $E_1(v)$ is 1. That is,

$$\sum_{e \in E_0(v)} ||w(e)||^2 = 1, \quad \sum_{e \in E_1(v)} ||w(e)||^2 = 1 .$$

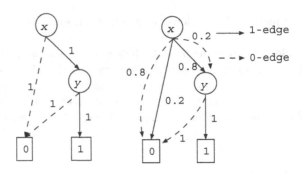

Fig. 1. probabilistic branching programs that compute $f = xy$ with error rate 0 and 0.2 respectively

The edge weight $w(e)$ represents the amplitude with which, currently in the node v, the edge will be followed in the next step.

Nodes are divided into the three sets of the accepting set (Q_{acc}), the rejecting set (Q_{rej}) and the non-halting set (Q_{non}). The configurations of P are identified with the nodes in $Q = (Q_{\mathrm{acc}} \cup Q_{\mathrm{rej}} \cup Q_{\mathrm{non}})$. A superposition of a QBP P is any element of $l_2(Q)$ (the space of mappings from Q to $\mathcal{C}$ with l_2 norm). For each $q \in Q$, $|q\rangle$ denotes the unit vector that takes value 1 at q and 0 elsewhere.

We define a transition function $\delta : (Q \times \{0,1\} \times Q) \longrightarrow \mathcal{C}$ as follows:

$$\delta(v, a, v') = w(e) ,$$

where $w(e)$ is the weight of the a-edge ($a = 0$ or 1) from a node v to v'. If the a-edge from v to v' does not exist, then $\delta(v, a, v') = 0$. We define a time evolution operator as follows:

$$U_\delta^{\boldsymbol{x}} |v\rangle = \sum_{v' \in Q} \delta(v, x(v), v') |v'\rangle ,$$

where $\boldsymbol{x}$ denotes the input of a QBP, and $x(v)$ denotes the assigned value in $\boldsymbol{x}$ to the labeled variable of the node v. If the time evolution operator is unitary, we say that the corresponding QBP is well-formed, that is, the QBP is valid in terms of the quantum theory.

We define the observable $\mathcal{O}$ to be $E_{\mathrm{acc}} \oplus E_{\mathrm{rej}} \oplus E_{\mathrm{non}}$, where

$$E_{\mathrm{acc}} = \mathrm{span}\left\{|v\rangle \,|\, v \in Q_{\mathrm{acc}}\right\}, E_{\mathrm{rej}} = \mathrm{span}\left\{|v\rangle \,|\, v \in Q_{\mathrm{rej}}\right\} ,$$

$$E_{\mathrm{non}} = \mathrm{span}\left\{|v\rangle \,|\, v \in Q_{\mathrm{non}}\right\} .$$

A QBP reads n inputs and returns a boolean value as follows: The initial state $|\psi_0\rangle$ is the source $|v_{\mathrm{s}}\rangle$. At each step, the time evolution operator is applied to the state $|\psi_i\rangle$, that is, $|\psi_{i+1}\rangle = U_\delta^{\boldsymbol{x}} |\psi_i\rangle$. Next, $|\psi_{i+1}\rangle$ is observed with respect to $E_{\mathrm{acc}} \oplus E_{\mathrm{rej}} \oplus E_{\mathrm{non}}$. Note that this observation causes the quantum state $|\psi_{i+1}\rangle$ to

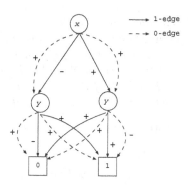

Fig. 2. A quantum branching program that computes $f = x \oplus y$ with no error. The weight of each edge is $\frac{1}{\sqrt{2}}$ or $-\frac{1}{\sqrt{2}}$, and only signs are put on the figure.

be projected onto the subspace compatible with the observation. Let the outcomes of an observation be "accept", "reject" and "non-halting" corresponding to E_{acc}, E_{rej} and E_{non} respectively. Until "accept" or "reject" is observed, applying the time evolution operator and observation is repeated. If "accept" ("reject") is observed, boolean value 1 (0) is returned.

We say that a QBP P computes a function f (with error rate $1/2 - \delta$) if P returns the correct value of f for any inputs with probability at least $1/2 + \delta$ ($\delta > 0$).   □

We show an example of a QBP in Fig. 2. To check well-formedness, we introduce the following theorem.

Theorem 1.

A QBP P is well-formed.

$\Leftrightarrow$ *for any input x, the transition function δ satisfies the following condition.*

$$\sum_{q'} \overline{\delta(q_1, x(q_1), q')} \delta(q_2, x(q_2), q') = \begin{cases} 1 & q_1 = q_2 \\ 0 & q_1 \neq q_2 , \end{cases}$$

where $\overline{\delta(q, a, q')}$ denotes the conjugate of $\delta(q, a, q')$.

Proof. It is obvious since U_δ^x is unitary if and only if the vectors $U_\delta^x |v\rangle$ are orthonormal.   □

Definition 3. *The Language Recognized by a Branching Program*

A language L is a subset of $\{0, 1\}^*$. Let the n-th restriction L^n of a language L be $L \cap \{0, 1\}^n$. A sequence of branching programs $\{P_n\}$ recognizes a language L if and only if, there exists $\delta (> 0)$, and the n-input branching program P_n computes the characteristic function $f_{L^n}(x)$ of L^n with error rate at most $1/2 - \delta$ for all $n \in \mathbf{N}$, where

$$f_{L^n}(x) = \begin{cases} 1 & (x \in L_n) \\ 0 & (x \notin L_n) . \end{cases}$$

□

Definition 4. *Bounded-Width Branching Programs*

For a branching program P, we can make any path from the source to a node v have the same length by inserting dummy nodes. Let the resulting branching program be P'. We say that P' is leveled. Note that P' does not need to compute the same function as P. The length of the path from the source to a node v is called the level of v. We define $\text{width}(i)$ *for P' as follows:*

$$\text{width}(i) = |\{v|\text{the level of } v \text{ is } i.\}| \ .$$

We define $\text{Width}(P')$ *as follows:*

$$\text{Width}(P') = \max_{i}\{\text{width}(i)\} \ .$$

We say that the width of P is bounded by $\text{Width}(P')$.

A sequence of branching programs $\{P_n\}$ is a bounded-width branching program if, for a constant w, $\{P_n\}$ satisfy the following condition.

$$\forall P \in \{P_n\}, \text{ The width of } P \text{ is bounded by } w \ .$$

$\square$

We also call a sequence of branching programs "a branching program" when it is not confused. We denote a bounded-width QBP and a bounded-width PBP as a bw-QBP and a bw-PBP respectively.

Definition 5. *Ordered Branching Programs*

Given a bounded-width branching program, we can make it leveled as shown in the above. For a given variable ordering $\pi = (x_{k_1} \leq x_{k_2} \leq \ldots \leq x_{k_n})$, if the appearances of the variables obey the ordering π, that is, x_{k_i} precedes x_{k_j} $(i < j)$ on any path from the source to a terminal node, and the labeled variables to all the nodes at the same level are the same, we say that the branching program is ordered. $\square$

3 Comparison of the Computational Power of Ordered bw-QBPs and Ordered bw-PBPs

In this section, we show that ordered bw-QBPs can compute some function that ordered bw-PBPs cannot compute. We define the function HALF_n and the language L_{HALF}.

Definition 6. *The Function HALF_n and the Language L_{HALF}*

We define $\text{HALF}_n : \mathbf{B}^n \longrightarrow \mathbf{B}$ *as follows:*

$$\text{HALF}_n(x_1, \ldots, x_n) = \begin{cases} 1 & |\{x_i|x_i = 1\}| = \frac{n}{2} \\ 0 & \text{otherwise} \end{cases} \ .$$

In the following, we denote the variables on which HALF_n depends as $X = \{x_i | 1 \leq i \leq n\}$. We define L_{HALF} as follows:

$$L_{\text{HALF}} = \left\{ x \mid x \in \{0,1\}^k, \text{HALF}_k(x) = 1 \right\} \ .$$

$\square$

3.1   Ordered bw-QBPs That Recognize L_{HALF}

In quantum computing, different computational paths interfere with each other when they reach the same configuration at the same time. In [3], a quantum finite automaton is constructed so that, only for inputs that the quantum finite automaton should accept, the computational paths interfere with each other.

In this paper, we modify this technique for quantum branching programs, and construct a quantum branching program that recognizes the language L_{HALF}.

Theorem 2. *Ordered bw-QBPs can recognize L_{HALF}.*

Proof. To show that ordered bw-QBPs can recognize L_{HALF}, we construct an ordered bw-QBP that computes HALF_n for any n. Figure 3 illustrates the QBP. We define the set of nodes Q as follows:

$$
\begin{aligned}
Q = \ &\{v_{\text{s}}, v_1, v_2, v_3, v_{\text{acc}}, v_{\text{rej1}}, v_{\text{rej2}}\} \\
&\cup \{v_{(i,x_k)} | x_k \in X, 1 \leq i \leq 3\} \\
&\cup \{v_{(i,x_k,j,\text{T})} | x_k \in X, 1 \leq i \leq 3, 1 \leq j \leq i\} \\
&\cup \{v_{(i,x_k,j,\text{F})} | x_k \in X, 1 \leq i \leq 3, 1 \leq j \leq 3-i+1\} \ .
\end{aligned}
$$

The labeled variable to the node $v_{(i,x_k)}$, $v_{(i,x_k,j,\text{T})}$, and $v_{(i,x_k,j,\text{F})}$ is x_k. The labeled variable to the node v_{s} is x_1. The labeled variable to the node v_1, v_2, and v_3 is x_n.

We define the accepting set (Q_{acc}), the rejecting set (Q_{rej}), the set of 0-edges (E_0), the set of 1-edges (E_1) and the weights of edges ($w(e)$) as follows:

$$
Q_{\text{acc}} = \{v_{\text{acc}}\}, Q_{\text{rej}} = \{v_{\text{rej1}}, v_{\text{rej2}}\} \ .
$$

$$
\begin{aligned}
E_0 = \ &\big\{(v_{\text{s}}, v_{(i,x_1)}) \,|\, 1 \leq i \leq 3\big\} \\
&\cup \big\{(v_{(i,x_k)}, v_{(i,x_k,1,\text{F})}) \,|\, 1 \leq i \leq 3, 1 \leq k \leq n\big\} \\
&\cup \big\{(v_{(i,x_k,j,\text{F})}, v_{(i,x_k,j+1,\text{F})}) \,|\, 1 \leq i \leq 3, 1 \leq j \leq 3-i, 1 \leq k \leq n\big\} \\
&\cup \big\{(v_{(i,x_k,3-i+1,\text{F})}, v_{(i,x_{k+1})}) \,|\, 1 \leq i \leq 3, 1 \leq k \leq n-1\big\} \\
&\cup \big\{(v_{(i,x_n,3-i+1,\text{F})}, v_i) \,|\, 1 \leq i \leq 3\big\} \\
&\cup \big\{(v_i, v_{\text{acc}}), (v_i, v_{\text{rej1}}), (v_i, v_{\text{rej2}}) \,|\, 1 \leq i \leq 3\big\} \ .
\end{aligned}
$$

$$
\begin{aligned}
E_1 = \ &\big\{(v_{\text{s}}, v_{(i,x_1)}) \,|\, 1 \leq i \leq 3\big\} \\
&\cup \big\{(v_{(i,x_k)}, v_{(i,x_k,1,\text{T})}) \,|\, 1 \leq i \leq 3, 1 \leq k \leq n\big\} \\
&\cup \big\{(v_{(i,x_k,j,\text{T})}, v_{(i,x_k,j+1,\text{T})}) \,|\, 1 \leq i \leq 3, 1 \leq j \leq i-1, 1 \leq k \leq n\big\} \\
&\cup \big\{(v_{(i,x_k,i,\text{T})}, v_{(i,x_{k+1})}) \,|\, 1 \leq i \leq 3, 1 \leq k \leq n-1\big\} \\
&\cup \big\{(v_{(i,x_n,i,\text{T})}, v_i) \,|\, 1 \leq i \leq 3\big\} \\
&\cup \big\{(v_i, v_{\text{acc}}), (v_i, v_{\text{rej1}}), (v_i, v_{\text{rej2}}) \,|\, 1 \leq i \leq 3\big\} \ .
\end{aligned}
$$

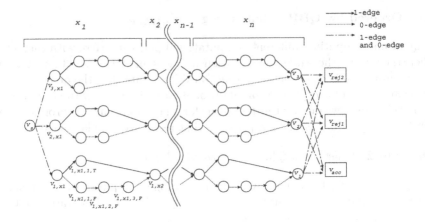

Fig. 3. a QBP that computes HALF$_n$

$$w((v_\mathrm{s}, v_{(1,x_1)})) = w((v_\mathrm{s}, v_{(2,x_1)})) = w((v_\mathrm{s}, v_{(3,x_1)})) = \frac{1}{\sqrt{3}} \ ,$$

$$w((v_1, v_\mathrm{acc})) = w((v_1, v_\mathrm{rej1})) = w((v_1, v_\mathrm{rej2})) = \frac{1}{\sqrt{3}} \ ,$$

$$w((v_2, v_\mathrm{acc})) = \frac{1}{\sqrt{3}} \ , \ w((v_2, v_\mathrm{rej1})) = \frac{1}{\sqrt{3}}exp\left(\frac{2\pi i}{3}\right) \ ,$$

$$w((v_2, v_\mathrm{rej2})) = \frac{1}{\sqrt{3}}exp\left(\frac{4\pi i}{3}\right) \ , \ w((v_3, v_\mathrm{acc})) = \frac{1}{\sqrt{3}} \ ,$$

$$w((v_3, v_\mathrm{rej1})) = \frac{1}{\sqrt{3}}exp\left(\frac{4\pi i}{3}\right) \ , \ w((v_3, v_\mathrm{rej2})) = \frac{1}{\sqrt{3}}exp\left(\frac{8\pi i}{3}\right)$$

The weights of the other edges are all 1.

Adding some more nodes and edges, each node of the QBP can be made to have

1. one incoming 0-edge and one incoming 1-edge with the weight of 1,
2. or, three incoming 0-edges and three incoming 1-edges with the same weights as between (v_1, v_2, v_3) and $(v_\mathrm{acc}, v_\mathrm{rej1}, v_\mathrm{rej2})$.

and have

1. one outgoing 0-edge and one outgoing 1-edge with the weight of 1,
2. or, three outgoing 0-edges and three outgoing 1-edges with the same weights as between (v_1, v_2, v_3) and $(v_\mathrm{acc}, v_\mathrm{rej1}, v_\mathrm{rej2})$.

In addition, incoming edges of each node can be made to be originated from the nodes labeled by the same variable. Then it is straightforward to see that this QBP can be well-formed by Theorem 1.

Given an input x, let the number of the variables in $X = \{x_1, \ldots, x_n\}$ to which the value 1 is assigned be k. Then the number of steps from $v_{(i,x_1)}$ to v_i is

$ik + (3 - i + 1)(n - k) + n$. Thus for any two distinct i and j $(1 \leq i, j \leq 3)$, the number of steps from the source to v_i (v_j) is the same if and only if $k = n/2$. Therefore the superposition of this QBP becomes $\frac{1}{\sqrt{3}}|v_1\rangle + \frac{1}{\sqrt{3}}|v_2\rangle + \frac{1}{\sqrt{3}}|v_3\rangle$ after $3n + 1$ steps if $\text{HALF}_n(\boldsymbol{x}) = 1$. Since $U_\delta^{\boldsymbol{x}}(\frac{1}{\sqrt{3}}|v_1\rangle + \frac{1}{\sqrt{3}}|v_2\rangle + \frac{1}{\sqrt{3}}|v_3\rangle) = |v_{\text{acc}}\rangle$, this ordered bw-QBP returns 1 with probability 1 if $\text{HALF}_n(\boldsymbol{x}) = 1$. On the other hand, since $U_\delta^{\boldsymbol{x}}|v_i\rangle = \frac{1}{\sqrt{3}}|v_{\text{acc}}\rangle + \frac{e^{i\alpha}}{\sqrt{3}}|v_{\text{rej1}}\rangle + \frac{e^{i\beta}}{\sqrt{3}}|v_{\text{rej2}}\rangle$, this ordered bw-QBP returns 0 with probability 2/3 if $\text{HALF}_n(\boldsymbol{x}) = 0$. Therefore this ordered bw-QBP computes HALF_n with one-sided error.   □

3.2   Ordered bw-PBPs Cannot Recognize L_{HALF}

Theorem 3. *Ordered bw-PBPs cannot recognize L_{HALF}.*   □

To prove Theorem 3, we introduce the following definition and lemma.

Definition 7. *Variation Distance*

The variation distance of two probability distributions P_1 and P_2 over the same sample space I is defined as follows:

$$\frac{1}{2} \sum_{i \in I} |P_1(i) - P_2(i)| \ .$$

Similarly, we define the variation distance of two vectors $\boldsymbol{x}_1 = (a_1, \ldots, a_n)$ and $\boldsymbol{x}_2 = (b_1, \ldots, b_n)$ $(a_1, \ldots, a_n, b_1, \ldots, b_n$: real numbers) as follows:

$$\frac{1}{2} \sum_{1 \leq i \leq n} |a_i - b_i| \ .$$

   □

Lemma 1. *Let Γ^m be the set that consists of all probability distributions of m events, that is,*

$$\Gamma^m = \{(a_1, \ldots, a_m) \,|\, a_1 \geq 0, \ldots, a_m \geq 0, a_1 + \ldots + a_m = 1\} \ .$$

For a finite set $S \subseteq \Gamma^m$ and a constant δ $(\delta > 0)$, if the cardinality of S is sufficiently large, the following holds.

$$\exists D_1 \in S, D_2 \in S$$
(the variation distance of D_1 and D_2) $< \delta$

Proof. We prove by induction on m. It is obvious that the lemma holds when $m = 2$. We assume that the lemma holds when $m = k - 1$. We consider the case of $m = k$. We assume that the cardinality of $S \subseteq \Gamma^k$ is sufficiently large, and, for any two distinct $D_i, D_j \in S$, the variation distance of D_i and D_j is greater than or equal to δ. For a_i, we define $\Gamma^N(l \leq a_i \leq r)$ as follows:

$$\Gamma^N(l \leq a_i \leq r) = \left\{(a_1, a_2, \ldots, a_N) \,\middle|\, \begin{array}{l} a_j \geq 0(1 \leq j \leq N), a_1 + \ldots + a_N = 1, \\ l \leq a_i \leq r \end{array}\right\} \ .$$

Since the cardinality of S is sufficiently large, the following holds.

$$\exists c, i$$
$$S' = S \cap \Gamma^k (c \le a_i \le c + \delta/2), \text{and}$$
the cardinality of S' is sufficiently large.

We fix c and i in the following. We assume that $i = k$ without loss of generality. For $\boldsymbol{x} = (a_1, \ldots, a_k)$, let the least index of the minimum of $\{a_1, \ldots, a_{k-1}\}$ be s. For example, if $\boldsymbol{x} = (0.35, 0.3, 0.05, 0.05, 0.25)$, the minimum is 0.05 and the least index is $s = 3$. For $\boldsymbol{x}' = (a_1, \ldots, a_{s-1}, a_s + a_k - c, a_{s+1}, \ldots, a_{k-1})$, we define the function F as $F(\boldsymbol{x}) = \boldsymbol{x}'$. For any two distinct $\boldsymbol{x}_1 = (a_1, \ldots, a_k) \in S'$ and $\boldsymbol{x}_2 = (b_1, \ldots, b_k) \in S'$ (we assume that $a_k \ge b_k$ w.l.o.g.), since the variation distance of $\boldsymbol{x}_1$ and $\boldsymbol{x}_2$ is at least δ, the variation distance of the vectors $F(\boldsymbol{x}_1)$ and $F(\boldsymbol{x}_2)$ is at least $\delta - \frac{1}{2}((a_k - b_k) - (a_k - c) - (b_k - c)) = \delta - (b_k - c) \ge \delta - \delta/2 = \delta/2$. Thus $F(\boldsymbol{x}_1)$ differs from $F(\boldsymbol{x}_2)$ for any two distinct $\boldsymbol{x}_1, \boldsymbol{x}_2 \in S'$. Therefore $|S'| = |S''|$ if we define

$$S'' = \left\{ \frac{1}{1-c} \cdot F(\boldsymbol{x}) \,\middle|\, \boldsymbol{x} \in S' \right\}.$$

Note that, for a vector $\boldsymbol{x} \in S'$ and $F(\boldsymbol{x}) = (a_1, \ldots, a_{k-1})$, $a_1 + \ldots + a_{k-1} = 1 - c$ and $a_1 \ge 0, \ldots, a_{k-1} \ge 0$. Thus, for a vector $(b_1, \ldots, b_{k-1}) \in S''$, $b_1 + \ldots + b_{k-1} = 1$ and $b_1 \ge 0, \ldots, b_{k-1} \ge 0$. Therefore in the case of $m = k-1$, there is $S'' \subseteq \Gamma^{k-1}$ such that, for any two distinct $D_i, D_j \in S''$, the variation distance of D_i and D_j is at least $\frac{\delta}{2(1-c)}$, and the cardinality of S'' is sufficiently large. This is a contradiction.   □

We show the proof of Theorem 3 in the following.
(Proof of Theorem 3)
We assume that there is an ordered bw-PBP $\{P_n\}$ that recognizes L_{HALF}, that is, $P \in \{P_n\}$ computes HALF_n with error rate $1/2 - \delta$. We say that a PBP is in normal form when all the variables appear on any path from the source to a terminal node. We assume that P is in normal form with the ordering $\pi = (x_1 \le \ldots \le x_n)$ without loss of generality. Let $S_{\frac{n}{2}}$ be the set of the variables of the former half of the variable ordering, that is, $S_{\frac{n}{2}} = \{x_j | 1 \le j \le \frac{n}{2}\}$. When n is sufficiently large, there are sufficient number of assignments to the variables in $S_{\frac{n}{2}}$ such that, for any two distinct assignments, the weights of the assignments differ. Let D_a denotes the probability distribution for the nodes at which we arrive when we compute according to a. That is, $D_a(v)$ is the probability such that we arrive at the node v after we compute according to a. For sufficient number of assignments, there are sufficient number of corresponding probability distributions. Thus, since P is a bounded-width PBP, when n is sufficiently large, there are two distinct assignment, a_1 and a_2, to the variables in $S_{\frac{n}{2}}$ satisfying the following conditions by Lemma 1.

– The variation distance of D_{a_1} and D_{a_2} is less than δ.
– The weight of a_1 (the number of 1 in a_1) differs from that of a_2.

Let the a_{rest} be the assignment to the variables of the latter half of the variable ordering such that, for the complete assignment $a_1 \cdot a_{rest}$, $\mathrm{HALF}_n(a_1 \cdot a_{rest}) = 1$. Then if we compute according to $a_1 \cdot a_{rest}$, we arrive at the terminal node of 1 with probability at least $1/2 + \delta$. On the other hand, if we compute according to $a_2 \cdot a_{rest}$, we arrive at the terminal node of 1 with probability at least $1/2 + \delta - \delta = 1/2$. We show the reason in the following.

Let I be

$$I = \{i | D_{a_1}(i) > D_{a_2}(i)\} \ .$$

Since the variation distance of D_{a_1} and D_{a_2} is less than δ,

$$\sum_{i \in I} (D_{a_1}(i) - D_{a_2}(i)) \left(= \sum_{i \notin I} (D_{a_2}(i) - D_{a_1}(i)) \right) < \delta \ .$$

Thus comparing the probabilities with which we arrive at the terminal node of 1 computing according to $a_1 \cdot a_{rest}$ and $a_2 \cdot a_{rest}$, the difference is at most δ. Therefore we arrive at the terminal node of 1 with probability at least $1/2 + \delta - \delta = 1/2$ if we compute according to $a_2 \cdot a_{rest}$. However this probability must be less than $1/2 - \delta$. This is a contradiction. □

4   Conclusion

In this paper, we show that there is a function that can be computed by ordered bw-QBPs but cannot be computed by ordered bw-PBPs. This is an evidence that introducing quantum effects to a computational model increases its power.

It is still the future work to study what results we obtain if we remove the restriction of bounded-width and variable ordering. Since quantum computational models must be reversible, introducing classical "reversible branching programs" and comparing them with quantum branching programs can also be future work.

References

1. A. Ambainis and R. Freivalds, "1-way quantum finite automata: strengths, weakness and generalizations," Proc. 39th Symp. on Foundations of Computer Science, pp. 332–341, 1998.
2. L. Grover, "A fast quantum mechanical algorithm for database search," Proc. 28th Symp. on the Theory of Computing, pp. 212–219, 1996.
3. A. Kondacs and J. Watorus, "On the power of quantum finite state automata," Proc. 38th Symp. on Foundations of Computer Science, pp. 66–75, 1997.
4. C. Meinel, "Modified branching programs and their computational power," Lecture Notes in Computer Science 370, Springer-Verlag, Berlin, 1989.
5. P. Shor, "Algorithms for quantum computation: discrete logarithms and factoring," Proc. 35th Symp. on Foundations of Computer Science, pp. 124–134, 1994.

Author Index

Lecture Notes in Computer Science

For information about Vols. 1–1763
please contact your bookseller or Springer-Verlag

Vol. 1810: R.López de Mántaras, E. Plaza (Eds.), Machine Learning: ECML 2000. Proceedings, 2000. XII, 460 pages. 2000. (Subseries LNAI).

Vol. 1811: S.W. Lee, H.. Bülthoff, T. Poggio (Eds.), Biologically Motivated Computer Vision. Proceedings, 2000. XIV, 656 pages. 2000.

Vol. 1813: P.L. Lanzi, W. Stolzmann, S.W. Wilson (Eds.), Learning Classifier Systems. X, 349 pages. 2000. (Subseries LNAI).

Vol. 1815: G. Pujolle, H. Perros, S. Fdida, U. Körner, I. Stavrakakis (Eds.), Networking 2000 – Broadband Communications, High Performance Networking, and Performance of Communication Networks. Proceedings, 2000. XX, 981 pages. 2000.

Vol. 1816: T. Rus (Ed.), Algebraic Methodology and Software Technology. Proceedings, 2000. XI, 545 pages. 2000.

Vol. 1817: A. Bossi (Ed.), Logic-Based Program Synthesis and Transformation. Proceedings, 1999. VIII, 313 pages. 2000.

Vol. 1818: C.G. Omidyar (Ed.), Mobile and Wireless Communications Networks. Proceedings, 2000. VIII, 187 pages. 2000.

Vol. 1819: W. Jonker (Ed.), Databases in Telecommunications. Proceedings, 1999. X, 208 pages. 2000.

Vol. 1821: R. Loganantharaj, G. Palm, M. Ali (Eds.), Intelligent Problem Solving. Proceedings, 2000. XVII. 751 pages. 2000. (Subseries LNAI).

Vol. 1822: H.H. Hamilton, Advances in Artificial Intelligence. Proceedings, 2000. XII, 450 pages. 2000. (Subseries LNAI).

Vol. 1823: M. Bubak, H. Afsarmanesh, R. Williams, B. Hertzberger (Eds.), High Performance Computing and Networking. Proceedings, 2000. XVIII, 719 pages. 2000.

Vol. 1824: J. Palsberg (Ed.), Static Analysis. Proceedings, 2000. VIII, 433 pages. 2000.

Vol. 1825: M. Nielsen, D. Simpson (Eds.), Application and Theory of Petri Nets 2000. Proceedings, 2000. XI, 485 pages. 2000.

Vol. 1826: W. Cazzola, R.J. Stroud, F. Tisato (Eds.), Reflection and Software Engineering. X, 229 pages. 2000.

Vol. 1830: P. Kropf, G. Babin, J. Plaice, H. Unger (Eds.), Distributed Communities on the Web. Proceedings, 2000. X, 203 pages. 2000.

Vol. 1831: D. McAllester (Ed.), Automated Deduction – CADE-17. Proceedings, 2000. XIII, 519 pages. 2000. (Subseries LNAI).

Vol. 1832: B. Lings, K. Jeffery (Eds.), Advances in Databases. Proceedings, 2000. X, 227 pages. 2000.

Vol. 1833: L. Bachmair (Ed.), Rewriting Techniques and Applications. Proceedings, 2000. X, 275 pages. 2000.

Vol. 1834: J.-C. Heudin (Ed.), Virtual Worlds. Proceedings, 2000. XI, 314 pages. 2000. (Subseries LNAI).

Vol. 1835: D. N. Christodoulakis (Ed.), Natural Language Processing – NLP 2000. Proceedings, 2000. XII, 438 pages. 2000. (Subseries LNAI).

Vol. 1837: R. Backhouse, J. Nuno Oliveira (Eds.), Mathematics of Program Construction. Proceedings, 2000. IX, 257 pages. 2000.

Vol. 1838: W. Bosma (Ed.), Algorithmic Number Theory. Proceedings, 2000. IX, 615 pages. 2000.

Vol. 1839: G. Gauthier, C. Frasson, K. VanLehn (Eds.), Intelligent Tutoring Systems. Proceedings, 2000. XIX, 675 pages. 2000.

Vol. 1840: F. Bomarius, M. Oivo (Eds.), Product Focused Software Process Improvement. Proceedings, 2000. XI, 426 pages. 2000.

Vol. 1841: E. Dawson, A. Clark, C. Boyd (Eds.), Information Security and Privacy. Proceedings, 2000. XII, 488 pages. 2000.

Vol. 1842: D. Vernon (Ed.), Computer Vision – ECCV 2000. Part I. Proceedings, 2000. XVIII, 953 pages. 2000.

Vol. 1843: D. Vernon (Ed.), Computer Vision – ECCV 2000. Part II. Proceedings, 2000. XVIII, 881 pages. 2000.

Vol. 1844: W.B. Frakes (Ed.), Software Reuse: Advances in Software Reusability. Proceedings, 2000. XI, 450 pages. 2000.

Vol. 1845: H.B. Keller, E. Plöderer (Eds.), Reliable Software Technologies Ada-Europe 2000. Proceedings, 2000. XIII, 304 pages. 2000.

Vol. 1846: H. Lu, A. Zhou (Eds.), Web-Age Information Management. Proceedings, 2000. XIII, 462 pages. 2000.

Vol. 1847: R. Dyckhoff (Ed.), Automated Reasoning with Analytic Tableaux and Related Methods. Proceedings, 2000. X, 441 pages. 2000. (Subseries LNAI).

Vol. 1848: R. Giancarlo, D. Sankoff (Eds.), Combinatorial Pattern Matching. Proceedings, 2000. XI, 423 pages. 2000.

Vol. 1849: C. Freksa, W. Brauer, C. Habel, K.F. Wender (Eds.), Spatial Cognition II. XI, 420 pages. 2000. (Subseries LNAI).

Vol. 1850: E. Bertino (Ed.), ECOOP 2000 – Object-Oriented Programming. Proceedings, 2000. XIII, 493 pages. 2000.

Vol. 1851: M.M. Halldórsson (Ed.), Algorithm Theory – SWAT 2000. Proceedings, 2000. XI, 564 pages. 2000.

Vol. 1853: U. Montanari, J.D.P. Rolim, E. Welzl (Eds.), Automata, Languages and Programming. Proceedings, 2000. XVI, 941 pages. 2000.

Vol. 1855: E.A. Emerson, A.P. Sistla (Eds.), Computer Aided Verification. Proceedings, 2000. X, 582 pages. 2000.

Vol. 1857: J. Kittler, F. Roli (Eds.), Multiple Classifier Systems. Proceedings, 2000. XII, 404 pages. 2000.

Vol. 1858: D.-Z. Du, P. Eades, V. Estivill-Castro, X. Lin, A. Sharma (Eds.), Computing and Combinatorics. Proceedings, 2000. XII, 478 pages. 2000.

Vol. 1860: M. Klusch, L. Kerschberg (Eds.). Cooperative Information Agents IV. Proceedings, 2000. XI, 285 pages. 2000. (Subseries LNAI).

Vol. 1861: J. Lloyd, V. Dahl, U. Furbach, M. Kerber, K.-K. Lau, C. Palamidessi, L. Moniz Pereira, Y. Sagiv, P.J. Stuckey (Eds.), Computational Logic – CL 2000. Proceedings, 2000. XIX, 1379 pages. (Subseries LNAI).

Vol. 1866: J. Cussens, A. Frisch (Eds.), Inductive Logic Programming. Proceedings, 2000. X, 265 pages. 2000. (Subseries LNAI).